Joep Fai Fredde

Studies in Surface Science and Catalysis 152

FISCHER-TROPSCH TECHNOLOGY

Studies in Surface Science and Catalysis

Advisory Editors: B. Delmon and J.T. Yates
Series Editor: G. Centi

Vol. 152

FISCHER-TROPSCH TECHNOLOGY

Edited by

André Steynberg
P.O. Box 1
Sasolburg 1947
South Africa

and

Mark Dry
Catalysis Research Institute
Department of Chemical Engineering
University of Cape Town, 7701, South Africa

2004

ELSEVIER

Amsterdam – Boston – Heidelberg – London – New York – Oxford – Paris – San Diego
San Francisco – Singapore – Sydney – Tokyo

ELSEVIER B.V.	ELSEVIER Inc.	ELSEVIER Ltd	ELSEVIER Ltd
Radarweg 29	525 B Street	The Boulevard	84 Theobalds Road
P.O. Box 211, 1000 AE	Suite 1900, San Diego	Langford Lane, Kidlington,	London WC1X 8RR
Amsterdam, The Netherlands	CA 92101-4495, USA	Oxford OX5 1GB, UK	UK

First edition 0000
Reprinted 0000

British Library Cataloguing in Publication Data
A catalogue record is available from the British Library.

Library of Congress Cataloging in Publication Data
A catalog record is available from the Library of Congress.

ISBN: 0-444-51354-X
ISSN: 01672991

♾ The paper used in this publication meets the requirements of ANSI/NISO Z39.48-1992 (Permanence of Paper).
Printed in The Netherlands

FOREWORD

Fischer-Tropsch technology is unique in many ways. Discovered early in the last century along with many other bulk chemical technologies, its development has been hampered by the lack of a continuous market pull. Interest in Fischer-Tropsch technology as a source of alternate fuels has been in spurts, driven by wars, political situations and peaks in the price and availability of crude oil. There have been surges of interest in the thirties and during the Second World War, in the fifties, as a result of the oil crisis of the seventies and now again in the last seven years or so. It is just possible that the present interest in Fischer-Tropsch technology is sustainable and will result in the establishment of a GTL industry. The signs are good.

There is a solid realization that the gas reserves exist and that they are largely in areas where the normal pipeline and power uses are not feasible. There are several nations and geographical areas with the right mix of gas reserves, local infrastructure and intent to add to their existing portfolios. The qualities of Fischer-Tropsch products are excellent and their environmental properties are being recognized as very valuable in the ongoing drive towards cleaner fuels and engines. The perceptions of supply/demand balance and pricing of crude oil are beginning to create an economic incentive for the conversions of gas to liquid fuels. And, above all, technology improvements reported by many of the major players have resulted in large and unprecedented reductions in the cost of the technology. All of these positive factors combined lead one to an expectation that Fischer-Tropsch technology will at last take its place as a significant factor in the global energy scene.

While the beginnings of Fischer-Tropsch were in the search for ways to turn coal into a liquid transport fuel, from the above it is now likely that its future is as a means of monetizing natural gas. Gas to Liquids technology, or GTL as it is referred to, is a complex integrated set of technologies with Fischer-Tropsch at its heart. While conversion of both coal and natural gas to liquids has been practiced in numerous locations over the whole history of the technology, today the only commercial operations are in South Africa (both coal and natural gas) and in Malaysia. The operations provide the springboard from which large modern GTL plants can be built and operated. While large plants, which benefit from economy of scale, are usually studied and are considered to be most likely to be viable, the use of GTL in niche applications for recovery of associated gas should not be forgotten.

For GTL technology to be converted from a curiosity into a modern industry, it has to be practiced by several players in multiple plants in several geographical areas. The GTL fuels and indeed the chemical co-products need to be marketed globally. The extent of the challenge involved should not be underestimated. To achieve commercial success GTL plants need to be built at a scale that has not been done before. In many cases, technologies practiced at a size of a few hundred barrels per day need to be scaled up by a factor of 100 or more and then incorporated into complexes costing billons of dollars. The plants will be located in remote locations often with unique difficulties. Of the many organizations preparing to enter this field only one or two have any commercial experience with the technology. The risks are indeed daunting. They are particularly so for multiple gas based units and products not often handled in the past. Small mistakes, omissions and oversights during design can very quickly lead to lengthy and extremely costly delays.

The construction by Sasol and its partner in Qatar of the 34,000 barrel per day Oryx GTL plant is a first step down the road. The approval of this project and its financing by consortia of international banks has clearly demonstrated to the world that GTL is real. This is confirmed by the large number of announcements of even larger projects mostly in the Middle East. There is every reason to believe that the technology patented by two German scientists, Frans Fischer and Hans Tropsch so many years ago is now well on the road to be the basis for a new global industry.

Given the interest in GTL, I believe that this book is extremely well timed. Its contents will be valuable to large numbers of engineers and scientists from the major GTL players through to the engineering contractors and equipment suppliers involved in the industry. Many of the chapters will give valuable background to newcomers to the industry and will reinforce the knowledge of the experienced. A blend of technical history book, academic review and reference, practical design manual and record of commercial reality, it will be welcomed by all technical people in the field.

The book contains a summary and review of a large amount of published material supplemented by the authors' extensive experience and know-how. Sasol related background has clearly dominated the book but, with over 50 years of practical CTL, GTL and Fischer-Tropsch experience this is not surprising. Indeed a Fischer-Tropsch reference without Sasol would not be complete. The many co-authors are all well known in the industry and bring together a collective experience that will be hard to beat.

In conclusion I would like to express my hope that Fischer-Tropsch will find its place in the global energy world and that this book will play its part in helping reach this goal.

John Marriott – July 2004

PREFACE

Remote natural gas has traditionally been used to supply energy markets through Liquefied Natural Gas (LNG), and sometimes to produce chemical products e.g. methanol. The last few years have seen a marked increase in the interest for the application of Fischer-Tropsch (FT) technology for the conversion of natural gas into easily transportable liquid hydrocarbon products. Several commercial scale plants have been announced and some are in advanced stages of design or even construction.

The successful commercial application of FT technology requires a very broad range of expertise. This demands a cooperative effort. Perhaps this is why companies that are active in the field stress the need for teamwork. Although some of the major oil companies may be capable of putting FT facilities in place using mainly in-house technology, it is expected that the most cost effective plants will require cooperation between several technology providers.

As an example of this approach Sasol™ and ChevronTexaco have formed the Sasol Chevron global alliance in order to apply FT technology to the conversion of remotely located natural gas to liquid products. This is expected to place the Gas-to-Liquids (GTL) industry on a footing similar to the LNG industry and involves massive capital expenditure measured in billions of US dollars ($US X 10^9). The prime example of a major oil company using mainly in-house technology for their GTL projects is Shell.

In the case of Sasol Chevron, the ChevronTexaco partner is well versed with natural gas utilization and the recovery of value from the associated condensate and natural gas liquids (NGL's, i.e. C_3+ hydrocarbons). The alliance uses technology from Haldor-Topsøe to produce synthesis gas from the lean natural gas. The synthesis gas preparation requires oxygen prepared in large scale air separation units (ASU's). The ASU's are provided by a number of American and European companies that bid on a competitive basis for the supply of these units. The ASU is the single most expensive unit in the GTL complex. Sasol provides the FT technology and the primary liquid products are upgraded using ChevronTexaco hydroprocessing technology. ChevronTexaco has a well established global network for the marketing of the fuel and lubricant products while Sasol has a significant global presence in the marketing of relevant commodity chemicals.

For the Sasol Chevron projects, a diverse team at the Foster Wheeler office in Reading, England does the feasibility studies and the Front End Engineering and Design (FEED) and prepares bid packages to enable large engineering and construction firms to bid for the engineering, procurement and construction (EPC). Currently detailed engineering and procurement is being done in Rome, Italy by Technip-Coflexip for the ORYX GTL Limited plant in

Qatar (a joint venture between Sasol and Qatar Petroleum). Sasol initiated this project prior to the formation of Sasol Chevron.

Sasol uses Stone & Webster in Cambridge, USA to assist with the engineering for the FT unit. The design of the FT unit is based on technology that was developed and demonstrated in Sasolburg, South Africa, where the Sasol corporate R&D facilities are located. Sasol also has an alliance with IHI from Japan for the supply of the FT reactor vessels. Currently, the only vessel fabricators capable of delivering these large vessels on a competitive basis are all based in Japan and Korea. Based on Sasol's commercial experience with large and complex plants using FT technology, Sasol takes overall technical responsibility as the single point licensor in addition to licensing the FT unit.

Two significant questions may be posed:
- How did FT technology develop to the point where it is set to form the basis for a significant new industry in the 21st century?
- What scope is there for the further development and application of this technology?

If these questions are of interest then this book should provide some useful insight.

This book has two primary aims: firstly, it is a comprehensive work documenting the state of the art for FT technology. It will be of use to people active in advancing and implementing this technology. There are comprehensive references to patents and previous publications. (A useful website for the location of additional historical publications relating to FT technology may be found at http://www.fischer-tropsch.org).

Secondly, this book is suitable for use as a reference book for courses in the fields of chemistry and chemical engineering. The book provides an explanation of the basic principles and terminology that are required to understand the application of FT technology. Known and proposed applications for the technology are described.

The intended audience is therefore researchers active in the development of FT technology, engineers involved in the implementation of this technology and students in the fields of chemistry and chemical engineering.

Existing text books and reference works dealing with the theory and application of FT technology are now somewhat dated. No single source exists that covers past, present and future applications starting with an explanation of the basic principles. This is the first publication that allows students to understand both theory and application for modern FT technology. The book is also useful as a contemporary reference work for those wishing to study the history of the application of FT synthesis. It can be used to locate previously reported research work and to understand the context and significance of the cited research. The text identifies the aspects of the technology that are proprietary and describes those aspects that can be considered to be open art. The emphasis is on explanation of principles, limits and the reasons for compromise in this technology field.

As with other endeavours relating to FT technology, this book is the result of a cooperative effort. Although it is the result of a Sasol initiative, the intention is to serve the interests of the rapidly expanding industries based on the use of FT technology. An attempt was made to involve the world's leading academic authorities and industry experts.

On the subject of synthesis gas generation from natural gas there was no need to look further than Haldor-Topsøe who are at the forefront of technology developments in this field and have previously produced many excellent publications. Their contribution to Chapter 4 is greatly appreciated. In the case of the coal gasification route to synthesis gas it was decided to use in-house experts from Sasol since Sasol is the world's largest producer of synthesis gas from coal. My thanks go to Dr Martin Keyser and Retha Coertzen for their contributions.

Nobody is better suited to bridge the gap between academic endeavour and industrial application than Professor Mark Dry from the University of Cape Town and formerly the head of Fischer-Tropsch research at Sasol. Professor Dry assisted with the editing of all parts of this book and provided a great deal of guidance in determining the content of the book. Mark Dry is the sole author for Chapter 3 dealing with the chemical principles that need to be understood for FT applications and Chapter 7 dealing with FT catalysts.

Professor Hans Schultz, a world renowned FT researcher from the University of Karlsruhe, also provided guidance on the appropriate ordering of the material. Professors Michael Claeys and Eric van Steen both now at the University of Cape Town had previously worked closely with Hans Schultz and were recommended by him to contribute a chapter dealing with the understanding of FT chemistry at a fundamental level. The result is Chapter 8.

My main contribution is to be found in Chapter 2 dealing with the FT reactors with which I have had a 24 year love affair. I am also very grateful to Professor Krishna from the University of Amsterdam, The Netherlands, for reviewing the material on FT reactors in Chapter 2. Excellent research is done on fluidization technology in the Netherlands and there is good cooperation in this field between a number of Dutch and German universities. Professor Krishna has published extensively and is especially well known for his work on slurry bubble column reactor systems. Thanks to Dr Berthold Breman who heads the Sasol Technology Process Development group based at Twente University in the Netherlands for the material dealing with fixed bed reactors.

This Process Development group was founded recently by Dr Ben Jager (now retired) who was also responsible for establishing the Process Development group in Sasolburg in which I worked for most of my career.

Important contributions particularly in Chapter 2 were also made by Professor Burtron Davis from the University of Kentucky. He is at the forefront of research in the field of FT catalysis, particularly for iron based catalysts used in low temperature FT reactors and is also particularly knowledgeable on the history of FT technology. He is well versed in all research and technology development activities in the USA that relate to FT technology. Professor Davis also reviewed Chapters 1, 2, 3, 5, 7 and the gasification material in Chapter 4.

Dr Marco Frank from the Twente group and Dr Jannie Scholtz (FT Technology Manager at Sasol) provided some valuable insights into the thermodynamic constraints for FT applications that are discussed in Chapter 5.

Chapter 6 was produced primarily by in-house experts at Sasol. Arno de Klerk provided the material dealing with the upgrading of products from the Synthol process (i.e. high temperature FT, HTFT) while Luis Dancuart was mainly responsible for the section dealing with the low temperature FT (LTFT) products. Dr Rob de Haan provided some sections dealing with hydroprocessing catalysts. The contribution from Dr Jerome Meyer and Richard Moore from ChevronTexaco, after reviewing all the material relating to hydroprocessing, is also gratefully acknowledged.

Given that this book is the result of part time contributions from people employed to do other things it has taken some time to reach completion. There is no doubt still room for improvement. Some areas such as those dealing with FT catalysts and FT products could benefit from more in depth publications that will hopefully follow this book.

Many other Sasol employees provided some assistance with this book. In particular I would like to thank Dr Anton Vosloo and Dr Jannie Scholtz for reviewing all the material; Ed Koper who was a constant source of support (moral and financial) and Caleb Hattingh for helping with some of the illustrations and equations.

Finally I would like to thank my employer, Sasol Technology, for allowing me to spend company time on the production of this book, Sasol Chevron for their financial support and to my family for accepting my neglect while working on this project.

André P. Steynberg
Sasolburg, May 2004

CONTENTS

Chapter 1 Introduction to Fischer-Tropsch technology
A. P. Steynberg

Chapter 2 Fischer-Tropsch Reactors
A. P. Steynberg, M. E. Dry, B. H. Davis and B. B. Breman

Chapter 3 Chemical concepts used for engineering purposes
M. E. Dry

Chapter 4 Synthesis gas production for FT Synthesis
K. Aasberg-Petersen, T. S. Christensen, I. Dybkjær, J. Sehested,
M. Østberg, R. M. Coertzen, M. J. Keyser and A. P. Steynberg

Chapter 5 Commercial FT process application
M. E. Dry and A. P. Steynberg

Chapter 6 Processing of primary FT products
L. P. Dancuart, R. de Haan and A. de Klerk

Chapter 7 FT catalysts
M. E. Dry

Chapter 8 Basic studies
M. Claeys and E. van Steen

Studies in Surface Science and Catalysis 152
A. Steynberg and M. Dry (Editors)

Chapter 1

Introduction to Fischer-Tropsch Technology

A. P. Steynberg

Sasol Technology R&D,
P.O. Box 1, Sasolburg, 1947, South Africa

1. ORIENTATION

Welcome to the amazing world of Fischer-Tropsch (FT) science and technology. The emphasis for this book is on the practical application of this technology. It is simply not possible to divorce the application of the technology from an understanding of the science. As a result this is a field that requires close co-operation between scientists and engineers (both with a heavy bias towards the chemical discipline).

Fischer-Tropsch technology can be briefly defined as the means used to convert synthesis gas containing hydrogen and carbon monoxide to hydrocarbon products.

The hydrocarbon products are mostly liquid at ambient conditions but some are gaseous and some may even be solid. For the above definition the term 'hydrocarbons' includes oxygenated hydrocarbons such as alcohols. However, the sole production of an oxygenated hydrocarbon such as methanol is excluded.

Interest in Fischer-Tropsch technology is increasing rapidly. This is due to recent improvements to the technology and the realisation that it can be used to obtain value from stranded natural gas. In other words, remotely located natural gas will be converted to liquid hydrocarbon products that can be sold in worldwide markets. This is often referred to as the Gas-to-Liquids (GTL) industry. There is an alternative approach to get value from remote natural gas via the deployment of conventional liquefaction technology that produces liquefied natural gas (LNG). The LNG industry has grown rapidly and it is anticipated that the GTL industry may expand at an equal pace and even surpass the LNG industry. This will happen if it is demonstrated on a commercial scale that the economic incentives for GTL are better than those for LNG.

Although LNG and GTL use the same feedstock, they sell their products into very different markets. There is currently no shortage of remote natural gas so there is scope for both technologies to flourish. The countries with these gas resources will benefit from product diversification. One of the important potential advantages for the GTL route is the ability to diversify further from fuels to higher value chemical products.

Figure 1 A graphic of the GTL plant for the first Sasol-Chevron joint project at the Escravos delta in Nigeria

The most popular goal is to utilize abundant and low cost natural gas to produce 'clean' (low sulphur, low aromatics) middle distillates/fuels, with the main co-product being a paraffinic naphtha to be sold as steam cracker feedstock to make mainly ethylene and propylene. The option exists to also produce other higher value co-products such as detergent and synthetic lubricant intermediates as well as propylene and alpha olefins used for the production of polymers.

The GTL products compete with crude oil derived products so the future of this industry depends on the actual and perceived future prices for crude oil. The GTL industry expansion predicted above will not happen unless the crude oil price averages above \$20 per barrel. Existing producers will survive at crude oil prices as low as \$16/barrel but the industry will not be viable if prices below this level are sustained for long periods. At present, it seems likely that this new industry will flourish.

Some advantages of FT hydrocarbons compared to crude oil as a feedstock for fuel production are the absence of sulphur, nitrogen or heavy metal contaminants, and the low aromatic content. The kerosene/jet fuel produced has good combustion properties and high smoke points and the diesel fuel with its high cetane number can be used to upgrade lower quality blend stocks produced from crude oil. Linear olefins required in the chemical industry can be produced either directly in the FT process or by dehydrogenation of the paraffinic cuts.

Potential co-products may also utilise material and energy that is not directly associated with the FT hydrocarbons. For example, waste heat may be converted to electrical power for sale and nitrogen produced in the air separation unit may be used to make ammonia which may be further reacted with waste carbon dioxide to make urea. The various potential primary and co-products are discussed in some detail later. At this stage it is only worth noting that the technology has the potential to make large quantities of products with higher values than fuels and to provide the hub for a whole new petrochemical industry.

The FT technology is actually quite old technology with roots in coal utilisation. The technology was first applied in Germany in the 1930's and most of the potential products and co-products were soon identified. However, the early technology was expensive and inefficient and could not compete with cheap and abundant crude oil. Nevertheless, the technology continued to fascinate people and research and technology development even continued at times when no commercial applications seemed likely.

Some of the technology interest, development and application resulted from strategic and political rather than purely economic considerations. This is discussed briefly later in this chapter. Strategic considerations may well result in further applications for FT technology with coal as a source for the synthesis gas. This applies to places with abundant coal and a shortage of other sources of energy. These places include China, the USA, Australia, India and South Africa. Hence, coal conversion has not been neglected in this book.

It is not the purpose of this book to deal with the business issues related to the application of FT technology or to compare the merits of the various potential technology providers. Rather, it is to explain how the technology works and equip future practitioners in this field with relevant knowledge.

Figure 2 The Sasol Synfuels plant at Secunda, South Africa

2. INTEREST AND CHALLENGES

What are the characteristics of this technology field that make it so fascinating? There are several features which are discussed below.

The chemistry involved in the FT synthesis has been described [1] as "a surprising phenomenon in heterogeneous catalysis that attracts the interest of world experts: the gases CO and H_2 enter the reactor and a hydrocarbon liquid exits." The thermodynamically preferred hydrocarbon product is methane gas so it is surprising that higher hydrocarbons are the predominant products.

The performance of the FT synthesis depends on the gas composition (feed/product component partial pressures), catalyst formulation and operating temperature. The effects of these variables are interrelated. Furthermore, the catalysts used undergo chemical (and sometimes physical) changes during the FT synthesis that further complicates reactor design and optimisation.

As an example increasing temperature:
- favours increased methane formation,
- favours deposition of carbon and other catalyst deactivation mechanisms (particularly with iron based catalysts)
- and reduces the average chain length of the product molecules.

These effects are undesirable. On the other hand, the rate of reaction increases and the quality of steam produced by the reactor heat removal system improves which are both desirable. Hence the application of FT technology involves a number of compromises and optimisations.

FT technology deals with things over a huge range of physical sizes. State-of-the-art surface science techniques are used to examine the surface of the Fischer-Tropsch catalysts with dimensions measured in angstrom units. At the other extreme, the world's largest pressure vessels are used as reactors. These vessels have diameters of the order of 10 metres, heights exceeding 30 metres and they weigh around 2000 metric tons (tonnes).

Catalyst particles in the powders used with modern fluid bed reactors range in size from about 2 to 200 microns. The hydrodynamic behaviour inside these reactors is a fruitful field for scientific and engineering research. Fixed bed reactors with millimetre size catalyst particles are also used for FT synthesis. Even for these reactors, first used in the 1930's, it has been observed that the catalyst and reactor should be optimised in a combined fashion [2]. This is even more applicable for the fluid bed reactors.

To complicate matters further, the FT reactor/catalyst system cannot be designed in isolation. There is a strong interaction between the design of the FT reactor and the choice of technology to provide the synthesis gas (containing H_2 and CO) fed to the FT reactor. Often recycles are involved and various separation technologies may be applied to process the tail gases to generate these recycle streams.

Separation technologies are also important to isolate various product components from the broad range of liquid hydrocarbon products formed.

The FT process is fairly unique in the field of heterogeneous catalysis in that there is not a single desired product but the emphasis is rather on the avoidance of undesirable by-products. The separated primary products generally require further processing to produce the variety of desirable end product fuels and chemicals. The choice from the available range of FT technologies may well depend on the prices obtainable for the different potential final products.

Figure 3 A propylene distillation column is raised into position

Another facet that requires optimisation is the energy balance. Both the generation of synthesis gas and the FT synthesis involve tremendous amounts of heat exchange. The application of the FT synthesis deals primarily with the conversion of a source of fossil energy into more convenient, user friendly fuels. It is important, for both economic and environmental reasons, that the energy conversion should be as efficient as possible. It is also important to note that there are practical limits to the amount of money that can be spent in the pursuit of energy efficiency.

It has been noted [3] that all types of processes found in the hydrocarbon and chemical engineering world come together when applying the Fischer-Tropsch synthesis for natural gas conversion in a GTL plant, including:

- Gas processing
- Gas conversion to synthesis gas akin to methanol and ammonia plants
- Hydrocarbon synthesis i.e. Fischer-Tropsch reaction, in fact a polymerisation reaction
- Refinery operations such as distillation, hydroprocessing and other typical refinery processes
- Integrated utility systems incorporating large steam systems and low level heat
- Generation of speciality (chemical) products thus 'monetising' (i.e. adding value to) the unique properties of the raw FT product
- Power generation to provide electrical plant power along with drives for rotating equipment

It has also been noted that the utilities play a critical role and are so integrated in the process that one should be aware of this at all times.

Synthesis gas may alternatively be generated from solid and liquid feedstocks. The processing of these raw materials may then be added to the above list. The materials handling and unit operations associated with the manufacture of the FT catalysts used also cover a wide spectrum of technologies. It is apparent that almost every unit operation known to man in the fields of chemical or hydrocarbon engineering and materials handling are involved.

3. THE THREE BASIC PROCESS STEPS

Although the application of FT technology usually involves complex integration, it inevitably consists of three basic steps. These are:
- Synthesis gas preparation
- FT synthesis
- Product upgrading

3.1 Synthesis gas preparation depends on the feedstock

Synthesis gas is prepared from a carbonaceous feedstock. The only essential requirement is that the feed should contain carbon. The term carbonaceous implies the presence of carbon. It is also desirable that the feedstock should contain hydrogen since this will increase the efficiency with which it can be converted to hydrocarbon products. If the feed is deficient in hydrogen, hydrogen is obtained from water and energy is required to split the water molecule.

The most important hydrogen lean feedstock is coal. This is because coal is the world's most abundant fossil fuel resource. The process of converting coal (or other hydrogen lean feedstocks) to synthesis gas is known as gasification. To make a synthesis gas suitable for FT synthesis, coal is gasified with steam and oxygen. There are several types of coal gasification technology that may be considered. The selection of the most appropriate technology will depend mainly on the particular characteristics of the coal resource.

If coal is selected as a feedstock, then there will be a large amount of waste process heat generated. This process heat can be harnessed for the production of electrical power. Currently the direct production of electrical power is the most common use for coal. It has been shown that the co-production of hydrocarbons and electricity offers higher efficiencies than individual plants producing these products [4].

Coal conversion is much more expensive than natural gas conversion to synthesis gas. In recent times reserves of natural gas, be it as such or associated with crude oil, have increased and a significant portion of this gas has been called 'stranded' [1]. In other words it is not viable to transport the gas by pipeline to a place where it can be sold. Such stranded gas from remote locations is converted into shippable liquids. There are a number of other possible shippable products besides FT hydrocarbons. These are ammonia, urea, methanol and LNG. The markets for ammonia, urea and methanol are small compared to FT liquids and LNG and the economic incentives to produce these products are no better.

Conversion of methanol into hydrocarbon liquids makes no sense now that hydrocarbon liquids can be produced directly by FT synthesis at a cost that is similar to the methanol production cost. However, production of all the abovementioned products at the same remote site makes a lot of sense, provided that the gas resource is large enough to support this. With the exception of LNG all these products require the production of synthesis gas as a first step. The conversion of natural gas to a shippable liquid product via synthesis gas is classified as Gas-to-Liquids (GTL) technology. It is this GTL approach that holds the most promise for the application of FT technology.

The main component of natural gas is methane. The conversion of natural gas to synthesis gas is called methane reforming. The use of the term 'reforming' is unfortunate because it is not particularly descriptive and it can be confused with the reforming process used to produce petroleum based gasoline. Nevertheless, this terminology is entrenched through wide use so it will also be used in this book.

For methane reforming, as with gasification, the feedstock usually reacts with steam and oxygen to produce hydrogen, carbon monoxide and carbon dioxide. The cleaner nature of the methane feed is conducive to the use of catalysts to enhance the conversion process. Much less carbon dioxide is produced for reforming than for gasification. The use of oxygen in the reforming process is optional. In the absence of internal partial combustion the endothermic steam reforming reaction will require an external source of heat. Methane reforming involves equilibrium reactions so that total consumption of the reactants to produce products is not possible. The exception is oxygen which is totally converted when it is used in the methane reforming process. Excess steam exiting the reformer is removed by cooling and condensing to liquid water. Unconverted methane passes through the FT synthesis step as an inert gas and is then either burned as fuel or recycled as feed to the methane reformer.

Carbon dioxide may be removed from the synthesis gas for total or partial recycle to the reformer inlet. Preferably, it remains in the synthesis gas fed to the FT synthesis step. Carbon dioxide from the FT synthesis step may then be partially recycled to the reformer. A number of different process configurations have been proposed depending mainly on which methane reforming technology and which FT catalyst is selected.

3.2 FT synthesis depends on the feedstock and the desired products

Iron catalysts are better suited to use with coal derived synthesis gas (syngas) because cobalt catalysts are more expensive and it is difficult to prevent coal derived catalyst poisons from reaching the FT catalyst. In the case of high temperature coal gasifiers that produce a low H_2/CO ratio synthesis gas, use of iron catalysts in the wax producing low temperature Fischer-Tropsch (LTFT) process eliminates the need for an upstream shift reactor to increase the hydrogen content in the syngas. Nevertheless, the high temperature Fischer-Tropsch (HTFT) process may still be competitive in spite of the need to use an upstream shift reactor. This is partly because of the higher light olefin content in the HTFT hydrocarbon products that results in increased income from these higher value products. It is also partly because it is easier to achieve high conversions with a high hydrogen content syngas fed to a FT reactor operating at higher temperatures. However, with further processing the LTFT process is also capable of providing higher value products such as lubricant base oils and detergent feedstocks. If markets can be found for these products in large quantities, then the best choice may swing back to the LTFT route. This is even more likely if the FT tail gas is used to produce electricity for sale at a high price.

If coal is used as a feedstock, it is likely that lower conversions will be tolerated in the FT reactor compared to natural gas applications and that the bulk of the FT tail gas will be directed to a gas turbine (operating in combined cycle mode with steam turbines) to produce electrical power for export. These concepts will be discussed in more detail in Chapter 5.

If the syngas is derived from the reforming of stranded natural gas then high conversion levels to FT hydrocarbon products will be required. This is because it is typically not possible to use the remaining reactants for any other useful purpose.

It is generally easy to remove potential FT catalyst poisons from natural gas prior to the reforming step. The production of a syngas composition that is optimal for either LTFT using cobalt catalyst or HTFT iron catalyst is easily achieved without any syngas processing step other than cooling to condense out water. It is more expensive to use an LTFT iron catalyst since it is not possible to achieve high per pass conversions with this type of FT catalyst resulting in the need for expensive recycles or multiple reactor stages.

As with coal applications, the choice for a natural gas feed is between HTFT and LTFT (but now using cobalt catalyst rather than iron). The products are very similar for the LTFT iron and cobalt catalysts.

Four types of FT reactor systems may find commercial application in future. These are:
- HTFT two phase fluidized bed reactors using iron catalysts
- LTFT three phase slurry reactors using precipitated iron catalysts
- LTFT three phase slurry reactors using supported cobalt catalysts
- LTFT tubular fixed bed reactors for special circumstances.

3.3 Product upgrading involves mainly separation and hydroprocessing

Chapter 6 discusses the various possible final products that may be produced from the primary FT hydrocarbons and describes the upgrading processes used to make these products.

The primary liquid product upgrading will typically start with the removal of light hydrocarbons and dissolved gases to make the hydrocarbons suitable for atmospheric pressure storage. In other words the vapour pressure is adjusted to meet storage specifications. This will allow some buffer capacity between the upstream production and the downstream upgrading processes. Olefins may be removed from the straight run liquid products for use as chemical feedstocks. This is achieved by means of fractionation and extractive distillation.

Figure 4 Alpha olefins distillation columns at Secunda

Olefins may be oligomerised, alkylated or hydroformylated to produce special final products or streams that are mixed with the straight run liquids at an appropriate point in the process or at the final product blending. It may be necessary to remove oxygenated compounds prior to these olefin processing options, for example, using liquid-liquid extraction. The remaining material is generally converted to paraffins in a hydrogenation step and fractionated into naphtha and diesel and optionally a kerosene/jet fuel cut. The naphtha may be further upgraded using conventional refining processes to produce gasoline. For LTFT processes using cobalt catalyst the olefin content may not be high enough to justify olefin extraction or processing and then the upgrading only involves hydroprocessing and separation.

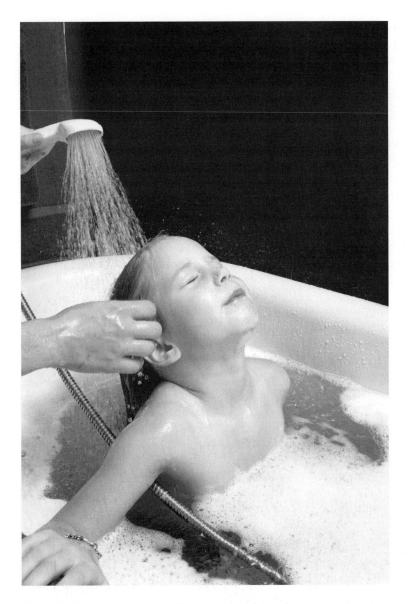

Figure 5 Shampoo made from FT derived olefins

The main reason for the hydrogenation step is to allow long term storage without discolouration or gum formation. The olefin content in the diesel cut may be low enough to be suitable for direct use as a diesel blend material. In this case chemicals may be added to enhance stability during storage. Other chemicals may be added to the diesel cut to improve properties such as lubricity.

The hydrogenation step may be combined with other processing steps that make use of hydrogen. For example hydro-isomerisation may be used to convert straight chain hydrocarbons to branched hydrocarbons. This is done to improve the cold flow properties of the diesel cut. The straight run material may also be combined with the FT wax and sent to a hydrocracker. The term hydroprocessing is then used to include all these possibilities i.e. hydrogenation, hydro-isomerisation or hydrocracking.

The wax product from the LTFT processes may be hydrocracked to provide further naphtha, diesel and optionally kerosene/jet fuel. Alternatively, special hydroprocessing technologies may be employed to also produce high quality lubricant base oils. Vacuum or short path distillation or supercritical solvent extraction may be used to separate wax or heavy lubricant base oils into the desired fractions or cuts.

The wax product may also be processed to make speciality wax products which may require that the wax is hydrogenated, hydro-isomerised or undergoes controlled oxidation. The market for the speciality wax products is small compared to the markets for lubricant base oils and very much smaller than the market for distillates that are produced by hydrocracking the wax.

Production of ethylene (ethene) and propylene (propene) is possible and the market for the use of these monomers to produce plastics is large. They command significantly higher prices than fuels. Butylene (butene) is also a useful petrochemical feedstock. These compounds may be recovered directly from the vapour product of the FT reactor by cooling to lower than ambient temperatures. The associated C_2 to C_4 paraffins may be separated and steam cracked to produce additional olefins. C_3 together with C_4 paraffins may be sold as LPG (liquefied petroleum gas) if there is no suitable cracker nearby. Here there may be some synergy with the paraffinic C_3 and C_4 material recovered from natural gas prior to the syngas generation step.

The naphtha cut discussed above may also be converted to light olefins. Due to the highly paraffinic nature of the naphtha, the olefin yields are higher than for crude oil derived naphtha. The naphtha is easily transported to places where existing naphtha crackers are located.

Oxygenated products are dissolved in the reaction water and are separated from the bulk of the water by distillation. The mixed oxygenates can then be used as a fuel or separated and processed to produce a wide range of chemical products. The scale of the plant generally determines whether the further processing is justified.

14

Figure 6 Ethyl acetate plant at Secunda

Figure 7   Sasol's n-propanol has captured 30 percent of the world market

In summary, the dominant future product is expected to be a high quality diesel motor fuel (also sometimes referred to as a compression ignition engine fuel). The primary FT hydrocarbons require hydroprocessing to produce this product.

There has been a gradual realization that diesel is the most desirable transportation fuel and diesel consumption is increasing at a faster rate than gasoline consumption. All FT processes are ideally suited for diesel production. Wax is easily cracked with a high selectivity to diesel using modern hydrocracking catalysts. Olefins in the FT naphtha are easily converted to diesel using modern oligomerisation catalysts. These diesel sources are then combined with the straight run diesel range material to produce similar amounts of diesel for all variations of the FT process. Thus future FT plants will most likely produce diesel fuel rather than gasoline. In addition this diesel fuel is essentially sulphur free and is cleaner burning than crude oil derived diesel.

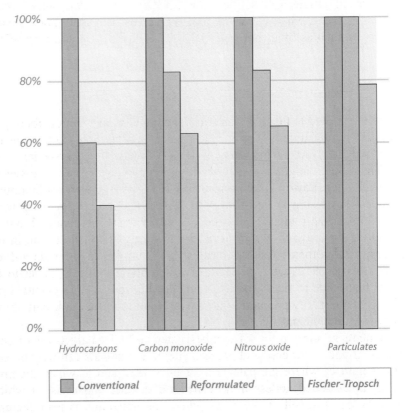

Figure 8 Results from compression ignition (diesel) engine tests carried out by South West Research Institute, San Antonio, Texas

It has been calculated that the production of the diesel rich product spectrum from a GTL plant has a lower greenhouse gas impact than the products from a typical oil refinery using the comparison method recommended by the ISO 14040 standard. GTL diesel is clearly superior when it comes to the local environmental impact at the places where the fuels are used.

Important secondary products are high quality lubricant base oils and olefinic hydrocarbons used in the petrochemical industry. Production of lubricant base oils involves fractionation under vacuum and hydro-isomerisation. Extractive distillation is used to separate the olefins from each mixed hydrocarbon cut while absorption or cryogenic technologies are required to separate the light hydrocarbons containing valuable light olefins from the FT gaseous outlet stream.

The diesel fuel product may be used as a blend material to enhance the properties of crude oil derived diesel. There are also some interesting synergistic effects when FT diesel fuel is blended with other non-conventional diesel fuels such as biodiesel or diesel derived from the processing of bitumen (tar sands).

4. PATENTS

Patents are treated in this book as a useful source of information about potential technology applications. It is not the purpose of this book to venture into the field of patent enforcement beyond the brief discussion in this section.

The product upgrading step is the least costly part of the three-step overall process. It does, however, attract the most attention in the patent literature. This may seem surprising since most of the processes used are well known in the petroleum refining industry. Some adaptation may be required but this is generally easy and well within the knowledge of the skilled person. In fact, the FT derived hydrocarbons are cleaner and simpler than crude oil derived hydrocarbons. The reason for this patent activity is that it is easier to enforce patents that relate to final products or the direct products of a patented process. This may be done by preventing the import and sale of the relevant products in countries where the patents are in force.

Patents relating to the FT synthesis step and FT catalysts are discussed in Chapters 2 and 7. Process and process integration patents can only be enforced in the countries where the patents have been filed and in which the processes (plants) are situated, unless they relate to a specific apparatus, in which case they may be enforced at the place where the apparatus is manufactured. FT catalyst patents can be enforced both where the catalysts are made and where they are used. Direct products from catalyst and process patents are protected.

The product and product upgrading process patents, on the other hand, are not discussed in any detail since they have little to offer in terms of understanding or advancing the application of FT technology. However, those intending to apply FT technology to produce new or speciality products will need to be familiar with the patent landscape in this field.

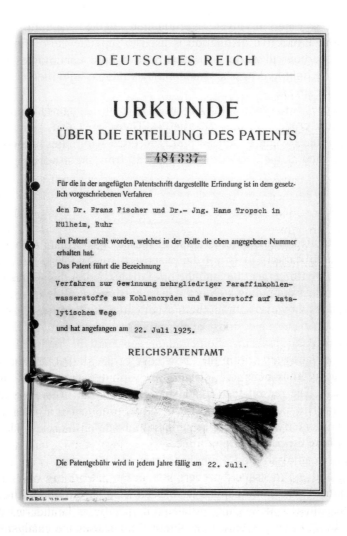

Figure 9 Patent granted to Fischer and Tropsch

It has also been necessary to restrict the discussion of patents relating to FT catalysts to only those that may be of commercial significance due to the large number of patents in this field. Some aspects of FT technology are kept as proprietary trade secrets so that a discussion of these aspects is not possible.

5. CATALYSTS

This book is confined to known catalysts of commercial significance of which there are three broad classes. These are:
- fused iron catalysts and
- precipitated iron catalysts,
- supported cobalt catalysts.

Each class of catalyst has an appropriate application. The techniques used to prepare these catalysts are discussed in Chapter 7.

Each of these catalyst types has scope for further improvement to decrease undesirable characteristics. These undesirable characteristics are often associated with catalyst changes that take place under synthesis conditions. An understanding of these changes is therefore important for attempts to enhance catalyst performance, so these changes are discussed in some depth in Chapter 7. Other improvements will relate to increasing catalyst activity, enhancing selectivity to desired products and inhibiting the formation of unwanted products, specifically methane. Catalyst development remains an area of ongoing research and there is still scope for further improvements. The preparation and characterisation of FT catalysts and their evaluation in laboratory reactors are topics that should be considered for future publications.

5.1 Fused iron catalysts

For the fused iron catalysts, alkali promotion is used to enhance catalyst activity and selectivity. In addition, structural promoters may be used to enhance the surface area of the final catalyst. These promoters are added into a molten bath of magnetite. The magnetite is then cooled to form a solid that is converted into a fine powder for use in fluidized bed reactors. The magnetite powder is first reduced with hydrogen to form the metallic catalyst before being loaded into the synthesis reactor.

In the synthesis reactor the core of the larger particles tends to revert to iron oxide (magnetite) and iron carbides are formed through the remainder of the catalyst. Free carbon tends to accumulate at grain boundaries particularly where promoter rich inclusions are found. This causes the catalyst particles to swell and break after some time under synthesis conditions. Periodic used catalyst partial unloading and reloading of fresh catalyst on-line is used. This keeps the catalyst physical properties and synthesis performance within desired ranges.

High alkali levels are desirable to decrease methane selectivity. There is a practical lower limit for the methane selectivity that can be achieved without causing operational problems in the fluidized bed reactors due to the formation of heavy hydrocarbons in the liquid phase. Other undesirable consequences associated with increased alkali content are an increased rate of carbon formation in the catalyst and increased levels of organic acids in the hydrocarbon products. High carbon formation rates result in increased catalyst consumption. Organic acids may cause corrosion problems and/or may be problematic in downstream refining operations. The lowest selectivity to methane achieved with the best catalysts is about 7% of the carbon in the hydrocarbon products.

The liquid products are highly olefinic. This makes fused iron the most desirable catalyst for the production of olefins for use in the petrochemical industry. Secondary reactions occur in which the light olefins, particularly ethylene, are hydrogenated and also converted to higher hydrocarbons. Promoter interventions may be desired to enhance the light olefin content by inhibiting these secondary reactions. Other promoter interventions may be aimed at decreasing the rate of carbon formation and/or organic acid formation. A more uniform distribution of promoters may also assist to achieve these objectives.

5.2 Precipitated iron catalysts

In order to further decrease the undesirable methane selectivity FT reactors containing hydrocarbons in the liquid phase can be used. The best catalysts can achieve methane selectivities as low as 3% of the carbon in the hydrocarbon products. This requires operation at lower temperatures than those used for the fused iron catalysts discussed above. In order to compensate for the decreased reaction rates at the lower operating temperature, higher catalyst surface areas are required. An inevitable consequence of these higher surface areas is weaker catalyst particles. Hence structural promoters and procedures applied in the catalyst preparation process to enhance catalyst strength are important issues.

The raw material requires to prepare these catalysts is usually iron metal which is more costly than the iron oxide used to prepare the fused iron catalysts. The metal is dissolved into an aqueous acidic solution (e.g. nitric acid) and promoters are added in the desired quantities. As for fused iron catalysts, alkali promotion is important. Copper is also typically added to enhance the reduction of iron oxide to in a subsequent catalyst reduction/conditioning step.

The catalyst is then precipitated from the acidic solution by the addition of a basic solution, for example, a solution containing sodium carbonate or ammonia. The precipitate is then filtered, washed, dried and formed into the shape required for use in the FT reactor. Extrusion techniques are typically used to provide the catalyst shapes used in fixed bed reactors. In the case of a slurry phase reactor application the drying and shaping steps typically occur simultaneously by making use of a spray drier to produce the catalyst powder. This powder should then be heat treated to improve the mechanical strength. Further catalyst treatment may take place in the FT reactor and/or in a separate reduction or conditioning reactor to produce a stable catalyst consisting of carbided iron.

The choice and level of promoters are important in producing a catalyst with a low selectivity to methane and a high selectivity to heavy hydrocarbon products with the desired olefin and oxygenate content in the products. Too high promoter levels have a negative effect on catalyst activity so an optimum level is sought. There are a great number of variables in the catalyst preparation procedure that influence the catalyst activity and the catalyst strength.

5.3 Supported cobalt catalysts

Modern cobalt catalysts are prepared by depositing the cobalt on a pre-shaped refractory oxide support. Typical support materials are silica, alumina, titania or zinc oxide or combinations of these oxides. The support would typically be prepared using a spray drier to provide the desired particle sizes for use in a slurry phase reactor. This might be followed by a classification step to refine the size distribution. Extrusion techniques will typically be used to shape the support for fixed bed applications.

The shaped support is heat treated to improve the mechanical strength. The control of the pore size in the support is an important factor in determining the amount of cobalt that can be placed on the support and the subsequent catalyst performance. Cobalt is then impregnated onto the support together with promoter metals such as lanthanum, platinum, palladium, rhenium and ruthenium. These metals are known to enhance the subsequent reduction step that provides cobalt metal on the catalyst surface. Zirconium has been reported to be beneficial for cobalt supported on silica. Metals such as lanthanum, ruthenium and rhenium have been reported to be effective to facilitate catalyst re-reduction. It appears, however, that promoters are not essential to produce a good supported cobalt catalyst but that effective reduction to metallic cobalt and the support geometry are important issues. The impregnated support is dried and then reduced using hydrogen at high temperatures.

The distribution of the cobalt crystallite size on the support is known to be important. Various methods have been described to control the cobalt crystallite size during the catalyst preparation procedure. Various support modification techniques may be necessary to improve catalyst stability under commercial reactor operating conditions.

Interestingly, supported cobalt catalysts can be classified as the oldest and the most modern commercial catalysts. The first commercial reactors used cobalt promoted with thoria and supported on kieselguhr. The modern catalysts are much more active as a result of improved understanding of the importance of the support geometry; the distribution of the cobalt on the support and enhanced cobalt reduction techniques.

Various different approaches have been adopted for modern reactor design using supported cobalt catalysts. When larger particles are used, for example in fixed bed reactors, it may be beneficial to only deposit the cobalt in a thin layer on the outside of the catalyst support. In this way significant intra-particle temperature and concentration gradients can be avoided. Continuous catalyst rejuvenation (or regeneration) is not possible with the fixed bed reactor design so catalysts are always designed for long term stability with this type of reactor. Continuous catalyst rejuvenation is sometimes proposed for slurry phase reactors. This is achieved by contacting the catalyst with hydrogen at an elevated temperature. In this way cobalt that has been oxidized under synthesis conditions is converted back to the metal. Heavy hydrocarbons that may foul the catalyst are simultaneously removed. This approach allows very small cobalt crystallite sizes to be used resulting in a very active catalyst. However, the first commercial slurry phase reactors using supported cobalt catalyst have been designed for a somewhat lower, but stable, catalyst activity.

The hydrocarbon products from supported cobalt catalysts comprise predominantly paraffins in contrast with the iron catalysts that usually produce olefins as the predominant product. With small catalyst particle sizes and CO rich synthesis gas, the olefin content in the product from supported cobalt catalysts in slurry phase reactors may be high enough to justify the use of the olefins for chemical applications. The primary hydrocarbon products are also highly linear. Compared to the LTFT precipitated iron catalyst the methane selectivity is higher and the selectivity to oxygenated hydrocarbons in the aqueous phase is lower. The best cobalt catalysts have a carbon selectivity to methane of about 5%.

An important characteristic for the best cobalt catalysts for natural gas applications is the absence of water gas shift activity so that significant amounts of carbon dioxide are not produced in the FT reactor. This advantage is only applicable to a comparison with the LTFT precipitated iron catalysts. For the

HTFT fused iron catalysts carbon dioxide can be consumed by the reverse water gas shift reaction, due to the higher operating temperature, which is advantageous. This characteristic allows fused iron catalysts to compete with cobalt catalysts for natural gas applications.

Cobalt catalysts are not suitable for coal applications due to the risk of catalyst poisoning from various impurities that are usually present in coal. These impurities are difficult to remove to the required low levels. The syngas compositions obtained from coal gasification are in any case generally considered to be better suited for processing by iron catalysts.

6. FISCHER-TROPSCH REACTORS

The different possible types of FT reactor are described in Chapter 2 along with some history associated with the development of these reactors. Sufficient information is given in Chapter 2 to understand their commercial application in an integrated plant which is discussed in Chapter 5.

After obtaining an appreciation for their commercial application, a more detailed understanding of the behaviour of the catalyst in the reactor from Chapter 7 will complete the knowledge needed to understand reactor design and modelling.

There have been some academic comparisons between the different types of reactor but these may sometimes be misleading. The reason is that there have been important modifications made to the initial simple reactor designs that considerably enhance their performance. These modifications may be proprietary to individual companies. The proprietary catalysts used by various companies are also not necessarily equivalent. The use of shaped catalysts that reduce pressure drop in fixed bed reactors and the use of liquid rather than gas recycle to control the heat release in these reactors are examples of relatively recent improvements.

Fluid bed reactors have an inherent advantage with higher heat transfer coefficients which is important due to the large amounts of heat that must be removed from the FT reactors to control their temperature. Fluid bed (also called fluidized bed) reactors may be two-phase (gas-catalyst) or three-phase (gas with catalyst suspended in a hydrocarbon liquid slurry). The three phase reactor is also known as a slurry phase reactor. It has been calculated that the heat transfer co-efficient for the cooling surfaces in a slurry phase reactor are five times higher than those for fixed bed reactors [2, 5, 6]. The magnitude of the heat transfer coefficient for a two phase fluidized bed reactor is similar to that for a slurry phase reactor. Until recently, there was a prejudice against running these fluid bed reactors at high velocities. A fluid bed reactor operating

Figure 10 The first truly commercial slurry phase reactor in Sasolburg shown at night.

with an inactive catalyst and a low gas velocity may cost more than an optimised fixed bed reactor with an active catalyst. However, the largest capacity proposed to date for a fixed bed reactor would produce about 8 000 barrels per day of product while fluid bed reactors producing 20 000 bbl/d are in commercial operation. This is an important advantage for large scale applications.

Figure 11 Model for the first SAS reactor (left) at Secunda replacing the CFB Synthol reactor to the right

It should also be remembered that the FT reactor itself is not a dominant cost component in the overall plant cost so the way in which it integrates into the process may be more important than the simple cost of the reactor vessel. The avoidance of excessive gas recycles is important.

It is not the purpose of this book to compare reactor technologies but rather to equip people, as far as possible, to prepare their own designs and draw their own conclusions.

Figure 12 Transport of one of the reactor sections to Secunda for the construction of a 10.7m diameter SAS reactor

7. HISTORICAL BACKGROUND

7.1 German roots for LTFT

Franz Fischer and Hans Tropsch were two chemists working at the Kaiser Wilhelm Institute for Coal Research in Mülheim, Ruhr. The institute was created in 1913, but the work on which Fischer and Tropsch were jointly engaged started producing results only in the 1920's. Their aim was to produce hydrocarbon molecules from which fuels and chemicals could be made, using coal-derived gas.

The first industrial FT reactor was the Ruhrchemie atmospheric fixed bed reactor (1935). It consisted of a box which was divided into sections by vertical metal sheets, the catalyst being loaded between sheets and tubes [1, 7]. They already used the best approach for heat removal from a FT reactor; the water cooling medium is kept at a controlled pressure and is converted to steam at a uniform temperature that corresponds to the boiling point of water at the prevailing pressure.

Figure 13 A meeting at the Kaiser Wilhelm Institute

Fischer and Pichler [1, 8] then developed the 'cobalt medium pressure synthesis' with the main products being middle distillates and wax. This was the most important technology used in Germany during the World War II.

Ruhrchemie AG, a company founded by Ruhr coal industrialists, envisioned the FT synthesis as an outlet for its surplus coke, and upon acquiring the patent rights to the synthesis in 1934, constructed a pilot plant in Oberhausen-Holten (Sterkrade-Holten) near Essen. After initial problems Oberhausen-Holten subsequently became the production centre for a standardized cobalt catalyst used in all the FT plants constructed later in the 1930's. The successful pilot plant research and development at Oberhausen-Holten was a major turning point in the FT synthesis [9].

By November 1935, less than three years after Germany's Nazi government came to power and initiated a push for petroleum independence, four commercial size Ruhrchemie licensed FT plants were under construction. Their total annual capacity was 100 000 – 120 000 metric tons (724 000 – 868 000 barrels) of motor gasoline, diesel fuel, lubricating oil and other petroleum chemicals. The motor vehicle products comprised 72% of the total. Other chemical products included alcohols, aldehydes, soft waxes (which when oxidized gave fatty acids used to produce soap and margarine), and heavy oil for conversion to a detergent known as Mersol [9].

Figure 14 Franz Fischer in his laboratory

All the plants were atmospheric pressure or medium pressure (5 – 15 atmospheres) syntheses at 180 – 200 °C, and used synthesis gas produced by reacting coke with steam in a water gas reaction and adjusting the proportion of carbon monoxide and hydrogen, using a cobalt catalyst (100 Co, 5 ThO_2, 8 MgO, 200 kieselguhr) that Ruhrchemie chemist Otto Roelen developed in 1933 – 38. This catalyst became the standard FT catalyst in Germany [9].

In addition to the nine Fischer-Tropsch plants built in Germany, there was one in France, one in Manchuria and two in Japan.

The 'iron medium pressure synthesis' was also invented by Fischer and Pichler [1, 10], commercialised by the Ruhrchemie and Lurgi companies (ARGE process) and established at Sasol™ in Sasolburg, South Africa in 1955. Similar technology, using modern supported cobalt catalyst and advanced fixed bed reactor design features, is used by Shell at a commercial GTL facility in

Bintulu, Malaysia which started up in 1993. This technology is soon to be the basis for a much larger Shell plant in Qatar.

These are all examples of what is now termed low temperature Fischer-Tropsch (LTFT) technology.

Figure 15 Hans Tropsch

7.2 American roots for HTFT

A two phase fluidized bed reactor using a fused iron catalyst operating at higher temperatures (now classified as HTFT) was developed by Hydrocarbon Research [1, 11, 12] and named the 'Hydrocol' process. A large scale Hydrocol plant operated during the years 1951-1957 in Brownsville, Texas.

It has been known for some time that natural gas is a better feed than coal for the production of hydrocarbon liquids. See, for example, a 1946 publication by R.C. Alden, director of research, Phillips Petroleum [13]. At that time two

American companies had announced their intentions to construct plants producing gasoline from natural gas. It seems that only the Hydrocol plant at Brownsville was eventually constructed. The Alden publication also pointed out the importance of economy of scale for successful GTL applications. So the use of HTFT fluidized bed technology for GTL processes has its roots in the USA. However, due to the fact that abundant crude oil was available and natural gas was close to markets where it could be sold at high prices, GTL applications were not economically viable in the USA. HTFT does form the basis for the world's largest GTL plant currently in operation in Mossel Bay, South Africa, producing gasoline, diesel, LPG and some oxygenated hydrocarbon products. This plant was fully commissioned and running at capacity in 1993. HTFT has yet to find widespread application outside South Africa.

In order to solve the perceived problems with the Hydrocol process, M W Kellogg developed the circulating fluidized bed (CFB) reactor for application by Sasol™ in 1955. Sasol applied this technology with a coal feed that was cheap and abundant in South Africa and located inland, far from any source of crude oil. Due to the initial problems with the CFB reactor the technology, the license was transferred to Sasol. The problems were eventually solved and the process was renamed as the 'Synthol' process. The mixture of hydrocarbons and oxygenated compounds obtained by Fischer and Tropsch in their early work was also given the name 'Synthol' as was a process proposed by an American company in the 1940's [14]. When Synthol is referred to in this book it means the Sasol Synthol™ process.

7.3 Modern commercial applications

The CFB Synthol reactors were used for the world's largest FT application by Sasol in Secunda, South Africa. Sasol subsequently replaced these reactors with more conventional fluidized bed reactors called 'Sasol Advanced Synthol' (SAS) reactors that are currently operating in Secunda. The performance of these reactors is far superior to the Hydrocol fluidized bed reactors. The largest version of the SAS reactor has a capacity of 20 000 barrels per day. The Secunda site has four of these reactors and another five reactors, each with a capacity of 11 000 barrels/day. Three CFB reactors still operate at the GTL facility in Mossel Bay, South Africa.

Figure 16 Even allowing for perspective, the goose-necked CFB Synthol reactors are clearly far larger than the SAS reactors seen on the left. Also the capacity of the SAS reactor on the extreme left is more than the combined capacity of the two CFB reactors seen on the right.

South Africa missed the opportunity to apply conventional fluidized bed HTFT reactors on three separate occasions. Firstly, the decision to use the M W Kellogg CFB reactors rather than the fluidized bed approach was taken in July 1950 shortly before the Hydrocol plant was operated. Secondly, when the Secunda plant design was initiated the appointed engineering contractor for the FT unit (Badger Engineers Inc.) proposed the use of the conventional fluidized bed approach but this was considered too risky given the tight project schedule and the massive capital investment. Thirdly, the decision to proceed with the Mossel Bay GTL project was taken a matter of months before the start-up of the first fluidized bed SAS reactor in Sasolburg in May 1989. The new technology was rightly considered by Sasol to not be sufficiently proven to licence at that time.

Opportunities to use a fluidized bed system, in the form of a slurry phase reactor, were also missed for the LTFT technology. A demonstration scale slurry phase reactor was operated in Germany in 1953, after the Sasolburg plant design had commenced. Secondly, Shell decided that it was too risky to use a slurry phase reactor for their Bintulu plant in spite of obtaining the intellectual property rights for a slurry phase system using cobalt catalyst that was patented by Gulf Oil in 1985.

1993 was an interesting year since it saw the commissioning of the Shell Bintulu GTL plant, the Mossel Bay GTL plant and the first truly commercial scale slurry phase reactor in Sasolburg

For the period from the mid 1950's to 1993 when the Shell plant was commissioned, the history of Fischer-Tropsch technology is synonymous with the history of Sasol. Much of Sasol's history is captured in a book by Johannes Meintjies [15] that mainly covers the period from 1950 to 1975. The Sasol pioneers searched worldwide and found the most advanced FT technology available at that time and Sasol produced the first FT products using both HTFT and LTFT iron catalyst technology in 1955.

7.3.1 Brief Sasol™ history

In the early 1930's a South African mining company, Anglovaal, in co-operation with the British Burmah Company, decided to establish a company to be called Satmar (short for The South African Torbanite Mining and Refining company). Their main objective was to mine and process oil-shale from the Ermelo district. The people assoviated with this development were A S Hersov and S G Menell on the Anglovaal side and A P Faickney on the British side. They lost no time in acquiring the right to use the Fischer-Tropsch process developed in Germany. One of Satmar's first employees was a young chemical engineer, J W (Johnny) van der Merwe. Later van der Merwe was in large measure responsible for the passing of the South African Liquid Fuel and Oil Act of 1947, and that same year went overseas to study the oil-from-coal industry for the South African government. He joined Sasol in December 1950, thus being one of the Sasol pioneers and eventually became joint general manager of Sasol.

In 1936 Dr H J van Eck, the later famous South African industrialist, became the consulting chemical engineer of Anglovaal. The main challenge for him was the possibility of an oil-from-coal industry being established in South Africa. In 1938 Van Eck appointed Etienne Rousseau as a research engineer. Rousseau was later regarded as the 'father' of Sasol. The wheels had begun to turn and Dr Frans Fischer himself visited South Africa. A German-American consortium already existed to develop the FT process further but as another world war was threatening, this co-operation ended and in 1937 fuel prices declined and the South African venture was temporarily halted.

Figure 17 Etienne Rousseau

In 1940 Van Eck left Anglovaal to become managing director of the South African Industrial Development Corporation (IDC), and in 1941 Rousseau left Satmar. Andrew Faickney encouraged by Hersov and Menell, and assisted by Frank Melville and John Beaumont 'carried the torch' as Rousseau once remarked. Menell, Melvill and Beaumont spent a considerable time in the USA during the war. The US Government was still very interested in starting an oil-from-coal venture.

In April 1950 Van Eck, as chairman of the IDC and on behalf of the South African government, went to New York to determine the status of the technology and to see whether finance would be available. On 25 May 1950, S G Menell of Anglovaal proposed to F J du Toit that an interim or action committee be formed with members representing both the government and

Anglovaal. To this committee Anglovaal would transfer all rights accruing from the licence agreements and all coal rights. On the same day Etienne Rouseau was thinking of a name for the project and suggested South African Synthetic Oil Limited (Suid-Afrikaanse Sintetese Olie Beperk) which could be shortened to SASOL. Someone objected to the word 'synthetic'. A 'Gallup poll' in the Greenside and Linden area of Johannesburg found that while people liked the short name 'Sasol', they too felt that 'synthetic' should be excluded from the long name. Thus the long name later became the South African Coal, Oil and Gas Corporation (Suid-Afrikaanse Steenkool, Olie en Gas).

The 'Interim Committee' was officially appointed on 2 June 1950, and its members were F J du Toit (chairman) Dr H J van Eck and Dr M S Louw (representing the IDC), S G Menell and A P Faickney (representing Anglovaal) and P E Rousseau (executive member). Its terms of reference were to consider the state of technical development of all aspects of the synthetic oil industry and generally to establish this new industry.

Figure 18 Sasol's first board of directors and the Zevenfontein farm house that served as Sasol's first offices. Back row from left to right: de Villiers, Rousseau and Louw. Front row from left to right: Faikney, du Toit and van Eck.

Van Eck, Faikney and Rousseau visited America and agreed that the Kellogg process was the best that America could offer and on 14 July they obtained confirmation from the M W Kellogg Corporation that they were prepared to negotiate the licensing of its know-how and patents and to assist in the design and erection of the plant.

On their way back Rousseau told his colleagues, while at a hotel in London, to wait there while he went to look at what the Germans were doing. This was South Africa's the first contact with Ruhrchemie since the outbreak of war. The men were impressed with the German approach. Following them home was a letter with an offer by Ruhrchemie Aktiengesellschaft, Oberhausen-Holten, and Lurgi Gesellschaft für Wärmetecknik GmbH, Frankfurt-am-Main, through an Arbeitsgemeinschaft (promptly abbreviated to 'Arge' at Sasol). The offer was for the designs and rights to operate plants for their Fischer-Tropsch process and the production of synthesis gas from coal.

The Interim Committee then recommended that a Government-sponsored company be formed and that a plant be erected immediately. On 26 September 1950 the South African Coal, Oil Gas and Gas Corporation Ltd., already called Sasol, became a fact as a public company. P E Rousseau was managing director and the members of the provisional directorate were F J du Toit (chairman), H J van Eck, A P Faickney and M S Louw. Staff appointments were A H Stander asassistant to Rousseau; A Brink, construction engineer; M W Neale-May and J W van der Merwe, chemical engineers; and D P de Villiers, company secretary.

That November, Rousseau, de Villiers and a patent attorney, Dr Hahn, who had had a pre-war connection with Fischer-Tropsch, went to Germany to negotiate various details of the Arge process, including license fees and terms for engineering services.

Of the five offers received from firms which had been invited to submit suitable proposals for the erection of a plant, three could be eliminated immediately and only Kellogg's and Arge's received further consideration.

In January 1951 Kellogg sent several top executives and Arge sent two men (Tramm and Herbert) to discuss their respective proposals. The Sasol team initially decided that, in view of the fact that the Kellogg proposal was much cheaper, they should favour Kellogg. However, as German experience and certain aspects of their process were attractive, they finally decided to adopt both processes. The plant would be two thirds American and one third German. The American engineers would be responsible for co-ordination and construction. The basic idea was now to combine the two processes. At first it was thought that they should work in parallel with combined services and utilities but the Americans and Germans came to agree that the two processes could be integrated and this turned out to be a very happy decision.

Members of the Construction Committee were Rousseau (chairman), Neale-May, Brink, van der Merwe, de Villiers and Stander (secretary). With great enthusiasm and the most meticulous eye for detail, this team and those who advised them began to bring Sasol to life.

The first main unit of the Sasol works to be started up was the power station, where the first boiler was fired on oil on 15 June 1954. Once coal was brought to the surface, the power station boilers were switched to coal and in the second half of 1954 the oxygen plant start-up commenced. When Etienne Rousseau fired the first boiler of the power station the first puff of smoke from the chimney was watched with interest by South Africans, Americans, Belgians, Swiss, Germans, Netherlanders and Danes as this was truly a multinational technical effort.

Historically, 1954 was one of the most important years for Sasol. The first coal was produced, the first smoke rose from the chimney stacks, the first oxygen was produced and the purple/blue flare was seen at the gasification plant as the first gasification test runs were conducted in November 1954. At that time Sasol's works manager was William Neale-May. On 23 August 1955 the first synthesis reaction was obtained in the Synthol plant and on 26 September the Arge synthesis was started.

The synthetic fuels and chemicals plant consisted of the following main sections: a steam and power generation plant, the oxygen plant, the gas generating plant, the gas cleaning plant (Rectisol), the Synthol plant, refining and work-up plants, the Arge synthesis plant with its product recovery and work-up plants, a gas reforming unit, a tar acids recovery plant, a tank farm and despatch installations and utility and auxiliary installations. A number of first-of-kind technologies were involved. The gas reforming unit was an early example of an open flame catalytic autothermal reformer. Sasolites were able to use their own petrol (gasoline) from 1 November 1955.

Thus was the beginning of Sasol in outline. On its creation, Minister Eric Louw made a point of expressing the sincere appreciation of the Government for the valuable contribution made by Anglovaal in bringing this national project to the stage of development where it could be taken over as a company sponsored by the South African government.

Dr Van Eck once remarked that South Africa had been extraordinary successful in the industrial field because those involved in the new ventures did not always know all the snags.

'Had Sasol been in any other country,' Dr Rousseau observed, '....I mean in a heavily industrialised country, people would certainly have walked off for other jobs. They would not have seen the thing through. But we virtually had a captive senior staff , and they *had* to see it through... When I walked through the works-office on a Friday afternoon , quarter to seven, the men were still there. It went on for years that they were at the factory every day, Saturdays and Sundays as well. I recall Neale-May and John Carr and Johnny van der Merwe coming to see me at half-past five one afternoon, having had three failures in getting a unit to work that day, and John saying to me, Etienne, do you think we're any good?" What a battle it was!'   Much of the later technology development required similar dedication.

Figure 19 Sasolburg taking shape in 1953

The Sasolburg facility would undergo many changes over the subsequent years. Today it produces only chemicals based on LTFT iron catalyst technology together with various gasification by-products and a world scale ammonia plant using hydrogen derived from the FT tail gas. Ammonia capacity was first added in 1963 using nitrogen that was being vented to atmosphere from the air separation unit and hydrogen extracted from the syngas. The period 1960-63 saw many changes in the growing Sasol. The nearby Fisons fertiliser factory was in production. Units were constructed to produce butadiene and styrene, the raw materials to be supplied to the Synthetic Rubber Company.

Then later a naphtha cracking plant was built to supply ethylene to AECI for the manufacture of plastics. AECI also purchased methane and ammonia for the production of cyanide for sale to the gold mining industry. There was also the chemical manufacturing firm, Kolchem, established jointly by Sasol and National Chemical Products (NCP).

In 1962 a major decision was taken to pursue further development of chemicals rather than increase production of liquid fuels. At that time abundant crude oil was available but there was a fast growing demand for feedstocks for

South Africa's secondary chemical industry. The petrochemical industry and the provision of synthetic rubber and fertiliser were considered to be strategically important.

Figure 20 The Sasolburg factory in recent times devoted wholly to chemicals production

An important new source of clean energy in the region was then provided in the form of pipeline gas. Town gas was fed into a high pressure pipeline and routed through the industrial heartland of the Vaal Triangle, Johannesburg and the East Rand. A separate company, known as Gascor, was formed in 1964 to sell the gas.

By 1975 Sasol was producing a wide range of products: petrol, diesel, kerosenes, light and heavy furnace oils, aviation turbine fuel, liquefied petroleum gas (LPG), bitumen, sulphur, industrial pipeline gas, ethylene, propylene, butadiene, styrene, methane/ethane, liquid and gaseous nitrogen, liquid oxygen, argon, carbon dioxide, olefins for biodegradable detergent manufacture, paraffin waxes, including high congealing-point waxes, oxidised waxes, waxy oils, ethanol, proponol, butyl and amyl alcohols, acetone, methyl-ethyl-ketone (MEK),aromatic/aliphatic solvents, creosote, road tar prime, coal tar pitch, tar acids, anhydrous ammonia, ammonium sulphte, nitric acid, limestone ammonium nitrate, ammonium nitrate solution and sodium nitrate.

As scientific advisor to Sasol, Prof. Helmut Pichler visited Sasolburg regularly from 1957 until a few years before his death in 1974. He was regarded as a Sasolite to such an extent that he was to receive Sasol's long service award – a golden tie pin – after he had been its consultant for ten years. Prof. Pichler was widely recognised as an expert in the field of petroleum-from-coal and had some 180 papers published on basic chemistry and chemical technology. His work had led to the successful application of the so-called medium-pressure synthesis, and he had done much of his research in collaboration with Dr Frans Fischer. Sasolites felt that they were using the Fischer-Pichler process rather than Fischer-Tropsch. In addition to several honorary awards conferred upon him in Germany, he was awarded an honorary doctor's degree by the University of Potchefstroom in South Africa in 1970.

The dream of the founders of Sasol, that a vast industrial complex would rise beside the oil from coal plant became a fact as time went by. The Fisons factory was commissioned in 1958 in order to make use of Sasol's ammonium sulphate. From 1964 they also used ammonia, ammonium nitrate and limestone ammonium nitrate. A second company arrived and later merged with Fisons to form the company known today as Fedmis and they were later joined by yet another fertiliser company, Omnia. AECI erected plants to produce chemicals and plastics from coal and raw materials provided by Sasol. The plastic making operation was taken over by Sasol many years later when the feed could be supplemented on a large scale with raw material from Secunda. Safripol manufactured high-density polyethylene and polypropylene. The Synthetic Latex Company and Orchem were established to serve the rubber industry. Karbochem operated an alkylate factory, producing raw materials for synthetic detergents, as well as plants producing carbon bisulphide and flotation agents for the mining industry. The Natref refinery was commissioned in 1971 and processed crude oil piped to Sasolburg from the coast. This then supplemented the synthetic fuels for the inland market.

Dr Adriaan Hendrikus (At) Stander was the first Sasol person appointed by Rousseau. As manager of research and development for several years he became a key figure in the Sasol organisation and was eventually given the responsibility for the establishment of the second oil-from-coal plant.

From 1970 a series of events on the international oil front showed signs of an impending crisis which was precipitated by the outbreak of the Yom Kippur War in October 1973. In 1971 Sasol reported that oil-from-coal had benefited from world-wide increases in the prices of petroleum products, but warned against undue optimism. In 1972 they sounded more hopeful. In 1973 they reported that in spite of a material improvement owing to sharp oil price increases and promising developments in the gasifier design, it would be unwise to erect a new oil-from-coal plant. In 1974 they disclosed that at the Government's request they had made a detailed study of the technical and financial implications of the establishment of a second oil-from-coal plant. On 5 December 1974 the Minister of Economic Affairs announced that the Cabinet had decided that a second Sasol plant must be built without delay to produce more than ten times the amount of fuel and petroleum products than the existing plant. It was based on the coal fields of the Eastern Transvaal Highveld, where Sasol had over the years acquired coal rights. No sooner was the second Sasol plant up and running when it was decided to duplicate it with a third plant of the same design next door at the new town named Secunda. Thus the world's single largest industrial facility was created while Dr Joe Stegmann was Sasol's managing director.

Figure 21 Map showing the locations of Sasolburg and Secunda

Figure 22 Secunda's second factory (Sasol Three, now Synfuels east) under construction

Doug Malan was given the responsibility to commission the Secunda plants and these plants came on-stream on schedule since mostly proven technology had been used. Nevertheless the commissioning of these enormous and complex plants was a major achievement. Doug Malan later also assisted with the commissioning of the Mossel Bay GTL plant as Sasol's senior representative at that plant site.

Figure 23 Sasol Synfuels west today previously known as Sasol Two

The capacity of the Secunda CFB Synthol reactors was three times larger than the Sasolburg reactors. Several design changes were implemented to eliminate problems that had been experienced in Sasolburg. Early in 1981 André Steynberg reported on the performance testing of the Sasol 2 Synthol units and concluded that they had met their design objectives. Thus the Badger Engineers under the leadership of Dave Jones had achieved an excellent scale-up and they were admirably supported with information and expertise from Sasol, notably that supplied by Terry Shingles. Over the years it was possible to substantially debottleneck the Secunda plants to a large extent due to innovations such as on-line catalyst renewal for the Synthol units. This was pioneered by André Steynberg on the Synthol units in Sasolburg in 1980. The replacement of the two stage cyclone system with single stage cyclones pioneered by Shingles and Steynberg in consultation with the cyclone vendor (Van Tongeren) also improved the Synthol reactor availability. Later further capacity increases were made possible by replacing the CFB reactors with SAS reactors. This was accompanied by focussed debottlenecking of the upstream synthesis gas preparation facilities and the creation of capacity in the downstream refinery by diverting primary products from fuels to chemicals.

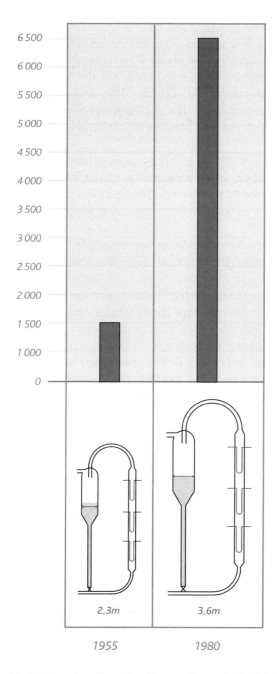

Figure 24 CFB reactors from Sasolburg to Secunda. Initial design capacities are shown. Final operating capacities were 2400 bbl/d in Sasolburg and 7300 bbl/d in Secunda.

Further refinements were made to the CFB reactor design for the Mossel Bay application. Due to the low methane content in the syngas at Mossel Bay the reactor capacity was increased by about 10% compared to the Secunda capacity. Steynberg and Shingles were responsible for the Mossel Bay reactor design which proved to be extremely reliable. The detailed design for the Synthol unit was once again done by Badger Engineers with supervision by Clive Jones from Sasol who ensured that the Synthol unit was brought into operation successfully.

Sasol was privatised in 1979 since it was anticipated that the new large-scale Secunda facilities would be commercially viable without government assistance. This indeed proved to be the case although from time to time, during periods of exceptionally low oil prices, the government supplemented the price paid for Sasol fuels.

In 1975 Johannes C (Jan) Hoogendoorn succeeded At Stander as manager of the Process Development department. The Research department was established in 1957 and Dr Len Dry was the manager for over 20 years during which time it grew in size continuously. Later under Dr Andries Brink the Process Development group moved from the Process Department to the R&D Department and, after the Secunda plants were up and running, Fischer-Tropsch technology development subsequently entered a period of rapid advancement from the early 1980's to the end of the 1990's. Both the Process and the R&D departments initially reported to Theo van der Pas. John Marriott soon took over from van der Pas during this exiting period for technology development and application. Van der Pas and Marriott had earlier played an important technology selection and implementation role for the plants established in Secunda and John Marriott was responsibly for the Sasol technology licence and technical assistance to the Mossel Bay GTL project. Dr Arie Geertsema took over from Andries Brink on his retirement during this period. In addition to the FT technology development Geertsema also drove the research into the extraction of chemicals from FT product streams with great vigour following the strategic guidance from Mr Paul Kruger, who was then the managing director. Emphasis was then given to the global application of FT technology for natural gas conversion and the participation in the global chemical industry under the current managing director Mr Pieter Cox. Dr Chris Reinecke the current R&D manager now heads a global R&D organization.

Jan Hoogendoorn had this to say about the Sasol 2 project at Secunda soon after it was announced: "Sasol began its operation in 1955 with a grass-roots plant in a sparsely populated farming area. Today it is surrounded by petro-chemical industries. It is logical to expect that history will repeat itself at the new plant in the Transvaal, which will also be a grass-roots operation." His prediction was indeed fulfilled.

To quote Meintjes [15]: "How does Sasol do it? This 'industrial whiz kid of South Africa' (*The Star*) has no magic wand after all. Its muscle has developed as a result of a variety of factors – relatively cheap coal, strategic and economic needs, acquired expertise and no-nonsense management approach." To this can now be added that Sasol management has the ability to make courageous decisions to apply new technology. This is based on a proven track record of successful implementation for first-of-a-kind technology.

Figure 25 A graphic representation of a slurry phase reactor used with the Sasol Slurry Phase Distillate process

7.4 Slurry phase synthesis

Early experiments with cobalt catalyst particles suspended in oil were performed by Fischer [1, 16]. The 'liquid phase synthesis' in a slurry bubble column reactor with a powdered iron catalyst and a CO rich syngas from coal was demonstrated by Kölbel et al at the Rheinpreußen Company in 1953 [1, 17]. This reactor operated at an intermediate temperature of 280 °C to produce mainly gasoline.

Modern slurry phase reactors also have some similarities to the oil-recycle process that was developed by the US Bureau of Mines in the 1940's. This was an early example of the so called 'liquid phase' Fischer-Tropsch synthesis [18, 19]. Pilot scale reactors were operated using both cobalt and iron catalysts. Surry phase reactors include ebullating bed reactors and slurry bubble column reactors in which catalyst is free to move while 'liquid phase' reactors includes, in addition, reactors containing hydrocarbon liquid with fixed catalyst beds.

In 1981, a slurry phase reactor pilot plant was operated by Sasol. This reactor was very similar to a pilot plant operated earlier by the US Bureau of Mines and used crushed Arge catalyst. At that time Mark Dry headed the FT research at Sasol and Terry Shingles was in charge of the pilot plants. A young engineer reporting to Terry Shingles, Anton Penkler designed the slurry phase pilot unit.

When the design work commenced on the demonstration scale fluidized bed HTFT reactor (i.e. the SAS reactor demo unit) also in 1981 both Mark Dry and Ben Jager (manager of the Process Development group) expressed the intention that this reactor would later be converted to demonstrate the slurry phase reactor approach. This intention was recorded in the meeting minutes by André Steynberg, a young chemical engineer in the Process Development group who later, in 1989, prepared the design for the modifications required to achieve this conversion.

In 1982 Dry [20, 21] proposed using slurry phase reactors to ultimately produce mainly diesel with naphtha as a significant co-product by using a scheme in which the reactor wax is hydrocracked. It was further proposed by Dry that the naphtha is a good feedstock for thermal cracking to produce ethylene. Gulf Oil [22], in 1985, proposed the use of a slurry reactor with a modern precipitated cobalt catalyst to produce mainly diesel as a final product.

In the 1980's the US Department of Energy (DOE) and the South African Council for Scientific and Industrial Research (CSIR) were actively promoting the use of slurry phase reactors for FT synthesis. In the case of the DOE, a pilot plant design was prepared in 1989 and a pilot plant was later operated; first as a methanol reactor; then using iron catalyst and finally using cobalt catalyst. In the case of the CSIR, useful hydrodynamic information was obtained using a large scale non-reactive mock-up using hot wax in 1989.

Figure 26 The demonstration reactor at Sasolburg used firstly to demonstrate SAS reactor operation then slurry phase reactor operation with iron catalyst and most recently the slurry phase reactor operation with supported cobalt catalyst.

Sasol now uses a commercial slurry phase reactor containing an iron based catalyst in Sasolburg to produce mainly wax. Rentech has also operated small scale slurry phase reactors (at very low gas velocities) using iron catalyst to produce diesel and naphtha final products [23]. Many other companies propose to use variations of this type of reactor, e.g. Statoil, ExxonMobil, Syntroleum, IFP/ENI/Agip and ConocoPhilips. These reactors are given various names e.g. 'slurry phase', 'slurry bubble column' and 'liquid phase'. Sasol currently operates the only truly commercial slurry phase reactor at Sasolburg. Construction has commenced for two reactors, each with a capacity of 17 000 bbl/d, to be installed by Sasol in Ras Laffan, Qatar. These Qatar reactors will use supported cobalt catalyst. Cobalt catalyst is typically used in the slurry phase reactor when the plant is fed by natural gas.

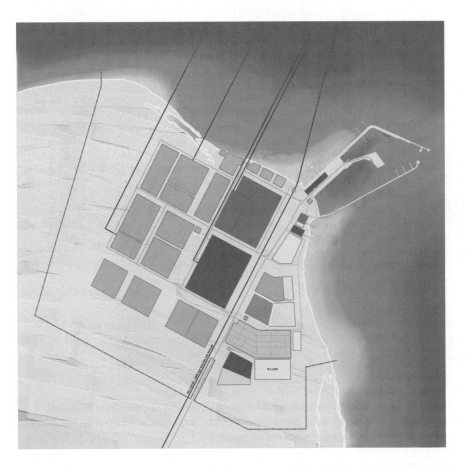

Figure 27 A schematic indication of the plans for the industrial site at Ras Laffan in Qatar with the site for the first Sasol GTL plant shown in light blue

Site preparation has also commenced for a second plant at Escravos, Nigeria to convert associated gas to diesel and naphtha using identical FT reactor designs to those used for the Qatar site.

Figure 28 The Escravos Delta in Nigeria, the site for the first GTL plant for Sasol-Chevron

When the feed is coal or some other carbon rich source, then an iron based catalyst may be preferred. Other special circumstances may influence the type of catalyst selected [24].

7.5 Modern reactor development achievements

It is not possible to do justice to all the people at Sasol and Badger (now Stone & Webster part of the Shaw Group) who contributed in important ways to the development of the new reactor technologies. For example, Tom du Toit an old hand at Synthol operations provided valuable input into the SAS reactor demonstration plant design. Roy Silverman (who was later responsible for running this unit) and Rodney Dry gave invaluable insight into the operation of fluidized systems when working under the direction of Terry Shingles. Silverman later joined Badger Engineers and gave valuable input for the design of modern slurry phase reactors using supported cobalt catalyst. The individuals responsible for the development of modern supported cobalt FT catalysts are named in the various published catalyst patents. Over many years, Mark Dry accumulated a wealth of knowledge on the behaviour of iron catalysts and their refinement to suit various operating conditions. Optimum reactor design and operation is not possible without this knowledge.

Figure 29 First oil from the first SAS reactor at Secunda

Evert Kleynhans made his mark with excellence in engineering supervision and commissioning. Dave Jones from Badger played an indispensable role in the direction of much of the development, scale-up and engineering work for the first-of-a-kind applications. Numerous Sasol managers were extremely supportive of and involved in these technology developments. Dr Ben Jager, who retired in 2003, was the manager responsible for process development during the entire period during which the major technology advancements were made and often provided valuable technical guidance.

Table 1

History of Sasol FT process development listing technical personnel responsible and key management decisions

Date	Event	Sasol	Badger[1]
1981	Design of nominal 1m diameter demonstration reactor for SAS technology	A Steynberg T Shingles	Y Yukawa
1981	Design and commissioning of a slurry phase reactor pilot unit	T Shingles A Penkler	
1983	Commissioning of SAS demonstration reactor	R Silverman T Shingles A Steynberg	Y Yukawa
1985	Engineering study for SAS reactor application	A Steynberg	Y Yukawa
1987	Decision to install commercial 5m diameter SAS reactor	Sasol management	
1988	Preparation of Mossgas Synthol CFB reactor design	A Steynberg T Shingles	Y Yukawa
May 1989	Successful start-up of 5m diameter SAS reactor	E Kleynhans	Y Yukawa
June 1989	Proposal to modify demonstration reactor to a slurry phase reactor	A Steynberg	
1990	Start-up of modified demonstration reactor as a slurry phase reactor	R Kelfkens A Steynberg	
Feb 1991	Risk assessment for the scale-up of the slurry phase reactor	R Kelfkens	D Jones Y Yukawa
July 1991	Decision to modify existing 5m diameter SAS reactor to a commercial slurry phase reactor	Sasol management	
1992	Mossgas start-up	C Jones D Malan	H Doleman Y Yukawa
1992	Process Design Basis set for 8 meter diameter SAS reactor at Secunda	A Steynberg	
May 1993	Completed plant modification and start-up of 5m diameter slurry phase reactor	R Kelfkens A Steynberg	Y Yukawa L Bonnell
June 1995	Start-up of 8m diameter SAS reactor at Secunda	E Kleynhans	Y Yukawa
March 1996	Kick-off meeting to replace all CBF reactors with SAS reactors in Secunda including four 10.7m diameter reactors	E Kleynhans A Steynberg	Y Yukawa M Lee
Oct 1996	Completion of Generic Sasol Slurry Phase Distillate Process Optimisation Study based on newly developed cobalt catalyst	J Price A Steynberg A Vosloo	S Pal R Silverman J Williams
May 1997	Completion of optimization study for the Slurry Phase Distillate Process using Haldor-Topsoe Autothermal Reformers	J Price M Coetzee A Steynberg	
1997	Design Basis set for Qatar plant	H Nel A Steynberg	R Silverman M Lee
Manager - Process Development during entire period		- B Jager	
Manager - FT Research during entire period		- M Dry	

[1]Badger Engineers later became Raytheon Engineers & Constructors then The Washington Group International and now Stone & Webster, part of the Shaw Group

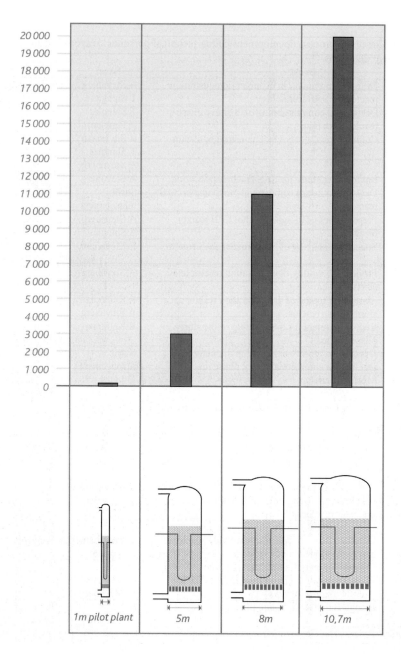

Figure 30 Illustration of the SAS reactor scale-up

Figure 31 Transportation of SAS reactor sections for erection at the Secunda site.

Figure 32 SAS reactor under construction at Secunda

Figure 33 In little more than a year, seven SAS reactors were erected at Secunda and successfully commissioned

The other companies that have worked on the development of slurry phase reactor technology will appreciate the major achievement to go from the design of a slurry phase demonstration reactor in 1989 to the operation of a commercial scale unit in 1993. André Steynberg, Ben Jager and Renus Kelfkens were awarded the South African Institution of Chemical Engineers (SAIChE) gold medal award for this achievement. Ben Jager and André Steynberg, were previously awarded this same medal for the development of the SAS reactor four years earlier along with Terry Shingles and Mark Dry from Sasol and Yoshio Yukawa from Badger Engineers.

8. PIONEER TECHNOLOGY DEVELOPMENT

The application of Fischer-Tropsch synthesis on a commercial scale builds on developments in other technology fields.

The initial synthesis gas feed was derived from coke-oven-off-gases that were available in excess from the German steel industry. Later, the very successful Lurgi gasifier was developed to produce synthesis gas from coal. This type of gasifier is capable processing low quality coal that is not useful for other purposes. The design of this type of gasifier has been gradually improved with time and the maximum capacity for a single gasifier has increased substantially.

Modern carbon conversion plants may now consider higher grade carbonaceous feeds such as low ash coal and petroleum coke using the newer high temperature entrained flow gasifiers. The Texaco gasifier is currently the most successful version. These modern gasifiers have been applied to the production of electrical power using steadily improved technology known as integrated gas combined cycle (IGCC) technology. This technology can now be considered for the combined production of hydrocarbons and electricity with synergistic benefits. Depending on the coal properties, modern Lurgi gasifiers may still be the gasification technology of choice. Gasification technology is discussed in Chapter 4 and the applications for coal derived synthesis gas are discussed in Chapter 5.

In the late 1980's it was realised that the lowest cost source of synthesis gas would be from remote natural gas. The Shell plant in Malaysia uses non-catalytic partial oxidation reforming while the Mossel Bay plant uses a two step approach of tubular steam reforming followed by autothermal reforming. Methane reforming technology has gradually improved over the years particularly with respect to the burner design and the ability to make larger capacity units. It has been discovered only very recently that the oxygen fired autothermal reformer (ATR) using low steam to carbon ratios of 0.6 or less is the technology of choice for large scale synthesis gas production.

Catalytic tubular steam reforming is still competitive for small scale chemical or refinery applications especially if hydrogen or a hydrogen rich gas is required but there is no version of steam reforming that can compete with ATR for the large plants that are necessary for FT applications to be viable. Catalytic or non-catalytic partial oxidation reforming is not significantly more expensive than ATR. Reforming technology is discussed in Chapter 4 and the importance of integrating the reforming and FT plant designs is discussed in Chapter 5.

FT technology development was preceded by other high pressure synthesis processes. The introduction of the Haber process for the high-pressure synthesis of ammonia provided Germany with a leadership role in high-pressure processes, especially those involving hydrogen. The importance and the scientific advances associated with the ammonia synthesis were recognized and Fritz Haber was awarded the Nobel Prize in 1918.

Figure 34 Fritz Haber (left) and Alfred Nobel (right)

After much research and development work, the first commercial success for the production of synthetic hydrocarbons from coal was in 1925 with the direct hydrogenation of coal (especially German brown coal) at high-pressure, high-temperature conditions. This produced liquids that could be utilized as transportation fuels. This was the Bergius Process and its discoverer, Friedrich

58

Bergius, was awarded the Nobel Prize in 1931. Bergius was led to this effort by his desire to be an entrepreneur and a desire to free Germany from dependence upon external oil supplies for their transportation fuels.

Figure 35 Friedrich Bergius

I.G. Farbenindustrie (Farben) purchased the rights to the Bergius process. Farben was attracted to this process, at least in part, because of the expertise they had developed with other high-pressure processes. By 1925, Carl Bosch had risen to a high leadership position in Farben following his successful efforts to commercialize first the ammonia synthesis process, discovered in 1908, and then the successful conversion of synthesis gas to methanol. In this work, Farben had become recognized for its ability to produce and utilize hydrogen at high pressures. For his leadership role in developing high-pressure processes and the materials of fabrication that was needed for the processes, Bosch was awarded the Nobel Prize together with Bergius in 1931.

So a number of other technology developments were prerequisites for the cost effective application of Fischer-Tropsch technology. Perhaps the most important has been the developments associated with the separation of relatively pure oxygen from air using cryogenic techniques. Essentially air is compressed, cooled and expanded to produce very low (cryogenic) temperatures. Liquid oxygen is then separated by distillation and pumped as a liquid to the pressure required. The oxygen is then vaporized and then preheated for use in the reformer. Efficient heat integration is important but the eventual design required is a compromise between high efficiency and low capital cost. Cost effective large scale units are now available from a number of reputable vendors such as Air Liquide, Air Products, Praxair, Linde and BOC. The combination of these air separation units with oxygen fired reformers has led to the cost effective preparation of synthesis gas from remote natural gas.

9. BASIC STUDIES

The aim of the basic studies is to understand the reactions that take place on the FT catalyst so that these reactions can be described accurately using predictive models. These models should then be capable of predicting changes in the consumption of reactants and the distribution of products with variables such as temperature, gas component partial pressures, catalyst promoter content and age dependant catalyst properties. This approach has been referred to as "kinetic modelling of FT synthesis" [1]. In engineering terminology 'kinetic modelling' is restricted to the prediction of the consumption of the reactants while the term 'selectivity modelling' is used to describe the prediction of the product distribution. Using a fundamental scientific approach both reactant consumption and product distribution are predicted simultaneously.

It is furthermore of prime importance to decrease the production of methane and it is hoped that a fundamental understanding of the methane formation mechanism will eventually lead to interventions that will decrease methane production. The reasons why the methane selectivity deviates from the theoretical prediction based on simple polymerization theory forms part of this fundamental understanding.

Other desirable interventions may be to increase the olefin or oxygenate content in the FT products in order to extract these components as chemical products that are more valuable than fuels.

Investigating FT activity and selectivity as a function of time has showed that the FT regime develops in a transient manner. The steady-state of FT synthesis is attained through the restructuring of the catalyst under the influence of reactants, products and intermediates. **Thus it can be said that the 'true catalyst' is formed and stable only in situ.** This may complicate the interpretation of laboratory results. There needs to be a verification that the catalyst studied at constant conditions of temperature and composition in the laboratory is the same as the catalyst in the industrial reactor where it is exposed to a range temperatures and reactant/product concentrations. Often the actual stable conditioned catalysts in the laboratory and industrial reactors are different even if they were prepared initially in the same way, particularly in the case of precipitated iron LTFT catalysts.

The composition of the hydrocarbons obtained by FT synthesis exhibits a distinctly regular pattern [1]. As a first approximation the molar amount of individual carbon number fractions declines exponentially with carbon number. This behaviour was originally noticed by Herrington [25]. It indicates a sort of polymerization kinetics. A mathematical model for homogeneous polymerisation had been formulated previously by Schulz [26] and Flory [27]. The FT polymerisation model was extended for chain branching by Friedel and Anderson [28]. In an ideal polymerisation reaction the chain propagation probability would be constant and independent of the carbon number. A logarithmic plot of product moles for each carbon number species versus the carbon number would then yield a straight line. This type of plot is now known as an ASF-diagram (Anderson-Schulz-Flory) and is commonly used to characterise FT synthesis products.

More recently, important contributions have been made to the understanding of the FT reactions particularly with regard to the secondary reactions that take place. The kinetic model developed by Schulz et al. was extended to account for several reactions of product desorption as olefins, paraffins, aldehydes and alcohols. The FT kinetic regime was termed 'non trivial surface polymerisation' in order to acknowledge its dual nature as a heterogeneous catalytic reaction and a polymerisation reaction [29-31]. The monomer is formed on the catalyst surface (from CO and H_2). The primary FT product distribution is determined by selective inhibition of the product desorption reactions. Thus the combination- (polymerisation-) reactions of surface species are faster than the desorption reactions.

A kinetic model was mathematically worked out in which the kinetic coefficients of individual reaction steps were defined as reaction probabilities of surface species. This selectivity model was linked to the rate of consumption of the CO and H_2 reactants. This then enables all the steps of the kinetic scheme to be calculated to predict the formation rate of each product compound. Using the model in reverse, it converts the experimentally obtained product distributions into the basic kinetic data which determine the product selectivity, e.g.:

- chain growth probability as a function of carbon number,
- chain branching probability as a function of carbon number,
- olefinicity per carbon number

FT synthesis on iron and cobalt catalysts have been compared using this model [32]. This revealed basic differences in kinetic behaviour.

Reversibility of olefin desorption, secondary olefin hydrogenation and olefin isomerisation have been introduced into the kinetic model together with chain length dependant solubility of olefins in the product liquid [33, 34].

Alternatively to effects of solubility, diffusion enhanced olefin re-incorporation in liquid filled catalyst pores has been suggested to cause carbon number dependencies of olefin to paraffin ratios as well as carbon number dependent probabilities of chain growth [35, 36], which cause 'positive deviations' from ideal ASF kinetics as often observed in ideal distributions. A positive deviation occurs when there is an increase in the probability of formation of a product species as the carbon number increases. The effect of solubility on the product distribution has however been demonstrated to usually dominate other effects such as product diffusivity [37].

Other researchers have modelled the deviations from ideal ASF kinetics for LTFT synthesis with the superimposition of two ideal distributions while postulating that chain growth on two different growth sites might occur or that there are two mechanisms occurring simultaneously [38, 39].

A recent review by Puskas and Hurlbut [40] examined the reasons for deviations from the ideal ASF plot. They challenged the 'evidence' for higher olefin incorporation into the FT products in secondary reactions and dispute the claims that these reactions are responsible for 'positive' deviations' from the ASF plot. In other words the product becomes heavier than predicted by a straight line ASF plot so that the actual plot curves upwards with increasing carbon number. Puskas and Hurlbut argue that a multiplicity of chain growth probability is the only reasonable cause for the reported 'positive deviations' in the C_6+ carbon number product range.

For the ideal ASF plot the product distribution is characterised by a single parameter, α which can be obtained from the slope of the plot and is a measure of the probability of chain growth. This type of ideal distribution can only be obtained if the kinetic environment is identical and constant at each catalytic site of the synthesis. In real reactors this is never achieved. The HTFT reactors in nearly isothermal fluidized beds come closest to this ideal situation. It has been shown that for these reactors α is not significantly affected by the gas composition changes through the reactor. For these reactors ideal ASF behaviour is closely approximated for the C_6+ products. For LTFT reactors, positive deviations are usually observed. At LTFT conditions, the gas composition and sometimes temperature profiles through the reactor result in a variation in the value of α as the reactants pass through the reactor. Thus a multiplicity of chain growth factors is produced. This results in a concave curvature for the ASF plot rather than straight lines. However, 'positive' deviations have also been observed in laboratory reactors with uniform temperature and composition so variations in conditions cannot explain all cases of concave curvature for the ASF plot.

It is also concluded that the case for 'negative' deviations cannot be regarded as being firmly established and that reported 'negative' deviations were most likely the result of analytical or sampling inadequacies.

In the case of a multiplicity of α values giving an ASF plot with positive curvature, it was found that this can be mathematically defined by two 'imaginary' α's and associated 'imaginary' product proportions. Thus actual C_{6+} product distributions can be predicted by fitting the two α parameters and another parameter, say β, that fixes the relationship between the two α's. Sasol researchers have been able to relate the parameters for this type of approach to catalyst formulations and operating conditions [41].

The above concepts are explained in detail in Chapter 8. Furthermore, the effects of reaction parameters (i.e. partial pressures of hydrogen, carbon monoxide and water; temperature and catalyst properties) on the individual reactions are discussed. This knowledge is considered to be crucial when either aiming for high C_{5+} selectivity or when aiming at the production of valuable chemicals from FT-synthesis, such as long chain alpha-olefins and oxygenates, where secondary reactions need to be forced back.

REFERENCES

[1] H. Schulz, Short history and present trends of Fischer-Tropsch synthesis, Applied Catalysis A: General 186 (1999) 3-12.

[2] J.J.C. Geerlings, J.H. Wilson, G.J. Kramer, H.P.C.E. Kuipers, A.Hoek and H.M. Huisman, Applied Catalysis A: General 186 (1999) 27.

[3] W. de Graaf and F. Schrauwen, Hydrocarbon Engineering (May 2002) 55.

[4] A. P. Steynberg and H.G. Nel, in ACS Spring National Meeting Symposium on Clean Coal Technology, K. Miura (ed.), American Chemical Society, New Orleans (2003).

[5] J.W.A. de Swart, PhD Thesis, University of Amsterdam, (1996).

[6] J.W.A. de Swart, R. Krishna and S.T. Sie, Studies in Surface Science and Catalysis, 107 (1997) 213.

[7] H. Pichler, F. Roelen, F. Schnur, W. Rottig and H. Kölbel, Kohlenoxidhydrierung, in Ullmanns Enzyklopadie der technischen Chemie, Urban a. Schwarzenberg, München-Berlin, (1957), 685.

[8] F. Fischer and H. Pichler, Brennstoff-Chemie 20 (1939) 41.

[9] A. N. Stranges, in AIChE 3rd Topical Conference on Natural Gas Utilization (Chen-Hwa Chiu, R.D. Srivastava and R. Mallison, Ed.), New Orleans (2003) 635-646.

[10] F. Fischer, and H. Pichler, Ges. Abh. Kenntn. Kohle 13 (1937) 407.

[11] P.C. Kieth, Gasoline from Natural Gas, The Oil and Gas Journal, May 16 (1946).

[12] J.H. Arnold and P.C. Kieth, in ACS meeting, Advances in Chemistry, (1951) 120.

[13] R.C. Alden, Conversion of Natural Gas to Liquid Fuels, The Oil and Gas Journal,

November 9 (1946) 79-98.

[14] B.H. Weil and J.C. Lane, Introduction, in Synthetic Petroleum from the Synthine Process, Rensen Press Division, Chemical Publishing Co., Brooklyn, New York, 1948, p.4.

[15] J. Meintjes, Sasol 1950-1975, Tafelberg Publishers Ltd., Cape Town (1975).

[16] F. Fischer and O. Roelen and Feiβt, Brennstoff-Chemie 13 (1932) 461.

[17] H. Kölbel and P.Ackermann, Chem. Ing. Techn. 28 (1956) 381.

[18] B.H. Davis, Overview of reactors for liquid phase Fischer-Tropsch synthesis, Catalysis Today 71 (2002) 249-300.

[19] H.H. Storch, N. Golombic and R.B. Anderson, The Fischer-Tropsch and related synthesis, Wiley, New York,1951.

[20] M.E. Dry, Sasol's Fischer-Tropsch experience, Hydrocarbon Processing (1982) 121-124.

[21] M.E. Dry, The Sasol route to fuels, Chemtech (1982) 744-750.

[22] H. Beuther, T.P. Kobylinski, C.E. Kibby and R.B. Pannell, South African Patent No. ZA 855317 (1985).

[23] Rentech Inc. website, http://www.rentechinc.com.

[24] G.G. Oberfell, Utilization of Natural Gas in the U.S., National Petroleum News, 38, 1 (1946) R-46.

[25] E. F.G. Herrington, Chem. Ind. 65 (1946) 346.

[26] G.V. Schulz, Z. Phys. Chem. B 30 (1935) 379.

[27] P.J. Flory, J. Am. Chem. Soc. 58 (1936) 1877.

[28] R.A. Friedel and R. B. Anderson, J. Am. Chem. Soc. 72 (1950) 1211, 2307.

[29] H. Schulz, K. Beck and E. Erich, Stud. Surf. Sci. Catal. 36 (1988) 457.

[30] H. Schulz, K. Beck and E. Erich, Fuel Proc. Techn. 18 (1988) 293.

[31] H. Schulz, K. Beck and E. Erich, in M. Phillips and M. Ternan (eds.), Proc. 9th Congress on Catalysis, vol. 2, Calgary 1988, The Chemical Institute of Canada, Ottawa (1988) 829.

[32] H. Schulz, E van Steen and M. Claeys, Stud. Surf. Sci. Catal. 81 (1994) 455.

[33] H. Schulz and M. Claeys, Appl. Catal. A: 186 (1999) 71.

[34] H. Schulz and M. Claeys, Appl. Catal. A: 186 (1999) 91.

[35] R.J. Madon, S.C. Reyes and E. Iglesia, J. Phys. Chem. 95 (1991) 7795.

[36] E. Iglesia, S.C. Reyes, R.J. Madon and S.L. Soled, Adv. Catal. 39, 2 (1993) 221.

[37] E.W. Kuipers, I.H. Vinkenberg and H. Oosterbeek, J. Catal. 152 (1995) 137.

[38] G.A. Huff and C.N. Satterfield, J. Catal. 85 (1986) 370.

[39] J. Patzlaff, Y. Liu, C. Graffmann and J. Gaube, Appl. Cat. A: Gen. 186 (1999) 109.

[40] I. Puskas and R.S. Hurlbut, Catalysis Today, 84 (2003) 99-109.

[41] A.C. Vosloo, Die Ontwikkeling van 'n Teoretiese Selectiwiteitsmodel vir die Fischer-Tropscc sintese, PhD Thesis, University of Stellenbosch (1989).

Credits

Studies in Surface Science and Catalysis 152
A. Steynberg and M. Dry (Editors)

Chapter 2

Fischer-Tropsch Reactors

A. P. Steynberg[a], M. E. Dry, B. H. Davis and B. B. Breman

[a]Sasol Technology R&D,
P.O. Box 1, Sasolburg, 1947, South Africa

1. TYPES OF REACTOR IN COMMERCIAL USE

There are four types of Fischer-Tropsch (FT) reactor in commercial use at present. Three broad categories of catalyst are used in these reactors. The four types of reactor are:

- Circulating fluidized bed reactor
- Fluidized bed reactor
- Tubular fixed bed reactor
- Slurry phase reactor

The fluidized bed reactors operate in the temperature range 320 °C to 350 °C. This temperature range is 100 °C higher than the typical operating temperature range used with the reactors shown on the right hand side of Fig. 1 of around 220 to 250 °C. Hence the term high temperature Fischer-Tropsch (HTFT) used to describe the reactors on the left hand side and the term low temperature Fischer-Tropsch used to describe the reactors on the right hand side of Fig. 1.

The key distinguishing feature between the HTFT and LTFT reactors is the fact that there is no liquid phase present outside the catalyst particles in the HTFT reactors. Formation of a liquid phase in the HTFT fluidized bed reactors will lead to serious problems due to particle agglomeration and loss of fluidization, [1, 2]. It was postulated by Caldwell [3] that reactors will be prone to waxing if the Anderson-Schulz-Flory (ASF) plot intersects the vapour pressure plot. These concepts are explained in more detail in Chapter 3. The slope (α) of the ASF plot is a measure of the chain-growth probability for the production of hydrocarbon products. The catalyst and operating conditions may be selected to obtain the desired chain-growth probability (α) or, in other words, the desired product spectrum.

The four reactor types are illustrated in Fig. 1. :

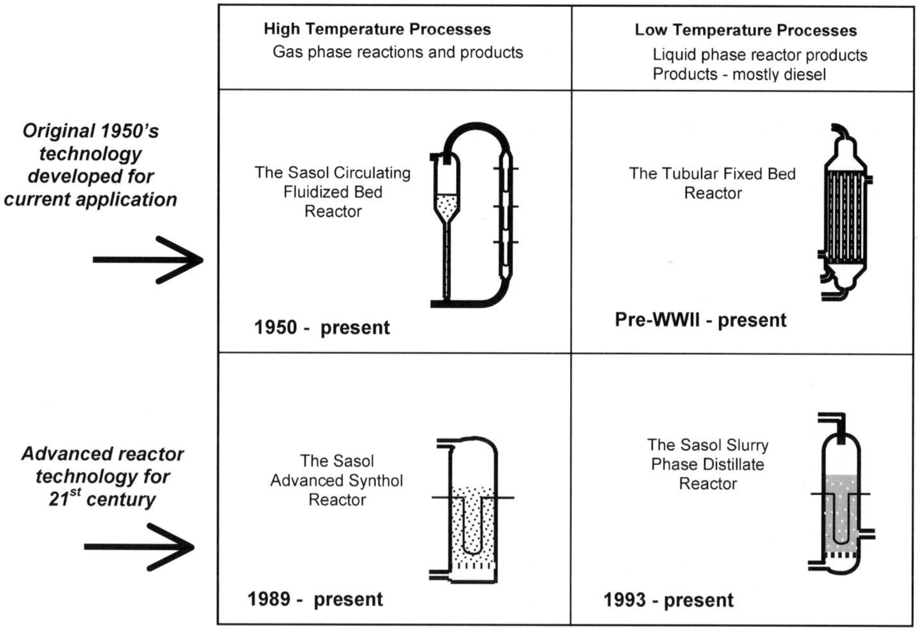

Figure 1. Types of FT reactor in commercial use

Table 1 shows the minimum temperature at which the reactor can operate to maximize a given hydrocarbon cut. For a given target α value, this determines the lower end of the feasible temperature range. This means that fluidized bed reactors cannot be used for maximized production of products heavier than the gasoline/naphtha cut [4]. At the upper end of the feasible temperature range, the typical catalysts used for fluidized bed reactors cannot operate much above 350°C without excessive carbon formation.

Table 1

Chain-growth probabilities (α) to maximize various product cuts and the minimum reactor temperature required to avoid a liquid phase [4]

Cut maximized (by mass fraction)	α	Minimum temperature to avoid liquid condensation (°C)
C_2 - C_5	0.5081	109
C_5 - C_{11}	0.7637	329
C_5 - C_{18}	0.8164	392
C_{12} – C_{18}	0.8728	468

The LTFT reactors are shown on the right hand side of Fig. 1. Heavy hydrocarbons in the form of liquid wax are present in these reactors.

Either precipitated iron catalysts or supported cobalt catalysts may be used in LTFT reactors. The choice of catalyst will be discussed in more detail later. At this point, note that commercial scale reactors exist or are under construction using both types of catalyst in both types of LTFT reactor.

Fluidized bed reactors are subdivided into two-phase (solid and gas), HTFT and three phase (solid, liquid and gas), LTFT systems. When the main objective is the production of long chain waxes the LTFT process is used and either multi-tubular fixed bed or three phase fluidized bed slurry reactors can be considered. For the HTFT process where alkenes and/or straight run fuels are the main products then only two phase fluidized systems are used.

Both types of HTFT reactor, shown on the left and side of Fig. 1, are also currently in commercial operation. The circulating fluidized bed (CFB) reactors are in use at the world's largest GTL plant in Mossel Bay, South Africa. The Sasol Advanced Synthol (SAS) reactors are used at the world's largest synthetic hydrocarbon plant which is based on coal derived synthesis gas in Secunda, South Africa. This is the reactor of choice for future HTFT applications.

Compared to many industrial operations the FT reaction is highly exothermic. The average heat released per 'CH$_2$' formed is about 145 kJ [5]. This is an order of magnitude higher than typical catalytic reactions in the oil refining industry. Any increase in the operating temperature of the FT synthesis will result in an undesirable increase in the production of methane and may result in catalyst damage. For HTFT in particular, high temperatures result in excessive carbon deposition. It is therefore very important that the rate of heat transfer from the catalyst particles to the heat exchanger surfaces in the reactor is high in order to maintain near-isothermal conditions inside the catalyst beds.

1.1 Fixed bed reactors

Vertical spaced packed bed and radial flow reactors with cooling between the beds are not satisfactory because of the negative effects of temperature rises within each individual adiabatic bed. The preferred fixed bed reactor type is multi-tubular with the catalyst placed inside the tubes and cooling medium (water) on the shell sides. Having a short distance between the catalyst particles and the tube walls (by using narrow tube diameters) and operating at high gas linear velocities, to ensure turbulent flow, greatly improves the transfer of the heat of reaction from the catalyst particles to the cooling medium. In order to achieve high percent conversions of the fresh feed gas it is common practice to recycle a portion of the reactor tail gas. This practise also of course increases the linear velocity through the reactor and hence further increases the rate of heat transfer. Recycle of liquid hydrocarbon product is also known to improve the temperature profile in the fixed catalyst bed.

The smaller the catalyst pellets or extrudates used the higher are the conversions achieved (Chapter 7). The combination of narrow tubes, high gas velocities and small particles, however, will result in unacceptably high differential pressures over the reactor. This will increase gas compression costs and also could cause disintegration of weak catalyst pellets. Catalyst loading and unloading may also become troublesome with narrow tubes. For all the above reasons compromises have to be made between the opposing operating and design factors. The activity of the catalyst employed also has to be taken into account. For iron based catalysts 5cm ID tubes are satisfactory but with the more active cobalt catalysts it is more difficult to control the bed temperatures. Thus the more active the catalyst the narrower the tubes should be.

Because of the abovementioned compromises that need to be made there will inevitably be radial as well as axial temperature gradients in the reactor tubes. In the case of iron based catalysts the axial gradient will be much more marked than in the case of cobalt based catalysts because for iron catalysts the rate of the FT reaction decreases much faster with bed length. As a result of this only a portion of the catalyst bed will operate at the optimum temperature, the other sections being either at lower or higher temperatures. Multi-tubular reactors are usually not suitable for high temperature FT operations. For iron based catalysts for instance, carbon deposition occurs at higher temperatures [6] and this will result in catalyst swelling and blockage of the reactor tubes. Large multi-tubular reactors can consist of thousands of tubes. This results in high construction costs. Such reactors are very heavy and this limits the size to which they can be scaled up as transportation can become the limiting factor. The design of the tube-sheets (the plates at either end through which the tubes pass) becomes very challenging for large scale reactors.

Despite the above disadvantages these reactors do have some advantages. They are easy to operate. There is no equipment required to separate the heavy wax products from the catalyst since the liquid wax simply trickles down the bed and is collected in a downstream knock-out pot. For slurry bed reactors additional equipment is required to achieve the complete separation of the finely divided catalyst from the liquid wax. The most important advantage, for the fixed bed multi-tubular reactor, is that the performance of a large scale commercial reactor can be predicted with relative certainty based on the performance of a pilot unit consisting of a single reactor tube.

For coal derived syngas, temporary slippage of catalyst poisons such as H_2S through the gas purification section is likely to occur. In the case of multi-tubular reactors this will result in only the upper sections of catalyst being deactivated leaving the balance of the catalyst bed relatively unscathed [7]. In slurry phase reactors all of the catalyst in the reactor will be deactivated to some extent. However, the catalyst in the slurry phase reactor can be replaced on-line while expensive and lengthy downtime is required to replace catalyst in the multi-tubular reactors. If a full load catalyst change is required for the slurry phase reactor this can be achieved relatively quickly. The slurry phase reactor generally copes well with poisons provided procedures are put in place to avoid extended periods of contact with highly contaminated gas. Expensive cobalt catalyst should preferably not be used with coal derived synthesis gas no matter which type of reactor is selected.

Five multi-tubular ARGE reactors, jointly developed by the two German firms, Ruhrchemie and Lurgi, were installed at the Sasolburg plant in the mid 1950's [6]. They have diameters of 3m and each contains 2050 tubes, 5cm ID and 12m long. Sasol currently uses extruded iron based catalysts in these reactors. These reactors are still currently in operation. A diagram of an ARGE reactor is given in Fig. 2.

After knock-out of the liquid wax, oils and water a portion of the tailgas is recycled to the reactor. This increases the percent conversion (on the fresh feed basis) and also increases the gas linear velocity through the catalyst bed thereby improving the heat transfer rate. The production capacity of each reactor is about 500 barrel/day (21×10^3 tonne per year). The reactors operate at approximately 2.7 MPa and 230°C.

In studies carried out in 5 cm diameter pilot plant fixed bed units at Sasolburg it was demonstrated that the production per tube could be increased by operating at higher pressures and simultaneously feeding proportionately more gas thereby maintaining the same linear velocity through the catalyst bed. Tests were carried out up to 6 Mpa and it was found that neither the percent conversion nor the carbon number distribution of the hydrocarbon products was altered. Based on this a 4.5 MPa unit was installed at Sasolburg in 1987 and it performed as predicted [7].

The four multi-tubular reactors at the Shell Bintulu plant, also designed by Lurgi, have much wider overall diameters, estimated at about 7m. Cobalt catalysts are used. Cobalt catalysts are generally more active than iron catalysts and unlike iron the activity remains high as the reactants are converted and product water is generated. Hence the tubes must have smaller diameters than those of the Sasol reactors in order to have a higher rate of heat exchange to thereby control the temperature. The original capacity of each Shell reactor of about 3000 barrel/day (125×10^3 tonne per year) has now been increased with single reactor capacities up to 8000 barrel/day now claimed for slightly larger reactor diameters. This is apparently due to new generation catalysts with higher activity. Further capacity increases may be feasible with even more active catalysts.

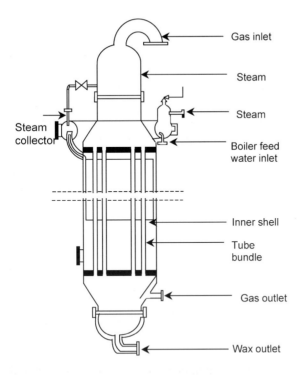

Figure 2. ARGE Reactor

1.2 Slurry phase reactors

In the 1950's and 1960's various sized slurry reactors were tested in Germany, England and the USA [6] but the space velocities used were all very low and so the performance at gas velocities applicable at likely commercial situations could not be judged [8]. Of these units the one developed by Kölbel [9] was by far the largest, with a 1.5m diameter and about 7.7 m bed height and a working volume of 10m^3. The slurry reactor system was considered to be suitable for the production of wax at low temperature FT operation since the liquid wax itself would be the medium in which the finely divided catalyst is suspended. However, a practical and efficient means of separating the product wax from the catalyst is an essential requirement that was not developed until much later. A suitable system was first demonstrated by Sasol in 1990.

When the Sasolburg plant was designed in the 1950's the multi-tubular ARGE reactors were chosen for wax production. Although these reactors operated very well, and continue to do so, it was decided in the late 1970's to evaluate the slurry system on pilot plant scale. The results obtained in 5cm ID units in the Sasol research department are summarised in Table 2 [6].

Comparisons were made between the fixed and slurry bed systems for wax production at the temperatures normally used in the fixed bed units. In both cases the precipitated iron catalyst used in the commercial multi-tubular reactors was employed, the only difference being the size of the catalyst particles. From Table 2 it can be seen that the hard wax selectivity and the percent ($H_2 + CO$) conversion obtained in the slurry phase reactor were as good, if not better, than those obtained in the fixed bed unit. This was despite the fact that the catalyst loading was three-fold lower in the slurry phase reactor. The reason for the higher activity per mass of catalyst is the much smaller particle size used in the slurry phase system. The rate of the FT reaction is pore diffusion limited even at low temperatures and hence the smaller the catalyst particle the higher the observed activity

Comparisons were also made between the slurry and the two phase (gas-catalyst) reactor systems at the higher temperatures normally used for the production of gasoline. In these runs the fused catalyst used in the commercial HTFT reactors was employed. For these tests the slurry bed gave the same product selectivities but the percent ($CO + CO_2$) conversion was considerably lower than that obtained in the two-phase fluidized bed unit. In this set of runs the slurry system did not have the advantage of smaller catalyst size to compensate for the lower loading. (For both cases the same size catalyst was used.) For the slurry test at 324 °C the wax inside the reactor was continuously being hydrocracked and so make-up wax was added daily. Such high temperature slurry phase operation is therefore not practical or viable.

The low temperature slurry system, however, was very promising. Further development was delayed for quite some time because of the problem of satisfactorily separating the wax produced from the very fine and friable precipitated iron catalyst used by Sasol. In 1990 an efficient separation device was successfully tested in a 1m ID demonstration unit and in 1993 a 5m ID, 22m high commercial unit was brought on-line [10, 11] and has been operating successfully ever since. The capacity of this reactor is about 2500 barrel/day (100×10^3 tonne per year), which equals the total production of the five 3m diameter multi-tubular fixed bed ARGE reactors. Slurry bed reactors with a capacity of at least 20 000 barrel/day (850×10^3 tonne per year) are feasible [11] and in fact Sasol reactors with a capacity of 17 000 barrel/day are currently under construction for application in Qatar.

Table 2
Comparison of fixed, slurry and 2 phase fluidized beds [6]

Reactor bed type	Fixed	Slurry	Fluidized	Slurry
Catalyst type	Precipitated		Fused	
Particle size	2.5 mm	40-150 μm	<70 μm	<40 μm
Fe loaded (kg)	2.7	0.8	4.2	1.0
Expanded bed height (m)	3.8	3.8	2.0	3.8
Average bed temperature (°C)	230	236	323	324
Recycle to fresh feed ratio	1.9	1.9	2.0	2.0
Total gas linear velocity (cm/sec)	36	36	45	45
Fresh feed conversion %				
$CO + H_2$	46	49		
$CO + CO_2$			93	79
Selectivity (C atom basis)				
Methane	7	5	12	12
Gasoline	14	15	43	42
Hard wax (BP > 500°C)	27	31	0	0

In principle, the Qatar reactor capacity could be more than doubled with more advanced catalysts by using a combination of decreased recycles and higher gas velocities. The slurry phase reactor is depicted in Fig. 3.

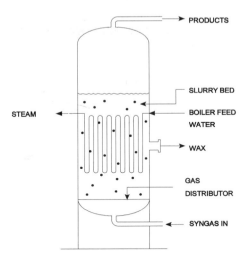

Figure 3 Slurry phase reactor

ExxonMobil has also developed a slurry bed reactor operating with a cobalt based catalyst. Successful demonstration runs in a 1.2m (4 ft) diameter unit were carried out [12]. The capacity of the unit was about 200 barrel/day (8.5×10^3 tonne per year). Other companies, notably Statoil, Conoco,

IFP/ENI/Agip and Syntroleum are also pursuing the slurry phase reactor approach using cobalt catalyst.

As previously mentioned fluidized systems have the disadvantage that even low levels of catalyst poisons, such as H_2S in the feed gas, would result in some deactivation of all the catalyst in the reactor. In the case of slurry systems with their relatively low catalyst loadings this problem becomes even more serious. (Very high catalyst loading causes a rapid increase in the viscosity of the slurry leading to negative effects for heat and mass transfer.) On-line removal of deactivated and addition of fresh catalyst can of course solve this problem. The behaviour and characteristics of slurry systems under churn-turbulent flow conditions has been reviewed recently by Sie and Krishna [15].

Based on experience in Sasolburg, the advantages of the slurry phase reactor over the multi-tubular reactor for iron catalysts are [11, 13]:

- The cost of a reactor train is only 25% of that of the alternative multi-tubular reactor system with the same capacity (Far less difference for Co catalysts).
- Pressure drop over a slurry reactor is much less, typically 0.1 MPa versus 0.4 MPa for a multi-tubular unit decreasing the gas compression cost.
- Because of the lower catalyst loadings in slurry bed reactors the catalyst consumption per ton of product is four times lower than in fixed bed units.
- The slurry bed is more isothermal and so it can be operated at a higher average temperature resulting in improved reactant conversion.
- On line removal and addition of catalyst allows longer reactor runs at higher average conversions. However, if the catalyst lifetime is long enough this advantage falls away [14].

No direct comparison has been published for cobalt catalysts. There should be far less advantage for the slurry phase reactor if the fixed bed reactor uses a high activity, long life cobalt catalyst particularly if the synthesis pressure is increased. For the fixed bed reactor the shell design is determined by the pressure of the steam produced in the cooling system and the thicker walls for the catalyst containing tubes at higher pressures are desirable to add rigidity to these small diameter tubes. The higher heat transfer coefficient at higher pressures results in a smaller deficit relative to the slurry phase reactor.

1.3 Two phase fluidized bed reactors (HTFT)

As discussed in the foregoing sections neither fixed nor slurry bed reactors can be utilised in high temperature FT processes for the production of light alkenes and/or gasoline. For this application two-phase fluidized bed reactors are the units of choice. There are two types of fluidized bed reactors currently in commercial use, the turbulent or fixed fluidized bed (FFB) and the circulating fluidized bed (CFB). The turbulent reactors have been named Sasol Advanced

Synthol (SAS) reactors by the developers of this advanced technology. Figs. 4 and 5 illustrate the two types of reactors.

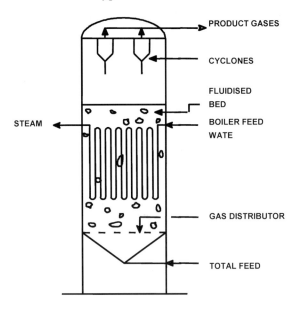

Figure 4 Sasol Advanced Synthol (SAS) reactor

Due to the high degree of turbulence in fluidized beds, they exhibit very high rates of heat exchange. This means they can cope with the large amounts of reaction heat released at high conversions with high feed gas throughputs that can be achieved at high operating temperatures. Despite this, the beds are virtually isothermal, temperature differences between the bottom and top of the reactor being only a few degrees. It is important, however, that process conditions must be such that the selectivity of long chain hydrocarbons is limited to ensure that excessive condensation of liquids in the pores of the catalyst does not occur. If the outer surfaces of the particles are wetted agglomeration of the finely divided particles, typically 5 to 100μm, would occur resulting in de-fluidization of the catalyst bed and the unit would then cease to function.

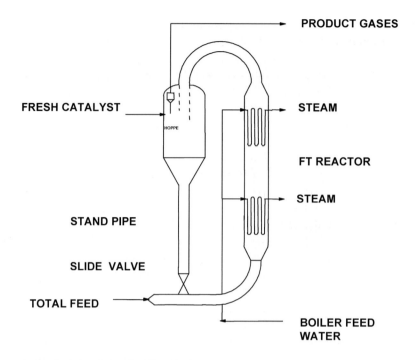

Figure 5 Synthol circulating fluidized bed (CFB) reactor

1.3.1 Circulating fluidized bed reactor (CFB reactor)

The circulating fluidized bed reactor design chosen by Sasol for the original plant at Sasolburg in the 1950's were CFB's. The gas linear velocities in CFB reactors are three to four times higher than in the turbulent fluidized beds. These reactors had been developed by Kellogg and had only been tested in a 10 cm ID pilot plant [6, 8]. For the Sasolbrg plant this was scaled up to 2.3 m ID, 46m high units. After many teething troubles several design as well as catalyst formulation changes were made and these resulted in satisfactory performances of the units. The reactors were re-named Synthol reactors and were operated successfully for thirty years.

The Synthol reactors in Sasolburg operated with iron catalyst at about 340°C and 2 MPa. At recycle to fresh feed ratios of about 2 the initial (CO + CO_2) conversion was between approximately 80% and 90% depending on the feed gas composition and flow rate. The initial design capacity of the first Kellogg two reactors installed in Sasolburg was about 1 500 barrel/day per reactor but later improvements implemented by Sasol increased the capacity to about 2 500 barrel/day (100 x 10^3 tonne per year). The aerated catalyst (dense phase fluidized) flows down the standpipe, the rate being controlled by the slide valve. The catalyst is swept up the reaction section (lean phase fluidization) by the feed gas which has been preheated to about 200°C. The average voidage here is approximately 0.9. The heat exchangers remove about 40% of the heat of reaction, the balance being absorbed by the feed gas and products [6]. The catalyst and gas disengage in the wide settling hopper and the aerated catalyst drops down into the standpipe. The bulk of any catalyst fines still entrained in the gas is knocked out in the cyclones and returned to the standpipe. Even though the efficiencies of the cyclones are in excess of 99% some fines still pass through and these are removed in the heavy oil scrubbers immediately downstream of the reactors.

To achieve a high conversion rate it is necessary to have a high catalyst loading in the reaction zone. However the loading, i.e. the pressure drop over the reaction zone, must not exceed the pressure drop over the standpipe. Should this happen the feed gas will pass up the standpipe, the cyclones will become choked with catalyst and massive losses of catalyst will occur. As no heat exchange takes place in the standpipe there will be a temperature runaway there and the catalyst will be damaged. Thus to avoid the feed gas going up the standpipe it is essential that the differential pressure over the standpipe should always exceed that over the reaction section [8, 16, 17]. With iron catalyst operating at 340°C a significant amount of carbon is continuously deposited within the catalyst and this causes particle disintegration which results in an increase in the rate of loss of fines through the cyclones. In addition, the deposition of carbon results in a decrease in the density of the catalyst particles. At a fixed gas linear velocity the lighter the particles the more rapidly they will be transported upwards by the gas and less catalyst back-mixing will occur. This means that the catalyst loading in the reaction zone decreases which results in a decrease in conversion rate with time on stream. The lower catalyst density of course also lowers the differential pressure over the standpipe. This makes it impossible to increase the catalyst loading on the reaction side in order to compensate for the loss of catalyst activity and mass in the reaction zone with time on stream [8]. To counter this on-line catalyst removal and addition of fresh catalyst is practised.

The second generation CFB reactors installed in the Secunda and Mossel Bay plants were larger and operated at about 2.5 MPa thereby increasing the capacity three fold to about 8 000 barrel/day (330 x 10^3 tonne per year) [6, 17]. The heat exchangers and the standpipe slide valve were also improved. By 1999 all the Secunda CFB reactors were replaced by the even higher capacity SAS reactors. Currently the only CFB reactors operating are those at Mossel Bay.

1.3.2 Turbulent fluidized bed reactor (SAS reactor)

The two 5m diameter FT reactors in the Brownsville, Texas plant were fixed fluidized bed (FFB) units [18]. They initially were plagued by low conversion attributed to poor catalyst fluidization. These problems were apparently overcome but the plant was shut down in the mid 1950's due to a sharp increase in the price of the natural gas from which the FT syngas was produced. The long term reliability of these FT reactors was thus not proven. It took over 30 years before this type of reactor was again used commercially.

As already stated CFB reactors were installed initially at three Sasol FT plants. The Sasol R&D pilot plants which were used to develop improved catalysts and to study process variables, however, were small scale FFB units operating in the slugging mode. These reactors performed very well and were easy to operate so it was decided in the late 1970's to further investigate their potential. A large Plexiglass cold model was constructed and used to investigate various gas distribution systems, the objective being to improve the quality and uniformity of fluidization and to investigate potential scale-up issues. Various gas distribution nozzles were tested and one recommended by Badger was found to perform well. A 1m diameter fixed fluidized bed (FFB) demonstration reactor, later named Sasol Advanced Synthol (SAS), designed by Badger, was subsequently built and came on line in 1984. It operated with the same catalyst and process conditions used in the commercial CFB reactors. The product selectivities were similar but the conversions were higher. In 1989 a 5m diameter, 22m high commercial SAS unit came on stream in the Sasolburg plant and it met all expectations [19]. The capacity of the reactor was 3500 barrel/day (145×10^3 tonne per year). Over the period 1995 to 1999 the sixteen second generation CFB reactors at the Secunda plant were replaced by eight SAS reactors. Four of these had diameters of 8m with capacities of 11 000 barrel/day (470×10^3 tonne per year) each and four of 10.7m diameter each with a capacity of 20 000 barrel/day (850×10^3 tonne per year). This increased the Secunda plant's capacity from 5.1×10^6 to 7.1×10^6 tons per year.

The key advantages of the SAS reactors over the CFB reactors are:
- The construction cost is 40% lower mainly because the reactors are physically much smaller. The SAS reactor is of similar size to the settling hopper of the CFB reactor. Furthermore, the support structure is much simpler and only costs about 5% of the support structure of a CFB reactor.
- Because the reaction section is much wider, more coiling coils can be installed increasing its capacity. Thus either more syngas can be fed by increasing the flow or by increasing the operating pressure at fixed gas linear velocity. Pressures up to 4 MPa are feasible [17].
- In CFB reactors only a portion of the catalyst charge participates in the FT reaction at any moment while in the SAS reactor all of the catalyst is involved. This results in higher conversions.
- As previously mentioned, carbon deposition on the catalyst results in lower catalyst loadings in the CFB reaction zone resulting in declining

conversions with time on stream. If allowance is made for bed expansion in the SAS reactors this problem is of less significance. For this reason a lower rate of on-line catalyst removal and replacement by fresh catalyst is required to maintain high conversion thus lowering overall catalyst consumption.

- The gas and catalyst linear velocities and the pressure drop across the reactor are much lower for the SAS reactors than the CFB's. The gas compression costs are consequently lower. Because the catalyst is very abrasive the narrower sections of the CFB reactor are ceramic lined and regular maintenance is essential. This problem is absent in the lower velocity SAS reactors and this allows longer on-line times between maintenance inspections, leading to higher production rates and lower maintenance costs.

2. HISTORICAL DEVELOPMENT

Much of the relevant history of the development of FT reactors has been covered in the preceding description of the types of reactor that are currently in commercial operation. The two phase HTFT technology development had its origins in the USA but was perfected in South Africa and has already been adequately described for the purpose of this book. The historical development of the LTFT reactors has been recently reviewed by Professor Burtron Davis [20] in a paper with the title "Overview of reactors for liquid phase Fischer-Tropsch synthesis". What follows is essentially an extract from this review.

2.1 Early developments

In viewing the reactors described for this early period, one should keep in mind that the Rheinpreuβen reactor capacity of about 80 bbl/day, was large for its time. However, today one thinks of 10-20,000 bbl/day reactor capacities as a minimum size so that a design that was reasonable to consider for this period may be impractical when projected to the reactor size needed today.

2.1.1 Work in Germany

Conversions in the liquid phase were initiated in Germany shortly after the discovery of the reaction. The developments progressed from the initial fixed-bed effort at atmospheric pressure to work in a bubble column reactor operated at atmospheric pressure to a large medium-pressure pilot plant that was operated during the 1940-1950 period. The German efforts were extensive and involved work by several companies. These companies added to the complexity faced by a reviewer because of their joint ventures with organizations located outside Germany.

2.1.1.1 Kaiser-Wilhelm-Instituts für Kohlenforschung (KWI)

The Institute, founded in 1913 with Franz Fischer as the director, was established to concentrate its efforts on fundamental studies of coal and its conversion. The onset of World War I within months following Fischer's arrival and Germany's need for synfuels led the institute toward more practical accomplishments. Both direct and indirect coal liquefaction studies were undertaken. The indirect approach eventually led to the low-pressure synthesis of predominantly hydrocarbon products, the Fischer-Tropsch synthesis (FTS). The exothermicity of the reaction and the need to provide a uniform temperature in the reactor were quickly recognized.

Fischer and co-workers turned to synthesis in the liquid phase soon after they discovered the FTS reaction. Fischer and Peters [21] evaluated the use of a bubble column reactor that was operated at atmospheric pressure for the hydrogenation and oligomerization of acetylene and for the hydrogenation of carbon monoxide using a supported nickel catalyst. They indicated that the use of the catalyst in liquid media for these exothermic reactions is advantageous for laboratory and commercial operations because it makes possible close control of temperature and eliminates overheating as is encountered in fixed-bed reactors. In November 1930 the pilot plant staff attempted to run the FTS in the liquid phase, and they were successful in solving the heat flow problem. Fischer and Küster [22] reported results to identify the influence of temperature and pressure on the synthesis. A horizontal stirred silver-lined autoclave was utilized for these studies. The authors found that higher temperatures were needed for liquid phase synthesis because there was no hot spot as was the present for fixed-bed operations. They utilized a catalyst with a composition of 9Co:2Th:1Cu:0.25Ce with an equal mass of Kieselgur as support. However, the typical catalysts of that day had an activity that was too low to have practical commercial interest. They also reported that pressures near atmospheric were preferred since 'Synthol' (alcohols, acids, aldehydes, ketones, etc) was formed at higher pressures.

Fischer and Pichler [23] indicated that four approaches were suitable for removing the heat of reaction and for maintaining a uniform temperature in the reactor: (1) circulating oil outside the tube with the catalyst, (2) suspending the catalyst in oil, (3) circulating superheated water outside the catalyst space (e.g., 180°C and 10 atm. (1.01 MPa)) and (4) the suspension of the catalyst in superheated water. Interestingly, they referred to middle-pressure synthesis not as Fischer-Tropsch synthesis but the Fischer-Pichler middle-pressure synthesis. They utilized four catalysts (Ni, Co, Fe and Ru) in a study of middle pressure synthesis in an aqueous phase utilizing a stirred-horizontal autoclave, similar to the one in Figure 3, to effect the synthesis. They found that only Co and Ru were suitable for synthesis since Ni formed the carbonyl and was carried from the vessel while the iron catalyst was too active for the water-gas-shift reaction. The best catalyst was Ru, which was active below 200°C. The amount and kind of products for both Ru and Co were similar to those of the dry synthesis. It was noted that water was effective in removing the heat of reaction and in providing a uniform temperature; this was based on the fact that the fixed bed catalyst in a reaction tube of the diameter they used would have not produced useful products (i.e., higher hydrocarbons). They indicated that work in the aqueous phase was not suited for scale-up because:

1) the results are not better than operating in the dry phase,
2) a larger reactor volume is needed,
3) the apparatus is more expensive due to the need for an acid resistant reactor liner,
4) considerable expenditure of energy is necessary for stirring to provide the vigorous agitation and
5) the reaction products are not easily removed from the reactor.

Thus, Fischer and co-workers were successful in using liquid-phase synthesis to control the temperature but the reaction velocity with their catalysts was too slow to be of commercial interest.

2.1.1.2 I. G. Farbenindustrie A. G. (Farben)

Farben utilized two general types of liquid-phase operations: (1) a fixed bed of solid catalysts with liquid and syngas circulating concurrently over the catalyst and (2) a suspension of fine catalyst particles due to the fine bubbles formed when the syngas passed through a ceramic bottom plate with slurry circulation through an external vessel for heat removal. In some versions, they also utilized stirring as a means to maintain the catalyst suspension. Only the fixed-bed version developed beyond the laboratory scale.

The foam process was one variation of the liquid phase synthesis [24-29]. During 1939 to 1944 Farben was developing a liquid-phase operation in which iron powder, prepared from iron carbonyl, was mixed with oil and the gas contacted with the liquid suspension [30, 31] using a process scheme as outlined in Fig. 6.

Michael [31] reports that laboratory reactors of 3 m height and 6.6 and 16.5 cm i.d. were utilized. These reactors were fitted with an internal central tube along the axis of the reactor. Because the gas bubbles rose in the annular space between the reactor wall and the inner tube, the density of the slurry in the annular space was lower than the essentially gas free slurry inside the smaller tube. The difference in density caused rapid circulation of the slurry within the reactor. In today's language, the inner tube would be a downcomer tube. When they constructed a pilot plant reactor identical to the laboratory reactors except that the volume was 300 L (2.0 m diameter), difficulties were encountered. If the circulation was interrupted for some reason, the catalyst would settle and, upon re-starting, the gas flow was insufficient to cause the catalyst to become uniformly suspended. The concentration of the catalyst at the lower portion of the annular space of the reactor provided a density that was greater than the slurry in the downcomer so that circulation could never be re-started. For this reason, a pump was installed and, for convenience of maintenance, placed outside the reactor vessel as shown in Fig. 4. A reactor of this design with a 0.5 m diameter and 8 m tall (1.5 m³ volume) was constructed and operated satisfactorily.A larger 1.5 m diameter and 8 m tall (14 m³ volume) reactor was constructed but was not operated because of a shortage of synthesis gas because of wartime problems.

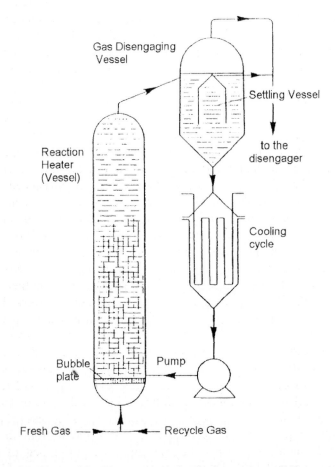

Figure 6 Schematic of the FT reactor unit for the foam process [31].

It was found to be difficult to operate the suspended catalyst because the formation of high molecular weight products of limited solubility caused the catalyst particles to agglomerate and settle [24, 32]. Frequently they experienced thick deposits of iron catalyst around the wall of the reaction vessel, and especially in the upper portions of the reactor.

2.1.1.3 Duftschmid process.

One Farben development [33-35] consisted of an oil recycle process in which a cooling oil was passed concurrently with the synthesis gas over granules of an iron catalyst (fused or sintered iron that was doubly promoted with aluminum and potassium oxides, or with titanium, manganese and potassium oxides). Cooling was effected by recycling the oil heated by the exothermic reaction through an external heat exchanger. The process was operated at a pressure of 20 to 25 atm (2.02-2.53 MPa) and a temperature of 260 to 300°C in the first stage, and 280 to 330 °C in the second stage if one was used. The throughput of synthesis gas (H_2/CO = 0.8 for this example) was controlled to yield about 0.5 kg of total product per litre of catalyst per day. The yield of C_3+ hydrocarbons was about 150 grams per cubic metre of synthesis gas (H_2/CO = 1 for this example) and was distributed as follows: 16% C_3 and C_4 (85% olefins), 40% gasoline boiling to 200°C (50% olefins), 20% gas oil (25% olefins), 20% paraffin wax and 4% alcohols, largely methanol and ethanol. The crude gasoline had a research octane number of 62 to 65 and the gas oil a cetane number above 70. Some of these patents were assigned to the Standard Catalytic Company. The Standard Catalytic Company was established by six U.S. petroleum companies to develop processes for the production of synthetic fuels, and included an agreement with Farben. Following World War II, legal actions divided the holdings of the Standard-Farben company. Jersey received complete possession of 544 patents and ownership of Standard Catalytic Company. Jersey retained half-ownership of another 254 patents.Benson et al. [36] report that after the preliminary work by F. Duftschmid, et al., as described above, a pilot plant with a reactor of 0.2 m diameter and 6 m height (about 0.75 m^3) was constructed during 1936 and 1937 and that it was operated, first at pressures of 100 atm (10.1 MPa) and then at 15 to 20 atm (1.5 to 2.0 MPa). In 1938 a larger plant with a 0.5 m diameter and 6 m height (4.7 m^3) was operated. The recirculation rate was high, 9-12 reactor volumes of catalyst/oil was recirculated each hour [37]. At the end of the war they were planning a 40,000 ton/year (860 bbl/d) plant for this process [37].

The Duftschmid process was considered to be different from the one developed by Ruhrchemie (described below) since the entire bed of solid catalyst is immersed in oil [36]. However, Farben obtained a patent [38] in which oil was sprayed onto the catalyst in a fixed bed. The liquid passed in the same direction as the gas. They indicated that, while it was desirable to always have a thin layer of liquid present on the surface of the catalyst, it may be advantageous to temporarily interrupt the trickling in order to cause a brief increase in the temperature of the catalyst, in effect, a catalyst rejuvenation step.

An early German liquid-phase reactor involved an arrangement of trays, each containing catalyst particles, that were contained in a cylindrical vertical reactor that was fitted at the top with a reflux condenser and an arrangement for removal of liquid at the bottom of the reactor [35, 39]. Experiments were conducted in a 7-tray reactor, 4.5 cm diameter and 60 cm tall, with up-flow of the reactant gas.

2.1.1.4 Ruhrchemie

A general license for the FT process developed at KWI was assigned to Ruhrchemie AG in Oberhausen-Holten on October 27, 1934. Independent of this licensing agreement, Ruhrchemie had started construction of a pilot plant on their own and had acquired the services of Otto Roelen, who was glad to leave the KWI since his relationship with Fischer was not entirely satisfactory [20]. Roelen is credited with much of the experimental success at both KWI and Ruhrchemie. In addition to his excellent experimental capabilities, Roelen had outstanding scientific and technical abilities. His keen intellectual curiosity led him to discover the Oxo process by concluding correctly and in contrast to the views of others that the recirculation of ethene to the FT reactor led to the formation of additional C_3-products, especially C_3-oxygenates.

Their reactors had a vertical tube 6 m in height that was inside a larger tube where oil circulated to remove the heat of reaction. The reactor volume of one was 55 L and a second reactor had 85 L of slurry volume. Because significant quantities of oil were carried over, even with a wider head at the top of the larger reactor, a reflux condenser was operated to return heavier oil to the reactor. Heavy products were withdrawn from the reactor through a filter stick immersed in the slurry. Initially a ceramic plate was used to generate small gas bubbles of feed gas, but the catalyst settled and plugged the plate when the operation had to be interrupted, even temporarily. The plate was replaced by a simple copper capillary tube which never blocked and, based on the conversion, operated to produce fine bubbles of syngas just as well as the ceramic plate.

They concluded that for commercial scale operation, it would be better to limit the conversion in the stage I reactor and to use reactors in series. This operational mode was claimed to limit methane formation and extend catalyst life.

The usage ratio (the ratio H_2/CO converted), when CO conversion was in the 70-75% range, as when operating a single stage, was usually 1.4-1.5; however, if the reactor was operated at lower conversion levels the usage ratio could be reduced to about 0.9. With two stages together, the usage ratio was 1.24-1.28, which was essentially that of the feed syngas. They had an objective of making the feed and usage ratio the same for each reactor in the series, but apparently never attained this goal.

Ruhrchemie [40] worked to develop a modified liquid-phase process in which finely divided cobalt-thoria-magnesia-kieselguhr catalyst was suspended in oil boiling between 240 and 300°C. During the synthesis, water was injected into the slurry where it vaporized to maintain control of the slurry temperature. The presence of a high water partial pressure, as must be present in this operation, implies that oxidation of their cobalt catalyst did not occur to an extent that it was a problem. An alternative explanation is that the activity of the catalyst was so low that the change could not be detected. With $H_2/CO = 2$, 10 atm. (1.01 MPa) and 2.5 L/gCo/hr at 190-210°C, 172 g of liquid and solid hydrocarbons per cubic meter were produced, with 90% of the liquid product boiling below 300°C. By the end of the war, the work had not progressed to the large-scale pilot plant stage [41].

Ruhrchemie workers also developed a version of the oil recirculation process. Their process utilized a fixed bed of catalyst that was sprayed continuously with an oil that was added through a device that was designed to provide a uniform distribution of oil across the radial dimension of the catalyst. The intent was to have the oil maintain the catalyst free of wax and to remove the heat of reaction by the latent heat of vaporization of the added oil [37]. It was not made clear how the added oil could wash the wax from the catalyst through the bottom of the reactor and, at the same time, be evaporated to maintain the temperature.

2.1.1.5 Summary of early German reactors

Sie and Krishna [15] provide the following summary of the early German reactor designs:

1) Fixed-bed reactor with internal cooling operating at high conversion in a once-through mode. The catalyst was packed in a rectangular box and water-cooled tubes fitted with cooling plates at short distances were installed in the bed to remove the reaction heat. This type of reactor was applied in the atmospheric synthesis process ('Normaldruck Synthese').

2) Multi-tubular reactor with sets of double concentric tubes in which the catalyst occupied the annular space, surrounded by boiling water. This type of reactor was applied with gas at medium pressure in a once through mode ('Mittedruck Synthese').

3) Adiabatic fixed bed reactor with a single bed, large recycle of hot gas whichwas cooled externally ('IG-Farben/Michael Verfahren').

4) Fixed-bed reactor with multiple adiabatic beds, inter-bed quenching with cold feed gas, recycle of hot gas and external cooling ('Lurgi Stufenoven').
5) Adiabatic fixed-bed reactor with large recycle of heavy condensate passing in upflow through the bed. The liquid recycle stream was cooled externally ('BASF/Duftschmid Verfahren').
6) Slurry reactor with entrained solid catalyst, large recycle of hot oil and external cooling ('BASF Schaumverfahren').

The above reactors are mainly of historical interest since they offer limited scope for large scale conversion. The commercially applied reactors mentioned above under 1 and 2 had very small production capacities by current standards, viz., of the order of 15 bbl/day. At the low gas velocities associated with once-through operation at relatively low pressures and temperatures, heat transfer rates from the bed to the cooling surface are so low that a very large cooling area is required, which is a severe limitation for further scale-up.

The other reactors with external cooling need very large recycle streams to take up and transport the generated heat out of he reactor. This gives rise to high pressure drops and very high energy consumption for gas or liquids circulation, if the reactors were to be applied on a large scale.

2.1.1.6 Summary of later German developments

The following summary was provided by Sie and Krishna [15]:
Developments in the period shortly after World War II (in some cases based on concepts generated somewhat earlier) led to reactors with increased potential for large scale application. The main ones are:

1) A multi-tubular fixed bed reactor operated with gas recycle at moderate per pass conversion, instead of once through operation aiming at maximum conversion as in the earlier mentioned 'Mittedruck Synthese'. This reactor, applied in the 'Arge Hochlast Synthese' developed by Lurgi GmbH and Ruhrchemie A.G., had a production capacity of about 400 bbl/day. This substantially increased the production rate compared to previous commercial fixed-bed reactors (by a factor of about 25). This is a result of higher temperatures and pressures, a more even reaction rate profile over the reactor length and improved heat removal as a result of higher gas velocities. A commercial plant based on the Arge process was installed by Sasol in Sasolburg, South Africa in the 1950's.
2) Slurry reactor in which synthesis gas is contacted in a bubble column with a slurry of fine catalyst suspended in liquid. In the process developed by Rheinpreussen AG and Koppers GmbH in the early 1950's, reaction heat is removed internally by cooling pipes immersed in the slurry [42-45].

The development studies culminated in the operation of a semi-commercial reactor. For this reactor a high (about 90%) conversion of carbon monoxide has been reported when operating in a once through mode with a low H_2/CO ratio feed gas at a superficial gas velocity of about 0.1 m/s [44].

2.1.1.7 Rheinpreußen (Rheinpreussen)

This company began work on the liquid-phase synthesis about 1937. Most of the work was under the direction of H. Kölbel and the work conducted up to 1945 was summarized by him [37]. The composition of most of the catalysts used by Kölbel and co-workers in this early work, as well as in the later pilot plant studies, had a composition of Fe:Cu:K_2O = 100:0.2-0.5:1.0 and was activated 10-20°C above the synthesis temperature. They used a suspension of 10-20% Fe, a syngas with H_2/CO = 0.5 and a space velocity of 75 h^{-1}, a very low value. The early experiments were conducted in a vessel with 15cm ID and 3-4 m height with cooling coils either inside or outside. They reported a usage ratio of 0.5, as it was reported that no water was present in the products. In this and other work, Kölbel reported that diesel and the product olefins, and even paraffins, reincorporated extensively to produce heavier products [46-48]. Excluding the mass of product formed from the diesel recycle fluid, 180 g product was formed from a m^3 of synthesis gas. With this catalyst they reported that methane was not formed. The synthesis products were obtained by withdrawing oil and catalyst, filtering, and returning the remaining catalyst slurry to the reactor, all operations being conducted under synthesis gas [37].

Kölbel and Ackermann [49, 50] obtained patents for an apparatus for carrying out gaseous catalytic reactions in liquid medium. The reactions included Fischer-Tropsch synthesis. It was stated that "the [Fischer-Tropsch] synthesis according to known processes is feasible without trouble in a reaction space of up to 20 cm [7.9 in] diameter. With increasing horizontal diameter of the reaction space the amount of gas conversion decreases and it is always more difficult to maintain a constant gas conversion. The larger the horizontal diameter of the reaction space, the more the liquid substance leans toward changing from stationary state to a state of vertical rotation. [Vertical rotation occurs when] the liquid flows downwardly along the surface of the wall and flows along the bottom to the middle of the reaction space, whereby it is drawn out by the gas bubbles leaving at the middle of the bottom. The compressed central gas stream flows along with the liquid toward the top whereby the firmly compressed gas bubbles combine to form large elongated gas bubbles. Only at the upper reversal point of the liquid, in the vicinity of the surface of the column of liquid, does the gas spread out horizontally across the transverse section and the large gas bubbles are partially decomposed."

As Kölbel and Ackermann illustrate, the liquid flows upwardly in the interior of the reactor tube and flows downward along the wall side of the reactor. They indicated several disadvantages that accrue from this situation, including:

1) decreased gas conversion,
2) occurrence of secondary reactions,
3) increased catalyst damage, and
4) increased catalyst aging.

These authors state that, for reactions like Fischer-Tropsch synthesis where the gas composition changes with conversion and where the products admix with the unconverted gas, "...the gas conversion should be as complete as possible on passing the gas through once..." The authors indicate that, prior to their patent, none of the available operations permit "...maintaining the liquid medium and the suspended catalyst stationary and nevertheless permitting the gas bubbles in uniform size and distribution to pass vertically through the liquid medium at equalized velocity..." The authors indicate that the disadvantages of the liquid and catalyst circulation can be overcome in a cylindrical reactor with a horizontal diameter of more than 30 cm and up to 3 m or more and more than 1.5 m in height, and a gas head space above the liquid at least as large as the reactor diameter. To provide within the reaction zone a stationary catalyst and liquid condition, the large reactor shell is subdivided into similar, vertical shafts which are open at top and bottom that have liquid-tight casings and a diameter of at least 5 cm. The shafts should terminate above the expanded liquid level; i.e., in the free-board gas space. Because of the flows in the bottom of the reactor, it is stated that each of the shafts, in the centre as well as at the wall region, receive the same amount gas. While there is circulation of the liquid in the sump (bottom of reactor), "...there are formed in the shafts extremely stationary liquid columns whose expansion depends on the amount of gas."

The authors indicate that in certain cases, as in the Fischer-Tropsch reaction, it may be an advantage to allow the temperature to rise toward the top of the reactor, and provisions are described which would allow for this to occur. Thus, as the partial pressure of the reactants decrease, the higher temperature will be able to compensate, completely or partially, by having the rate increase to compensate for the partial pressure decrease.

In spite of patenting the concept of using vertical baffles to inhibit backmixing, there is no indication that this was ever applied.

Kölbel and co-workers operated a large demonstration plant with a reactor that was 1.55 m in diameter and 8.6 m in height. Until the start-up of the slurry reactor by Sasol, the Rheinpreussen-Koppers demonstration plant was the largest slurry reactor that had been operated successfully. Kölbel states that at the time that most work was conducted using the demonstration plant (1952-53) the operation was confined almost exclusively to the production of gasoline [51].

The results of the operation of this plant and the smaller laboratory scale slurry phase reactor produced data that have become the 'standard' that is used to compare with other slurry phase studies. As indicated above, a typical catalyst used by Kölbel in this plant would have a composition of $Fe:Cu:K_2O = 100:0.1:0.05-0.5$; thus, it would be consistent with the objective of producing gasoline range material, and not high molecular weight reactor-wax.

Table 4

Operating data and results of 'liquid-phase synthesis' for one-step operation with a single passage of the gas over iron catalysts (from Ref. 51)

	Demonstration Plant (a)	Laboratory Plant (b)
Effective reaction space (volume suspension including dispersed gas) (L)	10,000	6
Catalyst (kg Fe)	800	0.4
Synthesis gas pressure (bar)	12	11
Synthesis gas (volume ratio, CO:H$_2$)	1.5	1.5
Quantity of synthesis gas (Nm^3/hr)	2,700	1.3
Linear velocity of the compressed gases at operating temperature referred to the free reactor cross section (cm/sec)	9.5	3.5
Total CO + H$_2$ used (Nm^3/hr)	2,300	1.1
Per m^3 of reaction chamber (Nm^3/hr)	230	183
Per kg of Fe (Nm^3/hr)	2.6	2.45
Average synthesis temperature, °C	268	266
CO conversion, %	91	90
CO + H$_2$ conversion, %	89	88
Synthesis products referred to CO + H$_2$ used:		
Hydrocarbons C$_1^+$ (g/Nm^3)	178	176
C$_1^+$ + C$_3$ (g/Nm^3)	12	11
C$_3^+$ (g/Nm^3)	166	165
O-containing products in the synthesis water (g/Nm^3)	3	2
Space-time yield of C$_3^+$ products including O-products in 24 hr (kg/m^3 of reaction chamber)	930	740

At the conversion level shown in Table 4, only 178 g of hydrocarbons were produced per m^3 gas. From the original paper in German, it is not possible to tell whether this volume of gas refers to the amount of gas fed or to the amount of gas converted. Even if it is taken as the amount fed, at the 90% conversion level, more than 178 g. of hydrocarbons should have been produced. For example, in the Mobil runs more than 200 g of hydrocarbons was produced. Sasol workers indicate that they could not repeat Kölbel's results in their early studies [10]. Kölbel ct al. report that through polymerization of lower olefins, about 18 g of alkylate gasoline can be produced for each m^3 of syngas that was converted.

When the alkylate gasoline was mixed with the reformed gasoline (112 g/Nm^3 CO+H$_2$), 130 g/Nm^3 CO+H$_2$ of finished gasoline could be produced. For a CO conversion of 91%, the H$_2$ + CO conversion was 89%; the feed gas ratio was H$_2$/CO = 0.67. With this gas ratio the only way that Kölbel could have obtained such similar high CO and CO + H$_2$ conversions would be to operate so that the single pass conversion was 50-60% and to recycle the unconverted gas.

It has not been widely appreciated that much of the work that Kölbel reports has been conducted under conditions designed to produce gasoline; in this mode the demand on wax/catalyst separation is minimal. Thus, much of Kölbel's work can be viewed as being conducted under conditions that make the operation of a slurry reactor much easier than the current goal of operating to maximize the reactor-wax fraction to subsequently hydrocrack to produce diesel fuel.

Kölbel stressed that the low viscosity and surface tension of the liquid was crucial for maintaining the small bubble size needed to maintain gas-liquid mass transfer. Kölbel maintained the view that it was necessary to establish upper limits upon the solids content of the slurry in order to maintain a low viscosity. Kölbel and coworkers obtained very low methane for the distribution of the other products that they reported. In spite of the question regarding mass balance, Kölbel and co-workers provided a strong scientific and engineering basis for further work.

It seems that Kölbel and Ackermann anticipated that the slurry phase reactor may be applied with catalysts other than iron; to higher H$_2$/CO ratio syngas and to wax production. The following extracts from a 1954 patent assigned to Rheinpreussen [52] are relevant to modern slurry phase reactors:
"Within a further embodiment of the invention, we find it possible to so adjust the conditions that our novel process may be adapted to any particular synthesis mixture whether rich in hydrogen or rich in carbon monoxide. When utilizing in the iron contact synthesis a mixture rich in hydrogen the water produced in the reaction enters with part of the carbon monoxide into a water gas equilibrium reaction which results in the partial removal of carbon monoxide thus

withdrawing it from the hydrocarbon synthesis and causing lower yields of synthesis products.

We have discovered that it is possible to substantially eliminate this disadvantage and to effect, even with gases rich in hydrogen, a practically complete utilization of the carbon monoxide with a maximum yield of synthesis products per cubic metre of synthesis gas used. This result is obtained by effecting a reduced gas-catalyst contact period which may be obtained by so adjusting the rate of flow of the synthesis gas through the suspensions that the same remain in individual contact therewith for only a relatively short period of time. The suspension of catalyst, however, is then repetitiously contacted with the synthesis gas whereby each individual contact is at a relatively high velocity or rate of feed through the suspension while the aggregate of the contacts per time unit fall within the general rate of gas flow limitations specified in accordance with the invention." **(Thus high gas flow with recycle operation is taught.)**

"Thus for instance gases rich in hydrogen (for example $2H_2$: $1CO$) are run at a velocity or rate of flow through the suspension up to fifteen times as high as that employed for gases rich in carbon monoxide (for example $1H_2$: $2CO$). This fifteen times velocity applies to each individual contacting, and the gases having passed the suspension at that rate of flow, are then re-contacted with catalyst suspension for a sufficient number of times within the general aggregate limits of rate of flow in accordance with the invention until substantially all the carbon monoxide is utilized. This repetitious contacting may be done by either recycling the emerging gases in each case through the same suspension at the higher rate of flow mentioned or by passing these gases into and through successive stages of a multiple stage synthesis unit." **(Thus multiple reactor stages are taught.)**

"When recycling part or all of the tail gases in accordance with this procedure, we have found it of advantage to add a certain amount of fresh synthesis gas to the mixture. Depending upon conditions affecting the conversion ratio H_2:CO such as pressure temperature, **nature of catalyst** and rate of fresh feed, we have found it of advantage to select the ratio of recycle or tail gas to fresh feed gas about 2 to 5 times as high as the volume ratio H_2:CO in the fresh feed gas.

When using multiple stage synthesis, the process involves a passage of the catalyst suspension from stage to stage to meet in each stage fresh synthesis gas.

We have also sometimes found it of advantage to add a suitable additive affecting surface tension of the oil in the catalyst suspension.

In the process between 180 and 195 g of hydrocarbons are formed out of the normal cubic metre of applied $H_2 + CO$, the properties of which are variable within a wide margin, depending on operating and catalyst conditions permissible within the scope of the invention. For instance, products can be produced which predominantly consist of C_3-C_8 olefins with a considerable percentage of iso-hydrocarbons, or hydrocarbons predominantly solid at normal temperature both with a high and low degree of branching and both with a high and low olefin content."

2.1.2 The Netherlands

International Hydrocarbon Synthesis Co (N. V. Internationale Koolwaterstoffen Syntheses Maatschappij, The Hague, The Netherlands) was reported to employ a form of the recycle technique that utilized vapour-phase synthesis in the first stage and this stage was followed by a liquid-phase operation that utilized a catalyst suspended in oil. The two reactors were separated by a product condensation unit [30]. However, the reference given for this work appears to apply only for an oil recirculation liquid-phase process that has the features of those described above [53]. This company obtained many patents covering FTS during the 1930s but only a limited number apply to liquid-phase synthesis.

2.1.3 United Kingdom

The Fuel Research Station at Greenwich first utilized fixed bed reactors that had a reaction space up to 50 litres volume and in 1947 began work on the fluidized bed reactor. In 1949 work began on the liquid-phase technique. Hall et al. [54] compared fixed bed, fluid-bed and liquid-phase reactors and in 1952 concluded that the maximum selectivity and flexibility could be obtained with the liquid-phase system. The decision was made to construct a pilot plant, based primarily on the fact that the liquid-phase reactor was the only one that could be operated with the CO-rich gas produced in a slagging coal gasifier. Pilot-scale operations were initiated at Greenwich and transferred to Warren Spring Laboratory in 1958. A plant was to produce up to 385 L of product/day and to operate at pressures up to 20 atm (2.02 MPa) and reaction temperatures to 300°C was operated. The reactor had a 9.75 inch (23.5 cm) internal diameter and was 28 ft tall (8.5 m), allowing a reactor volume of 273 L. Initial problems were encountered in the settling method they used for wax separation from the catalyst slurry and from carbon formation which increased the viscosity of the reactor contents to the point of near-gellation.

The operation of the British plant was terminated about the time that they had solved most of the operating problems and considered themselves to be at a point where they could operate to produce reliable data. The operation became a casualty due to the discovery of a plentiful supply of petroleum in the Middle East. Low catalyst activity and rapid catalyst aging were problems that limited the usefulness of the data produced during the period of operation of the plant.

Dreyfus, a British citizen who assigned U.S. patents to the Celanese Corp., described a liquid phase synthesis for Fischer-Tropsch synthesis [55, 56]. In his version a catalyst was immersed in a liquid medium which boils at about the temperature of the desired reaction. Syngas is bubbled through the liquid containing the catalyst and fitted with a reflux condenser to maintain the liquid phase. While the inventor indicated that the reaction could be conducted at high pressure, he indicated that it had the greatest value when applied to the synthesis of hydrocarbons under atmospheric pressure. Cobalt catalysts were utilized at a temperature of about 180-200°C and an iron catalyst at a temperature of about 250°C. The example utilized a cobalt catalyst that was promoted with thoria. Excess liquid products are drawn off by an overflow at the desired level, fitted with a liquid seal.

Dreyfus also described the use of liquid-phase synthesis in the range of sub-atmospheric to about 5 atmospheres (5.05 MPa) of pressure [56, 57]. A feature of this patent was the use of a metal active for Fischer-Tropsch synthesis that was present in a reduced form and at a low concentration. A salt of a reducible metal (especially Co, Ni or Fe) that is soluble in the reaction medium was indicated.

2.1.4 United States

2.1.4.1 U.S. Bureau of Mines

The U.S. Bureau of Mines constructed and operated a pilot plant that utilized oil-recycle in a reactor that was 3 in. (7.6 cm) and 8 ft. (2.44 m) tall [36]. After several runs during 1946-47 with co-current downflow of gas and liquid over cobalt catalysts, a run was made with the cobalt catalyst completely submerged in liquid and co-current upflow of gas and liquid. After it had been demonstrated that the unit could be operated with the cobalt catalyst, the later runs were made with an iron catalyst. The composition of the cobalt catalyst was $Co:ThO_2:MgO:kieselguhr = 100:7:12:200$. Since the catalyst bed occupied about 0.25 ft³ (0.026 m³), the GHSV was about 800 to 1600 h⁻¹ during the reduction at 360°C. Following reduction, nitrogen replaced hydrogen and the pressure was increased to 40 psig (0.28 MPa). Oil was then admitted to the reactor. After the oil was added, syngas flow (GHSV = 100) was started. The temperature was slowly increased from 150°C while increasing the pressure in small increments, which decreased the rate of evaporation of the cooling oil. The temperature was maintained below 175°C, maintaining a gas contraction of less than 50%, during 48 hours. The temperature was increased to 180°C during the next 24 h., after which the induction was considered complete. The conditions were then adjusted as required to obtain maximum productivity.

The U.S. Bureau of Mines also operated a larger 8 inch (20.3 cm) diameter reactor in the oil-recycle mode [58]. These units were operated with a precipitated and a fused iron catalyst that has a very low activity compared to the high surface area precipitated iron catalyst. The fused catalyst was used because it was hard and seemed to have the physical strength needed for slurry operation. Some experimental operating problems made it difficult to maintain constant temperature during significant portions of the runs. While it was demonstrated that this mode of operation was viable, little else was obtained that merit further consideration here.

The following summary was provided by Sie and Krishna [15]:

"Three-phase fluidized bed (ebullated bed, also called ebullating bed) reactor in which a packing of larger catalyst particles (e.g. 8-16 mesh) is expanded by cocurrent upflow of oil and gas. The process studied by the US Bureau of Mines features circulation of oil for attaining sufficiently high liquid velocities and has therefore been referred to as the oil circulation process. Process development studies were carried out in a 3 gallon-per-day pilot reactor of 3.2 m length and 7.5 cm diameter and also in a 1 barrel-per-day reactor of 20 cm diameter with a bed height of 2.4 m."

Storch [58], on page 420, describes further development of the slurry process by the Bureau of Mines in 1946. Some physical measurements were made in lab equipment to determine the gas velocities necessary to keep different concentrations of catalyst in suspension. A pilot plant was built and operated (See Fig. 7). A slurry depth of 10 ft was used in a 3 in diameter cylindrical reactor that was attached to an external downcomer to provide slurry circulation. The top of the reactor expanded into a 6 in freeboard section that contained a cooling coil to provide reflux to the top of the reactor. An in line filter draw-off was provided on the external downcomer pipe.

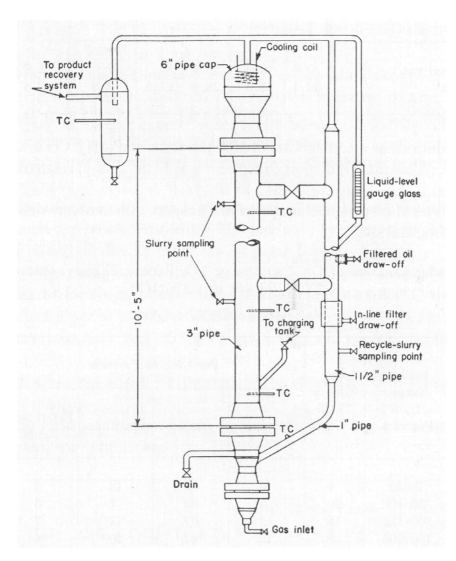

Figure 7 Bureau of Mines slurry phase reactor

2.1.4.2 Standard Oil (Jersey) (now ExxonMobil)

In the 1920s it became necessary for Farben to seek funding from partners that were located outside of Germany. One company that was targeted by Bosch, then head of the company, was Exxon. When the top management of Exxon visited the research facilities where Farben was working on the Bergius process and other high pressure hydrogenation and hydroconversion processes, they were sufficiently impressed to enter quickly into a joint effort that was centred on hydrogenation. In the early 1930s Exxon abandoned research on hydrogenation.

Exxon investigated the Fischer-Tropsch hydrocarbon synthesis process, developed by Ruhrchemie, A.G., which converted brown coal into liquid fuel. In 1938 and 1939, patents for this process outside Germany were transferred by Ruhrchemie to Hydrocarbon Synthesis Corporation, in which Standard Oil Development [now ExxonMobil] took 680 shares, Shell and Kellogg 425 each, and I. G. Farben 170. Both Great Britain and France considered the building of plants using this synthetic process for providing aviation gasoline, but they had been unable to accomplish anything by the time the war broke out in 1939.

It was recognized soon after the discovery of the Fischer-Tropsch synthesis that a reactor employing the catalyst in direct contact with a cooling oil was an attractive approach to maintain temperature control of the exothermic reaction. U.S. patents were obtained by Farben, Standard Oil Development Company and Standard Catalytic Company.

A procedure designed to facilitate removal of the reaction product was to pass the preheated synthesis gas into a 50-bubble tray tower at 105 m^3/hr. A suspension containing about 0.34 kg of finely divided nickel catalyst activated with magnesium and aluminum oxides per 3.8 L of sulfur-free paraffin was charged through the tower at the rate of 20 L/min. The reaction produced about 385 L of liquid hydrocarbons for each 105 m^3 of gas charged [59].

2.1.4.3 Texaco (now ChevronTexaco)

Most of Texaco's attention during this period was directed toward fluid-bed operations such as was utilized at the Brownsville plant. The Brownsville plant was developed by a group of companies and utilized a fixed-fluid bed reactor. The driving force for this plant was Dobie Keith who founded Hydrocarbon Research Inc.. The scale of the operation was small by today's standards: 900 bbl/day of diesel, 200 bbl/day of fuel oil and 300,000 lb/day of chemicals [60].

Texaco obtained a patent which provided some insight into the operation of a slurry bubble column reactor [61]. While the process equipment and its operation are described in much detail, neither experimental data nor examples are given. The reactor can be viewed as a grouping of tubes inside a shell. The catalyst slurry is circulated through the tubes while the cooling fluid is circulated in the space between the tubes and the shell.

Moore cautioned that the height/cross-sectional area should be large and provide an example where the height may range from about 10 to 40 feet (3.1 to 12.2 m) and have an inside diameter of 1 to 6 inches (2.54 to 15.2 cm). Moore taught that the flow of gas should be sufficient such that the 'flooding velocity' is exceeded and that diluent gases may be used to accomplish this objective.

The gas feed rate may be adjusted to give turbulent flow within the reactor tubes. The recycled slurry and dispersed gas flow can be adjusted to give rise through the reactors in 'tubular flow'. Presumably, tubular flow can be considered to be equivalent to plug flow, as this term is used today.

A variety of catalyst formulations were presented, including supported or unsupported cobalt and iron. Separation of wax from the slurry was accomplished external to the reactor and could be operated so that separation was accomplished continuously or by filtering the entire catalyst inventory at one time. The entire recovered catalyst could be regenerated and returned to the reactor or, as an alternative; regeneration could be effected by discontinuing the CO feed while continuing the H_2 feed to the reactor while the system was held at some elevated temperature.

In another variation, the liquid-phase synthesis takes place in a reactor similar to the one in the above patent but materials are added to alter the surface tension to permit the formation of a froth or foam [62]. This was claimed to increase the rate of synthesis by materially increasing the area of the liquid-gas interface. In one variation, water is introduced into the reactor above the entrance of the catalyst/slurry oil entrance. The steam will evaporate and form foam which will fill the reactor. The foam, after leaving the reactor, is broken and the slurry is separated from the products and recycled.

Another variation of the Texaco liquid-phase synthesis [63] is similar to the Duftschmid process. A solvent is injected between the catalyst layers of alternating catalyst and non-catalyst solids. Above each stage in the reactor, there is a device for injecting a fluid which, by partial evaporation, effectively removes the heat of the synthesis reaction.

A later Texaco patent (1948) describes many features found in modern slurry phase reactor designs [64]. The catalyst particles form a suspension or slurry in the liquid. The reactant gases are dispersed in the bottom of the liquid column and rise there through undergoing reaction as they do so. A stream of this liquid (slurry) is continuously drawn off from the upper portion of the liquid column, and either cooled or heated, as the case may be, and then returned to the lower portion of the liquid column. The liquid is thus continuously recycled through the reaction tower to provide a means for controlling the temperature of the reaction.

The unreacted gases together with vaporous products of the reaction, after rising through the liquid column are continuously removed from the top of the tower. In addition, the invention permits employing catalyst in relatively fine form. A catalyst useful for the hydrogenation of carbon monoxide to produce liquid hydrocarbons comprises cobalt, iron or nickel, together with a promoter on a supporting material. Gas recycle after cooling, condensing and separating liquid products is clearly described. Part of the heat of reaction is carried out as sensible heat of the gases and hydrocarbons leaving the top of the reactor. Provision may be made for recycling a substantial amount of the unreacted gas and in that way facilitate temperature control. Hydrocarbon products may also be recycled. Thus the effluent reaction mixture may be recycled in substantial amount such that a relatively low conversion per pass is obtained.

2.1.4.4 Gulf

Cornell and Cotton [64] patented a reactor that was dramatically different from the others described above. First, the reactor was oriented with its length horizontally rather than vertically. This reactor configuration would have intellectual appeal, especially where the height of a reactor is important, as it may be if the reactor is located on a barge. However, it appears that the operation of this plant in a steady-state condition would be an exceptionally demanding task.

2.2 Recently patented features

Table 5 provides a summary of recent patents relating to various features of slurry phase reactor design. From this Table it can be deduced that companies that have recently been active in FT slurry phase reactor decelopment include Statoil, ExxonMobil, Sasol, ENI/Agip/IFP, Conoco, BP/Davy, Syntroleum and Shell.

Two approaches are followed with respect to the way in which the catalyst is used in the reactor. Continuous catalyst rejuvenation or regeneration is used in one approach. The other approach is based on the attainment of a lower but stable catalyst activity. Slurry mixing to suspend the catalyst and attain a uniform reactor temperature is an important consideration for the reactor design. Gas mixing is relevant to reactors that target very high per pass conversions and is thus more important for the continuous catalyst regeneration approach. Separation of wax product from the catalyst is another important aspect for the design of modern slurry phase reactors. Other issues are effective gas distribution, entrainment separation, cooling methods and methods to start-up the reactor without damaging the catalyst.

Some of the patented features from Table 5 provide important advantages in comparison to a simple bubble column design but the features in many of the listed patents do not provide any significant advantages. Some would not even be considered for commercial application even if they were 'open art' options. The remainder of this chapter should help the reader to distinguish the useful features from those that offer little or no value.

Table 5 Patents relating to the design of slurry phase reactors

Patent No.	Assignee	Brief Description	Feature
US 5,387,340	Syncrude	Internal filter	Catalyst Separation
US 5,527,473	Syncrude	Internal filter	Catalyst Separation
EP 0627959	Statoil	Internal filter	Catalyst Separation
US 5,422,375	Statoil	Internal filter	Catalyst Separation
US 5,407,644	Statoil	Internal filter	Catalyst Separation
US 6,069,179	Statoil	Internal filter	Catalyst Separation
US 5,844,006	Sasol	Internal filter	Catalyst Separation
US 5,599,849	Sasol	Internal filter	Catalyst Separation
US 2002/0128330	Texaco	Internal filter	Catalyst Separation
US 5,770,629	Exxon	External filtration	Catalyst Separation
WO 00/043098	ExxonMobil	Internal filter	Catalyst Separation
WO 97/31693	Shell	External filtration	Catalyst Separation
US 6,096,789	Agip/IFP	Hydrocyclone	Catalyst Separation
US 6,217,830	N. Carolina U.	Solvent	Catalyst Separation
US 6,462,098	Sasol	Optimum catalyst size	Cat. Sep./Catalyst Size
ZA 855317	Shell (Gulf)	Optimum catalyst size	Catalyst Size
US 3,901,660	Hoechst	Downcomer/filtration	Mixing/Catalyst Separation
US 5,811,469	Exxon	Downcomer/filtration	Mixing/Catalyst Separation
US 5,157,054	Exxon	Second solid improves suspension	Mixing
US 5,252,613	Exxon	Secondary suspension fluid	Mixing
US RE37,229	Exxon	Downcomer	Mixing
US 5,332,552	Exxon	Draft tube	Mixing
US 5,348,982	Exxon	Slurry bubble column	Mixing
US 5,382,748	Exxon	Downcomer with gas disengaging	Mixing
US 5,866,621	Exxon	Gas and solids reducing downcomer	Mixing
US 5,962,537	Exxon	Multizone downcomer	Mixing
US 6,090,859	Exxon	Small catalyst particle addition	Mixing
US 6,201,031	Sasol	Multiple downcomer	Mixing
US 5,827,902	Agip/IFP	Multistage bubble column reactor	Mixing
US 5,869,541	Agip/IFP	Static mixer	Mixing
US 20030109590	ENI/Agip/IFP	Controlled liquid mixing	Mixing
US 5,961,933	IFP	Liquid recirculation	Mixing
US 6,348,510	ENI/Agip/IFP	Slurry bubble column with liquid upflow	Mixing

Patent No.	Assignee	Brief Description	Feature
US 2003014543	Conoco	Well-mixed gas phase	Mixing
WO 01/94499	BP	Liquid circulation system	Mixing
WO 01/94500	BP	External slurry recycle and gas recycle	Mixing
WO 02/26667	Davy	Slurry circulation	Mixing
WO 02/096836	BP/Davy	Continuously stirred reactor	Mixing
WO 02/096833	BP/Davy	Tubular loop reactor with high shear mixing zone	Mixing/Mass Transfer
WO 02/096837	BP/Davy	High shear mixing zone	Mixing/Mass Transfer
WO 02/097011	BP/Davy	Gas/catalyst contact before high shear mixing zone	Mixing/Mass Transfer
US 5,384,336	Exxon	Tube side slurry	Cooling/Mixing
US 5,409,960	Exxon	Pentane coolant	Cooling
WO 2002100981	ExxonMobil	Ribbed cooling tubes	Cooling
RU 2156650	G.K. Boreskova Inst of Catalysis	Condensate return	Cooling
ZA 2002/9907	Sasol	Cooling system configuration	Cooling
WO 02/096841	BP/Davy	Introducing liquid coolant into reactor with high shear mixing zone	Cooling
WO 02/097010	BP/Davy	External cooing of slurry to high shear mixing zone	Cooling
EP 824961	Shell	Sparger	Gas Distributor
US 5,905,094	Exxon	Grid	Gas Distributor
US 20030027876	ExxonMobil	Grid	Gas Distributor
ZA 2002/9581	Sasol	Sparger	Gas Distributor
US 4,589,927	Battelle Development Corporation	Hybrid bubble column/ebullating bed	Bubble Column/Ebullating Bed
US 4,256,654	Texaco	Iron catalyst ebullating bed	Ebullating Bed
US 5,776,988	IFP	Ebullating catalytic bed	Ebullating Bed
US 4,139,352	Shell	Gas-liquid separator	Entrainment Solution
US 5,866,620	Shell	Freeboard scrubber	Entrainment Solution
US 6,265,452	Sasol	Distillation trays	Entrainment Solution

Patent No.	Assignee	Brief Description	Feature
US 6,403,660	Sasol	Water removal – internal	Water Removal
US 20030134913	Conoco	Water removal - external	Water Removal
US 20030125397	Conoco	Water removal - external	Water Removal
US 20030027875	Conoco	Staged reactors	High Gas Flow
US 4,547,525	Exxon	Co-fed olefins	Olefin Feed
WO 02/096839	BP/Davy	Paraffins converted to olefins for recycle to reactor	Olefin Feed
US 4,626,552	Exxon	Process for start-up	Start-up Procedure
US 6,512,017	Sasol	Handling of a catalyst	Start-up Procedure
PCT/1803/00450	Sasol	Process for start-up	Start-up Procedure
WO 02/059232	Syntroleum	Process for start-up	Start-up Procedure
WO 02/096834	BP/Davy	Method to start-up reactor with high shear mixing zone	Start-up Procedure

3. DESIGN APPROACH FOR MODERN SLURRY PHASE REACTORS

3.1 Text book teaching

The starting point should be previous text books that deal with the subject of slurry phase reactor design for FT synthesis. A significant book is 'Bubble Column Reactors' by Professor W-D Deckwer edited by R.W. Field [66]. This book was originally published in German in 1985 and an English version was later published in 1992. It is perhaps significant that the first patents for modern supported cobalt catalysts appeared in 1985 [67] which is the reason why this book deals mainly with the design approach for iron catalysts. The content draws heavily on the experience from the Rheinpreußen demonstration reactor that was 1.55m in diameter and 8.6m high. This reactor was fed with coal derived synthesis gas with a H_2/CO ratio of 0.67 at an intermediate temperature with the objective of producing mainly gasoline.

Another relevant textbook: "Gas-Liquid-Solid Fluidization Engineering" by Professor L-S Fan [68] was published in 1989. A short six page section covers the topic of Fischer-Tropsch synthesis. To quote from this section, "One of the most important, and perhaps the most and longest studied application of three phase-phase fluidized systems is that concerning the hydrogenation of carbon monoxide by the Fischer-Tropsch process in the liquid phase." Reference is made to previous literature reviews published between 1977 and 1984 that all make use of the Rheinpreußen plant as their benchmark. According to this author, development on FT synthesis effectively ceased in the 1960's with the availability of inexpensive crude oil supplies but revived in the 1970's with the advent of unstable and rapidly rising crude oil prices. A notable omission by this author is a reference to the ongoing R&D work at Sasol in South Africa which continued from the 1950's to the present day.

Fan classifies gas-liquid-solid fluidized systems into 16 different variations. The types of system that are most often proposed for FT applications are the slurry bubble column and the ebullated (or ebullating) bed. According to Fan, the term 'ebullated bed' is commonly used in industry to describe what he calls a 'three-phase fluidized bed' where particles are in 'ebullation' induced by gas-liquid phases. The term is ascribed to P.W. Garbo in a patent by Johanson (1961) to describe a gas-liquid contacting process. According to Fan, the three-phase fluidized bed reactor was first used in 1968 for hydrotreating resids in the H-Oil process and slurry bubble columns have been used for hydrogenation reactions since the 1950's.

To quote further from Fan: " For gas-liquid upward flow with liquid as the continuous phase, fluidized bed and slurry bubble column systems are the two most extensively investigated three-phase systems involving a solid phase in suspension. A diverse range of operating conditions have been labelled slurry bubble column operation; indeed, the operating conditions for 'slurry' systems are loosely defined." Typical operating ranges for three-phase fluidized beds and slurry bubble columns overlap but the terminal velocities of the particles used are higher for the three-phase fluidized bed (ebullated bed) systems than for the slurry bubble column systems. Thus larger particles will be used for the ebullated bed but when using either terminology the solids particle size should be mentioned. In general the particle sizes for ebullated (or ebullating) beds will be above 100 microns (μm) preferably above 350 microns while those for slurry bubble columns will be less than 350 microns.

In this book the term 'slurry phase reactor' is used to include all types of gas-liquid-solid reactors that have freely moving solid particles.

There is now reasonable consensus that catalyst particle sizes between about 10 and 200 microns will be optimal for Fischer-Tropsch synthesis [68, 69 and 70]. The use of larger particle sizes is sometimes proposed, probably due to perceived difficulties with designing or operating slurry bubble column systems [71, 72]. In particular the separation of the hydrocarbon wax product from the catalyst may be thought to be problematic by some. However, Sasol has solved this problem using inexpensive proprietary separation techniques. This has been applied commercially for both iron and cobalt catalysts. Other companies notably ExxonMobil, Statoil, ENI/IFP/Agip and Conoco claim to have suitable separation technology but have not yet applied their technology in commercial operations.

In one of their patents [73], ExxonMobil teach the use of a simple slurry bubble column reactor with a supported cobalt catalyst to produce wax. This patent is somewhat confusing since it describes a non-existent problem of keeping catalyst in suspension while passing gas through a slurry bubble column at a velocity between 2 and 25 cm/s. Keeping catalyst in suspension was known to be easily achieved [67, 74]. They teach an upper size limit of about 50 microns [70], so that solids settling behaviour is determined by Stokes Law, and a lower size limit preferably above 30 microns to avoid solids separation difficulties. They teach no liquid up-flow over and above that induced by the withdrawal of liquid products. This approach was also used with iron catalyst for the Rheinpreußen reactor. Since solids separation was not an important issue for the gasoline operating mode, catalyst size degradation was not considered to be problematic for the Rheinpreußen operation. Some Air Products researchers working under US Department of Energy (DOE) sponsorship proposed the use of very small particles of less than 5 microns [75] but this provides no benefit for synthesis performance and the separation of solids from the liquid product will be difficult and costly.

In Chapter 1 of his book, Deckwer begins with a description of 'the simple bubble column' as discussed at the beginning of the previous paragraph and then proceeds to describe various modified bubble column designs and bubble columns with directional liquid circulation (both internal and external). The modified bubble columns include:

- a cascade of distributors within a single column that divide the column into a number of stages;
- a packed bed
- a static mixer in the slurry bed
- use of vertical baffles to segregate the bed.

It is also significant that all these options for modified bubble columns are illustrated with liquid up-flow. There are several subsequent patents that use these concepts, in some cases together with some additional features and in other cases with questionable validity.

The most commonly proposed approach targets particles in the range from about 30 to 300 microns for a reactor system in which the particles are uniformly fluidized by an upward liquid velocity. There are several patented techniques to achieve this result [70, 76-83]. This is somewhat surprising since this general design approach had been proposed as early as 1948 [84]. There are some proprietary features which provide important advantages, for example, the use of staged internal circulation devices (downcomers) [78]. Although the basic concepts for the design of modern slurry phase reactors are well known, there are clearly some pitfalls when it comes to practical implementation. It will be preferable to use designs that have been implemented successfully on a commercial scale or at least proven on a demonstration scale (e.g. in reactors with diameters of around 1 m or more).

A 1956 patent assigned to Koppers Company, Inc. [85] probably comes closest to describing most of the features that are now proposed for use with modern slurry bubble column reactors. This patent describes the following features:

- Enhanced internal slurry circulation
- Internal cooling
- Multiple banks of internal coolers using vertical tubes with water as cooling medium
- Staged reactors with product knock-out between stages
- Oil reflux to uniformly wash the walls of the reactor freeboard zone to avoid catalyst deposits on these walls
- External liquid recirculation to generate an upward liquid velocity through the slurry bed
- High shear mixing zone to generate small gas bubbles with a high interfacial area to enhance mass transfer
- High aspect ratio and relatively high gas velocity for that time

This patent describes a 1 m diameter and 18 m high reactor with a gas flow of more than 4000 m^3/h at a reaction pressure of 20 atm. (which corresponds to about 13 cm/s gas velocity assuming standard conditions for the stated flow). It seems that internal slurry circulation using downcomers is now a more popular approach than external circulation but the principles remain the same. The use of the high shear mixing zone would only be beneficial for catalysts with very high activity. This high catalyst activity may perhaps be attained using continuous catalyst rejuvenation or with future catalyst advances. The high shear mixing zone concept would certainly not have been particularly beneficial with the catalysts available at the time that this patent was conceived.

With the modern approach, which is accompanied by high liquid wax yields from the reactor, it is important to avoid particle degradation so as to facilitate the catalyst/liquid separation. It is significant that only Sasol [69] has claimed to have completely eliminated catalyst degradation for their commercial supported cobalt catalyst.

As with all fluidized bed systems, the selection and design of the gas distributor is particularly important and this is emphasised in the Deckwer book. Not mentioned is the importance of avoiding practical problems such as catalyst attrition; catalyst settling in stagnant zones and blocking of gas distributor nozzles. Successful gas distributor designs are usually closely guarded trade secrets although some approaches have been patented as can be seen from Table 5. Here again, the general principles are well known but there are some pitfalls for the design and scale-up of gas distributors that make it preferable to rely on a design that has been applied successfully in a large scale commercial reactor.

An understanding of the mixing behaviour in slurry phase reactors is very important for design and scale-up. Deckwer devotes an entire chapter to this subject with the title: "Dispersion in Gas-Liquid Flow". This will be discussed in more detail when considering mathematical models for slurry phase reactors. Deckwer uses the term 'dispersion' to describe all random spreading processes in the various phases including back-mixing. Dispersion is usually simplified as one-dimensional axial dispersion. To quote Deckwer: "As relations for mass transfer are generally a linear function of concentration, one dimensional models are normally adequate for bubble column reactors even when strong radial functions are in evidence. Radial dispersion phenomena only affect reactor conversion and selectivity under conditions of pronounced non linearity (reaction not first order, heat effects) plus highly irregular addition of the reactants at the same time."

Two extremes of mixing are perfect mixing and plug flow. The reactor volume may, in the presence of a considerable amount of dispersion (axial mixing), become many times greater than that for plug flow, especially when conversion is high. Deckwer proceeds to provide mathematical methods to describe axial dispersion in the gas and liquid (or slurry) in terms of a dispersion co-efficient for these phases which is incorporated into a Bodenstein number that characterises the mixing in each phase. The Bodenstein number is a special type of Peclet number that is often used to describe axial mixing in so-called axial-dispersion models. The higher the dispersion co-efficient and the lower the Bodenstein (Bo) or Peclet (Pe) number, the greater is the degree of mixing. For a given dispersion co-efficient, Bo (and thus the effect of mixing) decreases with increasing column length and phase velocity.

Another modelling approach that gives the same numerical result is to describe mixing in terms of a number of perfectly mixed cells in series. In this case the degree of mixing is described by the number of cells in series which can be related to a particular Bodenstein number from the axial-dispersion model.

Deckwer recommends certain correlations to predict the liquid and gas phase dispersion coefficients. The dispersion coefficients for both phases increase as a function of gas velocity and column diameter. However, as Saxena [86] later pointed out (1995), these correlations neglect the effect of column internals. In a recent paper by Forret, Schweitzer, Gauthier, Krishna and Schweich [87], the influence of vertical tubes on the liquid phase mixing has been investigated in a 1 m diameter column. This study shows the limitations of the one-dimensional axial dispersion model for this case. Due to the highly exothermic nature of the FT reaction, FT reactors are densely packed with cooling pipes and other internals are often used to enhance slurry phase mixing in order to promote uniform temperatures in the reactor. As a result literature correlations should not be used to design FT reactors. This is the reason why experience with large scale reactors is essential for the accurate prediction of reactor performance particularly at high conversions. The larger the range of diameters available, the greater is the confidence in the predictions. Non-reactive mock-up columns can also be used to obtain mixing data.

Saxena introduces the term 'baffled slurry bubble column' to describe the slurry phase reactors with internal surfaces which would typically be used for Fischer-Tropsch synthesis. According to Saxena "Most of the available literature concerns unbaffled slurry bubble columns and only limited investigations have been conducted on baffled columns." Saxena further states that "O'Dowd et al. concluded on the basis of experiments that the use of baffles in a slurry column in general provides a more uniform axial solids concentration than an unbaffled column for otherwise identical operating conditions. The effect was more pronounced at lower gas velocities (less than 0.10 m/s)."

Care must be taken in accepting the above conclusion by O'Dowd as being universally applicable since this result may well only apply to the diameter of the columns used for the experiments.

An important observation by Saxena is that measured heat transfer coefficients at gas velocities higher than 0.16 m/s are almost constant. Thus for higher velocities of industrial interest it is only necessary to make a single heat transfer measurement for design purposes. He also points out that the nature of the liquid (or slurry) phase is important and that it is adequately characterised only by its viscosity value.

The following extract from the Saxena review is also relevant:

"Most of the bubble column data have been generated for the semibatch mode of operation in which only the gas phase is in continuous flow. It is desirable to produce data for a cocurrent gas and liquid or slurry phase, i.e., for a continuous mode of operation at different liquid- or slurry-phase velocities. This would be particularly useful for actual industrial units which operate only in continuous mode. Separation of catalyst particles from the liquid products and their recycle to the bubble column reactor is a challenging problem, particularly when the particles are small and in the micron-size range."

It is unclear why Saxena considered the separation of liquid to be necessary for recycle to the bubble column. Downcomers (internal or external) or draft tubes can easily be used to achieve the desired slurry recycle without the need for liquid separation.

To continue the Saxena quote: "Uniform distribution of catalyst particles in the bubble column is essential for good conversion of gaseous reactants.

Sufficient experience exists in the literature on this aspect and this has been reviewed in this article so that proper design calculations of velocities needed to suspend the particles in different fluids of known properties are possible. Fortunately, in most operations, the gas velocities are large enough so that the accomplishment of uniform suspension of catalyst particles is readily obtained. The task is still easier when the catalyst particles are small, as is usually the case."

Saxena, like Deckwer, relies on the use of the well known sedimentation-dispersion model to determine solids concentration profiles. This model was developed by Cova [88] and Suganuma and Yamanishi [89] and later refined by many investigators [90]. Both Deckwer and Saxena fail to mention (perhaps because it is obvious) a significant fact. If the upward liquid velocity exceeds the settling velocity of the particles then there can only be a uniform solids concentration so that the sedimentation-dispersion model is then no longer applicable. This is the likely situation for modern slurry phase reactor designs. Patent teaching from US 5,348,982 [73] that applies the sedimentation-dispersion theory would obviously not be relevant to such modified slurry bubble column designs.

Returning now to Deckwer, he states: "In some cases the relatively sophisticated mathematical dispersion models used for describing bubble column reactors may be replaced by more simple versions. When the aspect ratio (length/diameter) is small, the liquid phase residence time is similar to that of an ideal stirred tank reactor. On the other hand, with a large aspect ratio, the gas phase can be modelled as an ideal plug flow reactor .... However, as the treatment of bubble column reactor models will show, no general conclusions on the effect of dispersion on conversion levels and selectivity can be drawn. This is of particular relevance when attempting to define conditions under which the more simple models can be used. A simple specification of Bodenstein numbers or dispersion values is not enough in this instance. Dispersion effects are best considered in combination with mass transfer and reaction rates, thus becoming a function of conversion levels."

The previous paragraph is highly relevant to certain patents that attempt to draw general conclusions regarding slurry bubble column performance based on mathematical descriptions of the mixing behaviour alone. Simply put, such patents have no scientific basis. This applies to simple bubble columns; baffled bubble columns and to modified bubble columns.

Now considering modified bubble columns, some uncertainty regarding scale-up effects can be eliminated by using packing or bubble column cascades. Deckwer recommends that the liquid phase dispersion coefficients for packed co-current bubble column reactors can be calculated from Heilman and Hoffman [91] and Steigel and Shah's [92] recommendations. The relation derived by Blass and Cornelius [93] is recommended for working out the effective cell number for bubble column cascades. Dispersion coefficients in the cascade liquid phase should be based on Sekizawa and Kubota's correlation [94].

For modified bubble columns with directional liquid circulation which are most likely to be used for commercial scale FT reactors, Deckwer provides some guidance in his chapter 10. The cell model with back flow (CMBF) is the ideal modelling approach for modern slurry phase reactors rather than the axial dispersion models (ADM). Quoting Deckwer again: "The CMBF model has a number of advantages over the ADM [95, 96]. For example, dynamic calculations can be solved numerically, even for non-linear problems, as the CMBF constitutes a set of first order differential equations which represent an initial value problem and which can be numerically integrated without involving convergence problems. The CMBF is also more flexible than the ADM, flow ratios and geometric arrangements being relatively easy to model mathematically. Hence, this cell model can be successfully used for bubble columns in which the liquid phase does not enter the reactor at the top or bottom but at a random point along the column..... Bubble columns in which properties differ between sections and which incorporate non-constant phase hold-ups are also satisfactorily described by the CMBF."

Deckwer introduces the concept of a dimensionless factor M^* which describes the ratio between mixing time and characteristic mass transfer time. The concentration of dissolved gaseous reactant in the liquid phase does not become constant until $M^* \leq 10$ at which point a CSTR (perfect mixing) model can be used.

$$M^* = K_L a L^2 / \varepsilon_L E_L \qquad (1)$$

$K_L a$ is a measure of the mass transfer rate; L is the reactor length; ε_L is the liquid fraction in the bed and E_L is the liquid phase dispersion coefficient.

Calculation of the value for M^* shows clearly that an assumption of perfect mixing in the liquid phase can never be a suitable approach for a commercial scale slurry phase FT reactor that may have a diameter of up to about 10 m and a bed height of between 20 and 40 m. Common errors have been to underestimate the mass transfer rate by failing to consider the effect of rapid bubble coalescence and break-up; to overestimate the decrease in fractional liquid hold-up with increasing gas velocity and to use correlations for the liquid phase dispersion coefficient that are not appropriate for columns with densely packed internals.

Deckwer [74] defines a relative mass transfer resistance, β, which is the ratio of the mass transfer resistance to the sum of the kinetic resistance and the mass transfer resistance. Thus

$$\beta = R_m/(R_k + R_m) \tag{2}$$

Deckwer correctly concludes that mass transfer resistance is small in comparison with kinetic resistance. This is in spite of the fact that he was probably not aware of the factors that further reduce mass transfer resistance at high gas velocities.

According to Saxena, the above conclusion by Deckwer has been disputed and remains controversial. He further states "The magnitiude of β provides a quantitative indication of whether the reactor conversion and throughput are limited primarily by gas-liquid mass transfer or by chemical reaction, or whether both these resistances are significant. Based on a knowledge of β, one would determine the design modifications that would have the greatest impact on reactor performance. For instance, if severe mass transfer limitations exist, one might attempt to increase conversion or reactor throughput by increasing the gas-liquid interfacial area. This could be accomplished by modifying the gas distributor design to produce smaller bubbles, by optimizing reactor geometry to minimize bubble coalescence, or by providing mechanical agitation. Alternatively, if the overall reaction rate is found to be limited by the intrinsic kinetics, reactor performance could be improved by decreasing gas holdup, increasing catalyst activity, catalyst loading, or reactor temperature if possible."

Current research is focussed on increasing catalyst activity which is a clear indication that chemical reaction resistance is significant. Conversion performance is considerably enhanced with very high activity fresh catalysts and a decrease in conversion is clearly observable as the catalyst activity declines. Upper temperature limits are determined by product selectivity and catalyst deactivation considerations. Decreasing gas hold-up and increasing catalyst loading go hand in hand but there comes a point when the slurry viscosity increases rapidly so that mass and heat transfer suddenly become problematic.

3.2 Subsequent contributions

Subsequent to the Deckwer text book Krishna and his co-workers made important contributions to the model descriptions for slurry bubble columns operating at high velocities that are of commercial interest. The desired operating regime is known as the heterogeneous or churn-turbulent regime. In this operating mode fast rising gas bubbles pass through the slurry bed at velocities that exceed 1 m/s. The fast rising gas bubbles are called the dilute phase and it is reasonable to model the dilute phase as being in plug flow. Small gas bubbles, which may account for a significant fraction of the gas in the bed, tend to follow the slurry flow and to account for a negligible portion of the gas through flow. The slurry together with these small bubbles is called the dense phase and it can then be described in terms of its mixing behaviour by an ADM or CMBF model.

This modelling approach is analogous to the two phase model used for gas-solid fluidization and there are many similarities in the behaviour of the two phases in gas-solid heterogeneous fluidization and gas-liquid or gas-slurry heterogeneous fluidization. This analogy is very important when it comes to the collection of relevant data for successful scale-up. Sasol was in the fortunate position of having scaled up the gas-solid fluidized bed Sasol Advanced Synthol (SAS) reactors to diameters exceeding 10 m, at similar operating pressures and with similar internals, before the designs for large commercial slurry phase reactors were prepared. The first FT slurry phase demonstration performed at Sasol made use of a retrofitted SAS reactor so that the similarities were immediately clear.

The use of tail gas recycle has been a feature of all commercial FT reactor applications. Neither Deckwer nor Krishna and his co-workers deal with this feature. Recycle is important for practical commercial reactor designs. It may be possible to avoid recycle operation in future if advanced supported cobalt catalysts can be made more resistant to catalyst de-activation at high per pass conversions. An alternative approach is to take the view that recycle operation can be avoided by making use of on-line catalyst rejuvenation. Current commercial applications maintain a lower but stable catalyst activity and obtain the desired overall conversion by using a recycle design. This approach is inherently safer without any significant cost differences. The stable catalyst activity approach requires a recycle compressor to achieve 90% reactant conversions and may require a longer bed, depending on the ratio of the catalyst activities; on the other hand the catalyst rejuvenation approach requires hydrogen production facilities to make the rejuvenating gas at well as a place to contact this gas with deactivated catalyst.

Fox [97] has provided an open literature explanation of the benefit of using a recycle approach. He shows that there is a gain in reactor productivity with increasing recycle ratio so that there is an engineering evaluation to be made as to the best per pass conversion level to design for. He also concludes that per pass conversion levels over 90% should be avoided because of the sharp drop-off in space time yield (STY). He suggests that 80% conversion per pass may be a good compromise between recycle requirements and high productivity. From the diagram presented by Fox it seems likely that the optimum will more likely be found in the range from 60% to 80% per pass. Fox has correctly identified that Deckwer was misled in the choice of superficial gas velocities by the use of a simplified gas hold-up expression that gives much too high a gas hold-up (and therefore too low a catalyst hold-up) at superficial velocities above 4 to 5 cm/s. To quote Fox: "Operation at 0.15 m/s inlet superficial velocity and 35 wt% slurry concentration appears as feasible in a Fischer-Tropsch as in a methanol slurry reactor." Thus the prejudice against optimum operation at velocities above about 10 cm/s, created to a large extent by the numerous publications by Deckwer, was challenged by Fox in 1990. This work by Fox was done for the US Department of Energy (DOE), Pittsburgh Energy Technology Centre under Contract No. DE-AC22-89PC89867.

However, Fox also provides some misleading opinions. For example, he states that "The limiting factor on conversion in the slurry bubble column is backmixing, particularly in the liquid phase, which makes it necessary to use lower space velocities than would be required in a plug flow reactor." He also states, "The superficial velocity restriction on a slurry reactor makes it generally unsuitable for low conversion, high recycle operation." This latter statement appears to be an unjustified conclusion based on experience with methanol synthesis. Fox notes that when reaction rate controls his simplified model approach reverts to a CSTR model. Reaction rate does usually control for FT synthesis but in reality this means that the behaviour can be closely approximated by a model that correctly describes the mixing in the dense phase relative to the reaction rate. This should be clear from the preceding discussion concerning M^*. A possible exception is when very high catalyst activities are attained with the catalyst rejuvenation approach, then mass transfer may play a more significant role and the reactor heights needed may be less so that the negative effect of dense phase backmixing is enhanced. It should be recognised that the catalyst is less effectively utilised if backmixing and mass transfer resistance become significant issues.

Fox uses a very simple rate expression to describe the FT synthesis that can only be justified for low per pass conversion designs. If, as expected, the FT reactor is designed using an approach where the reaction rate is controlling then it is very important to use the correct rate expression.

Fox in Ref. 97 determined that a gas velocity of 11.5 cm/s (based on the empty reactor shell) would be optimal for a slurry bubble column operating at 28.3 atm using iron catalyst and low H_2/CO ratio syngas feed. The calculated bed height was 11.69 m to achieve a per pass conversion of 80% giving a syngas conversion of 95% with a recycle to syngas ratio of 0.264. At the time of this publication, it was well known that the desired hydrocarbon products were predominantly wax (chain growth factor at least 0.9 for C20+ hydrocarbons) and that the wax could be easily hydrocracked to yield predominantly diesel boiling range hydrocarbon product. Thus this design uses a higher pressure and lower operating temperature than the Rheinpreußen-Koppers demonstration reactor to enhance wax selectivity. The following is a comparison of the two designs for slurry bubble column reactors using iron catalyst and low H_2/CO ratio syngas feed :

Table 6
Slurry bubble column FT reactor designs

Application	Rheinpreußen - Koppers	US DOE (Fox)
Syngas H_2/CO ratio	0.67	0.67
Pressure (atm)	12	28.3
Temperature (°C)	268	257
Bed height (m)	7.7	11.7
Recycle ratio	0	0.264
Conversion %	90	95
Superficial gas velocity at operating conditions at reactor inlet based on empty shell (cm/s)	9.5	11.5

The reactor designs are surprisingly similar. The proposed gas velocity is only marginally higher. Although it has been speculated that recycle was used to achieve the claimed performance for the Rheinpreußen reactor it is generally assumed that this claimed performance was for once through operation. So with this assumption, it seems that by using recycle a slightly higher conversion is obtained for about the same syngas feed volumetric rate. It is well known that iron catalyst productivity increases in direct proportion to operating pressure and so the choice of higher pressure is not surprising. If the feed molar flow is increased in proportion to the increase in pressure then for the same amount of catalyst the conversion and gas velocity remain the same. In principle, even higher pressures would be desirable but this would depend on economic and technical constraints for the upstream gasification process since syngas compression will usually not be cost effective. As pressure and gas velocity are increased the gas hold-up increases so a greater reactor height is required to hold the same amount of catalyst. It is therefore not surprising that the later design has a height increase that is slightly more than the velocity increase.

What is not clear is why the reactor height was constrained to 11.7 m. There is no reason why even higher gas velocities could be used with further height increase to maintain the target conversion.

3.2.1 Gas hold-up prediction

The gas hold-up is an essential parameter that needs to be known in order to design the reactor because it not only determines the reactor pressure profile, it also determines the amount of catalyst which can be held in a given reactor volume. For two phase fluidization, the science has developed to the point where it is possible to predict the gas hold-up from the physical characteristics of the solids and the gas (See Section 5.1). This is not yet the case for slurry bed systems where the phase hold-up prediction is much more complex and is dependant on properties, such as the liquid phase surface tension, which are difficult to predict at the reactor operating conditions. There is also a strong influence of the operating pressure on gas hold-up for slurry bed systems. As a result, large pilot units are currently required to successfully design commercial slurry bed reactor systems. Such pilot reactors should have a diameter greater than at least 30cm and should preferably have a diameter of the order of 1m or larger. The sizes of the largest individual gas bubbles were measured in Sasol's 1m diameter pilot unit by making use of cross correlation from readings using nuclear density measurement devices located at different axial elevations.

There are no reliable published correlations or methods to predict gas hold-up without the aid of suitable experimental data. However, some good work has been done to understand the hydrodynamic mechanisms that influence gas hold-up. This fact should be kept in mind when reading the subsequent discussions.

The combination of hydrodynamic data from the nominally 1m diameter demonstration reactor, together with kinetic and selectivity data from laboratory gradientless slurry bed micro reactors have been combined into computer models which have been successfully applied to the prediction of the performance of commercial scale reactors [11].

The volumetric catalyst concentration C_{cat} is related to the gas hold-up via:

$$C_{cat} = \frac{W_{cat}}{V_R \left(1 - \varepsilon_{int\,ernals}\right)} = \frac{\left(1 - \varepsilon_G\right)\varepsilon_p}{\rho_p}$$

(3)

As a rule of thumb, the volumetric gas-liquid mass transfer coefficient for a slurry bubble column reactor is roughly proportional to the gas hold-up:

$$k_L a = C * \varepsilon_G$$

(4)

One of the first publications to suggest the simple Eq. (4), with $C = 0.5$, was by Vermeer and Krishna [98].

The earliest recorded formulation of a gas distribution theory is credited to Toomey and Johnstone [99] and is known as the two phase theory of fluidization. This two phase theory, developed to describe the gas distribution in conventional powder fluidized beds, can also be of use in describing the gas distribution in a slurry bed (also known as a slurry bubble column). Simply stated, the theory specified that all gas in excess of that needed to bring a fluidized bed to minimum fluidization conditions passes through in the form of bubbles. This simple theory is slightly modified by defining a dense phase which consists of the aerated powder for gas-solid fluidization or the slurry aerated by small bubbles in the case of a slurry bubble column. Gas in excess of that required to maintain the dense phase passes through in the form of bubbles referred to as the dilute phase. The gas hold-up in the dense phase can be determined by means of a bed collapse experiment [100]. In the case of slurries, there is no minimum fluidization velocity but there is a velocity at which a transition occurs between a homogeneous bubbly regime and the churn turbulent regime [66]. This transition determines the dense phase transition to a combination of dense and dilute phases as defined by this modified two phase theory. This approach was pioneered by Krishna et al [101] who presented this concept at an AIChE meeting in 1979 and first published it in 1981 [102].

It is important to differentiate the dense phase gas hold-up from the dilute phase gas hold-up because the dense phase gas hold-up is not affected by the column geometry while, for small diameter columns the dilute phase gas hold-up is determined mainly by the column geometry. For large diameter columns (larger than 1m) the dilute phase gas hold-up is surprisingly constant for all fluidized systems while the dense phase gas hold-up may vary widely depending on the powder or slurry properties. In the case of slurries prediction of the dense phase voidage (i.e. gas hold-up in the dense phase) are usually unreliable because of the sensitivity to the liquid surface tension which is often affected by small amounts of impurities.

Using the modified two phase theory, the dilute phase voidage can be described using the equation:

$$\varepsilon_b = (U - U_{df}) / U_b \tag{5}$$

Using the correlation proposed by Werther [103] of $U_b = (g\ d_b)^{1/2}$ and substituting into Eq. (5) gives:

$$\varepsilon_b = (U - U_{df}) / \phi (g\ d_b)^{1/2} \tag{6}$$

Following the Darton et al [104] approach, the diameter of a sphere having the same volume as the actual bubble, for dispersion heights exceeding h* (where h* is the height above the gas distributor where the bubbles reach an equilibrium size), is given by :

$$d_b = \alpha\ (U - U_{df})^{2/5}\ (h^* + h_o)^{4/5}\ g^{-1/5} \qquad \text{for } h^* \le h \le H \tag{7}$$

Ellenberger and Krishna [101] derived the following equation for $H \geq h^*$:

$$\varepsilon_b = (U - U_{df})^{4/5} / \alpha^{1/2} \; \phi \; g^{2/5} (h^* + h_o)^{2/5} \tag{8}$$

Substituting for d_b from Eq. (7) into Eq. (6) gives Eq. (8). This shows that the approach of Ellenberger and Krishna is consistent with the modified two phase theory.

It is also known that h^* is a function of $(U - U_{df})$. Once this functional dependence has been determined, equation (8) can be used to describe how the dilute phase gas hold-up, ε_b, varies with superficial gas velocity U (having also determined U_{df}).

Werther [103] proposed that for columns with diameters smaller than 1m, $\phi = \phi_o \; (D_T)^{2/5}$ which can be used with equation (8) to quantify the dependence of dilute phase gas hold-up on column diameter.

The state of the art as described above was not available during the commercialization of the slurry bed reactor technology for the Fischer-Tropsch application. Sasol was, however, aware of the importance of column diameter and hence was successful in scaling up from the 1m diameter Works Pilot Plant to the 5m diameter reactor commissioned in 1993. Scale-up to any other reactor diameter is now an easy task for Sasol.

The prediction of the dense phase mixing characteristics of the slurry and the influence of column diameter on this mixing behavior is also an important scale-up issue. As mentioned previously relevant information can be obtained from industrial scale baffled slurry bubble column and fluidized bed systems. This point has been emphasized by Krishna, Ellenberger and Sie [105].

3.2.1.1 Maximum stable bubble size

Wilkinson [106, 107] observed that both the maximum stable bubble size and the mean bubble size decrease with increasing gas density and decreasing surface tension. Letzel [108] explained this quantitatively by a Kelvin-Helmholz stability analysis and concluded that the mean large bubble rise velocity is inversely proportional to the square root of the gas density ($u_{lb} \sim (1/\rho_G)^{0.5}$). This is mimicked by the gas density correction factor DF used by Krishna [109, 110].

On basis of a fundamental mechanistic model including the effect of internal gas circulation (especially important at higher gas densities), Luo et al. [111] arrived at the conclusion that the maximum stable bubble size is inversely proportional to the square root of the gas density:

$$d_{b,\max} \approx 2.53 \sqrt{\left(\frac{\sigma}{g\rho_G}\right)} \quad \text{for gas-liquid systems} \tag{9}$$

$$d_{b,\max} \approx 3.27 \sqrt{\left(\frac{\sigma}{g\rho_G}\right)} \quad \text{for gas-slurry systems} \tag{10}$$

where:
$d_{b,\max}$ = maximum stable bubble size (m)
σ = surface tension (N m^{-1})
g   = acceleration by gravity (m s^{-2})
ρ_G = gas density (kg m^{-3})

As a consequence, the rise velocity of a bubble at its maximum stable bubble size is proportional to $(1/\rho_G)^{0.25}$ as can be deduced by substituting into the Taylor-Davies correlation to give:

$$V_{b,\max} \approx \sqrt{\left(\frac{C g^{0.5} \sigma^{0.5}}{2\rho_G^{0.5}}\right)} \approx \sqrt{\frac{C}{2}} \left(\frac{g\sigma}{\rho_G}\right)^{0.25} \tag{11}$$

$V_{b,\max}$ = single bubble rise velocity of a bubble at the maximum stable
         bubble size (m s^{-1})
σ =   surface tension (N m^{-1})
g =   acceleration by gravity (m s^{-2})
ρ_G = gas density (kg m^{-3})
C =   constant, C = 2.53 for gas-liquid systems, C = 3.27 for gas-slurry
        systems (-).

The actually prevailing bubble size distribution will be determined by the dynamic equilibrium between bubble coalescence and bubble break-up, through which the mean bubble size is lower than the maximum stable bubble size. The bubble size distribution is generally known to narrow with a smaller mean bubble size when the gas density is increased [106, 107, 111, 112], because bubble break-up is substantially enhanced by increased gas densities. Increasing the solid content of the slurry widens the bubble size distribution and shifts the mean bubble size to larger values [113].

3.2.1.2 Effect of column dimensions
Wilkinson [106] concluded that the only experimental data relevant for design of commercial scale bubble columns are those obtained in bubble columns with a diameter larger than the critical column diameter, an aspect

ratio H/D > 5 or H > 1 – 3 m and perforated plate or single nozzle with hole diameters in excess of 1 – 2 mm as gas distributor. The critical column diameter above which the gas hold-up is not influenced by the diameter is usually close to 15 cm and can be expected to decrease to some extent with increasing gas density and decreasing surface tension. The latter can be explained from the reduced wall effect on the rising gas bubbles with decreasing bubble size [114]. gives the following correlation to estimate the critical column diameter D_c:

$$D_c = 20\left(\frac{\sigma^2}{g^2(\rho_L - \rho_G)\rho_G}\right)^{0.25}$$

(12)

with:
D_c = critical column diameter above which there is no effect of the
    column diameter on the gas hold-up any more (m)
σ = surface tension (N m^{-1})
g = acceleration by gravity (m s^{-2})
ρ_G = gas density (kg m^{-3})
ρ_L = liquid density (kg m^{-3})

Eq. (12) gives a critical column diameter of 0.09 – 0.11 m for gas densities of 6 and 10 kg m^{-3} respectively.

For megascale FT reactors of the future, with reactor diameters of the order of 10 m, the H/D ratio is likely to be in the 3-4 range. Use of experimental data obtained in columns with H/D > 5 (quite easy to do this in the laboratory) is open to serious question. This point has not been sufficiently recognized by academics.

Porous plate distributors and perforated plate distributors with hole diameters < 1 mm are frequently applied in academic R&D, but give higher gas hold-ups and interfacial areas, especially at relatively low superficial gas velocities and/or low dispersion heights, for industrially relevant gas spargers. Care should be therefore taken not to use experimental data from laboratory columns equipped with porous plate distributors for the design of an industrial column.

3.2.2 Flow regimes

The above discussion applies to the heterogeneous or churn-turbulent flow regime of industrial interest for FT reactors. It is important to understand the boundaries of this flow regime to which the above discussion applies.

Figure 8(a) gives the generally used map of flow regimes in bubble columns [109]. The homogeneous regime occurs in bubble columns at relatively low superficial gas velocities. This homogeneous regime is characterized by relatively small gas bubbles, which are dispersed homogeneously throughout the liquid phase (no significant radial gas hold-up profiles). The transition from the homogeneous regime to the churn-turbulent regime occurs at the so-called 1st transition velocity. This churn-turbulent regime is characterized by the occurrence of two types of gas phases, i.e. the small bubble gas phase or dense phase and the large bubble gas phase or dilute phase. The small bubbles are homogeneously dispersed over the entire column, whereas the large bubbles rise preferentially away from the walls and other surfaces where slurry down flow occurs. The large bubbles transport the major amount of feed gas through the column at high superficial velocity. As a consequence, significant radial gas hold-up profiles may occur in the heterogeneous regime with relatively high gas hold-ups around the centerline of the column [66].

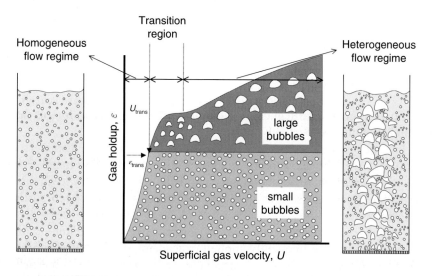

Figure 8(a) Homogeneous (bubble flow) and heterogeneous (churn-turbulent) flow regimes in gas-liquid bubble columns [109].

The heterogeneous regime prevails at superficial gas velocities in the range between the 1st transition velocity up and the 2nd transition velocity. Noteworthy, using chaos analysis of pressure fluctuation signals, Letzel [115] observed that an additional transition occurs within the boundaries of the heterogeneous regime for increased gas densities ($\rho_G > 5$ kg m^{-3}) and at a superficial gas velocity being typically 2-3 times the 1st homogeneous regime-heterogeneous regime transition velocity. This "pseudo" transition is explained from two classes of large bubbles, from which the sizes mainly overlap at low gas densities but are substantially different at increased gas densities.

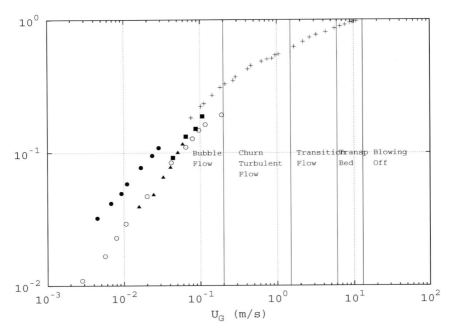

Figure 8(b) Flow regimes and gas hold-up as a function of superficial gas velocity for the air-water system at ambient pressure, temperature [116]. Column diameters: 0.02 (•), 0.075 m (+), 0.25 m (o), and 0.48 m (▲).

The heterogeneous regime is left and a gradual transition from a gas-in-liquid dispersion to a liquid-in-gas dispersion (phase inversion) occurs at superficial gas velocities above the 2nd transition velocity, see Fig. 8(b). [116, 117]. Notice from Fig. 8(b) and Table 7 that this phase inversion occurs smoothly over a rather broad range of superficial gas velocities. The on-set of the phase inversion – or transition flow regime is further characterized by gas hold-ups in excess of 0.65 – 0.7. The on-set of the phase inversion regime can be expected to occur at significantly higher superficial gas velocities for increased column diameters, because of a reducing effect on the gas hold-up. The upper value is not far from the interstitial liquid voidage of a dense 'packed bed' of spherical bubbles [118]. Increasing the superficial gas velocity beyond the upper limit of this transition flow regime results in the occurrence of a transported bed regime, where a true dispersion zone with a constant hold-up of dispersed liquid lumps occurs up to a certain column height above which the liquid hold-up (1 – gas hold-up) decreases exponentially with height [117].

The transition velocities depend strongly on the physical liquid properties (especially surface tension and viscosity), the gas density and the presence of catalyst particles in the liquid. Table 7 gives specific values for the air-water system in a 0.075 m ID diameter bubble column at ambient conditions [116]. Table 7 also gives the α-values in the empirical gas hold-up correlation $\varepsilon_G = K\, u_G^{\alpha}$ for the different flow regimes of air-water at ambient conditions. Notice that the transition regime for phase inversion starts for air-water at ambient conditions at a superficial gas velocity as high as 1.3 m/s with a corresponding gas hold-up in the range 0.6–0.7. As can be seen, the dependency of the gas hold-up on the superficial gas velocity becomes increasingly less pronounced with increasing gas velocity in the churn-turbulent regime.

Table 7
Superficial gas velocity (u_G) range, gas hold-up (ε_G) range and values of α in empirical gas hold-up correlation $\varepsilon_G = K\, u_G^{\alpha}$ for the different flow regimes of the air-water system in a 0.075 m ID bubble column.

Flow regime	Homogeneous bubble flow	Churn-turbulent flow	Phase inversion or transition flow	Transported bed
u_G (m s^{-1})	< 0.20	0.20 – 1.35	1.35 – 5.5	5.5 – 13.5
ε_G at regime transition	0.2 – 0.4	0.65 – 0.7	0.8 – 0.9	-
α in $\varepsilon_G = K u_G^{\alpha}$	0.60	0.32	0.22	0.11

3.2.3 Scale-up for slurry phase reactors

The following discussion is based on a paper prepared by Krishna for IFSA 2002, held in Randburg, South Africa and a publication in press by van Baten and Krishna [119, 120].

For successful scale up of the bubble column slurry reactor for Fischer Tropsch synthesis, a proper description of the hydrodynamics and transport phenomena is needed (gas and liquid hold-ups, gas and liquid (slurry) phase mixing, gas-liquid mass transfer, heat transfer to cooling tubes) as a function of reactor scale (column diameter and height) and operating conditions (superficial gas velocity, system pressure, slurry concentration,…). Most of the above mentioned hydrodynamic parameters are inter-related. A scale up strategy is presented using Computational Fluid Dynamics (CFD). Experimental results from non-reactive hydrodynamic studies have been used to verify the CFD model predictions.

CFD simulations were carried out in the Eulerian framework using both two-dimensional (2D) axi-symetric and transient three-dimensional (3D) strategies in order to describe the influence of column diameter on the hydrodynamics and dispersion characteristics of the bubble column slurry reactor for FT synthesis. The results demonstrate that there is a strong increase in liquid circulation with column diameter. It is concluded that the 3D Eulerian simulations can provide a powerful tool for the scale-up of bubble columns.

3.2.3.1 Introduction

For economic and logistic reasons Fischer-Tropsch conversions are best carried out in large scale projects and the capability of scaling up is therefore an important consideration in the selection of reactors for synthesis gas generation as well as in Fischer-Tropsch synthesis. It is now widely accepted that the slurry bubble column slurry reactor is an appropriate reactor type for large scale plants with a capacity of the order of 40,000 bbl/day or more. Typical design and operating conditions of a Fischer Tropsch slurry bubble column diameter for an optimally designed reactor can be obtained, for example, from the information given by Maretto and Krishna [121]:

- The column diameter ranges from 6 to 10 m
- The column height is in the range of 30 – 40 m,
- The reactor operates at a pressure of between 2 – 4 MPa,
- The reactor temperature is about 513 – 523 K,
- The superficial gas velocity is in the range 0.10 – 0.4 m/s depending on the catalyst activity and the catalyst concentration in the slurry phase,
- For high reactor productivities, the highest slurry concentrations consistent with catalyst handle-ability should be used. In practice the volume fraction of catalyst in the slurry phase, ε_s, is in the range 0.3 – 0.4,
- For removing the heat of reaction 5000 – 8000 vertical cooling tubes, say of 50 mm diameter and 150 mm pitch, will need to be installed.

The success of the process largely depends on the ability to achieve deep syngas conversions, say exceeding 95%. Reliable design of the reactor to achieve such high conversion levels, requires reasonable accurate information on the following hydrodynamics and mass transfer parameters:

- Gas hold-up
- Inter-phase mass transfer between the gas bubbles and the slurry (Editorial comment: only required for very high catalyst activities.)
- Axial dispersion of the liquid (slurry) phase
- Axial dispersion of the gas phase (Editorial comment: plug flow is usually a reasonable assumption.)
- Heat transfer coefficient to cooling tubes

Most of the above mentioned hydrodynamic parameters are inter-related. For a given column diameter the bubble rise velocity affects the gas hold-up and also determines the strength of the liquid circulations and, consequently the axial dispersion coefficient of the liquid phase. The distribution of bubble sizes and rise velocities determines the axial dispersion coefficient of the gas phase. The heat transfer to the cooling tubes is influenced by the renewal rate of the liquid film on the tube surface, which in turn is dictated by the bubble rise velocity. Increasing the column diameter has the effect of increasing the liquid circulations which enhances the bubble rise velocity; this impacts on all the hydrodynamic parameters. For a proper description of the hydrodynamics at different scales Computational Fluid Dynamics (CFD) in the Eulerian framework can be used. In the CFD model, the interphase momentum exchange, or drag, coefficient is obtained from experimental measurements on a relatively small scale.

In order to demonstrate this scale-up approach, experimental studies on slurry bubble columns in columns of 0.1, 0.19 and 0.38 m in diameter were carried out. These experimental results are compared with CFD simulations. Furthermore, CFD simulations for a 6 m column demonstrate the significant influence of scale on column hydrodynamics. (Editorial comment: for real applications the effect of column internals in the baffled columns typically used will need to be taken into account.)

3.2.3.2 Experimental set-up and results

Experiments were performed in polyacrylate columns with inner diameters of 0.1, 0.19 and 0.38 m. The gas distributors used in the three columns were all made of sintered bronze plate (with a mean pore size of 50 μm). The gas flow rates entering the column were measured with the use of a set of rotameters, placed in parallel, as shown in Fig. 9 for the 0.38 m column. This set-up was typical. Air was used as the gas phase in all experiments. Firstly, experiments were performed with paraffin oil (density, ρ_L = 790 kg/m^3; viscosity, μ_L = 0.0029 Pa.s; surface tension, σ = 0.028 N/m) as liquid phase to which solid particles in varying concentrations were added. The solid phase used consisted of porous silica particles whose properties were determined to be as follows: skeleton density = 2100 kg/m^3; pore volume = 1.05 mL/g; particle size distribution, d_p: 10% < 27 μm; 50% < 38 μm; 90% < 47 μm. The solids concentration ε_s, is expressed as the volume fraction of solids in gas free slurry. The pore volume of the particles (liquid filled during operation) is counted as being part of the solid phase. Further details of the experimental work are available elsewhere [109, 122-125].

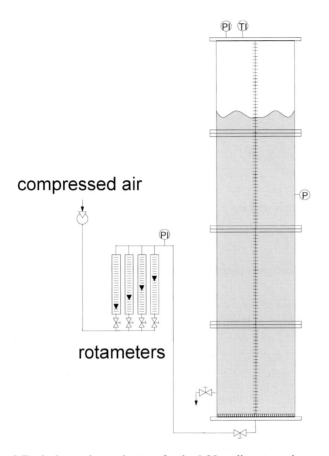

Figure 9 Typical experimental set-up for the 0.38 m diameter column.

The influence of the solids concentration on the total gas hold-up ε for varying superficial gas velocities are shown in Fig. 10 for the 0.38 m diameter column. It is observed that increased particles concentration tends to decrease the total gas hold-up, ε, to a significant extent. This decrease in the total gas hold-up is due to the decrease in the hold-up of the small bubbles due to enhanced coalescence caused due to the presence of the small particles. At low solid concentrations there is a pronounced maximum in the gas hold-up which is typical of the transition region. With increased solids concentration the transition occurs at a lower superficial gas velocity and the transition "window" reduces in size.

Fig. 11 presents a qualitative picture of the influence of gas velocity and particles concentration. At particles concentration exceeding 30 vol% the dispersion consists almost exclusively of fast-rising large bubbles, belonging to

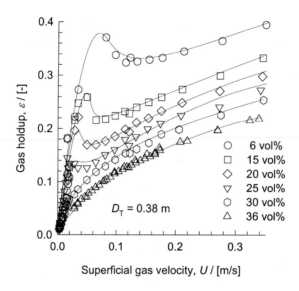

Figure 10 Influence of increased particles concentration on the total gas hold-up in 0.38 m diameter column. The liquid phase is paraffin oil containing varying concentrations of silica particles.

the spherical cap family. The gas holdup in concentrated slurries has been shown to have almost the same values as the gas holdup in a highly viscous liquid [123], such as Tellus oil with a viscosity of 75 mPa s. Furthermore, dynamic gas disengagement experiments [123] have established that for both Tellus oil and concentrated paraffin oil slurries the gas dispersion consists predominantly of large bubbles. Tellus oil can therefore be used to mimic the hydrodydnamics of a FT reactor with concentrated slurries.

Recent experimental study by Urseanu et al. [126] has shown that the influence of operating pressure on the gas holdup is negligible for high viscosity liquids and therefore we may conclude that the experimental gas holdups for Tellus oil at atmospheric conditions are representative of the FT reactor at a higher operating pressure. For a slurry concentration of 36 vol%, the gas hold-up decreases with column diameter; see Fig. 12. An exactly analogous dependence of ε on D_T has been observed for the highly viscous Tellus oil [123, 127]. With increasing column diameter, the liquid circulation velocities are higher, with the consequence that the bubbles tend to be accelerated leading to lower gas hold-up. This is evidenced by plotting the bubble swarm velocity, V_b calculated from $V_b = U/\varepsilon$ for the three columns; see Fig. 13. At low superficial gas velocities the bubble swarm velocity is practically the same for the three columns and $V_{b0} = 0.47$ m/s; this is indicated by the large filled circle in Fig. 12. $V_{b0} = 0.47$ m/s is used to determine the drag coefficient between the large bubbles and the liquid and thereby "calibrate" the CFD simulations.

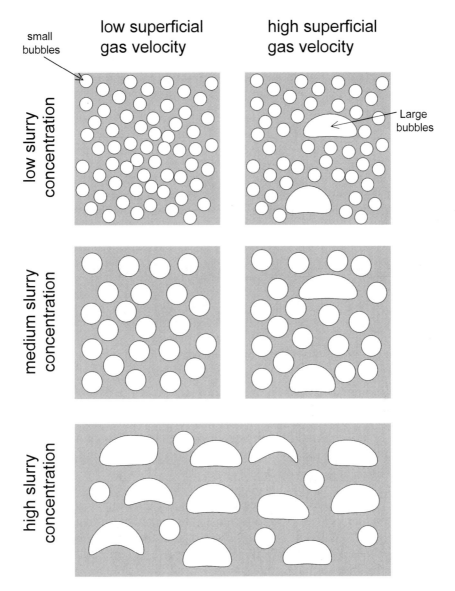

Figure 11 Qualitative picture of the influence of particles concentration and superficial gas velocity on bubble dispersions

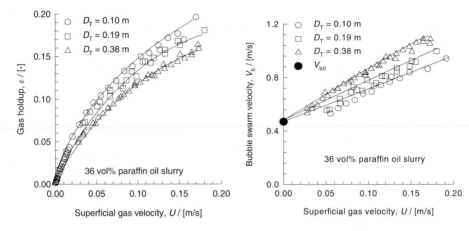

Figure 12 Influence of column diameter on the gas hold-up in 36 vol% paraffin oil slurry system.

Figure 13 Influence of column diameter on the average bubble swarm velocity in 36 vol% paraffin oil slurry system.

In contrast to the results shown above for viscous slurry systems, published experimental work on *air-water* systems [128, 129] show that the influence of column diameter on the *total* gas holdup is negligible. The rationalization of these observations is as follows. For air-water systems we have essentially a bi-modal bubble size distribution, with 'small' and 'large' bubble size populations [109]. With increased liquid circulations, the large bubbles that are concentrated in the central core tend to rise faster and there is a decrease in the large bubble gas holdup. The small bubbles however are predominantly present in the peripheral wall region [130]; with increased liquid circulations the small bubbles are dragged downwards in the wall region and this leads to *higher* small bubble holdup with increased column diameter. The *total* gas holdup is virtually unaltered with increasing column diameter. The situation with Tellus oil and concentrated slurries is quite different. In this case the holdup consists predominantly of large bubbles, and therefore increased liquid circulations leads to a decrease in the total gas holdup.

The liquid circulations tend to accelerate the bubbles travelling upward in the central core. When the bubbles disengage at the top of the dispersion, the liquid travels back down the wall region. Clearly, to describe the influence of liquid circulations on the gas hold-up, we need to be able to predict the liquid circulation velocity as a function of U and D_T. One measure of the liquid circulations is the velocity of the liquid at the central axis of the column, $V_L(0)$. Figure 14 (a) shows published data of Forret [128] and Krishna [131] on $V_L(0)$

for air-water systems for D_T in the range $0.1 - 1$ m. Also shown in Fig. 14(a) are the literature correlations for $V_L(0)$ of Riquarts [132]:

$$V_L(0) = 0.21 (g D_T)^{1/2} \left(U^3 \rho_L / g \mu_L \right)^{1/8} \tag{13}$$

and Zehner[133]:

$$V_L(0) = 0.737 \left(U \, g \, D_T \right)^{1/3} \tag{14}$$

The major uncertainty in extrapolating to say $D_T = 10$ m, for the FT reactor operating with concentrated oil slurries is self evident, especially in view of the fact that there are no experimental data for columns larger than 1 m in diameter. In this connection it must be remarked that the experimental work of Koide [134] and Kojima [135], carried out in a 5.5 m diameter column, are not usable for our purposes because the operation was restricted to superficial gas velocities below 0.05 m/s.

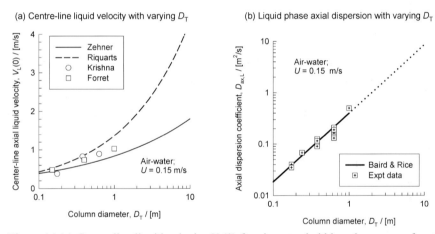

Figure 14 (a) Centre-line liquid velocity $V_L(0)$ for air-water bubble columns as a function of column diameter D_T. (b) Liquid phase axial dispersion coefficient $D_{ax,L}$ for air-water bubble columns as a function of column diameter D_T. Also plotted are the experimental data of Forret et al [128] and Krishna et al. [131].

With increasing liquid circulations, the dispersion (backmixing) in the liquid phase increases. For the FT reactor, we have conflicting requirements. In order to prevent hot spots and runaways, we would like the state of well-mixedness. However, from the point of view of achievement of high syngas conversions, we would like to have more staging of the liquid phase [136]. In any event, a good estimation of the axial dispersion coefficient of the liquid phase $D_{ax,L}$ is vitally important. Figure 14 (b) shows measured data [128, 137] on $D_{ax,L}$ for the air-water system for D_T in the range $0.1 - 1$ m. Also shown in Fig. 14 (b) is the Baird and Rice [138] correlation for $D_{ax,L}$:

$$D_{ax,L} = 0.35 D_T^{4/3} \left(g\, U \right)^{1/3} \tag{15}$$

The applicability of the Baird-Rice correlation for estimation of $D_{ax,L}$ for a FT reactor of 10 m, operating with concentrated oil slurries is open to question, as the data base used for setting up the correlation consisted largely of air-water experiments in columns smaller than 1 m in diameter, operating at atmospheric pressure. The influence of operating pressure on $D_{ax,L}$ has been investigated by Wilkinson [139] and Yang and Fan [140]; their results are however contradictory. While Wilkinson reports an increase of $D_{ax,L}$ with increasing pressure, Yang and Fan report a decrease of $D_{ax,L}$ with increasing pressure. In any event, the pressure effect on $D_{ax,L}$ may be expected to be small.

The major objective of the present work is to develop a strategy for obtaining information on gas holdup, liquid circulations and liquid dispersion for column dimensions and operating conditions relevant to the FT commercial reactor. Our approach relies on the use of Computational Fluid Dynamics (CFD) in the Eulerian framework, using both two-dimensional (2D) axi-symmetric and three-dimensional (3D) strategies. Firstly, we establish the ability of CFD simulations to reproduce the scale dependence portrayed in Figs. 11 and 13 for a concentrated oil-slurry system ($\varepsilon_s = 0.36$) in columns of 0.1, 0.19 and 0.38 m in diameter, using both 2D and 3D simulations. In the second campaign we use 3D simulations for columns of 0.38, 1, 2, 4, 6 and 10 m in diameter in order to establish the influence of D_T on ε, $V_L(0)$ and $D_{ax,L}$. For the latter campaign with varying D_T, the aspect ratios of the various columns were maintained above 5.

3.2.3.3 Development of Eulerian simulation model

For either gas or liquid phase the volume-averaged mass and momentum conservation equations in the Eulerian framework are given by:

$$\frac{\partial \left(\varepsilon_k \rho_k \right)}{\partial t} + \nabla \bullet \left(\rho_k \varepsilon_k \mathbf{u}_k \right) = 0 \tag{16}$$

$$\frac{\partial \left(\rho_k \varepsilon_k \mathbf{u}_k \right)}{\partial t} + \nabla \bullet \left(\rho_k \varepsilon_k \mathbf{u}_k \mathbf{u}_k - \mu_k \varepsilon_k \left(\nabla \mathbf{u}_k + \left(\nabla \mathbf{u}_k \right)^{\mathrm{T}} \right) \right) = -\varepsilon_k \nabla p + \mathbf{M}_{kl} + \rho_k \mathbf{g} \tag{17}$$

where, ρ_k, u_k, ε_k and μ_k represent, respectively, the macroscopic density, velocity, volume fraction and viscosity of phase k, p is the pressure, M_{kl}, the interphase momentum exchange between phase k and phase l and g is the gravitational force. On the basis of the hydrodynamic similarities between bubble columns operating with concentrated slurries and highly viscous liquids, we treat the slurry phase as a highly viscous liquid phase and use the properties of Tellus oil ($\rho_L = 862$; $\mu_L = 0.075$; $\sigma = 0.028$).

The momentum exchange between the gas phase (subscript G) and liquid phase (subscript L) phases is given by:

$$\mathbf{M}_{L,G} = \left[\frac{3}{4} \frac{C_D}{d_b} \rho_L \right] \varepsilon_G \varepsilon_L \left(\mathbf{u}_G - \mathbf{u}_L \right) \left| \mathbf{u}_G - \mathbf{u}_L \right| \tag{18}$$

where we follow the formulation given by Pan et al. [141]. We have only included the drag force contribution to $M_{L,G}$, in keeping with the works of Sanyal et al. [142] and Sokolichin & Eigenberger [143]. The added mass and lift force contributions were both ignored in the present analysis. For a bubble swarm rising in a gravitational field, the drag force balances the differences between weight and buoyancy and so the square bracketed term in Eq. (18) containing the drag coefficient C_D becomes [144]:

$$\frac{3}{4} \frac{C_D}{d_b} \rho_L = \left(\rho_L - \rho_G \right) g \frac{1}{V_{b0}^2} \tag{19}$$

where V_{b0} is the rise velocity of the bubble swarm at low superficial gas velocities (as indicated by the large filled circle in Fig.13). When the superficial gas velocity U is increased, liquid circulations tend to kick in and Eq. (18) will properly take account of the slip between the gas and liquid phases. Our approach is valid when the bubble size does not increase significantly with increasing U. Measurements of the mass transfer in slurries [145] show that $k_L a/\varepsilon$ is practically independent of U and this underlines the correctness of the assumption of a constant bubble size. It is important to note that we do not need to know the bubble diameter d_b in order to calculate the momentum exchange $M_{L,G}$.

For the continuous, liquid, phase, the turbulent contribution to the stress tensor is evaluated by means of k-ε model, using standard single phase parameters $C_\square = 0.09$, $C_{1\square} = 1.44$, $C_{2\square} = 1.92$, $\sigma_k = 1$ and $\sigma_\varepsilon = 1.3$. The applicability of the k-ε model has been considered in detail by Sokolichin and Eigenberger [143]. No turbulence model is used for calculating the velocity fields inside the dispersed bubble phases.

A commercial CFD package CFX, versions 4.2 and 4.4, of AEA Technology, Harwell, UK, was used to solve the equations of continuity and momentum. This package is a finite volume solver, using body-fitted grids. The grids are non-staggered and all variables are evaluated at the cell centres. An improved version of the Rhie-Chow algorithm [146] is used to calculate the velocity at the cell faces. The pressure-velocity coupling is obtained using the SIMPLEC algorithm [147]. For the convective terms in Eqs. (16) and (17) hybrid differencing was used. A fully implicit backward differencing scheme was used for the time integration.

(a) Grid for 0.1, 0.19 and 0.38 m dia. columns

(b) Grid for 6 m dia. column

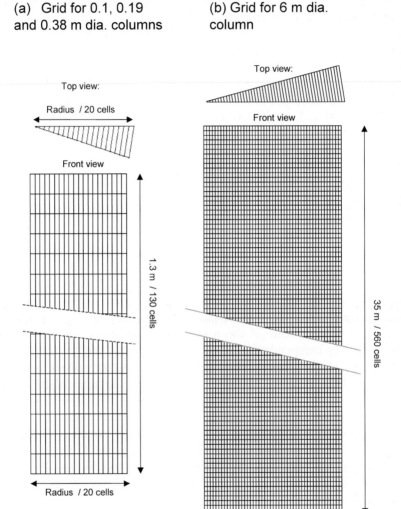

Figure 15 Grid used in the 2D cylindrical axi-symmetric Eulerian simulations.

A pressure boundary condition was applied to the top of the column. A standard no-slip boundary condition was applied at the wall. The physical properties of the gas and liquid phases are specified in Table 8. The details of the operating conditions and computational grids used in the various campaigns are specified in Table 9. For any simulation, the column was filled with liquid up to a certain height (as specified in Table 9) and at time zero the gas velocity was set at the final value at the bottom face. For 2D simulations, to prevent a circulation pattern in which the liquid flows up near the wall and comes down in the core, the gas was not injected homogeneously over the full bottom area. Instead, the injection of gas was performed on the inner 75% of the radius. The choice of aeration of 75% of the central core at the bottom is arbitrary but the results are only slightly different (within 10%) if the gas injection were taken to be 50% of the central distributor region. The gas aeration strategy is not very crucial to the results of 3D transient simulations as the flow is chaotic and the dispersion swishes from side to side.

Table 8
Physical properties of phases used in CFD simulations.

	Liquid (Tellus oil)	Gas (air)
Viscosity, μ / [Pa s]	75×10^{-3}	1.7×10^{-5}
Density, ρ / [kg/m^3]	862	1.3
Diffusivity of tracer, $Đ$ / [m^2/s]	1×10^{-9}	-

Table 9
Details of 2D and 3D simulation campaign. The RTD campaign was only carried out for the 3D simulations. Note that for the 0.38 m column, two sets of 3D simulations were carried out; these sets differ in their aspect ratio. When comparing the hydrodynamics of columns of diameters in the 0.38 – 10 m range, the simulation results obtained with the higher aspect ratio has been used.

	Column diameter, D_T / [m]	Gas velocity, U/ [m/s]	Column height, H_T/ [m]	Observation height for hydrodynamics, H_{obs} / [m]	Initial height of liquid in the column, H_0 / [m]	cells in radius	cells in height	cells in azimuthal direction	Total number of cells
2D	0.1, 0.19, 0.38	0.02, 0.05, 0.1, 0.15	1.3	0.9	0.9 - 1	20	130	-	2600
3D	0.38	0.02, 0.05, 0.1, 0.15, 0.2, 0.23	1.3	0.9	0.9	20	130	10	26,000
3D	0.38	0.15	2.66	1.52	1.76	15	133	10	19,950
	1.0	0.15	7.0	4.5	4.65	20	140	10	28,000
	2.0	0.15	14.0	6.0	9.3	30	210	10	63,000
	4.0	0.15	28.0	14.4	18.64	40	350	10	140,000
	6.0	0.15	42.0	24.0	28.0	50	420	10	210,000
	10.0	0.15	42.0	24.0	28.0	50	420	10	210,000

Typical time stepping strategy used was: 100 steps at 5×10^{-5} s, 100 steps at 1×10^{-4} s, 100 steps at 5×10^{-4} s, 100 steps at 1×10^{-3} s, 200 steps at 3×10^{-3} s, 1400 steps at 5×10^{-3} s, and all remaining steps were set at 1×10^{-2} s until real (in case of 2D) or quasi- (in the case of 3D) steady state was obtained. Quasi-steady state in 3D simulations was indicated by a situation in which all of the variables varied around a constant average value for a sufficiently long time period. In 2D simulations, true steady state was obtained in which none of the variables was subject to change.

To estimate the liquid phase axial dispersion, the final state of a hydrodynamics run was used to start a dynamic run in which a mass tracer is injected into the liquid phase near the top of the dispersion. The concentration of the mass tracer was monitored at two heights along the column, following a simulation technique described in earlier work [130]. The following equations are solved for the mass tracer:

$$\frac{\partial}{\partial t}\varepsilon_k\rho_k C_k + \nabla\cdot\left(\varepsilon_k\rho_k\mathbf{u}_k C_k - \mathcal{D}_k\varepsilon_k\rho\nabla C_k\right) = 0 \qquad (20)$$

Here, C_k is the concentration of mass-tracer in phase k and $\mathcal{D}_k$ is the diffusion coefficient of mass tracer in phase k (listed in Table 8). Since there is zero liquid throughput (liquid operates in batch), eventually all the mass tracer gets distributed equally along the liquid phase. For the mass tracer simulations, some smaller time steps were used to guarantee a smooth restart from the hydrodynamics run, and then time steps of 5×10^{-3} s were used for the 0.38 m diameter column and time steps of 1×10^{-2} s were used for all remaining column diameters.

The liquid phase axial dispersion coefficient was determined by a least-squares fit of the liquid-phase RTD curves at a distance L_i from the point of tracer injection [66]:

$$\frac{C_L(x,t)}{C_{L,0}} = 1 + 2\sum_{n=1}^{\infty}\cos\left(\frac{n\pi L_i}{L}\right)\exp\left[-D_{ax,L}\left(\frac{\pi n}{L}\right)^2 t\right]; \quad i = 1,2,3 \qquad (21)$$

Here, L is the total height of the dispersion, t is time and L_1 and L_2 are distances from the point of tracer injection along the dispersion height to the two monitoring stations (see Fig.16). An upper limit of $n = 20$ rather than infinity was found to be sufficiently accurate for the summation. The reference concentration $C_{L,0}$ was determined by the average concentration of all observation points at the end of the RTD simulation.

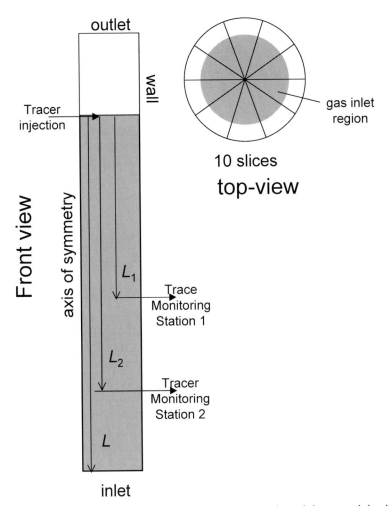

Figure16 Schematic showing the computational domain and the tracer injection and monitoring stations to determine $D_{ax,L}$.

All simulations were carried out on a set of five PC Linux workstations, each equipped with a single Pentium 4 processor. A single 3D campaign at $U = 0.15$ m/s on the 10 m diameter took more than 2 months to produce the hydrodynamics and RTD information. Further details of the simulations, including animations of column start-up dynamics are available on the web sites: http://ct-cr4.chem.uva.nl/viscousbc/ and http://ct-cr4.chem.uva.nl/FTscaleup/.

3.2.3.4 Simulation results for scale influence

In Fig. 17 the 2D simulations for the gas holdup ε are compared with the experimental data [123] for air – 36% paraffin oil slurry system. The 2D simulations are in reasonable agreement with the experimental results for all three column diameters, 0.1, 0.19 and 0.38 m, verifying the choice of the value of $\left[\dfrac{3}{4}\dfrac{C_D}{d_b}\rho_L\right]$ that was calculated taking the value of $V_{b0} = 0.47$ m/s, following Fig. 13 and Eq. 19 For the 0.38 m diameter column, we note that the 2D and 3D simulation results are close to each other for $U = 0.02$, 0.05 and 0.1 m/s. For $U = 0.15$ m/s, the ε predicted by the 3D simulation is slightly higher than that of the 2D approach. In order to understand the reason behind this, let us compare the dynamic behavior of the centre-line liquid velocity $V_L(0)$ for the 2D and 3D strategies as it approaches a steady state; see Fig. 18. The 3D simulations portray inherently chaotic behavior, with liquid sloshing from side to side; these effects, that are in conformity with visual observations, can best be appreciated by viewing the animations on our website: http://ct-cr4.chem.uva.nl/viscousbc/.

The 3D simulations were run for a sufficiently long period of time and the hydrodynamic parameters such as ε and $V_L(0)$ were determined by average over the time period where quasi-steady state prevails. The 2D simulations, on the other hand reach a constant steady-state. The time-average value of $V_L(0)$ for the 3D simulations, generally tend to be lower than the corresponding value for the 2D approach; see Fig. 19. The error bars in the $V_L(0)$ for the 3D simulations shown in Fig. 19 represent the standard deviations obtained from the transient $V_L(0)$ dynamics in Fig. 18. We also note from Fig. 19 that the differences in the 2D and 3D simulation results increase with increasing U. The experimental data [137] of $V_L(0)$ for air-Tellus oil are in better agreement with the 3D simulation results, emphasizing the superiority of the 3D approach. Fig. 20(a) shows the radial distribution of the axial component of the liquid velocity $V_L(r)$ obtained from 2D simulations for $D_T = 0.1$, 0.19 and 0.38 m and $U = 0.05$ m/s. We see that the liquid circulation velocities increase strongly with column diameter. At the centre of the column, for example the axial component of the liquid velocity $V_L(0)$ is 0.23 m/s in the 0.1 m diameter column; this value increases to 0.34 m/s in the 0.19 m column and to 0.47 m/s in the 0.38 m column. Since the drag coefficient between the gas bubbles and the liquid is the same for all column diameters, the rise velocity of the bubbles has to increase with increasing column diameter. Fig. 20(b) compares the $V_L(r)$ profile from 2D and 3D simulations for $U = 0.15$ m/s and $D_T = 0.38$ m. As has been remarked early in the context of Fig. 19 the 3D simulations predict a more realistic value of $V_L(0)$ in accord with experimental data [148].

133

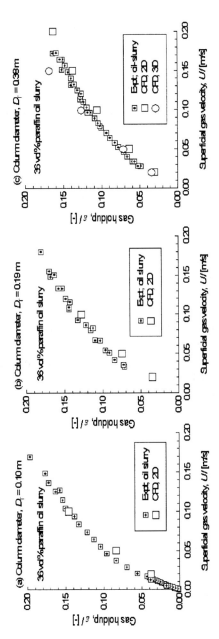

Figure 17 Data on gas hold-up as function of the superficial gas velocity U for columns of diameter D_T = 0.1, 0.19 and 0.38 m for 36 vol% paraffin oil system. Comparison with CFD simulations, both 2D axi-symmetric and 3D, with experimental data of Krishna et al. [119].

134

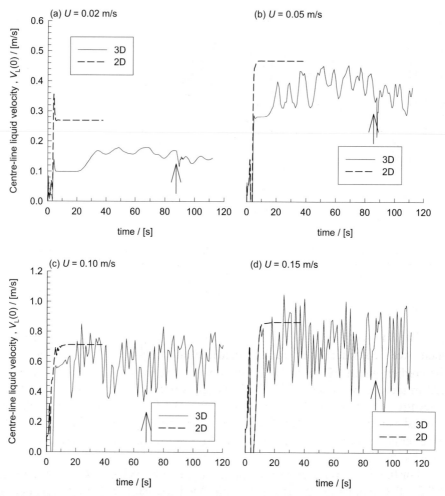

Figure 18 Transient approach to steady state (2D) or quasi-steady state (3D) for 0.38 m diameter column, operating at U = 0.02, 0.05, 0.1 and 0.15 m/s. Comparison of 2D axisymmetric and 3D simulation strategies. The arrows represent the time of injection of tracer in 3D simulations for determination of the liquid phase dispersion coefficient. Animations of column start-up dynamics are available on the web site: http://ct-cr4.chem.uva.nl/viscousbc/.

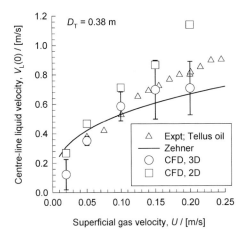

Figure 19 Data on centre-line liquid velocity $V_L(0)$ in 0.38 m diameter column as function of U. The experimental data with Tellus oil [137] are compared with both 2D and 3D simulations. The error bars on the 3D simulation data are obtained represent the standard deviations of the transient $V_L(0)$ data presented in Fig.18, obtained from the data set after the time indicated by the arrow mark.

The shortcomings of the 2D simulation strategy becomes more apparent when we consider the radial distribution of the gas holdup, $\varepsilon(r)$. The predictions of 2D and 3D simulations of $\varepsilon(r)$ for $U = 0.15$ m/s and $D_T = 0.38$ m are compared in Fig. 20(c). The 2D simulations predict an unrealistic off-centre maximum in the gas holdup, whereas the 3D simulations yield the classical parabolic holdup profile, often observed in practice [128, 148, 149]. The average gas holdup, ε, however, for 2D and 3D simulations are very close to each other.

For the FT slurry reactor the optimum operating value of U is in the range of 0.2 - 0.3 m/s, as discussed by Maretto and Krishna [121]. However, due to the fact that syngas is being consumed to form liquid product, the value of U at the outlet of the reactor is only 30-40% of the inlet value, depending on the conversion level. For a FT reactor operating at a value of $U = 0.25$ m/s at the bottom, the value of U at the top of the reactor will be reduced to about 0.10 m/s. Therefore, in the simulations for the FT reactor hydrodynamics as a function of scale (for column diameters $D_T = 0.38$ 1, 2, 4, 6 and 10 m; see also Table 9 for column heights used), it was decided to perform transient 3D simulations at $U = 0.15$ m/s, an average value between 0.1 and 0.25 m/s.

136

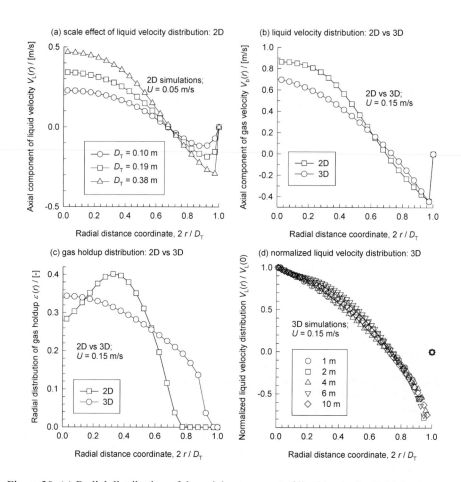

Figure 20 (a) Radial distribution of the axial component of liquid velocity $V_L(r)$ for operation at $U = 0.05$ m/s from 2D simulations of 0.1, 0.19 and 0.38 m columns.
(b) Comparison of 2D and 3D simulations of radial distribution of the axial component of liquid velocity $V_L(r)$ for operation at $U = 0.15$ m/s in 0.38 m column.
(c). Comparison of 2D and 3D simulations of radial distribution of the radial distribution of gas holdup $\varepsilon(r)$ for operation at $U = 0.15$ m/s in 0.38 m column.
(d) Radial distribution of the normalized axial component of liquid velocity $V_L(r)/V_L(0)$ for operation at $U = 0.15$ m/s from 3D simulations of 1, 2, 4, 6 and 10 m columns.

The transient dynamics of the centre-line velocity (monitored at the observation heights, H_{obs}, specified in Table 9) are shown in Fig. 21. It is apparent that with increasing scale both the magnitude of $V_L(0)$, and its fluctuation around the mean increases.

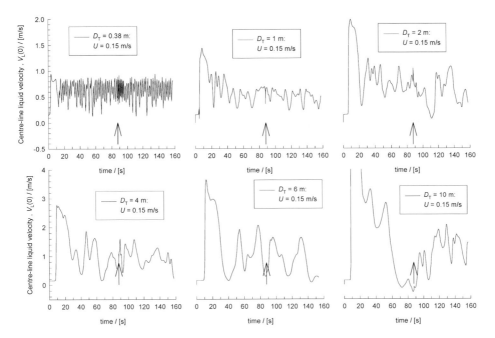

Figure 21 Transient approach to quasi-steady state (3D) for operation at $U = 0.15$ m/s in columns of 0.38, 1, 2, 4, 6 and 10 m diameter. The arrows represent the time of injection of tracer in 3D simulations for determination of the liquid phase dispersion coefficient. Animations of column start-up dynamics are available on the web site: http://ct-cr4.chem.uva.nl/FTscaleup/.

The hydrodynamic parameters were obtained by averaging over the time period during which quasi-steady state can be assumed to prevail. These time-averaged values of $V_L(0)$ are shown in Fig. 22(a), in which the error bars represent the standard deviations of the velocity fluctuations shown in Fig. 21. The $V_L(0)$ values, which increase with scale, appear to follow the trend predicted by the Zehner correlation [133], but the absolute values are significantly lower. Earlier work using 2D axi-symmetric simulations for scaling up bubble column reactors [109] had predicted $V_L(0)$ values conforming with the correlation of Riquarts [132] and significantly higher than found in the present study with realistic 3D simulations. The $V_L(0)$ predictions of this earlier work [109] are unrealistically high because of the artefacts introduced in the 2D approach as explained in the foregoing.

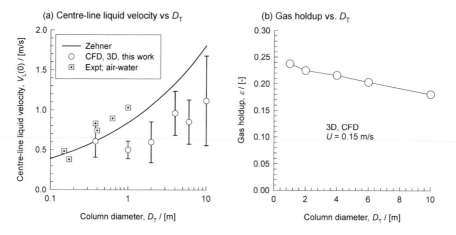

Figure 22 3D simulation data on (a) centre-line liquid velocity $V_L(0)$ and (b) gas holdup ε as function of D_T for operation at $U = 0.15$ m/s in columns of 0.38, 1, 2, 4, 6 and 10 m diameter. The error bars on the 3D simulation data in (a) represent the standard deviations of the transient $V_L(0)$ data presented in Fig. 21, obtained from the data set after the time indicated by the arrow mark. The continuous line in (a) represents the correlation of Zehner [133]. Also plotted in (a) are the experimental data of Forret et al [128] and Krishna et al. [131].

When compared with the air-water experimental data [128, 131] for $V_L(0)$, we see that the CFD predictions for the FT reactor are significantly lower, stressing the danger of using air-water information for scaling up. The main reason for the lower $V_L(0)$ predicted for the FT reactor, is due to the significantly lower gas holdup when compared to air-water systems. For air-water systems, operating at $U = 0.15$ m/s, we would have a significant fraction of the dispersion present in the form of small bubb. s; this small bubble population is virtually destroyed in concentrated slurries [150], leading to significantly lower gas holdups and $V_L(0)$ values.

When the liquid velocity profiles obtained from the 3D CFD simulations are normalized with respect to the centre-line velocity, the $V_L(r)/V_L(0)$ are practically independent of the column diameter. This is illustrated in Fig. 20(d) for the 3D simulation campaign at $U = 0.15$ m/s for various column diameters up to 10 m. The significance of the result portrayed in Fig. 20(d) is that the centre-line velocity can be taken to be a unique measure of the strength of liquid circulations.

An important consequence of the fact that the strength of the liquid circulations increase with increasing scale is that the gas holdup values are correspondingly lowered; this is shown in Fig. 22(b). We note that the gas holdup in the 10 m diameter FT reactor is 0.18, whereas for the 1 m column the value of $\varepsilon = 0.24$. A 20% decrease in gas holdup with increase of scale can have significant consequences for an FT reactor designed for high conversion targets.

A tracer is injected into the liquid phase near the top of the liquid dispersion, at the time step indicated by an arrow in Fig. 21 and the progression of this tracer is monitored at two stations along the height of the column (see Fig.16). The CFD simulations of the tracer RTD is then fitted with the model given by Eq. 21. Typical results comparison of the dimensionless RTD curves for the tracer are shown in Fig. 23 for (a) $D_T = 0.38$ m, and (b) $D_T = 2$ m columns, both operating at $U = 0.15$ m/s. We note from Fig. 23 that the tracer response is not smooth but oscillates. These oscillations are due to liquid sloshing from side to side causing a significant radial transport of the liquid tracer, as can be witnessed in the animations on the web site: http://ct-cr4.chem.uva.nl/FTscaleup/. In this context it is worth emphasizing that 2D simulations will yield a much lower value of $D_{ax,L}$ than 3D simulations because there is no mechanism for radial transport [130].

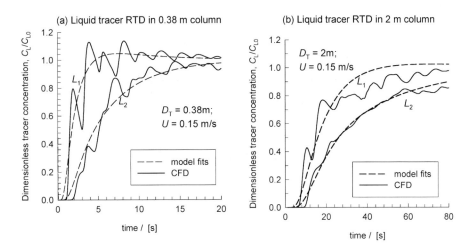

Figure 23 3D simulations of the dimensionless liquid tracer concentration measured at two different monitoring stations for (a) 0.38 m diameter column operating at $U = 0.15$ m/s, and (b) 2 m diameter column operating at $U = 0.15$ m/s. The dashed lines represent the fits of the two simulation data sets. Animations of liquid tracer dynamics are available on the web site: http://ct-cr4.chem.uva.nl/FTscaleup/.

Each of the tracer curves, such as those shown in Fig. 23 were fitted individually to obtain two different values of $D_{ax,L}$ for each run. Fig. 24 shows the results for the two campaigns with (a) varying U for $D_T = 0.38$ m and (b) varying D_T for $U = 0.15$ m/s. Also plotted in Fig. 24 are the experimentally determined $D_{ax,L}$ values for the air-water system [128, 137]. The $D_{ax,L}$ values

140

from our 3D simulations are lower than the experimental values for the air-water system, following the same trend as observed earlier for the $V_L(0)$ values in Fig. 22(a). The $V_L(0)$ value reflect the strength of liquid circulations and these directly influence liquid dispersion. The predictions of $D_{ax,L}$ using the Baird-Rice correlation [138] (shown by the continuous lines in Fig. 24), though representing the air-water data reasonably accurately, tend to be higher than the values for the FT reactor, operating with a concentrated oil slurry. We note that the simulated value of $D_{ax,L}$ for the 10 m diameter reactor is expected to be lie between 3 - 7 m^2/s, corresponding to a nearly well-mixed system.

Another important design parameter is the heat transfer coefficient to vertical cooling tubes, α ; this parameter is largely dictated by the surface renewal rate, which in turn is determined by the bubble rise velocity. Increased liquid circulation velocities with increasing scale will have the effect of enhancing the bubble rise velocity and improving the heat transfer coefficients. Therefore, CFD predictions of the $V_L(0)$ with scale allow better estimation of the heat transfer coefficient to vertical cooling tubes [151].

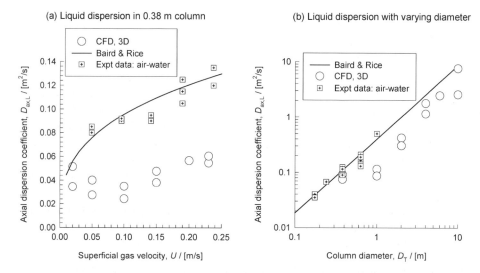

Figure 24 (a) Liquid phase axial dispersion $D_{ax,L}$ data obtained from 3D simulations of 0.38 m diameter column operating at $U = 0.02$ - 0.23 m/s.
(b) $D_{ax,L}$ data from 3D simulations for operation at $U = 0.15$ m/s in columns of 0.38, 1, 2 4, 6, and 10 m diameters. Note that the data for the 0.38 m diameter column in (b) was obtained at a higher aspect ratio than for the simulations for the same column shown in (a); details are given in Table 9. The continuous lines in (a) and (b) represent the correlation of Baird and Rice [138]. Also plotted in (b) are the experimental data of $D_{ax,L}$ of Forret et al [128] and Krishna et al. [137].

3.2.3.5 Conclusions from the CFD investigations

The following major conclusions can be drawn:

Both 2D and 3D simulations are able to provide a reasonable prediction of the average gas holdup for columns of 0.1, 0.19 and 0.38 m in diameter.

(1) The 2D simulations generally tend to predict a higher value of $V_L(0)$ than the 3D simulations. For very large diameter columns the 2D predictions of $V_L(0)$ are unrealistically high [109]. For scaling up to columns larger than 1 m in diameter, it is essential to resort to 3D simulations.

(2) The 2D predictions of the radial distribution of gas holdup show an unrealistic off-centre maximum in the gas holdup. The 3D simulations of yield the classical parabolic profile for the radial gas holdup distribution, typically found in experiments.

(3) For larger diameter columns, only 3D simulations are able to reproduce the chaotic hydrodynamics observed visually and only this strategy can yield reasonable values of $V_L(0)$ and $D_{ax,L}$ for large diameter FT reactors. 2D simulations of $D_{ax,L}$ give unrealistically low values because the radial dispersion contribution is absent [130].

(4) 3D simulations of the FT reactor of 0.38, 1, 2, 4, 6 and 10 m diameters show that the $V_L(0)$ values are lower than those predicted by the Zehner correlation [133], that is apparently only adequate to describe the air-water system.

(5) The axial dispersion coefficient of the liquid phase, $D_{ax,L}$, for the FT reactor show a similar trend as that for $V_L(0)$; these are lower than for air-water experiments. The predictions of $D_{ax,L}$ following the Baird and Rice correlation [138] tend to yield somewhat higher values than those obtained from CFD simulations.

It is concluded that 3D Eulerian simulations can provide a powerful tool for hydrodynamic scale up of bubble columns, obviating the need for large scale experiments on gas holdup, liquid velocity and mixing. Validation of the proposed scale up strategy is however desirable with experiments carried out in columns larger than say 2 m in diameter using an oil slurry.

3.2.3.6 Influence of operation at elevated pressures

The FT reactor operates at pressures of around 2 – 4 MPa and it may be important to take account of the influence of elevated pressures on the bubble column hydrodynamics; this influence can be very significant at lower slurry concentrations. [110, 152-156]. Consider the experimental data of Letzel et al. [152] for gas hold-up measured in a bubble column of 0.15 m diameter with the system nitrogen – water; see Fig. 25.

142

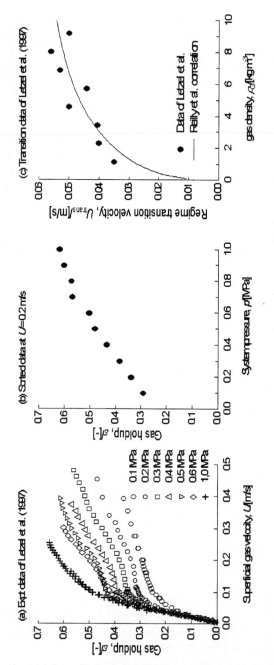

Figure 25 Experimental data of Letzel et al. [154] on pressure effect on gas holdup

For example, their data shows that for a superficial gas velocity U = 0.2 m/s, the gas hold-up ε increases from a value of 0.29 at p = 0.1 MPa to a value which is about twice as large for operation at p = 1 MPa; see Fig. 25 (b).

The observed increase in the gas hold-up with increasing pressures is due to two reasons. Firstly, increased pressure delays the onset of churn-turbulent flow, i.e. the superficial gas velocity at which regime transition occurs, U_{trans}, increases with increasing p; see Fig. 25 (c). The physical explanation for the delay in the regime transition with increased system pressure, which is equivalent with increasing gas density ρ_G, is to be found in the reduced probability of propagation of instabilities leading to delayed flow regime transition [157-159]. The correlation of Reilly et al. [160] adequately describes the increase of U_{trans} with gas density ρ_G (See Fig. 25 (c)). Using Kelvin-Helmoltz stability analysis, Letzel et al. [154] showed that there is a second influence of increased system pressures, that of decrease in the stability of large bubbles. The break-up of large bubbles, with increasing pressures, leads to a significant decrease in the rise velocity of the large bubble population.

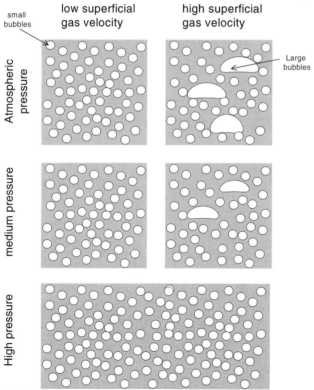

Figure 26 Qualitative picture of the influence of increasing pressure on the hydrodynamics of gas-liquid bubble columns (without the presence of suspended catalyst particles).

By comparing Figs. 10 and 18 we see that the influence of increasing amount of catalyst particles on the bubble hydrodynamics is opposite to the influence of increasing system pressures. While addition of catalyst particles promotes coalescence and increases the proportion of large bubbles, the influence of increasing pressure is to reduce the population and size of the large bubbles. The prediction of the hydrodynamics of slurry reactors operating at high slurry concentrations and at high pressures is, therefore, particularly difficult. The approach we suggest is to adopt CFD techniques wherein the inter-phase momentum exchange term is suitably modified to take the pressure effect into account. [156].

3.2.3.7 Interphase mass transfer

Measured experimental data on the volumetric mass transfer coefficient, $k_L a$, defined per unit volume of reactor, for air-water system in a 0.15 m diameter column is shown in Fig. 27(a) for varying U and operating pressures. [155] Increasing pressure also tends to increase $k_L a$, following the trend in the gas hold-up, see in Fig. 25. The volumetric mass transfer coefficient correlates with the gas hold-up ε; see Fig. 27(b). We therefore conclude that prediction of the gas hold-up is the key to the estimation of mass transfer. More recent work [161] has shown that scale effects are not significant. There is some experimental evidence in the literature to suggest that $k_L a$ values in concentrated slurries also conform to this correlation with the gas hold-up [109]. From the CFD simulation results presented in the foregoing, we should expect $k_L a$ to show a strong scale dependence. There is no experimental evidence in the literature to support this contention. There is a need for further studies on mass transfer in bubble columns.

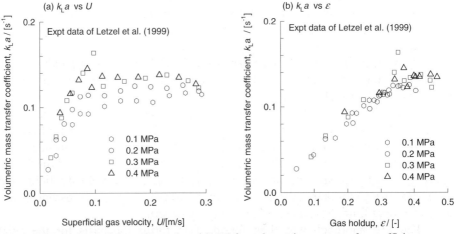

Figure 27 Experimental data of Letzel et al [155] for volumetric mass transfer coefficient.

3.2.3.8 Conclusions for scale-up strategies

A scale-up strategy for bubble column slurry reactor for Fischer-Tropsch synthesis using CFD as a pivotal tool has been presented above. For operation with concentrated slurries, with slurry concentrations in excess of 30 vol%, the dispersion consists almost exclusively of fast–rising large bubbles. By extrapolating the bubble swarm velocity data to low superficial gas velocities the slip velocity between the bubbles and the slurry phase can be determined. For the 36 vol% paraffin-oil slurry a value $V_{b0} = 0.47$ m/s is obtained; see Fig.13. This value of V_{b0} is used to estimate the drag coefficient C_D between the gas and the slurry phase using Eq. (19). Eulerian simulations of the slurry bubble column with varying diameters are then carried out by treating the slurry phase as a highly viscous liquid. The 2D CFD simulations are in very good agreement with the experimental results for gas hold-up in 0.1, 0.19 and 0.38 m diameter simple bubble columns.

Simulations for larger diameter columns required the use of 3D simulations to produce realistic results. The influence of various different internals in baffled bubble columns remains to be determined. If the slurry viscosity is not high enough to eliminate all the small bubbles then the influence of pressure may be significant. It may well be necessary to avoid the very high slurry viscosities associated with the loss of small bubbles in order to enhance both mass and heat transfer coefficients in the reactor. To date there is therefore still no reliable substitute for a relatively large scale demonstration reactor. With data from such a demonstration reactor with a diameter of the order of 1 m or less it is now possible to produce large scale reactor designs with confidence.

3.2.4 Modelling approaches for slurry phase reactors

Useful insight into the use of models in chemical engineering is provided by Levenspiel [162]. He starts with a quote from Denbigh's monograph [163] as follows:

"In science it is always necessary to abstract from the complexity of the real world, and in its place to substitute a more or less idealized situation that is more amenable to analysis."

In the first half of the twentieth century two flow models dominated for reactors, plug flow and mixed flow. Real reactors behave somewhere between these two extremes. According to Levenspiel the $100 approach to deal with this complexity is to evaluate the velocity field within the reactor and the $1000 approach is to evaluate the three-dimensional fluctuating velocity field, and then use the computer to tell you what would happen; in essence, use computational fluid dynamics. What an ugly procedure!

The $10 approach was then found. This forms the basis for the axial dispersion (AD) modelling approach. That was due to the genius of Danckwerts

[164]. He proposed a simple flow model to tell how a vessel acts as a chemical reactor. Introducing a pulse of tracer into the fluid entering the reactor and producing a concentration-time curve at the exit provides the information needed to determine how the reactor will behave. The curve is called the residence time distribution, or RTD curve.

The RTD curve can be used to predict how the reactor will behave – exactly for linear reaction kinetics, and as a close approximation for more complex reaction kinetics. The study of the RTD of flowing fluids, and its consequences is called tracer technology.

The problem with the RTD approach in the case of fluid bed reactors is that you would first need to build the reactor in order to do the tracer test to determine how the reactor will behave.

Although the word 'hydrodynamics' has been given other meanings it is now often used to refer to the study of the flow and mixing behaviour of the phases present in fluid bed reactors such as the slurry phase reactor that we now discuss.

Levenspiel ends his publication with the following proposed modelling strategy: always start by trying the simplest model and then only add complexity to the extent needed. This he refers to as the $10 approach or as Einstein said,

"Keep things as simple as possible, but not simpler".

Van der Laan [165] reported a model for FT synthesis in a slurry bubble column reactor (SBCR) using iron catalyst. His model exhibits well mixed liquid and two gas bubble regimes: small bubbles that are well mixed and large bubbles that exhibit plug flow behavior. Van der Laan also provides a summary of bubble column reactor models that others have utilized (Table 10). He concluded that the FT SBCR with is reaction controlled due to the low activity of the iron catalyst and the volumetric mass transfer coefficient of the large bubbles is enhanced due to frequent bubble coalescence and break-up.

It is somewhat surprising that only Inga et al appear to use the most appropriate modelling approach for use in the design of modern slurry phase reactors i.e. using plug flow (PF) for the gas (dilute) phase and the cell model with backflow (CMBF) for the dense (liquid) phase. Reasonable predictions can also be obtained using axial dispersion (AD) models if appropriate assumptions are used and this approach is certainly suitable for steady-state conversion prediction. The advantage in using the CMBF model approach is that it lends itself to predictions for the real life situation of imposed slurry flows and discrete heat removal surfaces at various locations inside the reactor.

Table 10
Comparison of Reaction Engineering Models for the Fischer-Tropsch Synthesis in Slurry Bubble Column Reactors (from Ref. 165)

Reference	Gas Phase	Liquid Phase	Catalyst Distribution	Energy Balance
Calderbank, et al.	PF	PF	uniform	isothermal
Satterfield, et al.	PF	PM	uniform	isothermal
Deckwer, et al.	PF	PM	uniform	isothermal
Deckwer, et al.	AD	AD	non-uniform	non-isothermal
Bukur	PF	PF,PM	uniform	isothermal
Stern, et al.	PF	PM	uniform	isothermal
Kuo	PF	PM,PF, AD	non-uniform	isothermal
Kuo	PF	PF	non-uniform	isothermal
Stenger, et al.	AD	AD	non-uniform	isothermal
Prakash, et al.	AD	AD	non-uniform	isothermal
Prakash	AD	AD	non-uniform	isothermal
De Swart and Krishna[1]	AD	AD	non-uniform	non-isothermal
De Swart and Krishna[2]	PF	PM	uniform	isothermal
Mills, et al.	AD	AD	non-uniform	non-isothermal
Inga, et al.	PF	MC	uniform	isothermal
Krishna, et al.[2]	PF	PM	uniform	isothermal
Van der Laan, et al.	PF	PM	uniform	isothermal

PF: plug flow; PM: perfectly mixed; MC: mixing cells, AD: axial dispersion.
1   Heterogeneous flow regime: large bubbles: PF, small bubbles and liquid: AD.
2   Heterogeneous flow regime: large bubbles: PF, small bubbles and liquid: PM.

Two recent publications use the AD model approach to give valuable insight into the behaviour and design considerations for slurry phase reactors. These are by Rados et al. [166], Chemical Engineering Laboratory, Washington University and by de Swart and Krishna [167], Department of Chemical

Engineering, University of Amsterdam. Rados et al. credit de Swart and Krishna with the first published dynamic model for a slurry phase FT reactor. The main objective for the work of Rados et al. was to develop a one dimensional dynamic model that properly accounts for the change in gas flow rate due to chemical reaction. The change in gas flow rate is accounted for from the overall gas balance. However, de Swart and Krishna already achieve this by making use of the contraction factor ALPHA, defined in Levenspiel [168].

The Rados et al. approach uses the premise that the design objective is to maximize conversion rather than the more realistic design objective to maximize reactor productivity. This leads to the common error of suggesting that "gas velocity should be as small as optimally possible in order to increase the reactant residence time but high enough to keep the flow in the churn turbulent regime insuring good mass and heat transfer". This is in essence the approach taught earlier by Deckwer. They used a low 5 m reactor height for their simulations which will inevitably lead to relatively low per pass conversion especially at high gas velocities. They compared their AD model prediction to a model using gas in PF and slurry well mixed and also to large bubble in plug flow with both small gas bubbles and slurry well mixed. In both cases the conversion predictions matched within 5%. The discrepancy can be expected to increase at higher per pass conversions. The spread in predicted conversions between ideal plug flow and fully mixed models is about 20%. This is the maximum deviation (for this set of operating conditions resulting in relatively modest conversion) in predicted conversion when the back-mixing modes for different phases are not well selected. They conclude the analysis as follows:

"The outcome of this comparison does not mean that the full AD model is better than some ideal reactor model. It simply shows that the full AD model is more versatile than the ideal reactor modelling approach. However, the success of the AD model depends on the accuracy of the correlations that are used for calculation of Peclet numbers in different phases. If these correlations are not accurate then the advantage of the AD model disappears. In this case, an ideal reactor model consisting of gas in plug flow and slurry well mixed for large diameter reactors or gas in plug flow and slurry in plug flow for very slender reactors should be used." This ideal reactor approach will lead to a conservative (i.e. oversized) large scale reactor design if it is based on data from a slender demonstration reactor.

Rados et al. also point out that, to take full advantage of the AD model, a more accurate and detailed kinetic scheme needs to be used. Dry [169] and Huff and Satterfield [170] showed that when hydrogen conversion is below 60% then first order kinetics for hydrogen is a good approximation. Hence it can be concluded that up to a conversion of 60% per pass the ideal reactor modelling approach proposed in the previous paragraph may suffice for engineering purposes. For higher per pass conversion designs a better understanding of the effect of scale-up on mixing and more refined kinetic expressions will be required.

De Swart and Krishna [167] report an elegant approach to slurry phase reactor modelling with clear reporting of the parameter values used. In analogy with Mills et al. [171], the mass and energy balances are applied over a differential element of the reactor using dimensionless equations. The complete reactor model is defined by four partial differential equations together with accompanying initial and boundary conditions. From their model predictions they come to the correct conclusion that the key for a successful scale-up procedure for the FT slurry reactor is the amount of backmixing in the liquid phase. They recommend that it is desirable to validate the correlations for reactor diameters up to 8 m and superficial gas velocities up to 0.5 m/s.

The four partial differential equations used are for:
1) the hydrogen balance for the 'large' bubbles
2) the hydrogen balance for the 'small' bubbles
3) the hydrogen balance in the liquid phase
4) the energy balance for the slurry phase

It was found that the small bubbles are almost in equilibrium with the liquid phase all over the reactor. This in essence means that the model could be further simplified to three differential equations without any significant loss of accuracy. The hydrogen balance for the small bubbles is not required. It is still necessary to quantify the gas hold-up attributable to 'small' bubbles since this needs to be subtracted from the active reactor volume since this portion of the 'dense' phase contains no catalyst and hence cannot participate in the conversion of reactants. Only negligible through flow is attributed to the small bubbles so they are essentially only using up space within the reactor from a modelling perspective. The small bubbles may well play a role in enhancing mass transfer between the large bubbles and the liquid but it is not necessary to capture this complexity in the model other than to select an appropriate value for the mass transfer coefficient or perhaps more correctly a suitable value for the 'effective' average bubble size. Mass transfer has been discussed recently by Krishna and van Baten [172]. Similarly to the situation for mixing, the accuracy of prediction for mass transfer becomes more important at higher per pass conversions.

The modelling approach used by de Swart and Krishna together with RTD data for the dense phase (liquid or solid) from various reactor sizes with suitable internals should provide all the necessary insight for accurate conversion prediction for future reactor designs.

More complex models may be required to accurately predict temperature profiles but the design objective will be to achieve a reactor that operates at a constant and optimum temperature throughout the bed (preferably also with a uniform catalyst concentration). If this objective is achieved then the steady state conversion prediction becomes even simpler. The energy balance in the slurry phase is then no longer required and the model reduces to two partial differential equations.

3.3 Recommended design approach

It is well known that it is less expensive to add reactor volume by adding reactor height rather than reactor diameter. **The first step should therefore be to select the maximum practical bed height.** This will obviously be higher than the 7.7 m bed height used for the Rheinpreußen demonstration reactor and will more likely be between 20 an 40 metres. Modern plants need to make maximum use of economy of scale to be economically viable which means that reactor diameters also need to approach maximum limits for cost effective vessel manufacture. Furthermore, there are limits to the maximum diameter that can be considered due to fabrication constraints. Vessel diameters much above 10 m will seldom be considered. Shop construction is usually less costly than field construction so that the maximum weight that can be transported in a cost effective way may constrain the vessel dimensions.

A suitable per pass conversion then needs to be selected. The upper limit per pass conversion may be limited by catalyst constraints which may limit the maximum tolerable water partial pressure (or perhaps the water to hydrogen ratio). In this case the partial pressure constraint sets the maximum allowable per pass conversion but below this constraint an economic optimisation is performed in order to determine the optimum recycle ratio.

There is a place for the use of both iron and cobalt catalysts in LTFT slurry phase reactors depending on the products to be produced. This is discussed in more detail in Chapter 5. If an iron catalyst is used then this is typically combined with another downstream gas processing operation because it is expensive to achieve high reactant conversions with LTFT iron catalysts. The exception is when the iron catalyst is fed with low H_2/CO ratio syngas in which case it is possible to target high per pass conversions. This is a special case because there is very little water product using this approach. If the syngas feed is based on Lurgi gasifier product or natural gas reforming then the syngas H_2/CO ratio will be at least 1.7. In this case water production is inevitable and the water partial pressure must be kept below 3 bar (300 kPa) in order to avoid rapid catalyst deactivation by oxidation. For this reason the per pass conversion is severely restricted and recycle and/or tandem operation is required to achieve reasonable overall conversions. For this reason and also to avoid unwanted carbon dioxide production, cobalt catalyst will normally be selected for natural gas applications as stated, for example, in Ref. 96. Iron catalysts would only be selected for natural gas applications if it is intended to derive value from the more olefinic hydrocarbon products and another high syngas conversion process is located downstream. The following discussion assumes natural gas feed to the process.

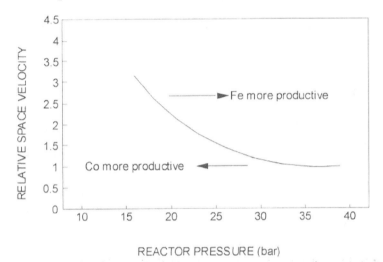

Figure 28 Productivity comparison between iron (240 °C) and cobalt (220 °C) based catalysts [173].

There is an operating conditions envelope in which either iron or cobalt catalysts are more productive. A comparison was performed by van Berge [173], using data obtained from the Sasol commercial precipitated iron catalyst and from the Sasol supported cobalt catalyst [174], and this is shown in Fig. 28. As expected, the iron catalyst is more productive at higher space velocities and operating pressures. The fact that cobalt catalysts are more productive at lower space velocities (higher water partial pressures) means that the cobalt is more productive under higher per pass conversion regimes.

Fig. 28 is perhaps slightly misleading when it comes to reactor design since this comparison assumes a constant reactor bed height. The y-axis should rather be recast as per pass conversion. The lower per pass conversion that favours iron catalyst can rather be attained by decreasing the reactor bed height instead of increasing space velocity. Thus for iron catalysts the reactor bed height will be constrained by the maximum desired water partial pressure at the reactor exit even when operating at the maximum feasible gas velocity. This means that **reactors using iron catalyst in water producing applications will have reactor heights that are less than the maximum constraint from fabrication considerations.**

A good starting point for modern cobalt catalysts is to select a per pass conversion of 60% and a feed to recycle ratio (internal recycle) of 1:1 and then to investigate whether lower recycle ratios are possible and desirable. This should provide an overall conversion of about 90%. In some cases the per pass conversion may be constrained by a water partial pressure limit. A 5 bar (500 kPa) water partial pressure has been mentioned as an upper limit for one application [175]. As will be seen later the overall conversion is also an optimisation variable for a natural gas conversion plant where there will typically also be a recycle to the methane reformer to form a gas loop. This is called an external recycle and ideally a reactor model should be incorporated into a gas loop model so that both the internal and external recycles are optimised simultaneously.

The gas velocity can be selected to achieve the desired per pass conversion. If the catalyst activity is very high then it is possible that the desired gas velocity exceeds a practical upper limit where the gas velocity - hold-up correlation changes form. In this case it will be necessary to target a lower bed height than that imposed by the fabrication constraints. **An iterative procedure will usually be required to find the optimum combination of per pass conversion and gas velocity.** Higher gas velocity will decrease the reactor cost per unit product but will give lower per pass conversions. Recycles can be adjusted in an attempt to improve the plant productivity but production will reach a maximum and capital cost a minimum at some combination of gas velocity and per pass conversion.

If the catalyst rejuvenation approach is used (without internal recycle or a water limit) then the same principle applies that the desired overall conversion and the catalyst activity will determine the gas velocity needed at the maximum bed height. If this velocity is excessive (i.e. for very active catalysts) then the bed height is decreased.

By now the need for a suitable reactor model should be clear. This model should:

- provide a realistic description of the mixing in the liquid phase.
- not underestimate the gas-liquid mass transfer coefficient
- not overestimate the influence of gas velocity on gas hold-up.

Mistakes on these three points prior to the work of Krishna et al. and the Sasol reactor designs have previously resulted in the gross underestimation of the maximum desirable gas velocity for slurry phase reactors.

An understanding is required of the relationship between the solids concentration in the slurry and the slurry viscosity. This will provide a curve that shows a rapid exponential increase in viscosity at higher concentrations. Some judgement is required to select the maximum concentration that avoids an excessively high slurry viscosity. This should be based on carefully compiled experimental data using actual catalyst. Literature correlations based on solids volume fractions may provide some guidance.

It is also important that the gas hold-up correlation is based on actual data at the selected slurry concentration. There is an advantageous relationship between gas hold-up and slurry concentration. As the concentration increases, the gas hold-up decreases. This means that as catalyst is added more liquid can be retained in the reactor which provides more space to add catalyst! This means that a surprisingly large amount of catalyst is loaded into the reactor in order to increase the slurry concentration.

Having set the catalyst concentration then the amount of catalyst in the reactor will depend only on the fraction of slurry in the reactor. For the fixed reactor dimensions selected at the outset, the catalyst mass in the reactor will decrease as the reactor gas velocity is increased which in turn increases the gas fraction in the reactor so that the slurry fraction decreases. In the churn-turbulent regime, gas fraction increases at a much slower pace than the gas velocity. Thus the gas velocity is increased until the target per pass conversion is matched. The gas velocity selected will depend on the activity of the catalyst; the feed gas composition; and the operating temperature and pressure. It is important that the rate equation used in the model correctly reflects the catalyst activity and the effect of these operating conditions. The higher the catalyst activity, the higher will be the velocity that will be selected. This approach assumes that there is no hydrodynamic constraint to the maximum velocity. If the velocity is high enough, the bed will be transported out of the reactor. Before this happens though the gas hold-up relation to gas velocity may change so that further increase in gas velocity is not desirable. However, for current catalyst activities it seems that the selected velocities will be well below the transition to a transported flow regime. On the other hand reactor designs with inlet velocities less than 30 cm/s should be a sign that the catalyst activity is not competitive. For example, Fig. 3 in a recent patent application (WO 03/052335, US 20030027875, see Table 5) shows the syngas per pass CO conversion decreasing to below 60% if the gas velocity is increased to above 30 cm/s. If this is based on actual operating data then it indicates that a catalyst with a low activity was used. It is also possible that the conversion was incorrectly extrapolated due to an overestimation of the gas mixing at high velocities.

Interestingly, a tubular loop reactor approach has been proposed in which long conduits (pipes) are used [176]. It would only be necessary to use this type of reactor to achieve a sufficient conversion at lower than transport velocities if the catalyst activity is very low so this type of reactor will most likely be considered for velocities above the transport velocity. A number of geometric configurations could be envisaged for such conduit reactors. It remains to be seen whether a tubular loop reactor approach will prove to be cost effective using highly active catalyst and high velocities which transport the slurry. The gas fraction increase after the transition to a transport regime has not been investigated in detail for the purpose of this book.

It seems likely that there will be a step change up in gas fraction after the transition to a transport regime. A gas velocity which gives a gas hold-up of 60% does not transport the slurry out of the reactor. If the pipe reactor gas hold-up is 80% then this will double the reactor volume required to hold the same amount of slurry. This can only be cost effective if the gas velocity is more than doubled. A gas slurry separation section will be needed for the pipe reactor approach which adds further cost.

Another approach to consider is to use reactors in series. With this approach, water is removed between stages to enhance the reactant partial pressures in the downstream reactor stage. The removal of other diluents between stages (such as carbon dioxide) may also be considered. This allows lower recycle ratios and increased steam production from the reactor heat removal system and decreased overall reactor volumes. These advantages may not necessarily compensate for the increased complexity of the reactor in series approach and detailed studies are required to determine the most cost effective design approach.

The staged introduction of recycle gas is a simple and effective method to decreasing the required reactor volume [177]. However, this approach requires a very good understanding of the reactor hydrodynamics.

Recently a multi-stage slurry reactor concept as described in the Deckwer text book has been proposed by Maretto, Krishna and Piccolo [178, 179]. The division of the reactor into 4 separate stages within the same vessel is recommended. Each stage requires its own independent cooling system with the percentage of the total duty being 33%, 31%, 25% and 11% in each of the respective stages from bottom to top. This approach decreases the uncertainty involved in scale-up and allows large scale reactors to be constructed without prior knowledge of the effect of scale-up on dense phase mixing. This is an important advantage for newcomers to the field of FT reactor technology. However, it may result an unjustified increase in cost due to the more complex reactor and cooling system design. For example, a reactor with a height of 32 m and a diameter of 8 m has an L/D ratio of 4 and the actual dense phase dispersion may not differ significantly from the recommended 4 mixing cells in series. This would then avoid the need to divide the column up into discrete compartments, each with its own cooling system and it should be possible to use a single steam drum for temperature control.

Furthermore, the muti-stage approach was not analysed with the possibility of using tail gas recycle. As a result it was concluded that the inlet superficial gas velocity should be restricted to 30 cm/s. However, the target overall conversion can be achieved by adding recycle and increasing the gas velocity to say 40 cm/s and as a result the reactor productivity is increased. As mentioned earlier, the target fresh feed conversion is the result of an optimisation exercise and may be less than the 95% chosen for this analysis, but it should be at least 90%. A fair comparison can only be made if the number of mixed cells in series required to model the dense phase behaviour is known. This is definitely more than a single mixed cell for large scale baffled slurry bubble columns of commercial interest.

Many FT reactor designs assume that the reactants in the syngas will be in stoichiometric balance (for example, the multi-stage reactor design discussed above). However, it is well known that the selectivity to the desired hydrocarbon products can be improved for cobalt catalyst if the syngas is slightly rich in carbon monoxide [180]. This will result in a lower target reactant conversion due to the excess CO but this is compensated by an increased selectivity to useful products. This is the subject for a further optimisation study that requires the reactor model to predict product selectivities simultaneously with the reactant conversion prediction. This is discussed in more detail in Chapter 5. After many years of comparison of model predictions with actual industrial data, Sasol can confidently use such models to determine optimum reactor designs for any desired reactor scale.

4. FLUIDIZATION FUNDAMENTALS AND THE SASOL ADVANCED SYNTHOL (SAS) REACTORS

Gas-solid fluidization finds wide application in the chemical industry. At the Sasol 2 and 3 plants in Secunda, dense phase turbulent fluidized bed reactors called Sasol Advanced Synthol (SAS) reactors are used to convert Syngas to liquid fuels and chemicals. This section covers the fundamentals of fluidization both in general terms and as they relate to the SAS reactors. Topics discussed include the relationship between bed pressure drop versus gas velocity, powder classification in terms of Geldart's powder classification diagram, fluidization regimes, fluidization regime transitions due to changes in powder classification, entrainment and the basic principles of cyclone and dipleg operation. This section is based on a paper presented by Sookai at the Industrial Fluidization, South Africa (IFSA 2002) Conference [181]. The standard text book on fluidization engineering is the book by Kunii and Levenspiel [118].

4.1 Powder classification based on fluidization properties

The effect of particle size and density on the fluidization behaviour of a particular powder is best described by Geldart's powder classification diagram shown in Fig. 29 [182].

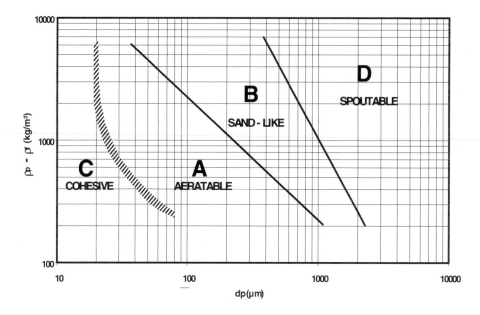

Figure 29 Geldart's powder classification diagram

According to this diagram particles are classified as being in either group A, B, C or D, depending on the particle and fluid density difference, and the mean particle size. Each group exhibits unique fluidization characteristics which forms the basis for distinguishing between groups. The group classification can therefore be used to identify the general "package" of fluidization characteristics of a particular powder without having to do experiments. For example, if a powder falls in group C then it would be known that these powders are cohesive, difficult to fluidize and are prone to gas channelling. However, if a powder has a group A classification, then it would be known that these powders display good fluidization properties, are easily aeratable, have a high dense phase voidage and limited bubble growth. Further characteristic fluidization properties are summarised in Table 11 [183].

The SAS reactors utilize a reduced and promoted iron oxide Geldart group A powder catalyst. However, depending on the process conditions, the group classification could potentially change. For example, the loss of fines can cause the powder to change to a group B classification or the accumulation of fines can cause the powder classification to change to group C. It will be shown later how this potential change in powder classification influences reactor operation.

Table 11
General fluidization characteristics of Geldart group A, B, C and D powders

Increasing particle size →			
Group	**C**	**A**	**B**	**D**
Typical solids	Flour, Cement	Cracking catalyst	Building sand, table salt	Crushed limestone, coffee beans
Most obvious characteristic	Cohesive, difficult to fluidize	Bubble free range of fluidization	Starts to bubble at min. fluidization velocity	Coarse solids
Bed expansion	Low when bed channels, can be high when fluidized	High	Moderate	Low
De-aeration rate	Initially fast, exponential	Slow, linear	Fast	Fast
Bubble properties	No bubbles, channels and cracks present	Bubble splitting and re-coalescence predominate. Bubble grow to a maximum size. Large wake	Bubble growth, no limit on size	No known upper size, small wake
Solids mixing	Very low	High	Moderate	Low
Gas backmixing	Very low	High	Moderate	Low
Slug properties	Solid slugs	Axisymmetric slugs	Axisymmetric and asymmetric	Horizontal voids, solid and wall slugs

4.2 Fluidization regimes

For fixed process conditions and particle properties, a fluidized bed operates in a particular fluidization regime. A fluidization regime transition can therefore occur in two ways, namely, (1) by changing the particle properties and (2) by changing the process conditions for example, the gas velocity. For simplicity the effect of gas velocity on fluidization regime transitions will be discussed here, but it will be shown later how a change in particle properties can also cause fluidization regime transitions.

158

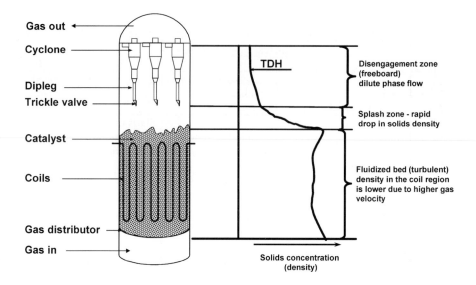

Figure 30 Schematic diagram of the SAS reactor showing internals and the variation of solids concentration with reactor height

Initially, as gas is first introduced into the bed of particles at a low velocity, it percolates through the void spaces between the particles. This is the **fixed bed regime** and particles display little or no movement. As the gas velocity is increased, the particles begin to move apart and the bed begins to expand, until a velocity is reached where the friction forces are balanced by the weight of the particles. At this stage the bed is just fluidized and the gas velocity at which this occurs is termed the minimum fluidization velocity (U_{mf}). Synthol catalyst, which has a wide particle size distribution, does not display a distinct U_{mf}. In this case the term full support velocity (U_{fs}) or complete fluidization velocity (U_{cf}) [184, 185] is used. This is the minimum velocity at which the bed is completely fluidized and is characterised by a bed pressure drop that is approximately equal to the weight of the bed (Fig. 31).

In Fig. 24, the curve ABCD shows schematically what happens to the bed pressure drop as the gas velocity is increased from zero for a narrow particle size distribution powder. Curve DCA shows what happens when the gas velocity is decreased after the bed has been fluidized.

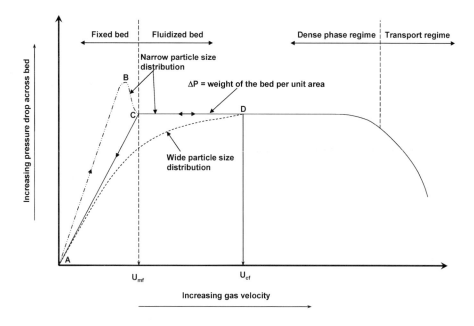

Figure 31 Schematic diagram showing the relationship between bed pressure drop and gas velocity for a narrow and wide particle size distribution powder

As can be seen for a powder having a narrow particle size distribution, the minimum fluidization velocity is clearly observable. This should be compared to curve AD which represents the bed pressure drop versus gas velocity for a powder having a wide particle size distribution (eg. Synthol catalyst). In this case a distinct U_{mf} is not observable. When the gas velocity is increased from zero, it will be noticed from Fig. 31 that curve ABCD is followed, and this shows that the pressure drop initially increases above that required to fluidize the bed (point B). The reason for this is that the powder is initially compact and in order to fluidize the powder, it is necessary to overcome both the weight of the bed and the compaction forces. Once the bed is fluidized, the dense phase can be easily 'deformed' hence there is no further increase in bed pressure drop with gas velocity. The bed pressure drop begins to decrease as the entrainment rate increases and the fluidization regimes change to the transport regime.

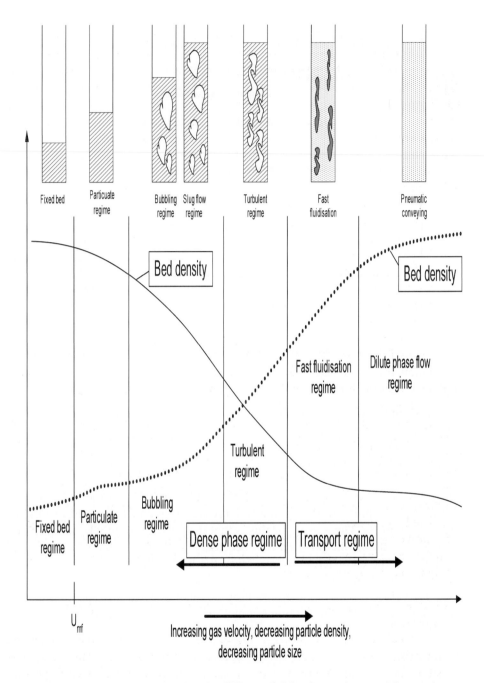

Figure 32 Schematic diagram showing the different fluidization regimes and factors that influence them

As shown in Fig. 32, increasing the gas velocity above U_{mf} causes the bed voidage (ε) to increase and this is reflected in an increase in the bed expansion. This is called the **particulate regime** and is of significance because in this regime the bed has the highest density whilst still maintaining fluid-like conditions. It will be shown later why this regime is important for dipleg operation. A further increase in gas velocity sees a fluidization regime transition to the **bubbling regime**. In this regime the bed consists of two phases namely a dense phase (mainly catalyst) and a bubble phase (mainly gas). The bed voidage increases further and the bubbles induce solid circulation, which makes the bubbling bed better mixed than the dense phase fluidised bed.

The powder group, reactor diameter, and particle and gas properties will determine the commencement of bubbling and the size of bubbles. For example, group A powders are known to have a high dense phase voidage and limited bubble growth. For these powders, it has been shown that an increase in fines causes an increase in the dense phase voidage while an increase in gas velocity causes an increase in bubble voidage [184, 186]. **Since conversion occurs predominately in the dense phase, group A powders are therefore ideal for catalytic reactors.**

While in the bubbling regime, an increase in gas velocity causes larger bubbles to form that move through the bed faster, which enhances the solids circulation rate. A further increase in gas velocity sees a further fluidization regime transition form the bubbling regime to the **turbulent regime**. In the turbulent regime distinct bubbles begin to disappear and the bed consists rather of regions of high density and low density particle clusters that dart to and fro. The bed is now in a high state of mixing and the entrainment of particles from the bed increases considerably, but a distinct bed surface is still visible. A further increase in gas velocity sees a transition to the **transport regime**, which consists of **fast fluidization** and **dilute phase flow** as shown in Fig. 32. In fast fluidization a distinct bed surface is not visible and solids have to be re-circulated to maintain a fast fluidised 'bed'. Reactors operating in this regime are known as Circulating Fluidized Beds (CFB). In dilute phase flow all particles are suspended and transportable and hence this regime is found in catalyst conveying applications.

4.3 Potential changes in powder classification during fluidized bed operation

A Geldart group A powder, for example, will display the fluidization characteristics of that group during operation (see Table 9). However, it is possible for a group A powder to change to either a Group B or C powder depending on the process conditions and the extent to which the particle properties change in situ. A group A powder can potentially change to a group

B powder as a result of the loss of fines, hence resulting in a coarser particle size distribution. Once the powder classification changes to group B, the powder then displays the characteristic fluidization properties associated with group B powders. Similarly it is also possible for a powder having a group A classification to change to one having a group C classification. This change can be brought about by the accumulation of fines, or by a decrease in particle density. Once the change has occurred, the powder will display the characteristic fluidization properties associated with group C powders. These changes in powder classification can cause fluidization regime transitions.

4.4 Potential changes in fluidization regimes due to changes in powder classification

When a powder undergoes a change in powder classification, the characteristic fluidization properties of the powder also changes. This can potentially change the fluidization regime during operation. For example, if under normal conditions a group A powder is operating in the particulate regime, a powder classification change to group B can result in the fluidization regime changing from particulate to packed bed flow and eventual de-fluidization. This occurs because a characteristic property of group B powders is that it de-aerates fast and does not display a particulate regime of fluidization. The other potential fluidization regime transition of interest can occur when a group A powder changes to a group C powder. In the fluidized state the group A powder may have been operating in the turbulent regime. When the powder classification changes to group C, the fluidization regime can potentially change to the fast fluidization regime. This change occurs because the accumulation of fines causes the bed density to decrease and it may decrease to such an extent that the bed becomes transportable, hence a transition to the transport regime. It will be shown later how these potential changes apply to the SAS reactors.

4.5 Entrainment

The terms entrainment, elutriation and carryover are often used interchangeably to describe the ejection of particles from the surface of a fluidized bed and their removal from the unit in the gas stream. Although there is no general agreement concerning the mechanisms by which entrainment occurs, bubbles (or gas voids) clearly have an important role to play. Particles are ejected into the freeboard either when the bubble 'bursts' at the bed surface or through the wake of the bubble. The entrainment rate of material from the bed depends on the rate at which bubbles "burst" at the bed surface, the bubble size, solids properties and in some cases the carrying capacity of the gas.

4.6 Cyclone and dipleg operation

The entrained solids are usually recovered in the reactor by passing the gas and solids through cyclones before the gas leaves the reactor. To improve the recovery efficiency of solids, several stages of cyclones could be used in series. However, in the SAS reactors single stage cyclones are used [187]. In the cyclones, the solids are separated from the gas by centrifugal forces. The separated solids flow down the cone of the cyclone and return to the reactor via the diplegs. During the separation process there is a pressure drop, which, results in the pressure inside the cyclone being less than the freeboard pressure. It is essential that there is a pressure recovery in the dipleg to return the solids to the reactor. This requires a fluidized head of solids as shown in Fig. 33.

As the solids flow down the dipleg the solid level builds-up and the pressure at the base of the dipleg increases. When this pressure is greater than the freeboard pressure, the trickle valve opens, causing the solids to flow out into the freeboard. Thus, the solids flow out of the dipleg is governed by the pressure balance shown in Fig 33.

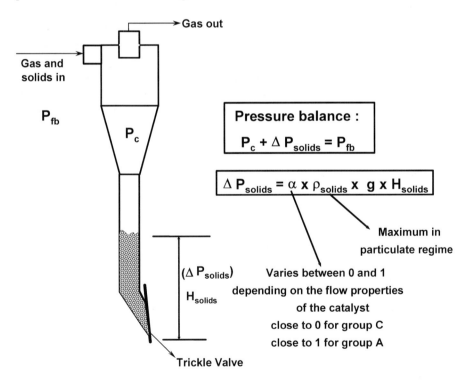

Figure 33 Schematic diagram showing the cyclone and dipleg

The pressure head developed by the fluidized solids in the dipleg depends on the catalyst properties as shown by the equation in Fig. 33. The flow coefficient (which varies between 0 and 1) depends on the fluidization properties of the catalyst. For example group A powders which have good flow properties will have a flow coefficient close to 1 whereas a group C powder will have a much lower flow coefficient. It will be shown in the next section (using the SAS reactor as an example) how this phenomenon can affect the dipleg operation.

4.7 The Sasol Advanced Synthol reactors

At the Sasol 2 and 3 plants in Secunda, dense phase turbulent fluidized bed reactors called Sasol Advanced Synthol (SAS) reactors are use to convert Syngas to liquid fuels and chemicals. At present there are five 8m diameter and four 10.7m diameter reactors being utilized. The fluidization related scale-up of these reactors is discussed elsewhere [188]. With reference to Fig. 30, the operation of the SAS reactors can be described briefly as follows. The gas enters at the bottom of the reactor and passes through a grid plate or gas distributor and fluidizes the catalyst bed (which operates in the turbulent fluidization regime). The gas distributor serves to distribute the gas evenly throughout the catalyst bed thus preventing preferential flow patterns and stagnant regions. It also acts as a support for the catalyst bed during bed slumps (no gas flow). In the fluidized bed, Fischer-Tropsch and side reactions occur, which results in a host of gaseous products. These reactions are exothermic, hence cooling coils are present in the reactor to remove heat (by generating steam in the coils). This together with the rapid mixing in the fluidized bed results in isothermal conditions. The product gas leaves the bed carrying some catalyst particles with it (entrainment). Therefore, before leaving the reactor, the gas passes through the cyclones where the entrained solids are separated from the gas. Solids collected in the cyclones are returned to the reactor via the diplegs.

It was noted earlier that in addition to gas velocity changes a fluidization regime transition can also occur as a result of a change in the particle density and size distribution. In the SAS reactors this transition is potentially possible because the Boudouard reaction, which is a side reaction to the Fischer-Tropsch reaction, results in the formation of free carbon [189]. The lay-down of this carbon on the catalyst improves the flow properties of the catalyst but it also lowers the bulk density of the catalyst and can lead to an increase in the fines fraction. This change in particle properties causes an increase in the bed voidage and if allowed to continue uncontrolled can result in a fluidization regime transition from the turbulent regime to the transport regime (fast fluidization). This must be avoided because the uncontrolled bed expansion and transport of catalyst causes the cyclones to block followed by a reactor shut down. An on-line (used) catalyst removal and (fresh) catalyst addition policy was implemented to maintain the fluidized bed density of the SAS reactor within certain limits to prevent the above occurrence.

For the successful continuous operation of the SAS reactors it is essential to maintain the pressure balance which causes catalyst to flow out of the dipleg into the reactor. Since the pressure build-up in the dipleg provides the principle driving force for solid circulation, it is essential to maintain a fluidization regime in the dipleg that maximises this pressure recovery. To achieve this, a Geldart group A catalyst operating in the particulate regime is required in the dipleg. With Synthol catalyst the change in powder properties could cause a powder classification transition from group A to group B. Because a group B powder de-aerates very rapidly (see Table 11), the flow regime in the dipleg can quickly change from dense phase flow to eventual de-fluidization. This would cause a reduction in the pressure recovery and can lead to eventual de-fluidization of the catalyst in the dipleg, causing the dipleg to block and eventually the reactor has to be shut down.

If the catalyst properties change such that there is a transition from group A to group C (e.g. through the accumulation of fines) the flow properties of the catalyst will also change. This would result in a lower flow coefficient and hence a higher catalyst level will be required in the dipleg to ensure that the necessary pressure recovery is established. The residence time of catalyst in the dipleg is increased and hence the risk of blocking the dipleg is also increased. To reduce the risk of the above potential dipleg blockages it is essential to provide constant aeration to the dipleg.

From the material presented in this section it should be clear that a proper understanding of the fundamentals of powder classification and the factors that influence fluidization regime transitions is important. This knowledge can be used to avoid potential problems in the operation of a fluidized bed. Operational procedures can be put in place to control the particle properties thereby ensuring sustained and successful commercial-scale fluidized bed operation.

5. DESIGN AND SCALE-UP OF FLUIDIZED BED FT REACTORS

An overview is provided of the most important considerations for the design and scale-up of catalytic fluidized bed reactors. Modern reactors used for high temperature Fischer-Tropsch synthesis are pushing the size and gas velocity limits for turbulent fluidized bed reactors operating at high pressures. Based on this experience, the recommended design and scale-up approach is explained. This overview is extracted from a paper by Steynberg at the Industrial Fluidization, South Africa (IFSA 2002) Conference [190].

The considerations for the design of catalytic fluidized bed reactors are in many cases different from those for fluidized reactors involving the processing of solids. Ideally for catalytic reactors the solids properties can be selected to give the best performance. If the solids properties change with time then an approach of on-line catalyst renewal may be adopted to keep the solids properties approximately uniform.

Typically the use of a fluidized bed is desirable because the chemical reactions are highly exothermic and high heat transfer coefficients are achievable with heat removal pipes in these reactors.

For optimum catalyst performance, small catalyst particle sizes are preferred. This avoids intra-particle concentration and temperature gradients and allows more effective use of the catalytic material. Typically the particle sizes will be less than 150 microns and the powder will be a type A powder according to the well-known Geldart classification [183] (See Fig. 29).

Fluidized bed reactors using a Geldart type A catalyst can be described using the two-phase model [99, 191]. This model divides the fluidized bed into bubbles and the expanded solids phase known as the dense phase (See Fig. 34).

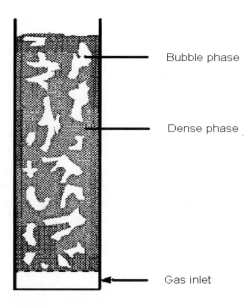

Figure 34   Two-phase model for fluidized bed

It is easier to predict the performance of type A solids with scale-up than the larger Geldart type B/D catalyst. A Fischer-Tropsch reactor using a type B/D catalyst was operated in Brownsville, Texas by Hydrocarbon Research

Inc.. The reactor performance was found to be significantly worse than the smaller scale units. This has been ascribed to a decrease in dense phase voidage with increasing column diameter [192] resulting in poor catalyst to gas contact (see Fig. 35). A further problem at Brownsville was the formation of fine carbon rich particles during synthesis referred to as 'bug dust'. This essentially resulted in a mixture of type C and type B solids and was no doubt particularly problematic for the ceramic filters used for gas-solids separation.

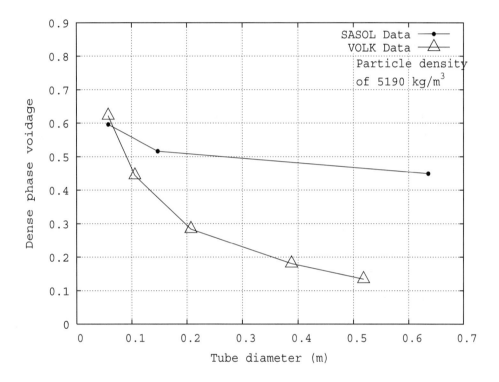

Figure 35 Influence of diameter on (dense phase) voidage [192]

It is well known that an increase in the level of fines in the type A catalyst will increase the dense phase voidage [186]. The appropriate definition of fines may differ depending on the particle density. For the relatively dense Synthol catalyst the fraction less than 22 microns is classed as 'fines'. For less dense catalysts it may be appropriate to use up to 44 microns as the upper size limit for 'fines'. See Fig. 36 for an example of the influence of the fines fraction on the dense phase voidage.

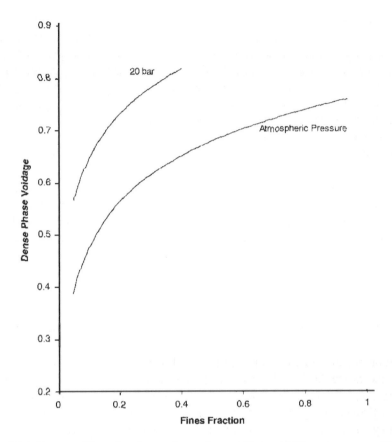

Figure 36 Influence of fines fraction on dense phase voidage at different pressures [193]

While good gas-catalyst contact is desirable, the dense phase voidage can be too high resulting in a decline in performance. This is because less catalyst can be accommodated in a given reactor volume and more gas will backmix with the solids resulting in lower reactant conversions. There is thus an optimum fines content in the catalyst particle size distribution.

5.1 Determination of voidage

It is necessary to determine the 'voidage' or space occupied by the gas in the fluidised bed in order to determine the amount of catalyst that can be accommodated in a given reactor volume. It is now well known that the volume occupied by bubbles depends on the column diameter for small diameter columns but is relatively constant at a given gas velocity for column diameters greater than 1m [186]. For large-scale columns, bubbles typically occupy about 10% to 20% of the bed volume so the amount of catalyst in the reactor is mainly determined by the dense phase voidage, which may exceed 70% [193]. Besides the fines fraction mentioned previously, the other parameter that influences the dense phase voidage is the gas density [193]. The dense phase voidage can be determined for the catalyst powder in question using relatively small scale equipment but it is important to cover the applicable range of gas densities. Dense phase voidage is also influenced by particle shape and experimental verification is still recommended rather than the use of published correlations.

It is a relatively simple matter to determine the dense phase voidage in a large-scale commercial reactor from a differential pressure measurement across a section of the bed if the bubble fraction, ε_b, and particle density are known. For a fluidized bed of height H, the bubble fraction is determined by:

$$\varepsilon_b = \frac{1}{H} \int_b^H \frac{(u - u_{df})}{\phi \sqrt{g d_B}} \, dh \tag{22}$$

The value for ϕ becomes constant for columns larger than 1m in diameter. There is a region of higher gas hold-up, which is dependant on the gas distributor design, up to about one metre above the gas distributor. At higher elevations a stable bubble size is typically attained. As the gas density increases, d_B will decrease and ø will increase so that ε_b remains approximately constant [194] at a given gas velocity. For a Fischer-Tropsch reactor the value of ø may be as high as 5. However, measurement of ε_b by a bed collapse experiment may be easier than finding the average bubble size, d_B, or the bubble rise coefficient, ø.

Newton and Johns determined d_B by measuring the number of bubbles using an X-ray imaging technique [195]. Using the Darton bubble growth model [104], Ellenberger and Krishna [101] derived the following equation to predict the bubble fraction (also referred to as bubble hold-up or dilute phase voidage):

$$\varepsilon_b = \frac{(u - u_{df})^{\frac{4}{5}}}{\alpha^{\frac{1}{2}} \phi g^{\frac{2}{5}} (h^* + h_0)^{\frac{2}{5}}} \quad for \ H \gg h^* \tag{23}$$

where α is a constant in the Darton bubble growth model, h_o is a parameter determining the initial bubble size at the gas distributor and h^* is the height above the gas distributor where the bubbles reach their equilibrium size.

Werther [196] proposed that for columns with diameters, D_T, smaller than 1m, $\phi = \phi_o (D_T)^{0.4}$.

5.2 Gas and catalyst mixing

Typically the superficial gas velocity, u, is much larger than the dense phase velocity, u_{df}, so most of the through flow of gas is by way of the bubbles and these can usually be considered to be in plug flow. It is the mixing behaviour of the dense phase that is of most interest. One way to characterise the dense phase mixing is with an axial dispension model. It has been found that the axial dispersion coefficient can be correlated with an expression of the form [197].

$$D_{ax} = \text{constant } V_b \, D_T \tag{24}$$

Where V_b is the bubble rise velocity and D_T is the column diameter. The bubble velocity depends only on the gas velocity for columns larger than 1m in diameter.

For large diameter columns it is inevitable that the dense phase will be highly backmixed. This is desirable from the point of view of attaining a uniform temperature profile but it becomes difficult to achieve high per pass gas conversions. This is even more of a problem when a reaction product has a negative impact on the reaction rate. This is the case with the water product from Fischer-Tropsch synthesis with an iron catalyst. High overall conversions may be attained by means of tail gas recycle or tandem reactors, or both, after water has been removed from the tail gas by cooling and condensation. If the reactor is designed with recycle and lower per pass conversions, then the design is much less sensitive to assumptions made with regard to mixing in the fluidized bed. In this case the assumption of plug flow for the gas phase and complete mixing for the dense phase is sufficient.

5.3 Mass transfer

In small scale fluidized beds, using small bed heights, the bubbles will probably not have grown to their equilibrium size and wall effects may constrain the maximum bubble size. Bubbles also move faster in small diameter columns. As a result the mass transfer between the bubbles and the dense phase is poorer in large commercial units compared to small-scale reactors.

The mass transfer for type A powders has been ascribed mainly due to cross flow so ignoring cloud boundary effects and diffusion, mass transfer is described by [198, 199]:

$$k_g a = 7.14 \, u_{df}/d_b \tag{25}$$

This is the main reason for the great deal of attention paid to characterisation of dense phase velocity and bubble size in the academic world. Research relating to bubble columns [113] has highlighted the need to consider the effect of the frequent bubble coalescence and break up. This tends to cause the effective bubble diameter to be much smaller than the actual average diameter. This results in significantly higher mass transfer predictions. It has also been stressed that the dense phase velocity may be significantly higher than the minimum fluidization velocity and that the dense phase velocity depends on the gas velocity, the mean particle size and the fines fraction in the solids [184]. Other physical properties such as particle shape and density also play a role.

5.4 Conversion prediction

With the assumptions of plug flow for the bubbles and perfect mixing in the dense phase, the following equation can be derived to predict conversion, χ, [199]:

$$\chi = \frac{(1 - v\exp(-NTU/v))NRU}{1 + NRU - v\exp(-NTU/v)} \qquad (26)$$

where v is the fraction of the total gas which flows up the reactor in the form of bubbles. For high capacity reactors it is often acceptable to set $v = 1$ so that the expression simplifies to:

$$\chi = \frac{(1 - \exp(-NTU))NRU}{1 + NRU - \exp(-NTU)} \qquad (27)$$

where $\qquad NTU = v\int_b^H \frac{k_g a}{V_b}\, dh = \int_b^H \frac{7.14u_{df}}{V_b d_b}\, dh \qquad (28)$

and $\qquad NRU = \frac{k_m \rho_b H}{(1 + \varepsilon)u} \qquad (29)$

where k_m represents a reaction rate expression (m^3 reactant converted per kg of catalyst per second); ρ_b is the bulk density and ε is the total voidage in the fluidized bed.

At this point it is impossible to overemphasise the importance of determining the correct value for the reaction rate constant and a suitable rate equation.

The overall conversion depends on the relative success of the catalyst researchers in increasing k_m and the fluidization experts in optimising the relative values for NTU and ε. The bed height must also be sufficient to accommodate the heat removal pipes.

It will further be noted that all scale-up risk is essentially eliminated if the demonstration reactor has a diameter of about 1m or more. This was confirmed by Sasol with scale-up from a 3ft diameter demonstration reactor to 5m, 8m and then 10.7m diameter reactors.

5.5 Practical considerations for the gas distributor design

The basic design concept is to use orifices to create sufficient pressure drop to evenly distribute the gas and avoid backflow. Quality control in the construction of the gas distributor is very important to avoid problems associated with gas maldistribution. It is also important that the pressure drop is still sufficient at the minimum flow turndown case, which is usually the total recycle case.

For many applications the avoidance of stagnant zones may be critical since this may lead to hot and cold regions with a variety of adverse effects. In the case of Fischer-Tropsch synthesis, this may lead to the progressive blockage of the gas distributor. For this reason it is desirable to design the gas distributor in a way that allows the backmixing dense phase to sweep catalyst from potential areas of laydown. This sweeping action improves with scale-up but may become so severe that the effect on the gas inlet jets needs to be considered.

Any form of valve design should generally be avoided since they are generally costly, ineffective and decrease the reactor on-stream factor. The gas inlet system must therefore be capable of tolerating some solids backflow during an interruption of the feed flow and these solids should then be easily returned to the bed on re-start.

A recent article by Pell [200] gives good guidance for the design of gas distributors.

5.6 The influence of internals

The cooling pipes will generally have a positive impact when scaling-up since they will to a certain extent dampen the relationship between reactor diameter and dense phase backmixing. It is unwise to rely on this dampening effect in the design of the reactor. In general, vertical cooling pipes will have minimal effect on the design of large-scale reactors but can have a significant effect for column diameters less than 1m.

5.7 The design of solids separation equipment

Both filters and cyclones may be considered for the solids gas separation. For filters, the gas flow and density determine the filter area required while the catalyst entrainment flux will determine the interval between blowbacks. Blowback system reliability may be critical. Cyclones are less expensive than filters and are potentially more reliable, if properly designed, due to the absence of moving parts. Erosion and dipleg blockages are common cyclone problems. Accurate prediction of solids entrainment flux is also vital for effective cyclone design. The method of Briens et al [201] is recommended to predict the entrainment flux. The use of single stage cyclones with dipleg purge gas is recommended for best reliability. Unlike filters, cyclones can never entirely avoid solids losses and it is not possible to retain entrained particles smaller than 5 microns.

It is easily possible to write an entire book on the proper design and maintenance of solids separation devices. It is important to choose an experienced equipment supplier and to ensure that there is good communication with the supplier so that your application characteristics and all possible operating modes are well understood. Proper inspection of the installed equipment prior to commissioning is also vital. Consultation with experienced operators or consultants is also recommended.

5.8 Gas velocity limits

There is a maximum velocity for which a stable fluidized bed can be maintained. This maximum velocity decreases as the pressure or gas density increases. According to published maps of acceptable operating velocity (See for example Fig. 37), the Sasol Advanced Synthol (SAS) reactors now operate very close to this maximum velocity.

The validity of these maps has not been verified at Sasol for gas densities equivalent to the 25 bar operating pressure used in the SAS reactors. Sasol will no doubt be pushing the gas velocity limits for future reactor designs.

To summarize, the reactor design steps are as follows:

1) Set a safe gas velocity upper limit.
2) Determine the bed height required to attain the conversion target.
3) Check that there is sufficient bed volume for cooling surfaces.
4) Consider all operating cases especially the minimum possible particle density.
5) Calculate the catalyst entrainment rate.
6) Consult solids separation vendors.

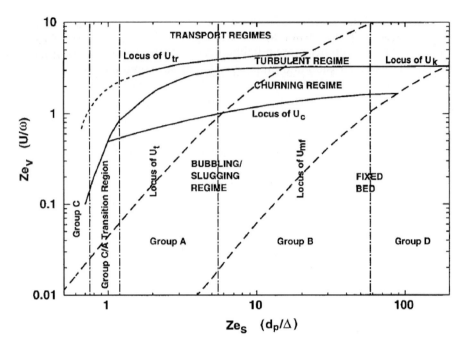

Figure 37 Unified flow regime map [202]

6. DESIGN AND SCALE-UP FOR MULTI-TUBULAR FIXED BED FT REACTORS

6.1 Text book teaching

In contrast to slurry phase reactor design for the FT-synthesis, no standard text books deal specifically with fixed bed reactor design for the Fischer-Tropsch synthesis. Chapter 11 of the well-known book "Chemical Reactor Analysis and Design" by G.F. Froment and K. B. Bischoff [203] deals with design of Fixed Bed Catalytic Reactors in a fundamental way without making specific reference to the FT-synthesis.

The low temperature FT-synthesis differs from typical fixed bed reactor processes such as ethylene oxide synthesis, carbon monoxide methanation, methanol synthesis, maleic anhydride synthesis, because a significant part of the FT-product is a liquid at the reaction conditions. As a consequence, the fixed bed reactor operates in a trickle flow regime with the synthesis gas - and the liquid product streams flowing co-currently downwards in the reactor. Typical volumetric liquid productivity is however such that the superficial liquid velocity is well below 1 mm/s (without liquid recycle), which ensures operation far from the trickle flow-pulsed flow transition as illustrated in Fig. 38. The effect of the liquid on hydrodynamics relative to 'dry' fixed bed operation is therefore small. The main effect is a somewhat increased pressure drop and slightly enhanced heat transfer, see Section 1.4 below.

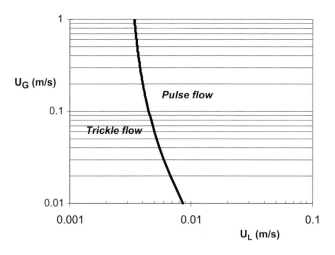

Figure 38 Flow map of trickle flow-pulsed flow regime transition in trickle bed reactors. Composed from Iliuta et al. [204], using representative physical properties for the LTFT synthesis and a particle diameter of 1.5 mm (sphericity factor = 0.85).

A major effect of the presence of the liquid is that the catalyst particle pores are liquid-filled due to capillary forces. Since the diffusion coefficients of the gaseous reactants hydrogen and carbon monoxide in this liquid medium are orders of magnitude lower than in the gas phase and pressure drop considerations dictate catalyst particle sizes to be 1 mm or larger, serious diffusion limitations occur in the fixed bed FT synthesis [205].

Keeping the above mentioned effects of the presence of a liquid in mind, the book of G.F. Froment and K. B. Bischoff [203] can serve as a valuable basis for design of fixed bed reactors for the low temperature Fischer-Tropsch synthesis.

6.2 Reactor modelling

One of the main advantages of the multi-tubular fixed bed reactor is that it provides modular scale-up from one representative tube on a pilot scale. Pilot-plant work is therefore relatively cheap and one can rely for an important part on experimental data for proper commercial design. Nevertheless, reactor modelling is indispensable in arriving at a thorough optimized commercial design, accounting properly for the combined action of heat removal, pressure drop, diffusion limitations and intrinsic kinetics.

Froment and Bischoff [203] discuss the various fixed bed reactor models in detail. Two main classes of models can be distinguished, i.e. pseudo-homogeneous models and heterogeneous models. The first class supposes physical and thermal equilibrium between the bulk gas phase and the catalyst particle surface, whereas the second class accounts for concentration – and temperature differences between bulk gas phase and catalyst particle surface as a results of inter-particle mass – and heat transfer resistances. Within each class of models, one- and two-dimensional sub-models exist with an increasing degree of complexity as axial mixing, radial mixing and intra-particle gradients are taking into account. Since the liquid-filled pores and particle sizes ≥ 1 mm guarantee relatively low volumetric reactions rates in the fixed bed FT-synthesis, inter-particle mass – and heat transfer resistances are insignificant and pseudo-homogeneous models can be used. Typically applied particle sizes (≥ 1 mm), tube diameters (1 – 2 inch), tube lengths (6 – 12 m) and superficial gas velocities (> 0.2 m/s) also imply that the Bo numbers are large enough to justify the assumption of plug-flow (no axial mixing).

Depending on the tube diameter, a one-dimensional model may suffice otherwise radial mixing has to be taken into account. There are substantial negative effects resulting from excessively high temperatures (high methane selectivity, carbon make and damage to expensive catalyst) so strict temperature control is necessary to avoid temperature excursions. This is much more a concern for Co-based catalysts than for Fe-based catalysts. Lower mean temperatures and narrower tubes are therefore applied for Co catalysts, radial temperature gradients are relatively small and a (pseudo) one-dimensional model suffices for reactor design purposes. Less stringent temperature control is necessary for Fe-based catalysts. Cheaper reactors with fewer, wider tubes operating at increased temperatures are therefore applied for Fe-based catalysts, e.g. the Arge reactors in Sasolburg that use 2" diameter tubes. Radial temperature gradients are then significant and homogeneous 2-dimensional models are needed to guarantee a reasonably accurate prediction of the reactor performance. As discussed above, intra-particle diffusion limitations need to be taken into account in both models. Wang et al. [206] recently reviewed modelling of the fixed-bed Fischer-Tropsch synthesis and concluded that the number of studies reported in open literature is rather restricted. As an omission, the modelling work of De Swart [113] is not referred to by Wang et al. Table 12 below summarizes Fischer-Tropsch fixed bed modelling studies.

Table 12
Overview of modelling studies available in the public domain on the LTFT-synthesis in multi-tubular fixed bed reactors.

Reference	Type of model	Catalyst	Tube diameter
Atwood and Bennet [170]	Heterogeneous one-dimensional Intra-particle diffusion limitations taken into account via analytical expression for effectiveness factor	Fe-based $d_p = ..$ mm	...
Bub et al. [171]	Pseudo-homogeneous two-dimensional. Effect of liquid on heat transfer and pressure drop neglected, insignificant pressure drop. No intra-particle diffusion limitations.	Fe-MnO $d_p = 2.9$ mm	20 mm
Jess et al. [172]	Pseudo-homogeneous two-dimensional Effect of liquid on heat transfer and pressure drop neglected, insignificant pressure drop. Intra-particle diffusion limitations accounted for by using lumped kinetics	Fe (ARGE-Lurgi-Ruhrchemie) $d_p = 2.5$ mm	41 – 90 mm
Wang et al. [169]	Heterogeneous one-dimensional Effect of liquid on heat transfer and pressure drop neglected Rigorous single particle model embedded to account for intra-particle diffusion limitations.	Fe-Cu-Zn	32 – 80 mm
De Swart [107]	Heterogeneous one-dimensional model Effect of liquid on heat transfer and pressure drop taken into account. Intra-particle diffusion limitations taken into account.	Co, $d_p = 2$mm	50 – 60 mm *

* Estimated from shell diameter and number of tubes (6.2 m, 8000 tubes)

Remarkably, only De Swart studied the LTFT-synthesis over a Co catalyst, despite its commercial relevance for GTL processes (Shell's SMDS process, [14]). This study is also most comprehensive in taking the effect of the trickling liquid flow on heat transfer and pressure drop into account. The study of Wang et al. is the only study which accounts for the effect of intra-particle diffusion limitations on product selectivity. All other studies ignore – or consider the effect of intra-particle diffusion limitations on Co conversion only, while it is known from Sasol's commercial experience that the effect on selectivity is at least as pronounced. Considering the above discussion, a successful FT reactor model for modern multi-tubular fixed bed reactors:

- is allowed to be of (pseudo) homogeneous class
- can be either one-dimensional (low axial temperature gradients, narrow tubes: Co) or two-dimensional (large axial temperature gradients, wide tubes: Fe)
- should embed some type of single particle model to account for the effect of diffusion limitations on productivity and selectivity.

6.3 Diffusion limitations

Commercial Sasol experience teaches that FT fixed bed reactor technology suffers from significant diffusion limitations for practical catalyst particle diameters (1 – 3 mm). The occurrence of diffusion limitations results in a reduced productivity but also, even more important, in a worsened selectivity (more methane, lighter product, less olefins), the latter due to a lower effective syngas pressure and an enrichment of H_2 relative to CO inside the particle [207]. Post et al. [205] studied quantitatively the effect of diffusion limitations in the Fe- and Co-based low temperature Fischer-Tropsch synthesis for particle sizes ranging from 0.22 – 2.6 mm. They showed that the effectiveness factor η_{FT}, defined as the ratio between the actual FT-rate and the FT-rate at bulk gas phase conditions, can be estimated with a reasonable accuracy from the following analytical expressions using pseudo-first order kinetics in hydrogen.

$$r_{FT} = k_{PS,FT} C_{H2,L} \tag{30}$$

$$\eta_{FT} = \frac{\tanh(M_T)}{M_T} \tag{31}$$

$$M_T = r_{P,eq} \sqrt{\frac{k_{PS,FT}}{\frac{\varepsilon_P}{\tau} D_{H2,L}}} = \frac{V}{A} \sqrt{\frac{k_{PS,FT}}{D_{H2,L}^{eff}}} \tag{32}$$

Within this framework of equations, r_{FT} represents the hydrogen consumption rate and the carbon monoxide consumption rate or the –CH_2– formation rate. The effective liquid phase diffusion coefficient of H_2 is related to the bulk liquid phase diffusion coefficient and the catalyst particle geometry via:

$$D_{H2,L}^{eff} = \frac{\varepsilon}{\tau} D_{H2,L} \tag{33}$$

Post et al. [205] found this effective diffusion coefficient to be typically in the range 1 x 10^{-9} – 2 x 10^{-9} m^2/s, which is said to correspond with ε/τ values of 0.2 – 0.4. As mentioned above, it is however as important to consider the effect of diffusion limitations on selectivity. Especially for the governing reaction kinetics over Co-catalysts, the effect of diffusion limitations on selectivity will occur before a noticeable effect on productivity is observed. Fig. 39 shows the effectiveness factor η_{FT} as a function of the 'equivalent' slab thickness $r_{p,eq}$ for various values of the pseudo first-order rate constant $k_{PS,FT}$. Here, $r_{p,eq}$ is defined as:

$$r_{p,eq} = \frac{V}{A} \tag{34}$$

with:
A = particle surface exposed to the bulk gas phase (m^2)
V = particle volume (m^3)

The equivalent slab thickness $r_{p,eq}$ for a sphere with radius r_p equals $r_p/3$. Fig. 39 is also valid for non-spherical catalyst particles, egg-shell type of catalyst particle, hollow cylindrical particles. Spherical egg-shell type of catalysts with radius r_p and an active catalyst layer of 0.2 times r_p have an equivalent slab thickness $r_{p,eq}$ of 0.5 x $r_p/3$.

Further improvement can be obtained for hollow cylindrical particles with a similar active catalyst layer (= wall of cylindrical particle). For such particles, the equivalent slab thickness $r_{p,eq}$ is equal to 0.25 x $r_p/3$ since the particle surface area exposed to bulk syngas is doubled by the hollow nature of the particles.

As can be seen from Fig. 39, applying these types of catalyst particles will results in a strong increase of the effectiveness factor when operating under serious diffusion limitations. The typically small catalyst particle size of 1 – 3 mm for LTFT complicates the manufacture of mechanically strong, hollow cylindrical catalyst particles (walls of only a few hundreds microns) and, therefore, egg-shell type catalyst particles are likely to be more practical for the LTFT-synthesis.

It is necessary to keep in mind that the hold-up of active catalyst phase in the packed catalyst bed decreases when applying egg-shell - and hollow cylindrical type of catalyst particles. In the above example, this fractional volume is ca. 0.5 x (1 - ε_B) for both the egg-shell – and the hollow cylindrical catalyst particles relative to (1 - ε_B) for regular catalyst particles. The overall effect of applying these types of catalyst particles on the volumetric reactor productivity is therefore marginal, since the volumetric reactor productivity is proportional to the product of the effectiveness factor and the hold-up of the catalyst phase.

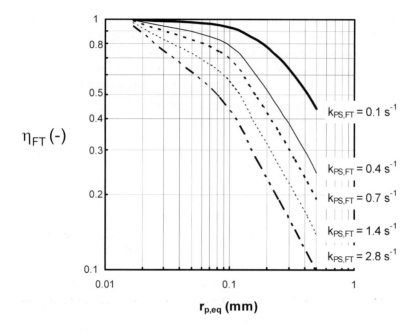

Figure 39 Calculated effectiveness factor, η_{FT}, as a function of the equivalent slab thickness for various value of the pseudo-first order rate constant of Eq. 34 [203]. Data used:
effective H_2 diffusion coefficient = 10^{-8} m^2/s
partial H_2 pressure in gas bulk phase = 23 bar
Henry coefficient of H_2 = 750 bar
molar liquid volume of pore liquid = 6 x 10^{-4} m^3/mol

The main incentives for using such special catalyst particles in the LTFT synthesis are therefore:
(i) to minimize the usage of precious metals (Pt, Rh, Pd) in the catalyst formulation
(ii) to obtain enhanced heat removal to the tube wall (see also section 6.5.1 below) and reduced pressure drops over the packed bed by applying larger effective particle sizes without excessively reduced effectiveness factors.

6.4 Effect of recycle

The effect of recycle has been discussed by Sie and Krishna [15]. Without recycle there is a strong radial temperature profile in the region near the inlet, since in this region the reaction rates are high because of the high partial pressures of the reactants. Further down the reactor tubes, rates are much lower as reactants are depleted and as a consequence radial temperatures are more even. The tube and catalyst dimensions have to be designed to cope with the temperature peaks so it follows that the larger part of the tube is over-designed.

More even axial profiles of reactant concentration and temperature are obtained when the conversion is restricted to say 30% per pass. If unconverted gas is recycled to enhance the overall conversion then gas velocities will be higher. The higher gas velocity has a beneficial effect on the effectiveness of heat removal. The Arge process derives its advantage over the earlier 'Mitteldruck' process from the application of gas recycle in combination with higher temperature and pressure. An enhancement of reactor capacity by a factor of 25 and a reduction of the cooling area by a factor of 12 was the result. The amount of catalyst and steel was lowered by a factor of about 7 [211].

An improvement of radial heat conductivity and heat transfer to the wall can not only be obtained by increasing the linear gas velocity in a gas-solid fixed-bed multi-tubular reactor, but also by operating in the presence of liquid [204, 212]. In the case of a Fischer-Tropsch reaction producing a relatively heavy product, the reactant stream is initially a gas that changes to a gas liquid mixture in the flow direction as condensable product is produced. In this situation, the effectiveness of heat removal will be lowest in the inlet region where the reaction rates are highest. By adding liquid, one can ensure that the whole tube including the most critical part operates in a trickle-flow mode, instead of only the bottom part.

6.5 Design equations
6.5.1 Heat transfer

The overall heat transfer coefficient in an equivalent pseudo-homogeneous one dimensional models follows from:

$$U = \cfrac{1}{\cfrac{1}{\alpha_w} + \cfrac{d_{tube,i}}{8\lambda_{ER}} + \cfrac{d_w A_i}{\lambda_w A_m} .. + \cfrac{A_i}{\alpha_o A_o}} \tag{35}$$

For a proper design, the heat transfer resistances of the tube wall and at the outside of the tubes are negligible. Numerous design equations exist for the heat transfer coefficient at the tube wall and the effective radial conductivity of the packed bed. Recommended design equations are:

$$\lambda_{ER} = \lambda_G \left(\left(\frac{\lambda_E^{\,0}}{\lambda_G}\right) + \left(0.197 + \cfrac{1}{1.9 + 0.0264 \cfrac{Re_L}{(1-\varepsilon_B)\beta_L} \cfrac{2}{3} \cfrac{\eta_L}{1e-3}} \left(\frac{2}{3} \frac{1-\varepsilon_B}{\varepsilon_B}\right) Re_L \; Pr_L \frac{\lambda_L}{\lambda_G} \right) \right) +$$

$$+ \; 0.095 \lambda_G \; Re_G \; Pr_G \tag{36}$$

$$\alpha_W = \alpha_{w,0} + 0.033 \frac{\lambda_G}{d_{p,sp}} \text{Re}_G \text{Pr}_G \qquad (37)$$

The static thermal conductivity of the packed bed, λ_E^0, depends among others on the thermal conductivity of the wax-filled catalyst particles and can be approximated with a reasonable accuracy from correlations as given among others by Froment and Bischoff [203]. The static contribution to the heat transfer coefficient at the tube wall, $\alpha_{w,0}$, depends on the catalyst particle properties, but also in a more complicated way on the voidage at the tube wall, the ratio of the particle diameter and the tube diameter and the absolute particle diameter. No consistent design equations are available to our knowledge for $\alpha_{w,0}$ and, when relevant, experimental determination is therefore recommended.

As can be seen from Eq. (36) and Eq. (37), heat transfer to the tube walls for a given tube diameter can be seriously enhanced by increasing the superficial gas velocity and by increasing the catalyst particle diameter. Further, the production of liquid product also contributes increasingly from tube top – to tube bottom to enhancement of heat transfer via the second term between the inner brackets in Eq. (36). If desired, this positive effect can be improved by recycling liquid product to the entrance of the tubes, resulting in an increased average Re_L number. This however imposes the technical challenge to distribute the recycle liquid evenly to all the thousands of tubes.

6.5.2 Pressure drop

The pressure drop can be calculated from the classic Ergun equation for packed beds:

$$\frac{\partial P}{\partial z} = -\left(150 \frac{(1-\varepsilon_B)^2}{\varepsilon_B^3 (d_{p,sp}\phi_p)^2} \eta_G u_G + 1.75 \frac{(1-\varepsilon_B)}{\varepsilon_B^3 (d_{p,sp}\phi_p)} \rho_G u_G^2\right) \qquad (38)$$

with:
$P =$ pressure (Pa)
$d_{p,sp} =$ equivalent spherical particle diameter (m)
$\phi_p =$ particle shape factor (-)
$\eta_G =$ gas viscosity (Pa s)
$\rho_G =$ gas density (kg m^{-3})
$u_G =$ superficial gas velocity (m s^{-1})

Since increasingly more liquid product is formed along the packed bed in the LTFT synthesis, the pressure drop will be somewhat higher than predicted from Eq. (38). This can become especially significant in case of recycling liquid product to the tube entrance to enhance heat transfer. Among others, Iliuta et al.

[168] give a detailed correlation to estimate the contribution of the liquid flow to the pressure drop.

6.6 Recommended design approach

In designing multi-tubular fixed bed reactors for the LTFT synthesis, one has to optimize the volumetric reactor productivity while meeting the following constraints:

(i)   reasonable per-pass syngas conversion

(ii)   a good product selectivity

(iii)   no excessive pressure drops over the catalyst bed

(iv)   no excessive temperatures within the catalyst bed (catalyst deactivation, deteriorated selectivity)

(v)   proper temperature controllability

Jess et al. [209] give as an important criterion for proper temperature control that the gradient $\Delta T_{max}\Delta T_{cool}$, representing the increase of the maximum temperature in the reactor per single degree of temperature increase of the cooling medium (generally: evaporating water), is smaller than 5.

To achieve an optimal design, the tube diameter, the catalyst particle diameter and the operating conditions (pressure, temperature and superficial gas velocity) can be varied. There is a big incentive to maximize the tube diameter to reduce reactor costs and to simplify reactor construction. The design is not an easy task since there is a compromise between the conflicting effects of catalyst particle diameter and operating conditions on the various constraints. For example, small catalyst particles favour a high utilization of the catalyst particles. This is desired in terms of obtaining a maximal volumetric productivity and good product selectivity. However, undesired side results of applying small catalyst particles are increased pressure drops, increased maximum temperatures and reduced temperature controllability (the latter two due to increased volumetric heat production and decreased heat transfer rates).

In general, high superficial gas velocities are needed to enable high volumetric reactor productivity while ensuring proper temperature control. As a consequence of a high superficial gas velocity, the syngas conversion per pass is only moderate for standard tube lengths and a significant gas recycle over the reactor is needed for high overall syngas conversions. Another reason to restrict the syngas conversion per pass is to avoid to high water partial pressures that may cause catalyst deactivation. Whenever gas recycle is mentioned in a Fischer-Tropsch reactor context it goes without saying that the gas is first cooled to condense water (and hydrocarbon liquids) before it is recycled.

Sending the combined tail gas of more parallel multi-tubular reactors (after cooling and liquid condensation and separation) to a shared second stage multi-tubular reactor, as applied by Shell in their Bintulu plant, is an important concept. This way, high overall syngas conversions (> 90 %) can be obtained in combination with high reactor productivity and a good product selectivity. To obtain similar high overall syngas conversions in a single stage of parallel reactors requires excessively high gas recycles, penalizing the reactor productivity (reduced throughput of fresh syngas feed) and the product selectivity (lower syngas pressures due to dilution with inerts and products). Since the manufacturing of syngas is the most expensive step in FT-based processes, the importance of high overall syngas conversions is self-evident.

From the above discussion, it is clear that pilot plant trials with a single representative tube and reactor modelling are indispensable for the design procedure.

NOMENCLATURE

C_D	drag coefficient, dimensionless
d_b	diameter of bubble, m
D_{ax}	axial dispersion coefficient, m^2/s
D_T	column diameter, m
g	gravitational acceleration, 9.81 m s^{-2}
h	height above the gas distributor, m
h_o	parameter determing initial bubble size, m
h^*	height at which bubbles attain their equilibrium size, m
H	fluidized bed height, m
$k_g a$	mass transfer coefficient, s^{-1}
$k_L a$	volumetric mass transfer coefficient, s^{-1}
k_m	pseudo first order reaction rate constant, m^3/kg cat.s
M	interphase momentum exchange term, N/m^3
NRU	number of reaction units, dimensionless
NTU	number of transfer units, dimensionless
p	system pressure, Pa
r	radial coordinate, m
t	time, s
u	velocity vector, m/s
U	superficial gas velocity, m s^{-1}
u_{df}	dense phase velocity, m s^{-1}
V_b	bubble rise velocity, m s^{-1}
$V_b(r)$	radial distribution of bubble velocity, m s^{-1}
$V_L(r)$	radial distribution of liquid velocity, m s^{-1}
V_R	reactor volume, m^3
V_{b0}	bubble rise velocity at low superficial gas velocities, m s^{-1}
$V_L(0)$	centre-line liquid velocity, m s^{-1}
W_{cat}	catalyst mass, kg
Ze_s	Zenz size number, dimensionless
Ze_v	Zenz velocity number, dimensionless

Greek:

α	constant in the bubble size correlation of Durton, dimensionless
ε	total gas hold-up (voidage), dimensionless
ε_b	fractional bubble hold up, dimensionless
ν	fraction of total gas flow in fluid bed which flows u[the reactor in the form of bubbles, dimensionless
ρ	density of phase, kg m-3
ρ_b	bulk density, kg/m^3
ρ_p	particle density, kg/m^3
ρ_f	fluid density kg/m^3
μ	viscosity of fluid phase, Pa s
σ	surface tension of liquid phase, N m-1
ϕ	Werther rise velocity constant, dimensionless
χ	fractional conversion of reactant, dimensionless
Δ	$[3\mu_f^2/4g\rho_f(\rho_p-\rho_f)]^{1/3}$, m
ω	$[4g\mu_f(\rho_p-\rho_f)/3\rho_f^2]^{1/3}$, m/s where μ_f = fluid viscosity, k/m s

Subscripts:

b	referring to bubbles	
G	referring to gas	
L	referring to liquid	
T	tower or column	
k,l	referring to phase k and l respectively	
U_{mf}	Minimum fluidization velocity	m/s
U_{fs}	Full support velocity	m/s
U_{cf}	Complete fluidization velocity	m/s
P_{fb}	Freeboard pressure	kPa
P_c	Pressure above solids level in dipleg	kPa
ΔP_{solids}	Pressure recovery in dipleg	kPa
ρ_{solids}	Fluidized/aerated density of solids in dipleg	kg/m^3
g	Gravitational constant	m/s^2
ε	Overall bed voidage	dimensionless
α	Flow coefficient	dimensionless

REFERENCES

[1] W.-D. Deckwer, F-T process alternatives hold promise, Oil Gas J. (1980) 198-213.

[2] M.J. Baird, R.R. Schehl, and W.P. Haynes, F-T processes investigated at the Pittsburgh Energy Technology Center since 1944, Ind. Engng. Chem. Prod. Res. Dev., 19 (1980) 175-191.

[3] L. Caldwell, in CENG 330, CSIR (South Africa) (1980).

[4] L. Caldwell, D.S. van Vuuren, On the formation and composition of the liquid phase in Fischer-Tropsch reactors, Chem. Eng. Sci., 41 (1986) 89-96.

[5] R.B. Anderson, in Catalysis, P.H. Emmett (ed.) Reinhold (1956) New York.

[6] M.E. Dry, in Catalysis Science and Technology, J.R. Anderson and M. Boudart (eds.), Springer Verlag, 1, 1981, p.159.

186

[7] M.E. Dry, Catalysis Today, 6, 3 (1990) 183.
[8] M.E. Dry, Applied Catalysis A: Gen., 138 (1996) 319.
[9] H. Kölbel, P. Ackermann and F. Engelhardt, Erdöl u. Kohle, 9 (1956) 153, 225, 303.
[10] B. Jager and R. Espinoza, Catalysis Today, 23 (1995) 17.
[11] R.L. Espinoza, A.P. Steynberg, B. Jager and A.C. Vosloo, Applied Catalysis: Gen., 186 (1999) 13.
[12] B.M. Everett, B. Eisenberg, R.F. Baumann, Conference on Natural Gas, Doha, Qatar, March (1995).
[13] B. Jager, Paper at Eurogas 96, Trondheim, Norway, June (1996).
[14] J.J.C. Geerlings, J.H. Wilson, G.J. Kramer, H. P. Kuipers, A. Hoek and M.H. Huisman, Applied Catalysis A: Gen. 186 (1999) 27.
[15] S.T. Sie and R. Krishna, Applied Catalysis: Gen., 186 (1999) 55.
[16] T. Shingles and A.F. MacDonald in Circulating Fluidized Bed Technology, P. Basu and J.F. Large (eds.) Pergamon press, N.Y., 1988, p. 43.
[17] A.P. Steynberg, R.L. Espinoza, B. Jager and A.C. Vosloo, Applied Catalysis A: Gen., 186 (1999) 41.
[18] P.C. Kieth, Gasoline from natural gas, Oil and Gas Journal, May 18, 45, (1946) 102.
[19] B. Jager, M.E. Dry, T. Shingles and A.P. Steynberg, Catalysis Letters, 7 (1990) 293.
[20] B. H. Davis, Overview of reactors for liquid phase Fischer-Tropsch synthesis, Catalysis Today (2002) 249-300
[21] F. Fischer and K. Peters, Brenst. Chem. 12 (1931) 286-293.
[22] F. Fischer and H. Küster, Brenstoff-Chemie, 14 (1933). 3-8.
[23] F. Fischer and H. Pichler, Brennstoff-Chemie, 20 (1939). 247.
[24] W.F. Faragher and W.A. Horne, Interrogation of Dr. Pier and Staff at I. G. Farbenindustrie A. G., Ludwigshafen and Oppau, Supplement II, U. S. Bureau of Mines I.C. 7376, August (1946).
[25] I. G. Farbenindustrie, Hydrocarbons, British Patent 449,274, June 24, 1936; British Patent 468,434, June 29 (1937).
[26] I. G. Farbenindustrie, Converting CO and H into Hydrocarbons, British Patent 516,352, Jan 1, 1940; British Patent 516,403, Jan 1 (1940).
[27] F. Duftschmid, E. Lindth and F. Winlder, Production of valuable hydrocarbons and their derivatives containing oxygen, US 2,159,077, May 23 (1939).
[28] W.F. Faragher and W.A. Horne, Improvements in the manufacture and production of hydrocarbons and their derivatives from CO and H_2, British Appl. 29636, Oct. 29 (1937).
[29] H. Tramm and W. Wischermann, Hydrogenation of CO in a liquid medium over suspended catalysts, German Patent 744,185, Jan. 26 (1944).
[30] H.V. Atwell, A.R. Powell and H.H. Storch, US Government Technical Oil Mission, Fischer-Tropsch Report One, TOM Report July 5, 1945, pp. 66, Office of publication board report 2051, 45 (1945).
[31] W. Michael, Brennstoff-Chemie, 37 (1956) 171.
[32] R. Holroyd, Report on Investigations by Fuels and Lubricants teams at the I.G. Farbenindustrie A.G. Works, Ludwigshafen and Oppau, Bureau of Mines IC 7375 (1946) 42.
[33] F. Duftschmid, E. Linckh and F, Winkler, Synthesis of hydrocarbons, U.S. Patent 2,287,092, June 23 (1942).
[34] F. Duftschmid, E. Linckh and F, Winkler, Synthesis of hydrocarbons, U.S. Patent 2,318,602, May 11 (1943).

[35] F. Duftschmid, E. Linckh and F, Winkler, Preparation of products containing valuable hydrocarbons or their derivatives, U.S. Patent 2,207,581, July 9 (1940).

[36] H.E. Benson, J. H. Field, D. Bienstock, R.R. Nagel, L.W. Brunn, C.O. Hawk, H.H. Crowell and H.H. Storch, Development of the Fischer-Tropsch Oil-Recycle Process, Bureau of Mines Bulletin 568 (1957).

[37] C.C. Hall and S.R. Craxford, Additional information concerning the Fischer-Tropsch process and its products, B.I.O.S. Final Report No. 1722, Item No. 22, British Intelligence Objectives Sub-committee.

[38] Great Britain Patent 464,308, April 15 (1937).

[39] F. Duftschmid, E. Linckh and F, Winkler, Production of valuable hydrocarbons and their derivatives containing oxygen, U.S. Patent 2,159,077, May 23 (1939).

[40] Ruhrchemie, Office of Publication Board Report No. 412, May 15 (1945).

[41] E. Reichl, U. S. Technical Mission in Europe; Technical Report No. 248-45, The Synthesis of Hydrocarbons and Chemicals from CO and H_2, September (1945).

[42] H. Kölbel, P. Ackermann and F. Engelhardt, Nue Entwicklungen zur Kohlenwasserstoff-Synthese, Erdöl und Kohle 9 (1956)153 225-303.

[43] H. Kölbel and P. Ackermann, Hydrogenation of carbon monoxide in the liquid phase, Proc. 3rd World Petroleum Congress, TheHague, Sect.IV, (1951) 2-12.

[44] H. Kölbel and P. Ackermann, Grosstechnische versuche zur Fischer-Tropsch synthese in flüssiger medien, Chemie-Ing-Technik 28 (1956) 381-388.

[45] H. Kölbel and R. Langheim, Carbon monoxide hydrogenation synthesis reactors, U.S. Patent 2,852,350, Sept. 16 (1958).

[46] H. Kölbel, Die Fischer-Tropsch-Synthese, Chemische Technologie, B and 3: Organische Technologie I, K. Winnacker and L. Kuchler, (eds.) Carl Hanser Verlag, Munchen, 1959, p. 495.

[47] H. Kölbel and P. Ackermann, Chemie.-Ing.-Techn., 28 (1956) 381.

[48] H. Kölbel and E. Ruschenburg, Brennstoff-Chem., 35 (1954) 161.

[49] H. Kölbel and P. Ackermann, Apparatus for carrying out gaseous catalytic reactions in liquid phase, U.S. Patent 2,853,369, Sept. 23 (1958).

[50] H. Kölbel and P. Ackermann, Apparatus for carbon monoxide hydrogenation, U.S. Patent 2,868,627, Jan. 13 (1959).

[51] H. Kölbel and M. Ralek, Catal. Rev.-Sci. Eng., 21 (1980) 225.

[52] H. Kölbel and P. Ackermann, U.S. Patent 2,671,103, March 2 (1954).

[53] International Hydrocarbon Synthesis Co., H_2-CO Synthesis, French Patent 860,360, Jan. 13 (1941).

[54] G.C. Hall, D. Gall and S.L. Smith, J. Inst. Petrol., 38 (1952) 845.

[55] H. Dreyfus, Improvements in or relating to the manufacture of hydrocarbons and other products from carbon monoxide and hydrogen, British 505,121, November 6 (1937).

[56] H. Dreyfus (assigned to Celanese Corp. of America), Production of organic compounds, U.S. Patent 2,361,997, November 7 (1944).

[57] British Patent 564,730, October 11 (1944).

[58] H.H. Storch, N. Golumbic and R.B. Anderson, The Fischer-Tropsch and related synthesis, John Wiley & Sons, New York, 1951.

[59] Standard Oil Development Co., Catalytic Apparatus and Process, British Patent 496,159, Jan 3 (1938).

[60] G. Weber, Oil & Gas J., March 24, (1949) 248.

[61] F.J. Moore, Methods of effecting catalytic reactions, U.S. Patent 2,440,109, April 20

(1948).

[62] M.M. Stewart, R. C. Garrett and E.E. Sensel, Method of effecting catalytic reaction between carbon monoxide and hydrogen, U.S. Patent 2,433,072, December 23 (1947).

[63] H.V. Atwell, Method for synthesizing hydrocarbons and the like, U.S. Patent No., 2,433,255, December 23 (1947).

[64] H.V. Atwell, U.S. Patent 2,438,029, March 16 (1948).

[65] P.W. Cornell and E. Cotton, "Catalytic reaction," U.S. Patent 2,585,441, February 12 (1952).

[66] W-D. Deckwer, Reaktionstechnik in Blasensaeulen, Otto Salle Verlag GmbH & Co, Frankfurt am Main Verlag Sauerlaender AG, Aarau, Switzerland (1985); W.-D. Deckwer, Bubble Column Reactors, (translated by Valeri Cottrell) R.W. Field (ed.), John Wiley and Sons, New York, 1992.

[67] H. Beuther, T.P. Kobylinski, C.E. Kibby and R.B. Pannell, Conversion of synthesis gas to diesel fuel in controlled particle size fluid system, South African Patent, ZA 855317, July 15 (1985).

[68] L.-S. Fan, Gas-Liquid-Solid Fluidization Engineering, Butterworth, U.S.A. (1989).

[69] A.P. Vogel, A.P. Steynberg and P.J. van Berge, Process for producing liquid and, optionally gaseous products from gaseous reactants, U.S. Patent, 6,462,098, October 8 (2002).

[70] C. Maretto, V. Piccolo, J.-C. Viguie and G. Ferschneider, Fischer-Tropsch Process, U.S. Patent, 6,348,510, February 19 (2002).

[71] P. Chaumette, P. Boucot and F. Morel, Process for converting synthesis gas into hydrocarbons, U.S. Patent, 5,776,988, July 7 (1998).

[72] W.C. Schlinger and W.L. Slater, Conversion of hydrogen and carbon monoxide into C1 – C4 range hydrocarbons, U.S. Patent, 4,256,654, March 17 (1981).

[73] E. Herbholzheimer and E. Iglesia, Method of operating a slurry bubble column, U.S. Patent, 5,348,982, September 20 (1994).

[74] W-D. Deckwer, Y. Serpemen, M. Ralek and B. Schmidt, Chem Eng. Sci., 36 (1981) 765.

[75] Air Products, Catalyst and reactor development for a liquid phase Fischer-Tropsch process, Final report for task 4 prepared for the US DOE under contract no. DE-AC22-8CPC30021 (1989).

[76] W.C. Behrmann, C.H. Mauldin and L.E. Pedrick, Hydrocarbon synthesis reactor employing vertical downcomer with gas disengaging means, U.S. Patent, RE37,229, June 12 (2001).

[77] M. Chang, Enhanced gas separation for bubble column draft tubes, U.S. Patent, 5,332,552, July 26 (1994).

[78] A.P. Steynberg, H.G. Nel and R.W. Silverman, Process for producing liquid and optionally, gaseous products from gaseous reactants, U.S. Patent, 6,201,031, May 13 (2001).

[79] D. Casanave, P. Galtier and J-C. Viltard, Process and Apparatus for operation of a slurry bubble column with application to the Fischer-Tropsch synthesis, U.S. Patent, 5,961,933, Oct. 5 (1999).

[80] W.C. Behrmann, C.H. Mauldin and L.E. Pedrick, Hydrocarbon synthesis reactor employing vertical downcomer with gas disengaging means, U.S. Patent, 5,382,748, Jan. 17 (1995).

[81] W.C. Behrmann and C.J. Mart, Gas and solids reducing slurry downcomer, U.S. Patent, 5,866,621, Feb. 2 (1999).

[82] S.C. Leviness, Multizone downcomer for slurry hydrocarbon syntheses process, U.S. Patent, 5,962,537, Oct. 5 (1999).

[83] J-M. Schweitzer, P. Galtier, F. Hugues and C. Maretto, Method for producing hydrocarbons from syngas in three-phase reactor, U.S. Patent Application, 20030109590, June 12 (2003).

[84] H.V. Atwell, Method of effecting catalytic conversions, U.S. Patent, 2,438,029, March 16 (1948).

[85] H. Lethäuser, W.L. Linder and E. Sattler, Apparatus for the production of hydrocarbons, U.S. Patent, 2,775,512, Dec. 25 (1956).

[86] S.C. Saxena, Bubble column reactors and Fischer-Tropsch synthesis, Catal. Rev.-Sci. Eng. 37 (1995) 227.

[87] A. Forret, J-M. Schweitzer, T. Gauthier, R. Krishna and D. Schweich, Liquid dispersion in large diameter bubble columns, with and without internals, Canadian J. Chem Eng, 81 (2003) 360-366

[88] D.R. Cova, Ind. Eng. Chem. Proc. Des. Dev., 5 (1966) 20.

[89] T. Suganuma and T. Yamanishi, Kagaku Kogaku, 30 (1966) 1136.

[90] S.C. Saxena, P.R. Thimmapuram, Rev. Chem. Eng., 8 (3, 4) (1992) 259.

[91] V.W. Heilmann and H. Hoffmann, Proc. 4th Euro. Symp.Chem. React. Engng., Pergamon Press, London (1971).

[92] G.J. Stiegel and Y.T. Shah, Can. J. Chem. Engng., 55 (1977) 3.

[93] E. Blass and W. Cornelius, Int. J. Multiphase Flow 3 (1977) 459.

[94] T. Sekizawa and H. Kubota, J. Chem. Eng. Japan, 7 (1974) 441.

[95] M. Popovic and W-D. Deckwer, Proc. Int. Congr., Contribution of computers to the development of chemical engineering and industrial chemistry, Paris, März 1978, Preprints vol. C, C-26-105, Soc. Chim. Ind. (1978).

[96] A. Chianese, M. C. Annesini, R. De Santis and L. Marelli, Chem. Eng. J., 22 (1981) 151.

[97] J.M. Fox III, Fischer-Tropsch reactor selection, Catalysis Letters 7 (1990) 281-292.

[98] D.J. Vermeer and R. Krishna, Hydrodynamics and mass transfer in bubble columns operating in the churn- turbulent regime, Ind.Eng.Chem. Process Design & Dev., 20 (1981) 475-482.

[99] R.D. Toomey. and H.F. Johnstone, Gaseous fluidization of solid particles, Chem. Engng. Prog., 48 (1952) 220-226.

[100] K. Rietma, in: Proc. Int. Symp. on Fluidization, Eindhoven (1967) 154.

[101] J. Ellenberger and R. Krishna, Chem Eng. Sci., 49 (1994) 5391.

[102] D. J. Vermeer and R. Krishna, Hydrodynamics and mass transfer in bubble columns operating in the churn- turbulent regime, Ind.Eng.Chem. Process Design & Dev., 20 (1981) 475-482.

[103] J. Werther, in: D. Kunnii and R. Toei (eds.), Fluidization IV, Engineering Foundation, New York (1983) 93.

[104] R.C. Darton, R.D. LaNauza, J.F. Davidson and D. Harrison, Trans. Inst. Chem. Engrs., 55 (1977) 274.

[105] R. Krishna, J. Ellenberger and S. T. Sie, Reactor development for conversion of natural gas to liquid fuels: A scale up strategy relying on hydrodynamic analogies, Chem. Eng. Sci., 51, 2041-2050 (1996).

[106] P.M. Wilkinson, Physical aspects of the scale-up of high pressure bubble columns, PhD thesis, University of Groningen, the Netherlands (1991).

[107] P.M. Wilkinson, L.L. Dierendonck, Pressure and gas density effects on bubble break-up and gas hold-up in bubble columns, Chem. Eng. Sci. 45 (1990) 2309-2315.

[108] M. Letzel, Hydrodynamics and mass transfer in bubble columns at elevated pressures, PhD thesis, Technical university of Delft, the Netherlands (1998).

[109] R. Krishna, A scale-up strategy for a commercial scale bubble column slurry reactor for Fischer-Tropsch synthesis, Oil Gas Sci. Technol., 55 (2000) 359-393.

[110] R. Krishna, M.I. Urseanu and A.J. Dreher, Gas hold-up in bubble columns: influence of alcohol addition versus operation at elevated pressures, Chem. Eng. Process., 39 (2000) 371-378.

[111] X. Luo, D.J. Lee, R. Lau, G. Yang and L. Fan, Maximum stable bubble size and gas hold-up in high-pressure slurry bubble columns, AIChEJ, Fluid Mechanics and Transport Phenomena, 45 (1999) 4.

[112] L.S. Fan, G.Q. Yang, D.J. Lee, K. Tsuchiya and X. Luo, Some aspects of high pressure phenomena of bubbles in liquids and liquid-solid dispersions, Chem. Eng. Sci., 54 (1999) 4681-4709.

[113] J.W.A. de Swart, Scale-up of a Fischer-Tropsch slurry reactor, PhD thesis, University of Amsterdam, the Netherlands (1996).

[114] G.G. Bartolomei and M.S. Alkhutov, Determination of the true vapour content when there is bubbling in the stabilization section, Thermal Eng., 14 (1967) 112-114.

[115] H.M. Letzel, J.C. Schouten, R. Krishna, et al., Characterization of regimes and regime transitions in bubble columns by chaos analysis of pressure signals, Chem. Eng. Sci. 52, 24 (1997) 4447-4459.

[116] R. Botton, D. Cosserat and J.C. Charpentier, Influence of column diameter and high gas throughputs on the operation of a bubble column, Chem. Eng. J., 16 (1978) 107-115.

[117] J.H. Hills, The operation of a bubble column at high throughputs 1. Gas holdup measurements, , Chem. Eng J., 12 (1976) 89-99.

[118] D. Kunii and O. Levenspiel, Fluidization Engineering, 2nd edition, Butterworth- Heinemann series in Chemical Engineering, 1991.

[119] R. Krishna, Scaling up bubble column slurry reactors, Industrial Fluidization South Africa (IFSA 2002) Proceedings of a conference on fluidization held in Johannesburg, South Africa, 20-21 Nov. 2002, SA Inst. of Mining and Metallurgy Symposium Series S31, A. Luckos and P. den Hoed, (eds.) 33.

[120] J.M. van Baten and R. Krishna, Eulerian simulation strategy for scaling up a bubble column slurry reactor for Fischer-Tropsch synthesis, Ind. Eng. Chem. Research, in press (2004).

[121] C. Maretto and R., Krishna, Modelling of a bubble column slurry reactor for Fischer-Tropsch synthesis, Catal. Today, 52 (1999) 279-289.

[122] R.Krishna, J. W. A. de Swart, J. Ellenberger, G. B Martina,. and C. Maretto, Gas holdup in slurry bubble columns: Effect of column diameter and slurry concentrations, A.I.Ch.E.J., 43 (1997) 311-316.

[123] R. Krishna, M.I. Urseanu, J.W.A. de Swart and J. Ellenberger, Gas hold-up in bubble columns: Operation with concentrated slurries versus high viscosity liquid, Can. J. Chem. Eng., 78 (2000) 442-448.

[124] R. Krishna, J.M. van Baten, M.I. Ursenu and J. Ellenberger, A scale up strategy for bubble column slurry reactors, Catal. Today, 66 (2001) 199-207.

[125] R. Krishna, J.M. van Baten, M.I. Urseanu and J. Ellenberger, Design and scale up of a bubble column slurry reactor for Fischer-Tropsch synthesis, Chem. Eng. Sci., 56 (2001) 537-545.

[126] M.I. Urseanu, R.P.M. Guit, A. Stankiewicz, G. van Kranenburg and J. Lommen, Influence of operating pressure on the gas hold-up in bubble columns for high viscous media, Chem. Eng. Sci., 58 (2003) 697.

[127] R. Krishna, J.M. van Baten, Scaling up bubble column reactors with highly viscous liquid phase, Chem. Eng. Technol., 25 (2002) 1015.

[128] A. Forret, J.-M. Schweitzer, T. Gautier, R. Krishna and D. Schweich, Influence of scale on the hydrodynamics of bubble column reactors: an experimental study in columns of 0.1, 0.4 and 1 m diameters, Chem. Eng. Sci., 58 (2003) 719.

[129] P.M. Wilkinson, A.P. Spek and L.L. van Dierendonck, Design parameters estimation for scale-up of high pressure bubble columns, A.I.Ch.E.J., 38 (1992) 544.

[130] J.M. van Baten and R. Krishna, Eulerian simulations for the determination of axial dispersion of liquid and gas phases in bubble columns operating in the churn-turbulent regime, Chem. Eng. Sci., 56 (2001) 503.

[131] R. Krishna, M.I. Urseanu, J.M. van Baten and J. Ellenberger, Influence of scale on the hydrodynamics of bubble columns operating in the churn-turbulent regime: experiments vs. Eulerian simulations, Chem. Eng. Sci., 54 (1999) 4903.

[132] H.P. Riquarts, Strömungsprofile, Impulsaustausch und Durchmischung der flüssigen Phase in Bläsensaulen, Chem. Ing. Techn., 53 (1981) 60.

[133] P. Zehner, Momentum, mass and heat transfer in bubble columns. Part 1. Flow model of the bubble column and liquid velocities, Int. Chem. Eng., 26 (1986) 22.

[134] K.Koide, S. Morooka, K. Ueyama, A. Matsuura, F. Yamashita, S. Iwamoto, Y. Kato, H. Inoue, M. Shigeta, S. Suzuki, T. Akehata, Behaviour of bubbles in large scale bubble column, J. Chem. Eng. Jpn., 21 (1979) 98.

[135] E. Kojima, H. Unno, Y. Sato, T. Chida, H. Imai, K. Endo, I. Inoue, J. Kobayashi, H. Kaji, H. Nakanishi, K. Yamamoto, Liquid phase velocity in a 5.5 m diameter bubble column, J. Chem. Eng., 13, Jpn. (1980) 16.

[136] C. Maretto, R. Krishna, Design and optimisation of a multi-stage bubble column slurry reactor for Fischer-Tropsch synthesis, Catal. Today, 66 (2001) 241.

[137] R. Krishna, M.I. Urseanu, J.M. van Baten, J. Ellenberger, Liquid phase dispersion in bubble columns operating in the churn-turbulent flow regime, Chem. Eng. J., 78 (2000) 43.

[138] M.H.I. Baird and R.G. Rice, Axial dispersion in large unbaffled columns, Chem. Eng. Jl., 9 (1975) 171.

[139] P.M. Wilkinson, H. Haringa, , F.P.A. Stokman L.L. Van Dierendonck, Liquid-Mixing in a Bubble Column under Pressure, Chem. Eng. Sci., 48 (1993) 1785.

[140] G.Q. Yang and L.S. Fan, Axial liquid mixing in high-pressure bubble columns, A.I.Ch.E.J., 49 (2003) 1995.

[141] Y. Pan, M.P. Dudukovic and M. Chang, Dynamic simulation of bubbly flow in bubble columns, Chem. Eng. Sci., 54 (1999) 2481-2489.

[142] J. Sanyal, S. Vasquez, S. Roy and M.P. Dudukovic, Numerical simulation of gas-liquid dynamics in cylindrical bubble column reactors, Chem. Eng. Sci., 54 (1999) 5071-5083.

[143] A.Sokolichin, and G. Eigenberger, Applicability of the standard k-epsilon turbulence model to the dynamic simulation of bubble columns: Part I. Detailed

192

numerical simulations, Chem. Eng. Sci., 54 (1999) 2273-2284.

[144] R. Clift, J.R. Grace, M.E. Weber, Bubbles, drops and particles, Academic Press, San Diego, CA, 1978.

[145] C.O. Vandu and R. Krishna, Gas holdup and volumetric mass transfer coefficient in a slurry bubble column, Chem. Eng. Technol., 26 (2003) 779.

[146] C.M. Rhie and W.L. Chow, Numerical study of the turbulent flow past an airfoil with trailing edge separation, A.I.A.A.J., 21 (1983) 1525-1532.

[147] J. van Doormal and G.D. Raithby, Enhancement of the SIMPLE method for predicting incompressible flows, Numer. Heat Transfer, 7 (1984) 147-163.

[148] R. Krishna, J.M. van Baten and M.I. Urseanu, Three-phase Eulerian simulations of bubble column reactors operating in the churn-turbulent regime: a scale up strategy, Chem. Eng. Sci., 55 (2000) 3275.

[149] A. Kemoun, B.C. Ong, P. Gupta, M.H. Al-Dahhan, M.P. Dudukovic, Gas holdup in bubble columns at elevated pressure via computed tomography, Int. J. Multiph. Flow, 27 (2001) 929.

[150] J.W.A. De Swart, R.E. van Vliet, R. Krishna, Size, structure and dynamics of 'large' bubbles in a two- dimensional slurry bubble column, Chem. Eng. Sci., 51(1996) 4619.

[151] G.Q. Yang, X. Luo, R. Lau, L.S. Fan, Heat Transfer Characteristics in Slurry Bubble Columns at Elevated Pressures and Temperatures, Ind. Eng. Chem. Res., 39 (2000) 2568.

[152] H.M. Letzel, J.C. Schouten, R. Krishna and C.M. van den Bleek, Characterization of regimes and regime transitions in bubble columns by chaos analysis of pressure signals, Chem. Eng. Sci., 52 (1997) 4447-4459.

[153] H.M. Letzel, J.C. Schouten, C.M van den Bleek and R. Krishna, Influence of elevated pressure on the stability of bubbly flows, Chem. Eng. Sci., 52 (1997) 3733-3739.

[154] M.H. Letzel, J.C. Schouten, C.M. van den Bleek and R. Krishna, Effect of gas density on large-bubble holdup in bubble column reactors, A.I.Ch.E.J., 44 (1998) 2333-2336.

[155] H.M. Letzel, J.C. Schouten, R. Krishna, and C.M. van den Bleek, Gas holdup and mass transfer in bubble column reactors operated at elevated pressure, Chem. Eng. Sci., 54 (1999) 2237-2246.

[156] R. Krishna and J.M. van Baten, Eulerian simulations of bubble columns operating at elevated pressures in the churn turbulent flow regime, Chem. Eng. Sci., 56 (2001) 6249-6258.

[157] R. Krishna, J. Ellenberger and D.E. Hennephof, Analogous Description of the 'Hydrodynamics of Gas-Solid Fluidized-Beds and Bubble-Columns', Chem. Eng. J., 53 (1993) 89-101.

[158] H.C.J. Hoefsloot and R. Krishna, Influence of Gas-Density on the Stability of Homogeneous Flow in Bubble-Columns, Ind. Eng. Chem. Res., 32 (1993) 747-750.

[159] R. Krishna, J.W.A. de Swart, D.E. Hennephof, J. Ellenberger and H.C.J. Hoefsloot, Influence of Increased Gas-Density on Hydrodynamics of Bubble- Column Reactors, A.I.Ch.E.J., 40 (1994) 112-119.

[160] I.G. Reilly, D.S. Scott, T.J.W. De Bruijn and D. MacIntyre, The Role of Gas-Phase Momentum in Determining Gas Holdup and Hydrodynamic Flow Regimes in Bubble-Column Operations, Can. J. Chem. Eng., 72 (1994) 3-12.

[161] C.O. Vandu and R. Krishna, Influence of scale on the volumetric mass transfer coefficients in bubble columns, Chemical Engineering & Processing, 43 (2004) 575-579.

[162] O. Levenspiel, Modeling in chemical engineering, Chem. Eng. Sci., 57 (2002) 4691-

4696.

[163] K. Denbigh, The thermodynamics of the steady state. Methuen's monographs on chemical subjects (1951) London.

[164] P.V. Danckwerts, Chem. Eng. Sci. 2 (1953) 1.

[165] G. van der Laan, Ph.D. Thesis, University of Groningen, The Netherlands (1999).

[166] N. Rados, M.H. Al-Dahhan and M.P. Dudukovic, Modeling of the Fischer-Tropsch synthesis in slurry bubble column reactors, Catalysis Today 79-80 (2003) 211-218.

[167] J.W.A. de Swart and R. Krishna, Simulation of the transient and steady state behavior of a bubble column reactor for Fischer-Tropsch synthesis, Chemical Engineering & Processing 41 (2002) 35-47.

[168] O. Levenspiel, Chemical Reaction Engineering, Wiley, New York, 1972.

[169] M.E. Dry, Advances in Fischer-Tropsch chemistry, Ind Eng. Chem.. Prod. Res. Dev. 15 (1976) 282.

[170] G.A. Huff Jr. and C.N. Satterfield, Intrinsic kinetics of the Fischer-Tropsch synthesis on a reduced fused-magnetite catalyst, Ind Eng. Chem.. Prod. Res. Dev., 23 (1984) 696.

[171] P.L. Mills, J.R. Turner, P.A. Ramachandran and M.P. Dudukovic, The Fischer-Tropsch synthesis in slurry bubble column reactors: analysis of reactor performance using the axial dispersion model, in A. Schumpe, K.D.P. Nigam (eds.) Three Phase Sparged Reactors, Chapter 5, Gordon & Breach, New York, 1996.

[172] R. Krishna and J.M. van Baten, Mass transfer in bubble columns, Catalysis Today 79-80 (2003) 67-75.

[173] P.J. van Berge, Stud. Surf. Sci. Catal., 107, Natural Gas Conversion IV (1997) 207.

[174] R.L. Espinoza, J.L. Visage, P.J. van Berge and F.H. Bolder, RSA Patent ZA 962759 (1995).

[175] A.P. Steynberg, A.C. Vosloo and P. van Berge, Handling of a catalyst, U.S. Patent 6,512,017, Jan. 28 (2003).

[176] J.J.H.M. Font-Friede, D. Newton and C. Sharp, Fischer-Tropsch process, PCT patent application, WO 02/096833, 5 Dec. (2002).

[177] A.P. Steynberg and B.B. Breman, U.S. Provisional Patent No. 60/471, 323 (2003).

[178] C. Maretto and R. Krishna, Design and optimization of a multi-stage bubble column slurry reactor for Fischer-Tropsch synthesis, Catalysis Today 66 (2001) 241-248.

[179] C. Maretto, R. Krishna and V. Piccolo, Hydrodynamics, design and scale-up of a multi-stage bubble column slurry reactor for Fischer-Tropsch synthesis, presented at 6th World Congress of Chemical Engineers, Melbourne (2001).

[180] P.C. Kieth and N.J. Peapack, Method of synthesizing hydrocarbons, U.S. Patent No. 2,276,274, March 17 (1942).

[181] S. Sookai, Fluidization fundamentals and the Sasol Advanced Synthol (SAS) reactors, Industrial Fluidization South Africa (IFSA 2002) Proceedings of a conference on fluidization held in Johannesburg, South Africa, SA Inst. of Mining and Metallurgy Symposium Series S31, A. Luckos and P. den Hoed (eds.) Nov. 20-21 (2002) 63.

[182] D. Geldart, Powder Technology, 7 (1973) 185-195.

[183] D. Geldart, Gas Fluidization Technology, 39, Wiley, J. & Sons, 1986.

[184] R.J. Dry, M.R, Judd and T. Shingles, Powder Technology 34 (1983) 213-223.

[185] T.M. Knowlton, Paper 9b, 67th annual meeting of AIChemE, Washington DC, December 1 to 5 (1974).

[186] R.J. Dry, Gas Distribution in Fluidized Beds of Fine Powders, Masters Thesis,

University of Natal, South Africa (1982) 50.

[187] T.Shingles, J. F. Kirsten and P. L. Langenhoven, Chemical Technology, March 2002, 11-13.

[188] S. Sookai, P.L. Langenhoven and T. Shingles, 2001 Scale-up and commercial reactor fluidization related experience with Synthol, gas to liquid fuel, dense phase fluidized bed reactors, In M. Kuauk, J. Li and W. C. Yang, (eds.) Fluidization X: Proceedings of the10th United Engineering Foundation Conference, New York, 2001, pp. 621-628.

[189] M.E. Dry, in J.R. Anderson and M. Boudart, (eds.) Catalyis Science and Technology 1: Chapter 4, Springer, Berlin, 1981.

[190] A.P. Steynberg, Design and scale-up of catalytic fluidized bed reactors, Industrial Fluidization South Africa (IFSA 2002) Proceedings of a conference on fluidization held in Johannesburg, South Africa, SA Inst. of Mining and Metallurgy Symposium Series S31, A. Luckos and P. den Hoed (eds.) Nov. 20-21 (2002) 81.

[191] J.J. Van Deemter, Mixing and contacting in gas-solid fluidized beds, Chem. Engng. Sci., 13 (1961) 143-154.

[192] R.W.Silverman, A.H. Thompson, A.P Steynberg, Y. Yukawa and T. Shingles, Development of a dense phase fluidized bed Fischer-Tropsch Reactor, K. Ostergaard and A. Sorensen (eds.) Fluidization V, Engineering Foundation, New York (1986) 441- 448.

[193] A.P. Steynberg, T. Shingles, M.E. Dry, B. Jager and Y. Yukawa, Sasol commercial scale experience with Synthol FFB and CFB catalytic Fischer-Tropsch reactors, Proc. 3rd International Conference on Circulating Fluidized Beds, Nagoya, Japan, P. Bassu, M. Horio and H. Hasitani (eds.) Pergamon Press, New York,1991, p. 527.

[194] A.P. Steynberg, Prediction of entrainment for a commercial scale high pressure fluidised bed, Preprints Fluidization VIII, International Symposium of the Engineering Foundation, Tours, France (1995) 785-790.

[195] D. Newton and D. Johns, Novel extensions to the Kunii and Levenspiel fluidized bed reactor model: comparison of model predictions with experimental data, Preprints Fluidization VIII, International Symposium of the Engineering Foundation, Tours, France (1995) 467-474.

[196] J. Werther, The influence of the distributor on gas-solid fluidised beds, Chem. Ing. Tech., 49 (1977) 777.

[197] R. Krishna, J. Ellenberger and S. T. Sie, Reactor development for conversion of natural gas to liquid fuels: a scale-up strategy relying on hydrodynamic analogies, Chem. Eng. Sci., 51 (1996) 2041-2050.

[198] J.F. Davidson, D. Harrison, R.C. Darton and R.D. La Nauze, Chemical Reactor Theory, A Review, L. Lapidus and N.R. Amundson (eds.) Prentice-Hall, Englewood Cliffs, N.J., 1977, p. 583.

[199] R. Krishna, Simulation of an industrial fluidised bed reactor using a bubble growth model, Chem. Eng. Res. Des., 66 (1988) 463.

[200] M. Pell, Understanding the design of fluid-bed distributors, Chemical Engineering, August (2002) 72.

[201] C.L. Briens, M.A. Bergougnou and T. Baron, Prediction of entrainment from gas-solid fluidised beds, Powder Technology, 54 (1988) 183.

[202] S.B.R. Karri, Powder & Bulk Solids Conference, Chicago (1989).

[203] G.F. Froment and K.M. Bischoff, Chemical Reactor Analysis and Design, 2nd edition, Wiley series in chemical engineering, John Wiley and Sons, New York, U.S.A., 1990.

[204] I. Iliuta, A. Ortiz-Arroyo, F. Larachi, B.P. Grandjean and G. Wild, Hydrodynamics and mass transfer in trickle-bed reactors: an overview, Chem. Eng. Sci., 54 (1999) 5329-5337.

[205] M.F.M. Post, A.C. van 't Hoog, J.K. Minderhoud and S.T. Sie, Diffusion limitations in Fischer-Tropsch Catalysts, AICHE Journal, July 1989, Vol. 35, No. 7.

[206] Y.N. Wang, Y.Y. Xu, Y.W. Li, Y.L. Zhao and B.J. Zhang, Heterogeneous modeling for fixed bed Fischer-Tropsch synthesis: Reactor model and its applications, Chem. Eng. Sci., 58 (2003) 867-875.

[207] K.P. de Jong, M.F.M. Post and A. Knoester, Deposition of iron from iron-carbonyls onto a working Co-based Fischer-Tropsch catalyst: The serendipitous discovery of a direct probe for diffusion limitations, Natural Gas Conversion V, Stud. in Surf. Sci.and Catal. (1998) 119.

[208] H.E. Atwood and C.O. Bennett, Kinetics of the Fischer-Tropsch reaction over Iron, Ind. & Eng. Chem. Proc. Des. Dev., 18 (1979) 163.

[209] G. Bub, M. Baerns, B. Bussemeier and C. Frohning, Prediction of the performance of catalytic fixed bed reactors for Fischer-Tropsch synthesis, Chem. Eng. Sci., 33 (1980) 348.

[210] A. Jess, R. Popp and K. Hedden, Fischer-Trospch synthesis with nitrogen-rich syngas: Fundamentals and reactor design aspects, Applied Catalysis A: Gen. 186 (1999) 321-342.

[211] H. Tramm, Technische und Wirtschaftliche Möglichkeiten der Kohlenoxyd-Oydrierung, Chem. Ing. Technik. 24 (1952) 237-332.

[212] D. Bode and S.T. Sie, Neth. Pat. Appl. 8500121, assigned to Shell Internationale Research Mij B.V., Jan. 18 (1985).

Studies in Surface Science and Catalysis 152
A. Steynberg and M. Dry (Editors)

Chapter 3

Chemical concepts used for engineering purposes

M. E. Dry

Catalysis Research Unit, Department of Chemical Engineering, University of Cape Town, Rondebosch, 7701, South Africa

1. STOICHIOMETRY

The term stoichiometry is commonly used to describe the way in which the components in a chemical reaction combine to form products. Thus in the case of the Fischer-Tropsch process the stoichiometry is primarily concerned with the ratio of consumption of hydrogen and carbon monoxide and in some cases also carbon dioxide. In this context, the H_2 to CO consumption ratio (or simply the consumption ratio); the H_2 to CO usage ratio (or simply the usage ratio) and the stoichiometric ratio (in the absence of CO_2 reaction) are synonyms. The simple terminology of the usage ratio is preferred in this text.

When CO_2 is a reactant, the stoichiometric ratio i.e. the ratio in which reactants are consumed, will involve CO_2. The H_2 to CO usage ratio may still be of interest but it is then no longer synonymous with the stoichiometric ratio.

The chemistry taking place in a Fischer-Tropsch reactor is complex but can be simplified into the following chemical reactions:

methane
$$CO + 3H_2 \rightarrow CH_4 + H_2O \tag{1}$$

heavier hydrocarbons
$$nCO + 2nH_2 \rightarrow (-CH_2-)_n + nH_2O \tag{2}$$

alcohols
$$nCO + 2nH_2 \rightarrow C_nH_{2n+2}O + (n-1)H_2O \tag{3}$$

water gas shift (WGS)
$$CO + H_2O \leftrightarrow CO_2 + H_2 \tag{4}$$

For the present discussion it is assumed that the desired products are the heavier hydrocarbons so that the predominating reaction is reaction (2). It must be kept in mind though that this reaction is a simplification of reality and that the ratio of consumption of carbon monoxide and hydrogen may vary significantly depending on the extent of the other reactions shown and the secondary reactions.

The water gas shift (WGS) reaction, reaction (4) above, may have a profound effect on the usage ratio. For the best cobalt catalysts, the extent of the WGS reaction is negligible and this reaction may then be treated as a one way reaction producing a small amount of carbon dioxide. In this situation carbon dioxide is typically treated as a carbon containing product. At the other extreme, the use of iron catalysts at the upper end of the operating temperature range results in the water gas shift reaction approaching equilibrium. In this situation, carbon dioxide is best treated as a reactant. The direction of the WGS reaction depends on the prevailing gas composition.

At one extreme with cobalt catalysts, the usage ratio is determined primarily by reaction (2) with a significant influence from reaction (1). The usage ratio is typically between 2.06 and 2.16 depending on the extent of methane formation; the olefin content in the longer chain hydrocarbons and the slight water gas shift activity.

At the other extreme, when the WGS reaction is in equilibrium, the combined usage ratio of hydrogen and carbon monoxide for the Fischer-Tropsch and water gas shift reactions together, is a moving target that depends on the feed gas composition. Although methanation (reaction (1)) tends to be more prevalent in these circumstances, the role of the WGS reaction is best explained by examining equation (4) together with the main reaction of interest i.e. equation (2). Dividing equation (2) by n and adding to equation (4) gives the most important equation that affects the usage ratio when there is a net formation of carbon dioxide:

$$2\,CO + H_2 \rightarrow \text{-}CH_2\text{-} + CO_2 \tag{5}$$

In this case the usage ratio is 0.5. As will be seen later, this is a good match with the synthesis gas composition produced by high temperature coal gasifiers. Consider now the reverse shift reaction i.e.

$$CO_2 + H_2 \rightarrow CO + H_2O \tag{6}$$

Adding this reaction to reaction (2) gives:

$$CO_2 + 3H_2 \rightarrow \text{-}CH_2\text{-} + 2H_2O \tag{7}$$

It is now necessary to consider carbon dioxide as a reactant and both equation (2) and equation (7) must be considered to determine a stoichiometric ratio. Thus when carbon dioxide consumption occurs and the reactants are in stoichiometric balance, two hydrogen molecules are consumed by each CO molecule and three hydrogen molecules are consumed by each CO_2 molecule when making 'CH$_2$' product. When the reactants are in stoichiometric balance, then the ratio $(H_2)/(2CO + 3CO_2)$ known as the Ribblett ratio, equals 1.

By examining various possible combinations of equations (2) and (7), it will be seen that the H_2 to CO usage ratio may vary from 2 with no CO_2 in the feed to infinity with no CO in the feed. However, the case with no CO_2 in the feed is likely to cause CO_2 to be a product rather than a reactant so that the usage ratio will then drop below 2.

Considering the idealized case of carbon as a feedstock which is gasified to produce syngas which subsequently undergoes Fischer-Tropsch synthesis, the overall equation may be represented as follows:

$$2C + \tfrac{1}{2} O_2 + H_2O \rightarrow \text{ - } CH_2\text{ - } + CO_2 \tag{8}$$

If more H_2O is used in the gasifier to increase the ratio of H_2 to CO in the syngas to make water in the Fischer-Tropsch reactor then there is a dual cost penalty of vaporizing water for the gasifier and condensing water after the Fischer-Tropsch reactor. The amount of CO_2 produced in the process of making the hydrocarbon product is not changed by adding water to both sides of equation (8). In fact the energy required to split the water molecule in the gasifier may have to be provided by 'burning' carbon to CO_2 so that excess CO_2 is produced over the stoichiometric requirements of equation (8). Neglecting energy considerations, the stoichiometric balance dictates that the best possible carbon efficiency when converting carbon to hydrocarbon products is 50%.

Considering the idealized case of methane as a feedstock, the overall reaction for the most efficient reformer design may be represented as follows:

$$CH_4 + \tfrac{1}{2}O_2 \rightarrow \text{ -}CH_2\text{- } + H_2O \tag{9}$$

This ideal partial oxidation process is never achieved due to the need to add some steam for practical reasons (See Chapter 4) and the need for energy to get the reactants to the operating temperature of the reformer. Furthermore, the WGS reaction is typically close to equilibrium at the exit of the methane reformer inevitably producing some carbon dioxide. If the Fischer-Tropsch catalyst does not have reverse WGS activity to consume this carbon dioxide then carbon will be wasted unless it is recycled to a reformer (or sent to a separate reformer). These concepts will be revisited in later chapters. A non-shifting Fischer-Tropsch catalyst can nevertheless be used to produce an efficient design for a plant using a methane feedstock since the benefits of using a shifting catalyst may be cancelled by other considerations.

A further point of interest is that the stoichiometric requirement for high temperature Fischer-Tropsch (HTFT) is almost identical to that used for the production of methanol from syngas. For methanol production the stoichiometric ratio (also called the stoichiometric number) is sometimes expressed in the form:

$$(H_2 - CO_2)/ (CO+CO_2) = 2$$

It is clear that this can be rearranged to Ribblett Ratio = 1 by dividing both sides by 2 and adding CO_2 to both the numerator and the denominator.

In future, the most efficient reformer designs are likely to use some of the waste heat from the oxygen fired reformer outlet gases to drive endothermic steam and/or CO_2 reforming reactions. This approach will be equally applicable for all processes converting methane to Fischer-Tropsch products or methanol. This is discussed in more detail in Chapters 4 and 5.

2. CONVERSION

Conversion performance relates to the consumption of reactants rather than the appearance of products. From the previous section on stoichiometry, it is clear that stoichiometric considerations may affect the relative amounts of the reactants that are consumed. It was also pointed out though that in all cases the rate of consumption of $CO+H_2$ is independent of the extent of the WGS reaction since CO and H_2 are on opposite sides of the equation:

$$CO + H_2 O \leftrightarrow CO_2 + H_2$$

Another useful concept is to express conversion in terms of the rate of consumption of $CO+CO_2$. This approach is valid even if CO_2 is produced rather than consumed. Conversion expressed in this way is also independent of the extent of the WGS reaction since CO and CO_2 are on opposite sides of the equation. This approach has the further advantage that the calculated amount of carbon consumed can be directly equated to the amount of carbon in the products. In order to calculate the amount of carbon in the products from the $CO+H_2$ conversion it is necessary to know the ratio of carbon to hydrogen and oxygen in the products. This is unfortunately different for various product selectivity scenarios.

For a non-shifting catalyst the small amount of CO_2 formed may be treated as a product. In this case the CO conversion can be directly equated to the amount of carbon in the 'products'.

In this book the term 'kinetics' is generally applied to the equations used to describe the rate of consumption of reactants and is thus related to the prediction of conversion performance.

3. SELECTIVITY

Having calculated the consumption of reactants, it is also necessary to describe the way in which the resulting products are distributed. The way in which the products are distributed is termed the selectivity performance.

A 'light' product distribution will generally have less long-chain hydrocarbons and more short-chain hydrocarbons than a 'heavy' product distribution.

It is convenient to express product selectivities as % carbon selectivities. For example the carbon selectivity of propylene will be:

$$\frac{\text{moles propylene produced} \times 3}{\text{total moles of carbon converted}} \times 100\%$$

The calculation of the total moles of propylene produced is then as follows:

$$CO + CO_2 \text{ in the feed gas x \% } \underline{CO + CO_2 \text{ conversion}} \text{ x } \frac{\text{propylene selectivity}}{3 \times 100}$$

and similarly for the other components.

It may be desirable to express selectivities as a mass % of the total products. Moles are easily transferred to mass using the component molecular mass. Carbon selectivities and mass selectivities produce similar numbers because the hydrocarbons are made up of 'CH_2' building blocks with a molecular mass of 14 so that the mass of the component is directly proportional to the carbon number of the component. The exceptions are the methane and low molecular mass alkane selectivities and especially the selectivities of the oxygenated compounds.

4. SYNTHESIS GAS COMPOSITION AND THE FT REACTIONS

The production of purified synthesis gas (syngas), which is suitable as a feed gas to the FT reactors, is a major cost factor. Whether coal or methane is used as raw material, syngas can account for up to about 70% of both the capital and operating cost, depending on the complexity of the overall downstream plant. Hence, it is very important that the conversion of the syngas to hydrocarbon products in the FT reactors is as efficient and as complete as possible i.e. that as much as possible of the reactants (CO, H_2, and CO_2) are consumed to provide useful products. The selectivity of the FT reaction should be controlled so as to minimise the production of undesired products such as methane.

In order to achieve the above objectives the syngas composition should match the usage ratio of the FT reactions. The usage ratio will depend on the overall product selectivity, which in turn depends on several other factors (Chapter 5). The effect of selectivity on the H_2 to CO usage ratio is illustrated in Table 1 for a few typical FT products.

Table 1
Usage ratio of FT reactions

FT product	Reactions	H_2 to CO usage ratio
CH_4	$CO + 3H_2 \rightarrow CH_4 + H_2O$	3
C_2H_6	$2CO + 5H_2 \rightarrow C_2H_6 + 2H_2O$	2.5
Alkanes	$n\ CO + (2n+1)H_2 \rightarrow C_n H_{(2n+2)} + n\ H_2O$	$(2n+1)/n$
Alkenes	$n\ CO + 2n\ H_2 \rightarrow C_nH_{2n} + n\ H_2O$	2
Alcohols	$n\ CO + 2n\ H_2 \rightarrow C_n H_{(2n+1)} OH + (n-1)\ H_2O$	2

For all alkenes and alcohols the usage ratio is 2.0, irrespective of chain length, but for alkanes the usage ratio decreases with increasing chain length. The extent to which the water gas shift reaction (WGS) occurs over the FT catalyst used is also important. For a cobalt catalyst, which has little or no WGS activity, the overall H_2 to CO usage ratio under typical FT syntheses conditions is between 2.05 and 2.15.

When precipitated iron based catalysts are used at low temperatures, the H_2 to CO usage ratio is lowered due to the simultaneous water gas shift (WGS) reaction. For an iron based catalyst operating in a fixed bed reactor at about 225°C the H_2 to CO usage ratio is approximately 1.65. For iron catalysts operating at higher temperatures, e.g. at 340°C, in fluidised bed reactors the WGS reaction is rapid and goes to equilibrium. This means that in practice CO_2 can also be converted to FT products via the reverse WGS reaction:

$$CO_2 + H_2 \rightarrow CO + H_2O$$

followed by the normal FT reaction. For CO_2 conversion the H_2 usage ratio is 3 for alkene formation, 4 for CH_4, etc. Thus for iron catalysts operating at high temperatures a high percentage conversion of all the reactants can be achieved provided that the Ribblett ratio $H_2/(2CO + 3CO_2)$ is about 1.05. It should be noted that in order to produce one unit of 'CH$_2$' three units of the sum of (H$_2$ + CO) is required irrespective of the route via which the 'CH$_2$' is produced.

$$2H_2 + 1CO \rightarrow \text{'CH}_2\text{'} + H_2O$$
$$1H_2 + 2CO \rightarrow \text{'CH}_2\text{'} + CO_2$$
$$3H_2 + CO_2 \rightarrow \text{'CH}_2\text{'} + 2H_2O$$
$$3CO + H_2O \rightarrow \text{'CH}_2\text{'} + 2CO_2$$

The formation of CO$_2$ via the WGS reaction is not necessarily a waste of syngas as the sum of (H$_2$ + CO) remains unchanged. Note that if CO$_2$ formation is a waste of CO then H$_2$O must be a waste of H$_2$.

The rates of the FT reaction will be affected by the partial pressures of hydrogen and of carbon monoxide in the syngas fed to the reactor. Using the kinetic rate equations presented later, it can be calculated that for iron catalysts the rate at the reactor entrance will be lower by a factor of about 1.3 if the H$_2$/CO ratio of the feed gas is lowered from 3 to 1.4. At the 50% conversion level the factor by which the rate is lowered increases to about 3.5. For the same two feed gas compositions, if a cobalt catalyst is used the factors by which the FT rates are lowered are about 2 and 4.3 respectively. In other words the reaction rate is more sensitive to H$_2$/CO ratio for cobalt catalysts than it is for iron catalysts.

5. CONVERSION AND SELECTIVITY EVALUATION

As stated in Section 4, because of the high cost of syngas production, it is important that the syngas be efficiently utilised. This requires that the conversion of the components of the syngas that can be converted to FT products is high and also that the selectivity of desired products is maximised. It might appear that the ideal situation would be that the composition of the syngas fed to the FT reactors should match the overall FT usage ratio of the reactants in order to achieve maximum utilisation of the syngas. It must, however, be borne in mind that, in practice, additional hydrogen is always required in several of the downstream product work-up operations (see Chapter 6). Thus it may be convenient that there is some excess of hydrogen in the FT synthesis loop from which the required hydrogen can be extracted. When using cobalt catalysts, which have little or no watergas shift (WGS) activity, the H$_2$ to CO usage ratio does not change with increasing FT conversion levels. Iron catalysts, however, are active for the WGS reaction, and when using these catalysts at the lower temperatures the overall usage ratio decreases as the conversion increases (see Chapter 4) and so this also should be borne in mind. When using iron catalysts at the higher temperatures it is in any event very desirable to have excess hydrogen in the FT reactor in order to minimise carbon deposition on the catalyst (see Chapter 7).

As a measure of the FT conversion activity the $\%(CO+H_2)$ or the $\%(CO+CO_2)$ moles converted over the reactor can be used. Note that the extent or direction of the WGS reaction does not affect the molar sum of either CO plus H_2 or of CO plus CO_2 leaving the reactor. In general the use of $\%(CO+CO_2)$ may be preferred for the following reasons:

When, as is normally practised, there is excess hydrogen present in the FT reactor then it is theoretically impossible to attain a high $\%(CO+H_2)$ conversion. This is not the case for the $\%(CO+CO_2)$ conversion. In this case the $\%(CO+H_2)$ conversion will not be related in a simple way to the total amount of the hydrocarbons produced (or to the selectivity of the products for which there are different C to H ratios). When the FT selectivity is calculated on a carbon atom basis then the amount of carbon atoms converted to FT products is required. In this case there is a direct link between the molar $\%(CO+CO_2)$ conversion and the calculation of the selectivity.

The FT selectivity can be calculated on a mass, carbon atom or molar basis. Since commercially the products are sold on a mass or volume basis the molar selectivity of the FT process is of little practical interest. For instance, to produce one mole of a high molecular mass wax would require the conversion of a very much larger amount of syngas than that required for one molecule of methane and so molar selectivity gives a distorted picture of syngas utilisation. For practical FT plants the bulk of the products are alkenes, alkanes, oxygenated hydrocarbons and, if the operating temperature is high, some aromatics and naphthenes. Consequently there is little difference between calculating the selectivity on a mass or carbon atom basis. From a theoretical point of view the selectivity on a carbon atom basis is preferred and all the selectivity data presented is on this basis. It should be borne in mind that for some low molecular mass products there can be a big difference between mass selectivity and carbon selectivity, e.g. for ethane (molecular mass 28) and acetic acid (molecular mass 60), both containing two carbon atoms per molecule.

The FT products are commonly analysed by gas chromatographic (GC) techniques and the results are calculated as mole percentages. As discussed previously the selectivity, on a carbon atom basis is easily calculated, By way of example, the fractional selectivity of butene produced in the FT reaction would be:

(moles butene produced x 4)/ (total moles of CO and CO_2 converted).

Table 2 shows the values for the percentage carbon atom selectivities for some typical commercial FT catalysts.

Table 2
FT product spectra (pressure 2MPa)

Catalyst	Cobalt	Iron	Iron
Reactor Type	Slurry	Slurry	Fluidized
Temperature °C	220	240	340
% Selectivities (C atom basis)			
CH$_4$	5	4	8
C$_2$H$_4$	0.05	0.5	4
C$_2$H$_6$	1	1	3
C$_3$H$_6$	2	2.5	11
C$_3$H$_8$	1	0.5	2
C$_4$H$_8$	2	3	9
C$_4$H$_{10}$	1	1	1
C$_5$-C$_6$	8	7	16
C$_7$-160°C	11	9	20
160-350°C	22	17.5	16
+350°C	46	50	5
Total water soluble oxygenates[1]	1	4	5
ASF alpha value[2]	0.92	0.95	0.7

[1] alcohols, aldehydes, ketones and acids dissolved in the water phase.
[2] ASF: Anderson Schulz Flory, probability of chain growth.

6. PRIMARY AND SECONDARY FT REACTIONS

The FT synthesis produces a wide range of hydrocarbon and oxygenated products. The formation of the various products appears to be controlled by mechanistic and kinetic factors and the product spectra are very different from what would be expected from thermodynamic considerations. Tillmetz [1] calculated that if a gas with a H$_2$/CO ratio of 1.0 were to go to complete equilibrium at 0.1 MPa and at a typical FT temperature then the main products would be methane, carbon dioxide and graphite and the amounts of higher hydrocarbons would be negligible. Similar calculations were presented by von Christoffel [2] for 600 K, 1.6 MPa and a gas with an initial H$_2$/CO ratio of 2.0. In Table 3 the results are compared with the actual selectivities typically found over iron catalysts under similar operating conditions. To facilitate direct comparison both sets of data have been normalised to 100 mass units of methane. It is clear that the FT reactions under normal operating conditions are nowhere near thermodynamic equilibrium, the observed C$_2$ and higher products are produced in huge quantities relative to thermodynamic expectation.

Table 3

Relative mass values of predicted product selectivities at 600K and 1.6 MPa [2] and those observed in actual FT reactors (iron catalyst)

	Predicted*[1] (No carbon deposition)	Typical observed*[1]
CH_4	100	100
C_2H_4	1×10^{-7}	40
C_2H_6	5.6×10^{-3}	40
C_3H_6	6×10^{-10}	120
C_3H_8	1.2×10^{-6}	20
C_5H_{12}	8×10^{-14}	20
C_2H_5OH	1.8×10^{-9}	20

*1 Both sets of data normalized to 100 mass units of CH_4

In Table 4 the typically obtained ratios of various FT products are compared with the calculated values assuming the relevant reactions went to thermodynamic equilibrium at the actual partial pressures of H_2, H_2O, and CO_2 at the reactor exit.

For the reaction class (A) Table 4 shows that whereas the alkenes should theoretically be virtually completely hydrogenated to alkanes they are in fact the dominant hydrocarbons produced. This indicates that alkenes must be primary FT products and that, if it occurs, their subsequent hydrogenation over iron catalysts in the gas atmosphere prevailing in the FT reactors must be slow. Even with cobalt catalysts, which are much more active for hydrogenation reactions than iron the alkenes exceed the alkanes for the lower molecular mass hydrocarbons (see Table 2). Thermodynamically the hydrogenation of ethene should be more complete than that of propene and in line with this it can be seen (from Table 4) that the observed alkene to alkane ratio for C_2 is indeed lower than for C_3. From C_5 to C_8 thermodynamics predicts that the ratio of alkene to alkane decreases somewhat, but at higher carbon numbers there should be little further change in the alkene to alkane ratio. In practice the observed ratios, although still much higher than predicted, do decline with increasing carbon number and only at very high carbon numbers do the ratios fall to low levels. For the data in Table 4, case (A), the hydrogen partial pressure was 0.6 MPa. At 510 K and at the much higher hydrogen partial pressure of 4.5 MPa in a FT reactor operating with an iron catalyst the alkene to alkane ratio shifts with carbon number as follows:

Carbon No	Alkene /alkane ratio
C3	1.4
C10	0.8
C20	0.3
C30	0.0

Table 4

Product ratios observed compared to thermodynamic equilibrium predictions

React -ion Class	Reactions	Temp. (K)	Ratios	Expected at Equilibrium [1]	Typical Observed [2]
A	$C_2H_4 + H_2 \leftrightarrow C_2H_6$	600	C_2H_4/C_2H_6	5.9×10^{-7}	2
	$C_3H_6 + H_2 \leftrightarrow C_3H_8$	600	C_3H_6/C_3H_8	1.7×10^{-5}	8
	$C_5H_{10} + H_2 \leftrightarrow C_5H_{12}$	600	C_5H_{10}/C_5H_{12}	1.3×10^{-5}	7
	$C_{10}H_{20} + H_2 \leftrightarrow C_{10}H_{22}$	600	$C_{10}H_{20}/C_{10}H_{22}$	1.1×10^{-5}	6
	$C_{20}H_{40} + H_2 \leftrightarrow C_{20}H_{42}$	600	$C_{20}H_{40}/C_{20}H_{42}$	1.1×10^{-5}	3
B	$C_2H_4 + H_2O \leftrightarrow C_2H_5OH$	600	C_2H_5OH/C_2H_4	0.0027	0.0045
		510		0.024	1.2
C	$C_2H_5OH + H_2 \leftrightarrow C_2H_6 + H_2O$	600	C_2H_5OH/C_2H_6	3.6×10^{-9}	0.6
D	$C_2H_5OH + H_2O \leftrightarrow CH_3COOH + 2H_2$	610	CH_3COOH / C_2H_5OH	0.14	0.14
		510		0.0009	0.26
E	$C_2H_5OH \leftrightarrow CH_3CHO + H_2$	600	CH_3CHO / C_2H_5OH	0.22	0.21
		510		0.0028	0.27
F	$1\text{-}C_5H_{10} \leftrightarrow 3Me\ 1\text{-}C_4H_8$	600	$3Me\ 1\text{-}C_4H_8 / 1\text{-}C_5H_{10}$	1.4	0.17
G	$2CH_3COOH \leftrightarrow CH_3COCH_3 + CO_2 + H_2O$	600	$C_3H_6O / (C_2H_4O_2)^2$	5.6×10^4	1.9×10^2
H	$CH_3COCH_3 + H_2 \leftrightarrow 2\text{-}C_3H_7OH$	610	iC_3H_7OH / C_3H_6O	0.2	0.2
		510		7	0.1
I	$C_7H_{14} \leftrightarrow C_7H_8 + 3H_2$	600	C_7H_8 / C_7H_{14}	2000	0.15

[1] At the relevant partial pressures at the reactor exit. [2] Iron based catalysts in all cases.

Table 5 also shows the shifts for several different reactor types and catalysts.

If ethene is deliberately added to the syngas fed to a FT reactor operating at 600 K with an iron catalyst, about 50% of the added ethene is hydrogenated to ethane [3]. However, if the liquid oil produced is recycled to the reactor at a ratio of (two recycled oil) to (one oil produced per pass) then there is hardly any change in the alkene content of the oil. If, however, an excess of octene is added (about 40 added to 1 normally produced) then about 15% of the added octene is hydrogenated to octane. Thus overall it can be concluded that the higher the molecular mass of the alkenes added to the feed gas the less likely it is to be hydrogenated in a secondary reaction. The explanation for these observations is provided in Chapter 8. (Note that in the absence of syngas, e.g. when an alkene is added to hydrogen flowing through the FT catalyst bed the alkene is extensively hydrogenated or hydrocracked, depending on the operating temperature.)

The situation, however, for the growing primary FT species on the catalyst surface is different. The higher the carbon number of these growing species the longer the time these species have spent linked to the catalyst surface sites and hence the greater the likelihood that these chains are more 'fully hydrogenated' species. This is in any event in keeping with thermodynamic expectations. Other factors may play a role but this is discussed in more depth in Chapter 8.

Table 5
FT hydrocarbon isomers

Catalyst Temperature °C	Cobalt 220	Iron 240[1]	Iron 340[2]
C_5-C_{12} Cut			
% Alkanes	60	29	13
% Alkenes	39	64	70
% Aromatics	0	0	5
% Oxygenates	1	7	12
C_{13}-C_{18} Cut			
% Alkanes	95	44	15
% Alkenes	5	50	60
% Aromatics	0	0	15
% Oxygenates	low	6	10
C_{24}-C_{35} Cut			
% Alkenes	Low	10	-
C_4 Cut			
% 1-butene	-	-	74
% Me-1-propene	-	-	8
C_6 Cut			
% 1-hexene	-	-	58
% Me 1 pentenes	-	-	24
C_{10} Cut			
% 1-decene	-	-	38
% Me 1 nonenes	-	-	20

[1] Slurry bed operation (LTFT), [2] Fluidized bed operation (HTFT)

For reaction class (B), Table 4 the observed ratio of ethanol to ethene is higher than predicted from thermodynamics at both the temperatures listed. This indicates that alcohols are not formed by the hydration of alkenes in secondary reactions. The reverse, the de-hydration of alcohols to alkenes, is however possible. The data for reaction (C) shows that at the hydrogen pressure in the reactor the ratio of ethanol to ethane is very much higher than expected. From cases (B) and (C) it thus appears that the alcohols too are primary FT products.

For reactions (D) and (E) it can be seen that ethanol, acetic acid and acetaldehyde are in thermodynamic equilibrium at 600 K and so these two interrelated reactions do take place. (This is confirmed by the experimental finding that when ethanol, acetaldehyde or ethyl acetate is added individually to the syngas being fed to an iron catalyst, additional amounts of the other two compounds are always found in the exhaust gas, i.e. the three compounds are readily inter-convertible.) At the lower temperature of 510 K, however, the ratios observed in normal FT synthesis are much higher than the expected ratios. This implies that acetic acid is not formed by subsequent oxidation of the primary ethanol by water, nor is acetaldehyde formed by de-hydrogenation of ethanol. The reverse, i.e. the hydrogenation of acetic acid or of acetaldehyde to form ethanol is, however, feasible. So it appears that aldehydes and acids too may be primary products.

It is well known that the FT hydrocarbons are predominantly linear. In accordance with this, the data for reaction (F) show that the ratio of branched to linear pentane is much lower than expected from thermodynamics. The reason is presumed to be that the FT mechanism favours the formation of linear products over the formation of branched products even though both may be primary reactions (see Chapter 8). In addition, the secondary isomerisation of the linear primary alkenes or alkanes is very probably a slow reaction at the normal FT temperatures. The branches are predominantly methyl branches. The degree of branching increases with the operating FT temperature. When using iron catalysts at high temperatures, e.g. at 320 °C, the degree of branching increases as the carbon number of the product increases [4]. When using iron catalysts for the production of waxes, i.e. at lower temperatures, e.g. at 220 °C, the degree of branching decreases with increasing chain length but as the temperature was increased beyond 220 °C the degree of branching increased [5].

Ketones and iso-alcohols only appear in the products in significant amounts when the FT process is carried out at higher temperatures. This could be because their formation as primary products is a slow process or that they are in fact secondary products. From the data presented for reaction (G), the ketonisation of acetic acid does seem to be feasible under FT conditions. This concept is apparently supported by the fact that as the FT synthesis temperature is increased from 310 to 380 °C (see Table 6) the acid selectivity decreases and the ketone selectivity increases up to about 360 °C after which it also decreases at higher temperatures [3]. The latter decrease could be due to the hydrogenation of the ketones to iso-alcohols. When acetic acid is deliberately added to the syngas fed to an iron catalyst at about 340 °C a large amount of additional acetone is produced. When feeding additional ethanol to the reactor, there was also a marked increase in acetone production [3]. As reaction (D), Table 4, is at thermodynamic equilibrium at 610 K then adding ethanol to the feed should increase the production of acetic acid and ketonisation of the latter could account for the increase in acetone production. Analyses of all the ketones present in the FT products shows that they are predominantly methyl ketones (e.g. methyl ethyl ketone, methyl pentyl ketone, etc). Also the dominant acid formed is acetic and the amount of higher molecular mass acids rapidly declines with increasing carbon number. It therefore follows that if ketonisation of this mixture of acids takes place then the methyl ketone would emerge as the dominant isomer for each carbon number ketone.

For reaction (H) it can be seen that at 610 K the hydrogenation of acetone to iso-propanol appears to have proceeded to equilibrium, but at the lower temperature of 510 K the ratio iso-propanol/acetone is much lower than expected. This suggests that the hydrogenation of the acetone was slow at the lower temperature and so this may indicate that iso-alcohols are secondary products.

Table 6
Influence of temperature on a Fe catalyst in HTFT operation [3]

Temp (°C)	Selectivity % (C atom basis)				Gasoline cut analysis		C_3H_6/C_3H_8 ratio
		Water soluble chemicals					
	CH_4	Alcohols	Ketones	Acids	Br Number[1]	% Aromatics	
310	10	2.3	0.4	0.3	109	4	6
330	14	2.3	0.8	0.4	94	8	10
350	17	1.6	1.1	0.2	92	10	9
360	20	1.1	1.3	0.2	93	13	8
370	23	0.8	1.2	0.1	88	18	6
380	28	0.5	0.8	0.1	85	26	4

1   The bromine number is an indicator of the amount of alkenes present

Below about 500 K the production of aromatics over iron catalysts is virtually zero but as can be seen from the results presented in Table 6 the aromatic content of the gasoline produced increases with temperature and at about 650 K the gasoline cut contains about 25% aromatics. As aromatisation is a highly endothermic reaction the increase in aromatics with increasing temperature is according to expectation. Note also that the Bromine Number (a measure of the alkene content) of the gasoline decreases as the aromatics increase. This could be in part due to the secondary hydrogenation of the alkenes and in part to their conversion to the thermodynamically much more stable aromatics. It is probable that alkenes and aromatics are formed on the catalyst surface from common precursors. The thermodynamically unfavoured mono-alkyl benzenes predominate and this fits the concept that aromatics are formed by linkage between the first and sixth carbon atom of the carbon chain followed by dehydrogenation to the aromatics. Note that in the gasoline cut there are also significant amounts of unsaturated ring compounds present which fits the proposed route to aromatics. The aromatics in the gasoline cut are predominantly C_7 to about C_{10} alkyl benzenes. Of the aromatics only about 1% is benzene which is in line with the known fact that the rate of aromatisation of hexane is very much slower than for C_7 and higher hydrocarbons. In Table 4, reaction (I), it can be seen that the thermodynamically expected ratio of toluene to heptene is very much greater than that obtained under FT conditions. Recycling the FT oil fraction to the reactor did not increase the aromatic content of the oil produced. If aromatisation was a secondary reaction then it could be expected that recycling of the oil would increase the aromatic content. It is thus possible that the formation of ring compounds and aromatics are primary reactions, which only occurs very slowly at 600 K. (Note that commercial reforming of gasoline is practised at about 800 K.)

At the hydrogen pressures existing in FT reactors the hydrocracking of longer chain hydrocarbons to shorter chain hydrocarbons is thermodynamically favourable but in practice the FT process at about 230 °C can make huge amounts of very long chain hydrocarbons, e.g. see Table 2. The 'hard wax' fraction (BP > 500° C) contains linear alkanes with carbon numbers well over 100. All this suggests that secondary hydrocracking hardly occurs, if at all.

It has been reported that with $Ni/Co/SiO_2$ catalyst hydrocracking was only observed when the FT conversion levels were very high [6]. Craxford found that wax hydrocracking was a slow reaction when CO was present [7]. Over Ru catalysts it was concluded that chemisorbed CO inhibited the hydrocracking of the FT products. Investigators at Sasol found that when the light oil produced over Fe catalysts at about 330 °C was recycled to the FT reactor there was no overall loss in oil production, nor was there any change in the molecular composition of the oil [3]. Thus there was no evidence of hydrocracking even at this high temperature. If, however, oil was passed over the same catalyst at 330 °C in the presence of only hydrogen then extensive hydrocracking took place.

Overall, it can thus be concluded that under normal FT conditions, using normal FT catalysts, minimal, if any, hydrocracking occurs. This probably is because once the primary hydrocarbons have de-sorbed from the catalyst surface they cannot readily re-adsorb as explained in Chapter 8. The only significant exception is ethene. Patents and theories that claim extensive re-adsorption of olefins should be viewed with scepticism. As will be explained in more detail in Chapter 8, the key to predicting the non-oxygenated FT product distributions is an understanding of the process of desorption from the catalyst surface.

7. FT PRODUCT DISTRIBUTIONS

Whatever the operating conditions the FT reaction always produces a wide range of hydrocarbon and oxygenated hydrocarbon products. Methane, which is an unwanted product, is always present and its selectivity can vary from as low as about 1% up to 100%. (The latter selectivity is obtained when using a strongly hydrogenating catalyst like nickel metal at high temperatures.) At the other end of the product spectrum the selectivity of long chain linear waxes can vary from zero to over 70%. Ruthenium operating at about 170 °C produces waxes with carbon numbers in the polyethylene range. The same catalyst at say 400 °C will produce mainly methane. The intermediate carbon number products between the two extremes are only produced in limited amounts. Thus it seems not to be possible to produce, on a carbon atom basis, more than about 18% C_2, about 16% C_3, about 42% gasoline/naphtha (C_5 to 200 °C BP) and about 20% diesel fuel (200 to 320 °C).

The spread in carbon number products can be varied by altering the operating temperature, the type of catalyst, the amount or type of promoter present, the feed gas composition, the operating pressure, or the type of reactor used. Whatever the process conditions there is always a clear interrelationship between all of the products formed. Fig. 1 illustrates this relationship between the lower molecular mass hydrocarbons produced with a fused iron catalyst operating at about 330 °C, and Fig. 2 shows the relationship between the higher molecular mass hydrocarbons produced over a precipitated iron catalyst operating at about 225 °C. ('Hard wax' is the hydrocarbon fraction boiling above 500 °C.) Fig. 3 shows how the selectivity of some different C_2 compounds varies with the methane selectivity. Considering only the alcohols produced it is also found that they range from methanol to high molecular mass linear primary alcohols and the relationships between them are similar to those illustrated in Figs. 1 and 2. The same holds for the aldehydes and acids.

For over eighty years the detail, on a molecular level, of the FT reaction has been a very controversial matter and several mechanisms, and variations thereof, have been proposed. The mechanistic proposals are discussed in Chapter 8. There is no doubt that the mechanism is of intriguing scientific interest. However, there is no convincing evidence to date that, based on the 'correct' mechanism, a catalyst has been developed that will markedly improve the three crucial factors of the FT process, namely, the lifetime, the activity and the selectivity of the products. This then remains as a challenge to those involved with catalyst R&D.

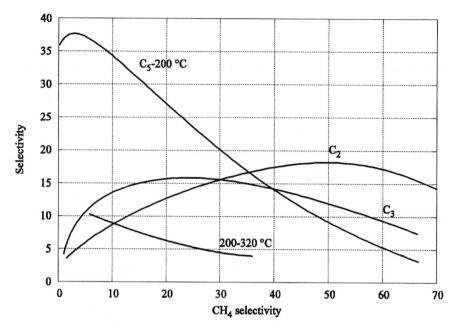

Figure 1 Hydrocarbon product selectivities relative to the methane selectivity

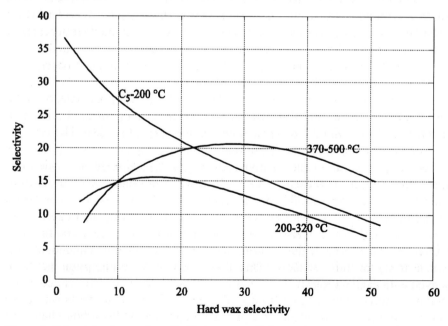

Figure 2 Selectivity for product distillation cuts relative to the hard wax selectivity

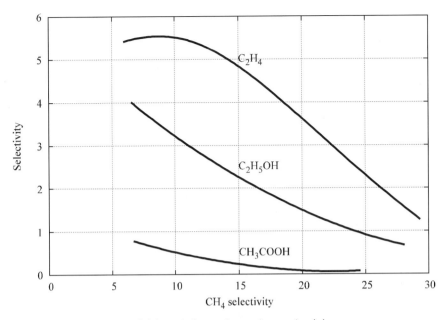

Figure 3 C₂ product selectivities relative to the methane selectivity

Common to all the proposed mechanisms is the assumption that chain growth occurs by a stepwise procedure and it is this aspect that will be dealt with here. The reaction sequences that are presented here should not necessarily be taken as support for one or other mechanism but rather to see whether they match up with the production of the wide range of products that are observed in practice.

For the sole purpose of explaining the hydrocarbon product selectivity let it be assumed that the basic building blocks, or 'monomers', of the FT reaction are the 'CH₂' units which are chemisorbed on the catalyst surface. The reaction sequences that could occur are illustrated in Fig. 4.

A 'CH₂' unit can either react with hydrogen to yield methane which will then desorb from the surface or it can link up with a another 'CH₂' unit to form an adsorbed 'C₂H₄' species. The latter unit now has three options. It can desorb to yield ethene, or it can be hydrogenated to produce ethane, or it can link up with another 'CH₂' monomer to produce an adsorbed 'C₃H₆' unit. The first two options are chain termination actions and their combined likelihood can be taken as the probability of chain termination whereas the third option would be the probability of chain growth, α (alpha). The reaction sequences can continue and thus hydrocarbons ranging from methane to high molecular mass waxes are produced, the higher the value of alpha the longer the hydrocarbon chains. In the steady state the concentration on the catalyst surface of each C_nH_{2n} species should be constant. If the probability of chain growth is α then the probability of chain termination is $(1-\alpha)$. According to this mechanism, the selectivity to methane must differ from that predicted by the chain growth probability α because it is not formed by the insertion of a 'CH_2' species but rather by the creation of this species. Also methane can only terminate as a paraffin while higher hydrocarbons can terminate as either olefins or paraffins.

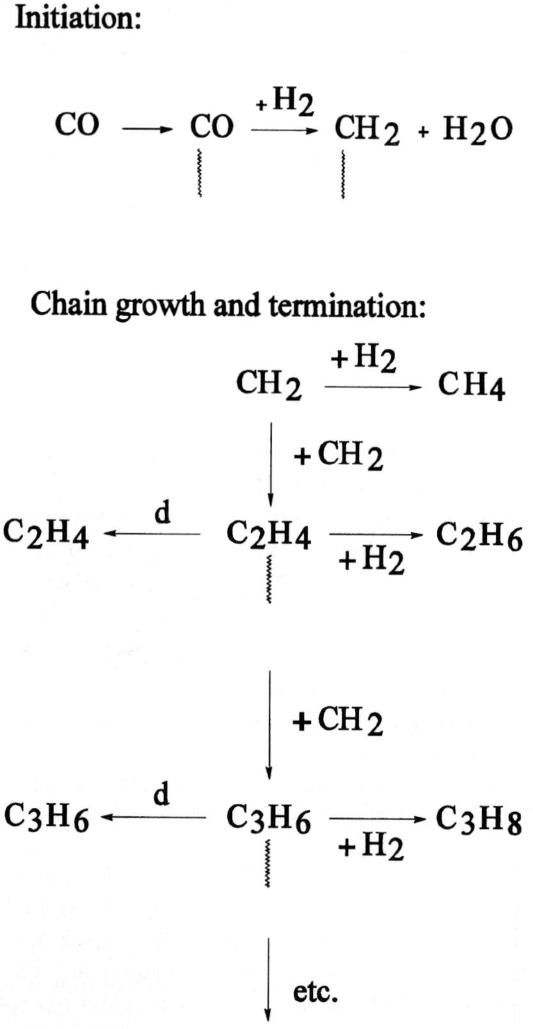

Initiation:

$$CO \longrightarrow CO \xrightarrow{+H_2} CH_2 + H_2O$$

Chain growth and termination:

$$CH_2 \xrightarrow{+H_2} CH_4$$

$$\downarrow +CH_2$$

$$C_2H_4 \xleftarrow{\;d\;} C_2H_4 \xrightarrow[+H_2]{} C_2H_6$$

$$\downarrow +CH_2$$

$$C_3H_6 \xleftarrow{\;d\;} C_3H_6 \xrightarrow[+H_2]{} C_3H_8$$

$$\downarrow$$

etc.

Figure 4

One useful approach is to distinguish between chain initiation and chain propagation. Even if chain initiation is started with both a single carbon atom initiator and a double carbon atom initiator (e.g. from re-adsorbed ethylene), the probability of chain growth for all higher carbon number species will still exhibit a constant chain growth probability if there are no other chain initiators. This will be explained in more detail in Chapter 8.

If it is assumed that the probability of chain growth is independent of chain length then the entire hydrocarbon product spectrum can readily be calculated for values of alpha from zero to one. The result of such calculations is illustrated in Fig. 5.

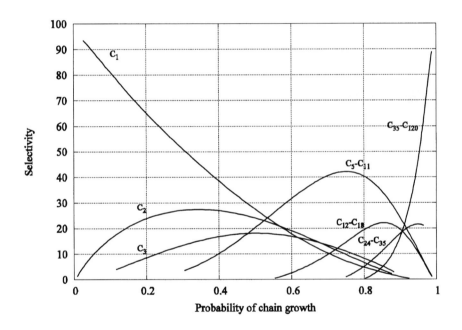

Figure 5 Product selectivity as a function of chain growth probability

These calculations show, as observed in practice, that between the two carbon number extremes all the product cuts go through maxima as the probability of chain growth increases. Comparing Fig. 5 with Figs. 2 and 3 it is seen that there is in fact a reasonably good agreement regarding the maximum amounts of the intermediate products that can be formed. One clear exception is for the C_2 products. Fig. 5 predicts a maximum of about 30% whereas in practice, over iron catalysts at least, the number never exceeds 20%. The apparent misfit of the probability of chain growth of the C_2 species relative to that of the C_3 and higher species is observed not only with iron catalysts but also with cobalt [8, 9] and with ruthenium catalysts [8, 10]. Some examples of the C_1 to C_4 selectivities are given in Table 7. For the high methane selectivity (50%) case the C_2 selectivity is higher than the C_3 and C_4 selectivities but lower than the C_1 selectivity, all of which is as expected from Fig. 5. Except for the one low methane selectivity (2.7%) case over an iron catalyst the C_2 selectivity for the remaining cases is lower than both the C_1 and the C_3 selectivities. This is not expected from Fig. 5.

Table 7
Examples of measured C_1 to C_4 product selectivities

Catalyst	Fe	Fe	Fe	Co	Co	Ru	Ru
C_1	2.7	11.0	50	12	8.4	3.7	14.8
C_2	3.6	10.8	17	2.1	1.9	1.2	3.5
C_3	4.6	14.3	12	3.0	4.7	4.4	3.8
C_4	5.2	12.5	6	3.8	4.0	3.8	3.8

Notes:
All selectivities in C atom %.
Sasol data include hydrocarbons and oxygenates.

The mathematical equation depicting the stepwise chain growth concept was developed by Herrington [11], Anderson [12] and earlier by Flory [13]. Assuming alpha to be independent of chain length the equation is:

$$Log(W_n/n) = n \log \alpha + constant$$

where, W_n is the mass fraction of the species with carbon number n. From the slope of the plot of log (W_n/n) against n the value of α is obtained. These plots are often referred to as ASF (Anderson, Schulz, Flory) distribution plots. It is usually found that over the carbon number range from 3 to about 12 the plots are linear which confirms that over this range the probability of chain growth is constant. With iron catalysts operating in the commercial HTFT gasoline operating mode there is only one alpha value, α_1 which is about 0.7. When considerable amounts of waxes are also produced then it is usually found that in vicinity of carbon number 12 the ASF plots shift to yield another straight line with a lower slope which translates to a higher chain growth probability (α_2) in the $C_{20}+$ wax carbon number range. With iron catalysts operating in the commercial LTFT wax producing mode the α_2 is typically 0.95 at the start of run. This double alpha effect is observed for both iron and for cobalt catalysts. It could be argued that there are not really two separate distinct alpha values, rather the value of alpha slowly increases with chain length [14] (See Fig. 6).

Under FT conditions the waxes are in the liquid state and so the pores of the catalyst are presumably filled with a vapour/liquid emulsion. It has been postulated that when liquids are present inside the catalyst pores the higher molecular mass products would have longer residence times within the catalyst particles. This could lead to re-adsorption of the primary alkenes and thus result in further chain growth, i.e., the probability of chain growth is increased. This process is referred to as alkene re-incorporation.

The more severe the diffusional restriction the longer the residence time of these long chain olefins and so the more likely their re-adsorption. However for chain growth to occur CO is also required at the relevant sites and this will not occur if the diffusion restrictions are too 'severe'. Thus an optimum diffusional situation is required for a high long chain hydrocarbon selectivity. Based on this concept Exxon developed an 'egg shell' cobalt catalyst for high performance in the FT reaction [26].

Workers at Shell have also studied olefin re-adsorption by comparing the FT product spectrum over a cobalt foil and Co supported on SiO_2. On the foil they found that olefin hydrogenation was the main secondary reaction which depended on chain length. This is in keeping with the commonly observed fact that the olefinity of FT products decreases with increasing chain length. In the case of the foil there was only a single alpha value. On the Co/SiO_2 catalysts, however, continued chain growth was the main secondary reaction of re-adsorbed olefins which was chain length dependent and the 'double alpha' effect was observed [56, 57].

With regard to the concept of alkene re-incorporation as an explanation of the double alpha phenomena it should be pointed out that iron catalysts appear to produce somewhat longer chain waxes than do cobalt catalysts. This could be due to the fact that Fe produces more alkenes than Co and thus there is a higher probability of re-incorporation of alkenes in the case of Fe catalysts. There is, however, disagreement about the factors that control residence times and alkene re-incorporation [24, 41, 42, 43, 44, 45, 46]. Three different reasons have been put forward to explain the longer residence times of the longer chain products inside the catalyst particles, viz., slower diffusion rates, higher solubility in the liquid phase and stronger physisorption of the longer chain products. These aspects have been discussed elsewhere [46].

It could be argued that the higher the molecular mass of the products the slower would they diffuse out of the catalyst particles. The longer the residence times the greater the likelihood that the primary alkenes will be hydrogenated to alkanes. This possibility would contribute to the finding that the longer the chain length the lower the alkene/alkane ratio of the products. In apparent support of this postulate it is found that when the same precipitated iron catalyst is used under identical FT conditions in a slurry and in a fixed bed reactor the same carbon number distributions are found. However, the alkene/alkane ratios are clearly higher for the slurry bed products [17, 23]. It should be noted that the actual size of the catalyst particles used in the slurry bed reactor is much smaller than that used in the fixed bed reactor. Thus the residence time in the fixed bed catalyst particles would be much longer which could result in a greater degree of secondary hydrogenation of the alkenes.

For alkene secondary hydrogenation to occur, the alkenes must be re-adsorbed onto the catalyst. The hard wax selectivities for the two reactor types, however, are the same. The question then arises that, if the longer residence times of the products in the larger fixed bed catalyst particles explains the higher degree of hydrogenation, why does this then not also lead to higher chain growth probabilities?

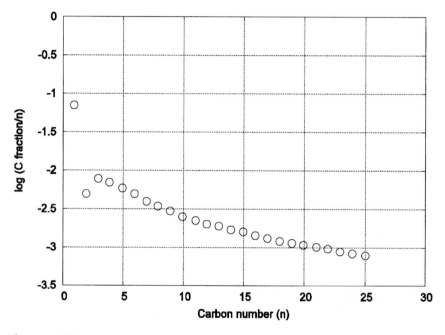

Figure 6 ASF plot

In the Fig. 6 example the value of α_1 is 0.84 and α_2 is 0.93. The point for the C_2 product invariably falls below the linear ASF plot and so, as previously pointed out, is out of line with prediction. Quite often the methane point falls well above the linear line, especially in some cases when cobalt catalysts are used. The implications are that the chain growth probability for the 'CH_2' surface specie is lower, and that of the 'C_2H_4' specie is higher, than for the higher carbon number surface species. It is possible that an isolated methylene unit (CH_2) on the catalyst surface is more likely to be hydrogenated to methane than are the adsorbed alkenes to be hyrogenated to alkanes. The lower than expected C_2 production, i.e. the higher probability of chain growth, may be due to the likelihood that an adsorbed methylene group can attach itself to an adsorbed ethene specie from either side of the latter. For C_3 and higher adsorbed alkene species, the methylene group, for steric hindrance reasons, can attach itself to the alkene specie more readily from one side than from the other [15]. These concepts are depicted in Fig. 7. This would result in a higher probability of chain growth for the adsorbed ethene species and thus in a lower probability of chain termination than that of the higher carbon number adsorbed alkene species.

H_2 ⟍   ⟋ CH_4

CH_2

|

An isolated CH_2 group is readily converted to CH_4

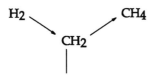

CH_2 ⟶ H_2C————CH_2 ⟵ CH_2

An adsorbed C_2H_4 unit can readily be attacked from either side by adsorbed CH_2 units

CH_2 H_2C————$CHCH_2R$

The adsorbed alkene unit can be attacked by an adsorbed CH_2 unit more readily from the one side than the other.

Figure 7

Alternatively, or in addition, it has been demonstrated that ethene itself can incorporate into growing chains on the cobalt catalyst surface [16, 17] and this effectively increases its probability of chain growth. It should, however, be noted that ethene incorporation over iron catalysts does not readily occur and despite this the C_2 selectivity misfit is still present. When compiling ASF distribution plots it should be born in mind that not only the hydrocarbon products should be considered but also the oxygenated products as well. For instance, of the alcohols, aldehydes and acids the C_2 compounds are produced in much larger amounts than the C_3 and higher compounds in each of the oxygenated types (see Table 10). If it is accepted that all the ketones are formed by ketonisation of the acids then all the acetic acid involved in these reactions should also be included in the C_2 fraction. If these C_2 oxygenates are not included in the total C_2 compounds then the C_2 point will fall even lower with respect to the linear portion of the ASF plot. On the other hand if taking all of the above into consideration actually brings the total C_2's in line as regards the ASF plot then the foregoing assumption that the adsorbed ethene species have a higher probability of chain growth, and of course the proposed explanations of this, could fall away.

Earlier studies reported that, as for hydrocarbons, there was a regular relationship between alcohols. The distribution curves for alcohols were similar to those of the (n+1) hydrocarbons [6]. However, whereas (on a carbon atom basis) ethanol always seems to be the major alcohol produced, the peak for the hydrocarbons can shift from C_3 to C_4 to C_5 etc depending on the probability of chain growth. When the FT operating conditions and the catalyst are geared at producing larger amounts of alcohols and alkenes it is found that when all the alkanes, alkenes and alcohols are evaluated separately the probabilities of chain growth (α_1) are fairly similar. By way of example all three would be 0.64 ± 0.03. The similarity suggests that the mechanisms by which alkenes and alcohols are produced are very probably the same except for the chain termination reaction step. It should be noted that under mild FT conditions, where secondary reactions are minimal, the alkenes are predominantly linear 1-alkenes and the alcohols are linear primary alcohols. Similarly, the aldehydes and acids too are linear. The chain growth reaction sequence for these compounds too could be the same as for alkenes and alcohols, except again for different chain termination steps.

The explanation for the double alpha phenomenon is discussed in more detail in Chapter 8 but it may be observed at this point that it appears to be related to the chain length dependant solubility in the liquid FT product. It may also be observed that for a stand alone Fischer-Tropsch catalyst it is never observed that α_2 is less than α_1. This can only happen if a second catalytic function is introduced into the FT reactor. This has never been achieved in practice without an unacceptable increase in methane selectivity.

In summary then, the ASF plots are useful to describe FT product selectivities but adjustments are required to predict the lower ethylene selectivity and higher methane selectivity that are usually observed as well as the duel α values observed for reactors that contain a liquid phase.

It will be useful to understand the reasons for the higher methane selectivity if this can lead to an intervention that decreases the production of this unwanted product. For HTFT where ethylene is a significant product, which is worth the cost of recovery, it is desirable to find ways to increase the ethylene selectivity. Given that the deviation from ASF theory differs for different catalysts, this may be possible.

Another question that arises is the chemical nature of the 'monomer' that adds on to the linear growing hydrocarbon chain attached to the surface. Clearly the very first reaction is the chemisorption of CO on to the catalyst surface. Does it, as such, insert into the growing chain, and is subsequently hydrogenated to CH_2, or does the CO first dissociate to C and O atoms on the surface and the C then hydrogenated to CH_2 which then inserts into the growing hydrocarbon chain? In the case of the latter concept the adsorbed O atoms could react with hydrogen to form water which certainly is the main 'by-product' of the FT reaction. From a thermodynamic point of view, however, the reaction of the surface O with CO to form CO_2 is the more favourable reaction. In spite of this little or no CO_2 is produced over cobalt catalyst at their normal FT operating temperatures. Either CO decomposition does not occur on the catalyst surface or the actual rate, due a big difference in activation energy, of hydrogen reacting with the oxygen atoms is much faster that that of CO reacting with the oxygen atoms. When only CO is passed over reduced cobalt or over reduced iron catalysts at about 220 °C the metal carbides are formed and CO_2 is produced, i.e., CO decomposition does occur on both metals. (The reaction is known as the Boudouard reaction.)

$$2CO \rightarrow C + CO_2$$

However, in the presence of syngas, i.e. CO and H_2, cobalt is not carbided and so even if the Boudouard reaction does occur to some extent the carbon produced is hydrogenated to 'CH_2'. In the case of reduced iron catalysts the iron is converted to carbide in the presence of syngas at 220 °C, but once the carbiding process is completed no further carbon build-up takes place. At high FT temperatures, e.g. 330 °C, after completion of the carbiding process the deposition of carbon continues and so 'free' carbon accumulates on the catalyst. This means that at the higher temperature carbon deposition rate, whether by the Boudouard reaction or by the decomposition of chemisorbed CO, exceeds its removal rate by hydrogenation.

On the assumption that it is unlikely that the FT chemical reaction processes over cobalt and iron catalysts are different the question then arises why cobalt catalysts produce little or no carbon dioxide whereas iron catalysts do always produce carbon dioxide as well as water? A probable explanation of this is that whereas iron is an active catalyst for the water gas shift (WGS) reaction cobalt is not. Note that for iron catalysts at low temperatures the WGS is far from equilibrium because there is much more water present than CO_2. This observation supports the concept that water is the primary product and carbon dioxide is the result of the subsequent WGS reaction and not the result of CO reacting with free O atoms on the catalyst surface. It could be argued that this casts doubt on whether the adsorbed CO molecule does in fact dissociate to C and O atoms on the catalyst surface under FT conditions although the preceding observations about carbiding and carbon deposition does support the CO dissociation concept.

Various possible reaction sequences required to form the 'monomer' are depicted in Fig. 8. Figs. 9, 10 and 11 show various possible chain growth and chain termination reactions which may account for the formation of the various primary FT products. In several of the reaction steps the involvement of surface oxygen atoms is depicted. These atoms could be the result of reversible dissociation of adsorbed CO, CO_2 or H_2O.

$$CO \leftrightarrow C + O \quad \text{[This is the initial reaction in Figure 8(b)]}$$
$$CO_2 \leftrightarrow CO + O$$
$$H_2O \leftrightarrow 2H + O$$

In the case of Fig. 9 where CO insertion is depicted as the initial step in further chain growth it is also possible that CO insertion is in fact a 'chain stopping' step and that subsequent reactions lead to the three primary oxygenated products as depicted in step (b). This could therefore mean that after CO insertion has occurred (i.e. step (b) or (c) in Fig. 9) then continued chain growth does not readily occur. If CO insertion is indeed the route to the formation of the oxygenated products then the selectivity of these products would be increased by having a high CO coverage on the catalyst surface. Higher temperatures are likely to enhance the rate of decomposition of adsorbed CO to C and O atoms and this would result in lower surface coverage by CO. Thus operating at lower temperatures should enhance alcohol selectivity. The data in Table 6 do show that the alcohol and acid selectivities did indeed increase as the temperature was lowered.

In view of the apparent correlation of acid selectivity with the partial pressure of carbon dioxide (see Section 9.3.1), reaction (d) represents a possible alternative route to acid formation. This is of course very speculative as the apparent link between CO_2 and acids may be coincidental and not real.

Note that the illustrations should be seen as possible overall reaction sequences and do not necessarily depict the actual chemical structures of the various surface complexes. Note also that in Fig. 8 that the three 'monomers' are chemically interrelated and so a mixture of all three further reaction sequences depicted in Figs. 9 to 11 could be taking place. In the latter Figs. when hydrogen is involved in the reactions the hydrogen is presented as two atoms of H, but they could equally well be represented by activated hydrogen molecules. Should the formation of the CH_2 monomer proceed via the reaction sequence depicted in Fig. 8(d) this may be the explanation why H_2O and not CO_2 is the primary product. Carbon dioxide formation would then be accounted for by the subsequent WGS reaction. This could explain why cobalt catalysts, being inactive for the WGS, make little CO_2, while iron catalysts, which are active for the WGS, can make copious amounts of CO_2.

As previously mentioned the hydrocarbon chains produced in the FT process are predominantly linear but as the synthesis temperature is raised the amount of branched chain hydrocarbons produced increases. The branches are predominantly methyl branches and mono- di- and tri-methyl etc. branched hydrocarbons are produced at the higher temperatures. The formation of methyl branches could be a primary surface reaction as depicted in Fig. 10(d), but if the rate of this reaction sequence is slow relative to that of Fig. 10(c) then this would explain why the hydrocarbon chains are predominantly linear. It is of interest to note that the amount of 2-methyl-1-alkene produced is less than the amounts of 3-(or 4-) methyl-1-alkene produced (see Table 9) From thermodynamics the expected ratio of 3-methyl-1-pentene to 2-methyl-1-pentene is 0.14 at 327 °C, whereas the observed ratio is about 2. So once again it appears that mechanistic and kinetic aspects determine the FT product spectra and not thermodynamics.

8. SELECTIVITY AT TYPICAL COMMERCIAL CONDITIONS

Irrespective of the operating conditions or of the type of catalyst used the FT process always produces a wide range of products. The products can consist of a mixture of linear and branched alkenes, dienes and alkanes, aromatics, saturated and unsaturated ring compounds and oxygenated products such as alcohols, aldehydes, ketones and carboxylic acids. In practice the actual spectra of products can vary widely. The control of the spectra will be dealt with in Section 9. Table 2 shows the carbon number distribution of some typical FT processes. For low temperature (about 200 to 240 °C) operations in fixed or slurry bed reactors the hydrocarbons range from methane to long chain hard waxes. The low temperature Fischer-Tropsch (LTFT) processes are usually geared at maximum wax production, while the high temperature (HTFT) operations are geared at the production of low molecular mass alkenes and liquid products in the gasoline (primarily) and diesel fuel range.

An important selectivity issue is the combination of values for alpha and operating conditions that will result in the formation of a liquid phase in the FT reactor. This would make it impossible to operate in the two phase mode used for the HTFT operation. It was postulated by Caldwell [18] that reactors will be prone to waxing if the ASF plot intersects the vapour/pressure plot. At 220 C, it was calculated that this will occur if alpha exceeds 0.653. The theoretical explanation was provided by Caldwell and van Vuuren [19]. At 280 C it has been calculated that waxing will occur if alpha exceeds 0.71 [20]. It is clear that the HTFT reactor that operates with an alpha value of about 0.7 is close to the practical upper limit for operation in a two phase fluidized bed.

Table 5 shows the breakdown of the various types of compounds present in several different liquid product fractions. Table 8 gives the alkene to alkane ratios of several carbon number products obtained over different catalysts at different temperatures. From Table 8 it can be seen that the HTFT operation at 340 °C produces more alkenes than the LTFT operations. Also, under similar operating conditions and in the same type of reactor, the iron catalyst produces more alkenes than the cobalt catalyst. In all cases the alkenes are predominantly linear 1-alkenes. The branched hydrocarbons are predominantly mono-methyl isomers. In the case of the HTFT process the degree of branching tends to increase with carbon number and the number of di- and tri-methyl branches per molecule also increases with carbon number. For the LTFT operations the hydrocarbons are predominantly linear.

Table 8
Ratios 1-alkene/n-alkane

Catalyst	Co	Fe (a)	Fe (a)	Fe (b)[1]	Fe (b)[2]	Fe (c)
Reactor Type	Slurry	Slurry	Fixed bed	Fixed bed	Fixed bed	Fluidized bed
Temp (°C)	220	230	230	240	240	340
Entry	1.9	1.9	1.9	2.0	5	5
H₂/CO ratio						
Carbon Number			Ratios 1-alkene/n-alkane			
C_2	0.05	0.4	0.2	0.5	0.2	1.5
C_3	1.4	3.9	1.2	1.5	0.5	6.5
C_4	1.4	3.2	0.9	1.2	0.3	7
C_6	1.0	2.1	0.9	1.0	0.2	7
C_{10}	0.9	2.0	0.8	0.4	0.1	6
C_{15}	0.4	0.2	0.05	0.1	0	2

*1 No recycle of tail gas (TG)   *2 TG recycled
(a) precipitated
(b) sintered
(c) fused

For the HTFT case the percentage of aromatics increases while that of the alkenes and oxygenated products decrease with increasing carbon number (Table 5). For the LTFT processes the products contain no aromatics. The alkene/alkane ratios depend on the type of catalyst used and on the operating conditions (Table 8). Comparing the cobalt and iron [Fe (a)] catalysts, both operating in slurry beds under fairly similar conditions, it can be seen that the cobalt catalyst produces a much more saturated product, i.e. the alkene/alkane ratios are lower. This is not unexpected as cobalt is a much more active catalyst for hydrogenation reactions. The two precipitated iron catalysts [Fe (a)] used in the slurry and fixed bed reactors were identical except for the average particle size, the latter being much larger than the former. Table 8 shows that for the iron catalyst in the fixed bed reactor the alkene/alkane ratios were clearly lower than for the same catalyst in the slurry reactor. Two factors could be involved in the explanation of this difference:

(1) The residence time of the products will be longer in the larger catalyst particles and hence the possibility of secondary hydrogenation of the primary alkenes to alkanes will be greater.

(2) Due to diffusion effects the ratio H_2/CO increases as the syngas penetrates into the porous catalyst particles and, for iron catalysts, the water gas shift reaction may also increase the hydrogen content.

The larger the catalyst particles are the greater is the increase in the average H_2/CO ratio inside the particles. Thus the fixed bed reactor, having much larger particles, produces a more saturated product. The effect of the H_2/CO ratio on the alkene/alkane ratios is demonstrated by the two runs with catalyst Fe (b), a sintered iron catalyst, both run in fixed bed reactors but having different H_2/CO ratios. The catalyst Fe (c) was a typical catalyst used in the HTFT process at 340°C and, as mentioned previously, large amounts of alkenes are produced. (The preparation of the various iron catalysts is described in Chapter 7.)

Table 9 shows the distribution of linear and methyl- branched alkenes in the C_4, C_5 and C_6 cuts produced over iron catalysts at about 340°C. The 2-methyl isomers are lower than the 3-methyl or 4-methyl isomers. Possible reasons for this are discussed in Chapter 8. As the carbon number increases the percentage of the 2-methyl isomer as well as its ratio, relative to the 1-alkene, decreases.

The amount of oxygenated hydrocarbons produced with iron catalysts depends on the type of catalyst and the process conditions used (see Tables 10 and 11).

Table 9

C$_4$, C$_5$ and C$_6$ alkene isomers (HTFT 340°C)			
Cut	Isomer	Mass %	Relative ratio
C$_4$	1-butene	74	1.00
	2-Me-propene	8	0.11
C$_5$	1-pentene	66	1.00
	2-Me-1-butene	6	0.09
	3- Me-1-butene	11	0.17
C$_6$	1-hexene	58	1.00
	2-Me-1-pentene	4	0.07
	3-Me-1-butene	10	0.17
	4-Me-1-butene	10	0.17

Table 10

Typical composition of water soluble oxygenated products with iron catalysts

Compound	LTFT[1] 230°C	HTFT[2] 340°C
Non-acid chemicals		Mass%
Ethanal (acetaldehyde)	0.5	2
Propanal	0.1	0.5
2 Propanone (acetone)	4	23
2 Butanone (MEK)	0.3	6
Methanol	24	0.5
Ethanol	45	40
1 Propanol	13	12
2 Propanol	1	5
1 Butanol	5	4
2 Butanol		1
2 Me 1 Propanol		1
Acids (mass % distribution)		
CH$_3$COOH		70
C$_2$H$_5$COOH		16
C$_3$H$_7$COOH		9
Acid content of water (mass %)	0.4	1.2

1 Precipitated iron catalyst
2 Fused iron catalyst

The percentage oxygenates of the total FT products is higher for the HTFT process than for the LTFT process. However, for the various cuts (the low molecular mass oxygenates dissolved in the condensed water phase and in the hydrocarbon fractions) the percent of primary alcohols of the total alcohols is higher for the LTFT fixed bed process than the HTFT fluidized bed process. Of particular note is the big difference in the methanol selectivity, the LTFT process having a much higher selectivity than the HTFT operation (Table 10). For the ketones, iso-alcohlols and acids the situation is reversed in that the

HTFT process makes more of these than the LTFT process. This is in keeping with the discussions in Section 6, namely that at the higher temperatures more of the acids are converted to ketones and more of the ketones are hydrogenated to iso-alcohols. Of the alcohols, aldehydes and acids produced the most abundant species on a carbon atom basis are the C_2 compounds, namely, ethanol, acetaldehyde and acetic acid. The most abundant ketone produced is acetone.

Under normal commercial FT operating conditions the amount of oxygenated compounds produced is low relative to the amount of hydrocarbons, especially so in the case of cobalt because it is a more active hydrogenating catalyst than iron. At low temperatures, e.g., at about 170 °C, cobalt can, however, produce large amounts of oxygenates [6, 21].

Table 11
Oxygenates in the naphtha and diesel cuts

Cut		LTFT Fixed Bed[2] 230°C	HTHT Fluidized Bed[3] 340°C
Naphtha	Total Oxygenates[1]	7	12
	% Alcohols	88	44
	% Ketones	8	42
	% Acids	4	14
Diesel	Total Oxygenates[1]	6	10
	% Alcohols	94	58
	% Ketones	5	28
	% Acids	1	14

[1] Carbon atom basis of total FT products
[2] Precipitated iron catalyst
[3] Fused iron catalyst

9. FT SELECTIVITY CONTROL

In Section 7 it was pointed out that selectivity on a carbon atom basis, is essentially a function of the probability of chain growth, α (alpha). Control of the product selectivity will therefore to a large extent be determined by the factors that influence the value of alpha. The main factors are the temperature of the reaction, the choice of metal catalyst used, the chemical and structural promoters added to the catalyst, the gas composition and, more, specifically, the partial pressures of the various gasses in contact with the catalyst inside the reactors. Overall, by manipulating these factors a high degree of flexibility can be obtained regarding the type of product and the carbon range thereof.

9.1 The influence of operating temperature

Irrespective of the type of metal catalyst used, whether iron, cobalt, nickel or ruthenium, raising the FT operating temperature shifts the spectra to lower carbon number products. Desorption of growing surface species is one of the main chain termination steps and since desorption is an endothermic process higher temperatures should increase the rate of desorption which would then result in a shift to lower molecular mass products. Putting it in another way, thermodynamically the formation of methane is much more favoured than the formation of higher molecular mass products at all FT operating temperatures. Therefore one should expect that as the temperature is raised, and the rates of all reactions subsequently increase, the situation should move towards that predicted by thermodynamics.

As the temperature is increased the system becomes more hydrogenating and so the ratio of alkenes to alkanes decreases, again as expected from thermodynamics. Probably for the same reasons the selectivity of the alcohols and of the acids decrease as the temperature is raised (see Table 6). The degree of chain branching increases as the temperature is raised, as one would expect from thermodynamics.

Tables 6 and 12 illustrate the influence of temperature on the commercial Sasol HTFT and LTFT processes respectively. Bear in mind that, as can be seen from Figures 1 and 2, the methane and the hard wax selectivities are good indicators of the overall product spectra. For both processes the overall selectivity shifts to lower molecular mass products and the alkene to alkane ratios decrease. In the LTFT process no aromatics are formed. For the HTFT process the amount of aromatics increases as the temperature is raised. As previously discussed note the inverse relationship between ketones and acids (Table 6) which agrees with the suggestion that ketones are formed from the ketonisation of acids (see Section 6). Note, however, that above 360 °C the ketones decrease again, probably due to hydrogenation to iso-alcohols and/or to alkanes, as one would expect from thermodynamics.

Table 12
Influence of temperature on an iron catalyst in LTFT operation [3]

Temperature °C	Hard wax selectivity[1] C atom %	% Alkenes in diesel cut
213	47	45
227	34	39
237	24	33
247	17	40

1. Hard wax is that fraction with BP > 500°C.

Cobalt catalysts are more active for hydrogenation reactions than iron catalysts and thus the products with cobalt are more saturated (see Tables 2 and 5). The methane selectivity also increases more rapidly with increasing temperature than is the case with iron catalysts. Early studies with cobalt catalysts found that when operating at low temperatures (165 °C) large amounts of oxygenates were formed, the oil fraction contained about 40 % oxygen containing compounds, while at 200°C only 1% was present [21], [22]. At low temperatures, about 170 °C, ruthenium catalysts produce very high molecular mass waxes (in the polyethylene range) [23, 24], whereas at high temperatures, about 400 °C, only methane is produced. Of the group VIII metals ruthenium is in fact the most active methanation catalyst [25]. Similarly nickel produces wax at low temperatures and only methane at high temperatures. (Most commercial methanation catalysts are nickel based.)

9.2 Catalyst metal type and promoters (chemical and structural)

Only the four metals, iron, cobalt, nickel and ruthenium have been found to be sufficiently active for application in the FT synthesis. The relative cost of these four metals is shown in Table 13. Note that the raw material used in the preparation of iron based FT catalysts is either scrap iron, steel works mill scale or iron ore of suitable purity, all of which means that the iron source is cheap relative to the alternative metals. While nickel was used in the early German investigations [6, 21] it was found to make too much methane at low synthesis pressures while at high pressures, where the methane production was lower, volatile nickel carbonyls were formed, i.e. nickel was lost from the reactors. The use of nickel was therefore abandoned. The price of ruthenium is very high and also the fact that the amount available is insufficient for large-scale industrial application rules it out as a viable option. This leaves only cobalt and iron as practical catalysts.

Table 13
Approximate relative cost of metals active for FT

Fe[1]	1
Ni	250
Co	1000
Ru	48000

1   Fe as scrap metal

9.2.1 Promotion of cobalt catalysts

Because of the high cost of cobalt the use of bulk cobalt catalysts (as employed in the pre-World war II FT plants) is not viable. It is important that the cobalt be supported on a suitable high area porous material under conditions that yield well dispersed, i.e. small cobalt crystals, thus resulting in a high exposed surface area per unit mass of cobalt metal. As discussed separately in this book, the cobalt crystals need to be larger than a certain minimum size as those smaller than the minimum are converted to inert oxide under normal FT conditions [26-30].

Unlike iron catalysts, cobalt based catalysts in general do not appear to be very sensitive to the chemical nature of promoters or supports. The early German studies with cobalt supported on kieselguhr showed that when operating at atmospheric pressure, promotion with alkalis such as potassium carbonate increased the selectivity of wax. However at pressures above 1.5 MPa, which is in any event preferred for industrial FT plants, the effect of alkali promotion was minimal [6]. Different investigators have reported conflicting results regarding the influence of promoters on the performance of cobalt catalysts. This could be due to different catalyst preparation techniques and /or different FT test conditions used.

At Sasol the effect of adding, individually, small amounts of the oxides of Cr, Mg, K and Th to an alumina supported Co catalyst was investigated. The FT reactions were studied at 1.8 MPa and about 220 °C and it was concluded that none of the additives had any beneficial effects [31].

Iglesia studied the FT performance of cobalt supported on different oxides, namely Al_2O_3, SiO_2 and TiO_2, at 2.0 Mpa and 200 °C [26]. A good correlation was found between the FT activity and the amount of metal area available. However, on the basis of unit metal surface area all three catalysts had the same activity, i.e. the same turn over frequency. It can therefore be concluded that the oxide supports had no chemical promoting effect.

Exxon found that the addition of low levels of ruthenium not only increased the rate of reduction of cobalt supported on either silica or on titanium oxide, but also increased the FT activity as well as the selectivity towards higher molecular mass hydrocarbons [32].

Goodwin reported that while Ru promotion of cobalt supported on alumina did enhance the rate of reduction and did increase the FT activity it had no effect on selectivity [33]. (The FT tests were carried out at 0.1 MPa, which was much lower than that used by Exxon.) Adding 1% Re to an alumina supported Co catalyst also was reported to enhance the FT activity without affecting the selectivity. The effect on titania or silica supported catalysts was, however, less marked [34].

The researchers at Trondheim reported that promoting alumina or silica supported cobalt with Pt enhanced reducibility as well as the FT activity without affecting the selectivity. The FT tests were carried out at 0.1 MPa, 200°C and a H_2/CO ratio of 7 [35]. It has been reported that the FT activity and the selectivity of silica supported cobalt was enhanced by lanthanum oxide [36] as well as by zirconium oxide [37]

Work carried out by Fischer in the 1930's showed that for precipitated cobalt supported on kieselghur addition of Cr_2O_3 or Al_2O_3 resulted in lower FT activities while the addition of MgO, MnO or ZnO had little or no effect on the FT performance [21].

For more information about the effects of various promoters and supports on the selectivity of cobalt catalysts see Chapter 7.

9.2.2 Promotion of iron catalysts

Contrary to cobalt catalysts, iron based catalysts are significantly affected by the chemical nature of promoters or supports. For all iron catalysts the promotion with the optimum amount of alkali, such as potassium salts, is vital for satisfactory FT activity as well as the required selectivity. Because the strong alkaline salts of K and Na have such a dominant influence on the FT selectivity it is debatable whether other promoters or supports present in the catalyst formulation have, in their own rights, any marked influence on selectivity. The presence of other compounds, whether added or present as impurities, can however have a marked influence on the effectiveness of the alkaline promoter. Thus if there is any silica present the alkali can react with it to form alkali-silicates which will be less basic (alkaline) than the 'free' alkali, e.g. as KOH or K_2CO_3 and hence the 'basicity' of the working iron surface would be lowered. Similarly the alkali could interact with acidic supports such as alumina to form less basic aluminates. Even if the supports or other high area promoters added do not chemically interact at all with the alkali they will effectively dilute the alkali in that the alkali will not only be distributed on the iron surface but also on the surface of the other components. Overall it means that the amount of alkali that has to be added in order to attain the required basicity of the iron surface needs to be adjusted to take into account the chemical nature and the surface areas of the other components present.

By way of example, the effect of various promoters on the performance of the two types of iron based catalysts used at Sasol is presented below. The catalyst used in the LTFT systems in the fixed and in the slurry bed reactors consist of co-precipitated Fe and Cu oxides which is then supported/bound by silica gel. A base stock of this material was used to study the effects of various promoters. In one series of tests the base stock was impregnated with equivalent amounts of different potassium salts and the FT performance of these catalysts then compared [3]. It was found that the carbonate, hydroxide, nitrate, fluoride, borate, oxalate, silicate and permanganate all yielded similar performances. The chloride, bromide, sulphate and sulphide salts all yielded lower activity catalysts. The poor performance of the later two salts was expected in view of the known fact that sulphur containing compounds are strong poisons for iron catalysts. In contradiction to the above results Hammer [38] reported that neither KCl nor KBr resulted in inferior catalysts.

In another series the same Fe/Cu/silica base stock was impregnated with equivalent amounts of the Group I alkalis, namely, Li, Na, K and Rb as the nitrates. The catalyst promoted with K had the highest FT activity. Equating the activity of the K catalyst to 100 units the activity of the Na and Rb catalysts were both about 90 units while that of the Li catalyst was much lower at about 40 units. As was expected from the relative basicities of the alkalis the wax selectivities increased in the order, Li, Na, K and Rb. The lower activity of the Rb case may possibly have been due to the fact that it produced a very high molecular mass wax. This wax, being present in the narrow catalyst pores, might have slowed up the diffusion rate of reactants and of products in and out of the catalyst particles and thus resulted in lower FT reaction rates. In another way of comparing Na with K promotion two catalysts were prepared as described in Chapter 7, but instead of using potassium water-glass, sodium water-glass was used as the means of adding the silica. The two catalysts, containing the equivalent amounts of alkali promoter were then tested for their FT performance. The activity of the Na- promoted catalyst was about 25% lower and the wax selectivity was about 20% lower than that of the K-promoted catalyst.

Table 14 illustrates the marked influence of the amount of potassium promoter on the activity and the selectivity of precipitated iron catalyst in the LTFT process [3]. Two sets of results are shown, one for unsupported Fe/Cu catalysts and the other for the supported Fe/Cu/silica catalysts. The well known shift to higher molecular mass hydrocarbons, i.e. higher chain growth probability, with increasing alkali is reflected by the increase in the hard wax selectivity. Note also the increase in the selectivity of the oxygenated compounds, namely the acids and alcohols. It has been reported previously that as the alkali level is increased the activity increases up to a certain level and then decreases at higher alkali levels [6]. In Table 14 this peak in activity was only observed in the unsupported series. In the supported series the alkali levels were apparently all beyond the peak level and thus the activity in this series decreased with further alkali additions. (In later studies it was confirmed that at lower alkali levels the activities were also lower.) As the wax selectivity increases with increasing alkali the average chain length of the waxes increases as previously discussed. This probably results in slower diffusion rates inside the wax-filled pores which results in lower overall reaction rates.

Table 15 shows the influence of changing the type and the amount of support material [3]. Two sets of data are presented. In the one set various individual supports were evaluated. The other set all contained the same amount of silica support, but in addition different individual other supports were also incorporated into the catalyst. (Note that kieselguhr is a natural form of silica of marine origin while the 'SiO$_2$' was added as K-water-glass as described in Chapter 7.)

As discussed previously, it should be noted that the results presented in Table 15 should be compared with caution since the key promoter, K$_2$O, was not necessarily at the optimum level for each different catalyst formulation. For a thorough comparison the alkali level should have been varied for each catalyst type and the best cases then compared. Nevertheless the impression gained from Table 15 is that using different supports, or incorporating different supports in addition to the 'standard' silica support, did not result in any marked improvement, relative to the standard silica supported catalyst, in either the wax selectivity or FT activity. Whether any of the other supports would improve the long-term stability of the catalyst is another matter, but this was not evaluated.

Table 14
Influence of alkali content on synthesis performance of precipitated iron catalysts [3]

Catalyst type	K$_2$O level[a]	Hard wax selectivity[b]	Water analysis wt%		Activity[a]
			Alcohols	Acids	
Unsupported Fe$_2$O$_3$	0	5		<1	26
	1.0	34		3	47
	1.6	41		8	50
	2.0	53		12	53
	3.0	63		19	40
Silica supported Fe$_2$O$_3$	12	18	2.5	0.2	112
	16	20	3.0	0.3	109
	21	30	3.4	0.3	85
	24	38	2.0	0.5	83
	32	44	2.0	0.7	75

a. relative quantities
b. wt% of hydrocarbon product boiling above 500°C

For the HTFT process (320 to 350 °C) Sasol to date uses catalyst prepared by fusing (melting) suitable iron oxides together with the required promoters (see Chapter 7). In this kind of catalyst, as is the case for typical ammonia synthesis catalyst, structural promoters such as the oxides of aluminium, magnesium or titanium is added to increase the surface area of the reduced catalyst. In addition these promoters minimise subsequent loss of surface area as the result of sintering with time on stream [3]. However, since diffusion control is involved in the overall FT kinetics, surface areas above a certain minimum probably do not result in increased activity. As for precipitated catalysts the added structural promoters and other compounds that may be present can have a significant influence on the effectiveness of the vitally important chemical promoter, namely, K_2O. Higher surface areas coupled with a fixed amount of alkali amounts to a lower surface basicity, i.e. a lower percentage of the surface iron atoms associated with the alkali promoter molecules. If the structural promoters or other compounds present, such as silica, chemically interact with the alkali then the basicity of the alkali could be lowered. The influence of additional silica on the FT performance of the catalyst is shown in Table 16.

Table 15

Influence of different supports on synthesis performance of precipitated iron catalysts [3]

Catalyst	SiO_2 g/100 Fe	Additional[a] support/g (100g Fe)$^{-1}$	K_2O[b]	Activity[b]	Wax Selectivity[b]
Unsupported Fe-Cu	0	0	0.6	34	25
Fe-Cu-Cr$_2$O$_3$	0	24	2.0	31	30
Fe-Cu-MgO	0	32	3.6	16	13
Fe-Cu-MgO	0	68	3.2	28	10
Fe-Cu-Al$_2$O$_3$	0	23	3.0	35	10
Fe-Cu-Al$_2$O$_3$	0	100	12	18	34
Fe-Cu-SiO$_2$[c]	24	0	10	45	34
Fe-Cu-ZnO	0	30	2	40	12
Fe-Cu-SiO$_2$-CaO	24	10 CaO[a]	10	38	31
Fe-Cu-SiO$_2$-Cr$_2$O$_3$	24	10 Cr$_2$O$_3$	10	42	30
Fe-Cu-SiO$_2$-Kieselguhr	24	20 Kieselguhr	10	31	29
Fe-Cu-SiO$_2$-Al$_2$O$_3$	24	10 Al$_2$O$_3$	10	41	25
Fe-Cu-SiO$_2$-V$_2$O$_5$	24	10 V$_2$O$_5$	10	41	31
Fe-Cu-SiO$_2$-ThO$_2$	24	10-ThO$_2$	10	4$^-$	31
Fe-Cu-SiO$_2$-MgO	24	10 MgO	10	3u	30
Fe-Cu-SiO$_2$-TiO$_2$	24	10 TiO$_2$	10	43	30
Fe-Cu-SiO$_2$[c]	24	O	10	45	34

a. Added in addition to the standard amount of SiO_2
b. Relative quantities
c. Standard reference catalyst

Table 16

Synthesis performance of promoted fused catalysts. SiO_2 varied with alkali content fixed. Fluidized bed at 320°C [3]

Alkali Type	SiO_2/Alkali molar ratio	$CO + CO_2$ conversion %	CH$_4$ selectivity	C_3H_6/C_3H_8 ratio
Na$_2$O	0.6	86	13	2.8
	2.1	83	13	2.4
	4.2	81	13	3.1
K$_2$O	0.8	85	10	6.1
	4.2	78	27	0.9

For both the alkalis, Na and K, as the silica level was increased the activity of the catalysts decreased, but more so in the case of the potassium promoted catalyst. In the case of the sodium promoted catalyst the influence of the silica on the selectivity was not really evident. However, for the stronger alkali (potassium) the silica markedly increased the methane selectivity and decreased the alkene to alkane ratio, both shifts indicating that the catalyst had become more hydrogenating because of the effective lowering of the basicity of the catalyst. A tentative reason for the smaller effect of silica on the performance of the sodium promoted catalyst is as follows. The sodium ions are capable of going into solid solution in the magnetite phase (see Chapter 7) and so not all of it will be associated with the silica in the final reduced catalyst. Since neither the silicon nor the potassium ions can go into solid solution in magnetite they will react to form silicates and so the basicity of the catalyst surface will be lowered.

Table 17 shows the effect of increasing the alkali levels on the FT performance of fused/reduced catalysts at high temperatures (330 °C) in fluidized bed reactors.

Table 17
Synthesis performance of promoted fused magnetite catalysts. Amounts and types of alkalis varied. Fluidized bed operation at 330 °C [3]

Alkali	Amount of alkali (relative units[a])	Activity $CO + CO_2$ conversion %	CH_4 selectivity[c]	Water acids selectivity[c]	NAC selectivity[b,c]	C_3H_6/C_3H_8 ratio
K_2O	0	63	41	0.05	0.3	0.2
	1.75	90	20	0.2	1.7	3
	2.1	90	15	0.3	2.6	5.7
	2.65	92	13	0.5	3.4	7.6
	3.4	94	12	0.7	3.8	11
	3.95	93	10	0.8	4.0	12
Na_2O	0	63	41	0.05	0.3	0.2
	2.3	78	15	0.15	2.7	1.8
	4.5	85	13	0.2	2.0	2.3
	6.8	84	14	0.2	2.5	2.0
	9.1	87	16	0.2	2.7	1.7

a. Alkali levels all expressed on molar basis.
b. NAC = Non Acid Chemicals (alcohols and ketones) in water phase.
c. All selectivities, % carbon atom.

For both alkalis the activity increases and the selectivity shifts to higher molecular mass products (as indicated by the decreasing methane selectivity), to more oxygenated products (acids, alcohols and ketones) and to higher ratios of alkenes to alkanes. Clearly, the effect of the stronger alkali, potassium, is much more pronounced than that of the weaker sodium. It also appears that beyond a certain level addition of more sodium has little further effect on the catalyst's performance.

Table 18 shows that when the fused/reduced catalysts are used at low temperatures (200 °C) in fixed bed reactors the FT activity decreases with increasing potassium promoter, as was observed for the precipitated/silica supported catalyst also operated at the lower temperature (see Table 14 and the related discussion).

Table 18
Synthesis performance of fused magnetite promoted with a fixed amount of Al_2O_3 and varying amounts of K_2O. Fixed bed operation at about 200 °C. [3]

K_2O content (relative units)	Activity (relative units)
0	64
0.4	60
0.8	55
1.1	50
2.6	29

It has been reported that the manner in which the potassium promoter was added to the catalyst used in a fluidized bed, whether by impregnation or as loose powder, had little effect on the performance of the catalyst in the FT reaction [39]. Investigations at Sasol revealed similar results. When potassium silicate was separately finely ground and the powder then added to a finely ground alkali-free fused magnetite the FT performance of this mixture was very similar to that of the catalysts prepared in the normal way. (Fusing the magnetite together with the promoters and then grinding the ingots to the required particle size distribution).

These results are not really surprising when the microscopic investigations reported in Chapter 7 are considered. It was found that essentially all the potassium in normally fused catalysts was present in the separate silicate glass inclusions and after milling a large portion of these inclusions was present as separate particles, i.e. the normal catalyst was in any event a heterogeneous mixture. It was also demonstrated that when alkali-free catalyst was loaded into a reactor and run for a few days the expected poor FT selectivity was observed. When some powdered potassium silicate was then added to the running reactor the selectivity of the catalyst the next day had improved markedly and was very similar to that of the catalyst prepared in the normal way. It appears therefore that under FT conditions the alkali silicate migrates over the surface from particle to particle. It nevertheless seems doubtful if this would result in a perfectly uniform distribution of the alkali over the iron surface and if this assumption is correct there is room for improvement regarding this aspect.

The following speculation is offered as an explanation of the observed effects of potassium promotion on the activity and selectivity shifts when the amount of alkali promoter is increased. The presence of potassium on the surface of reduced iron has been shown to increase the heat of chemisorption of CO [40, 41]. It is therefore expected that alkali promotion would increase the amount of CO adsorbed on the surface of the iron catalyst and consequently the surface coverage by the FT 'monomer' (e.g. 'CH$_2$') would be higher. Alkali being a strong base could act as an electron donor thus strengthening the bonds of the adsorbed CO and 'CH$_2$' units with the surface Fe atoms. The rate of the FT reaction may equate to the rate at which the monomers react, e.g. insert into the surface hydrocarbon chains. If the presence of alkali not only increases the strength of the carbon to metal bond in the case of the adsorbed monomer, but also does the same for the carbon to metal bond of the growing hydrocarbon chains, this would mean that the probability of desorption of the chains would be lowered (i.e. the probability of chain growth would increase). This would mean that as long as there is an ample amount of adsorbed monomers present on the surface both the rate of the FT reaction would increase and the selectivity would shift to higher molecular mass products.

At higher alkali coverage, however, the surface could be predominantly occupied by growing hydrocarbon chains and the coverage by monomers would be relatively lower. This would mean that chain growth would still continue, but at a lower rate because of the low availability of monomers, i.e. the rate of the FT reaction would decrease but the shift in selectivity to longer chain products would continue to increase. If CO insertion is the route to the formation of oxygenated products (see Fig. 9) and alkali increases the amount of CO adsorbed on the catalyst surface then this could explain why alkali also increases the selectivity of oxygenates.

The strength of adsorption of hydrogen on iron is known to be weaker than that of CO. If alkali does increase the coverage of the catalyst surface by growing chains and monomers there would be fewer vacant site available for the adsorption of hydrogen. The system therefore would be less hydrogenating and this could explain why alkali promotion increases the ratio of alkenes to alkanes (see Table 17).

9.3 Gas composition, partial and total pressures

Considering any of the various possible reaction sequences illustrated in Fig. 4 and in Figs. 8 to 11 it could be argued that the higher the CO partial pressure the higher the catalyst surface coverage by adsorbed monomers will be, whatever the chemical make-up of these monomers are.

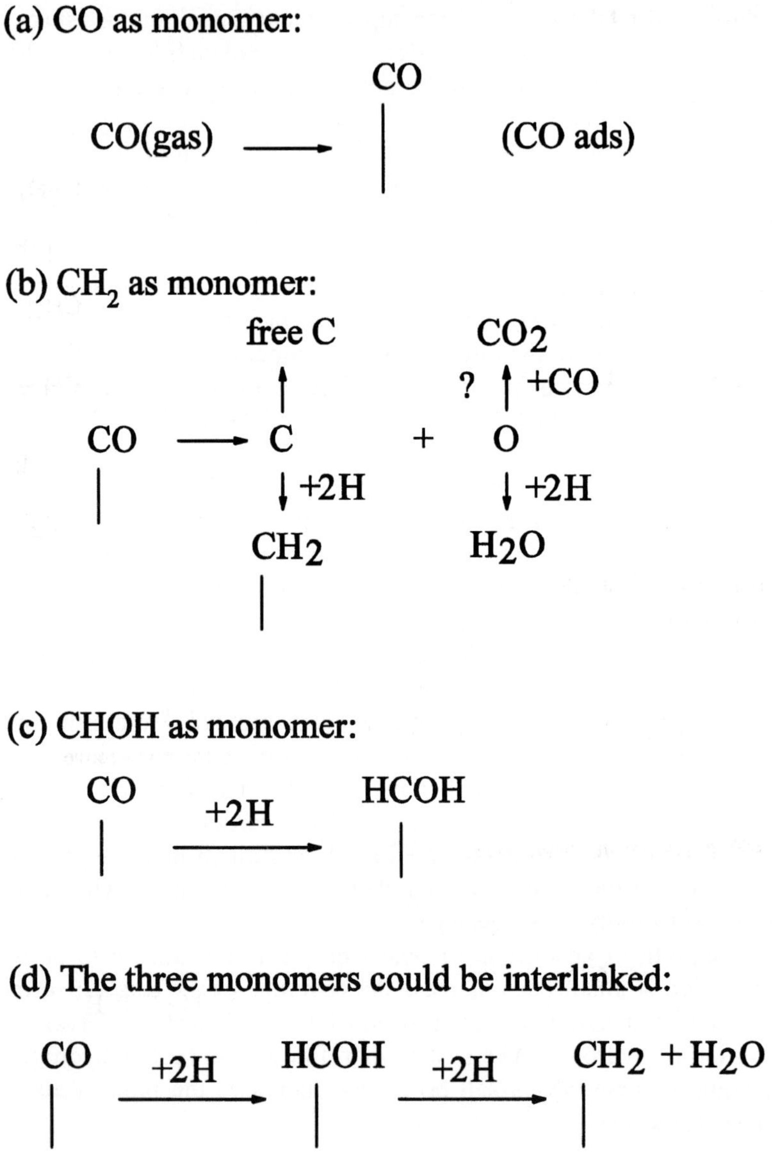

(a) CO as monomer:

$$CO(gas) \longrightarrow \overset{\displaystyle CO}{\overset{\displaystyle |}{}} \quad (CO\ ads)$$

(b) CH_2 as monomer:

$$
\begin{array}{ccccc}
 & & \text{free C} & & CO_2 \\
 & & \uparrow & & ?\ \uparrow\ +CO \\
CO & \longrightarrow & C & + & O \\
| & & \downarrow +2H & & \downarrow +2H \\
 & & CH_2 & & H_2O \\
 & & | & &
\end{array}
$$

(c) CHOH as monomer:

$$\overset{\displaystyle CO}{\overset{\displaystyle |}{}} \quad \xrightarrow{\ +2H\ } \quad \overset{\displaystyle HCOH}{\overset{\displaystyle |}{}}$$

(d) The three monomers could be interlinked:

$$\overset{\displaystyle CO}{\overset{\displaystyle |}{}} \quad \xrightarrow{\ +2H\ } \quad \overset{\displaystyle HCOH}{\overset{\displaystyle |}{}} \quad \xrightarrow{\ +2H\ } \quad \overset{\displaystyle CH_2 + H_2O}{\overset{\displaystyle |}{}}$$

Figure 8   Possible Monomers

CO as monomers (CO Insertion)

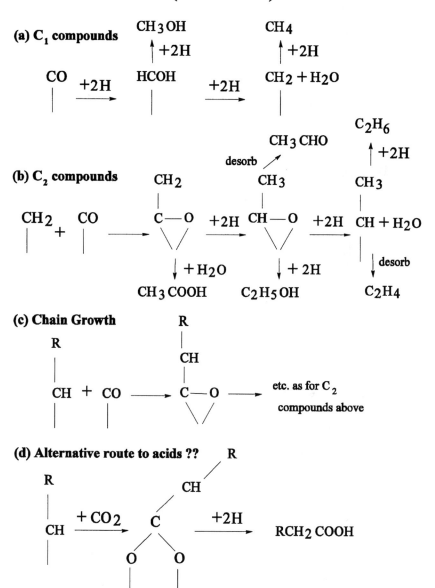

Figure 9   CO as Monomer

CH$_2$ as monomer

(a) C$_1$ compounds

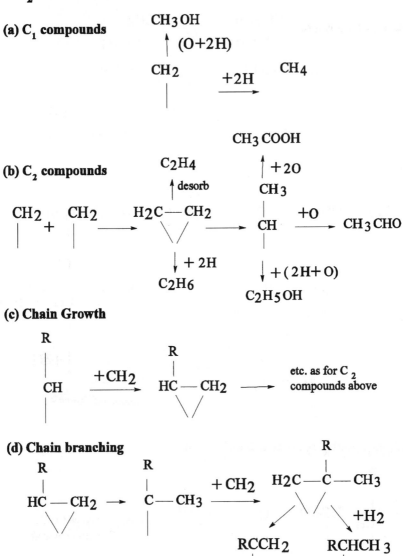

(b) C$_2$ compounds

(c) Chain Growth

(d) Chain branching

Figure 10   CH$_2$ as Monomer

CHOH as monomer

(a) C₁ compounds

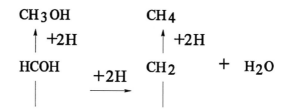

(b) C₂ compounds

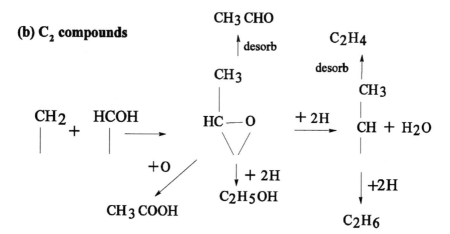

(c) Chain Growth

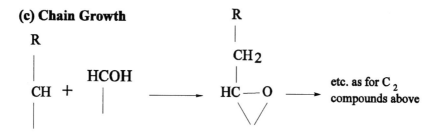

Figure 11   HCOH as Monomer

The higher the coverage by monomers the higher the probability of chain growth is expected to be. Two key steps leading to chain termination are desorption of the chains to yield alkenes and hydrogenation of the chains to yield alkanes. The lower the partial pressure of CO the lower the surface coverage by monomers and thus the higher the probability of desorption, i.e. of chain termination. The higher the partial pressure of hydrogen is the higher is the probability of chain termination by hydrogenation. Overall it could be argued that the higher the ratio H_2/CO the higher the probability of chain termination and thus the higher the selectivity of methane and the lower molecular mass hydrocarbons.

In an operating FT system the actual situation on the catalyst surface is probably more complex as the surface coverage by CO and H_2 could also depend on competition for vacant surface sites by other gases such as CO_2 and H_2O. As CO chemisorption is much stronger than H_2 chemisorption it is likely that the partial pressures of CO_2 and of H_2O will have a greater negative effect on the amount of hydrogen adsorbed than on the amount of CO adsorbed. The selectivity would then not simply be dependent on the simple ratio H_2/CO but on a more complex ratio, for example, such as

$$P_{H2}^{a}/(P_{CO}^{b} + P_{CO2}^{c} + P_{H2O}^{d}).$$

The actual partial pressures of the gases in contact with the catalyst at various positions along the catalyst bed will depend on several factors. The main contributing factors are the following; the composition and flow rate of the fresh feed gas; the H_2 to CO usage ratio of the FT reaction; the difference between the feed ratio and the usage ratio; the extent of conversion; the extent to which the water gas shift (WGS) reaction occurs and the amount of the exit gas (tail gas) that is recycled to the reactor. Recycling tail gas is a common practice as it results in higher overall conversion of the fresh feed and for fixed bed reactors it is required to assist with heat removal. Typically recycle is necessary because the per pass conversion is limited to avoid excessively high water partial pressures in the reactor.

For cobalt based catalysts the extent of the WGS reaction is almost negligible. Iron catalysts, however, are active WGS catalysts and so this can have a significant effect on the composition of the gas inside the reactor. At the lower temperatures, 210 to 240 °C, the rate of the WGS reaction is slow but nevertheless some CO_2 may be produced or consumed depending on the feed gas composition. At higher temperatures, above 300 °C, the WGS rapidly proceeds to equilibrium. In practice this means that when the conversion of CO via the FT reaction is high, i.e. the partial pressure is low, the reverse WGS reaction occurs, producing CO, which then reacts further in the FT reaction. By this means CO_2 is effectively converted to FT products. So it often occurs that CO_2 is produced at the entrance of the reactor and consumed near the exit.

9.3.1 Iron based catalysts

Using the Sasol pilot plants the influence of various total and partial pressures and of gas compositions on the FT selectivity was investigated. Typical industrial iron based catalysts used in the LTFT and HTFT processes were used. The operating temperatures were fixed at those used in the commercial processes. Tables 19 and 20 give the results for the LTFT and HTFT studies respectively. For the LTFT tests precipitated/supported iron catalyst was used and for the HTFT tests fused catalysts were used (their preparations are described in Chapter 7).

Table 19
Reaction over fixed bed precipitated iron catalyst, 225°C [3]

		Reactor entry pressure bar abs.			H_2/CO ratio	Selectivity % C atom			Diesel alkenes %
	Total	P_{CO2}	P_{H2}	P_{CO}		Hard wax	NAC[a]	Acids[b]	
A	7.8	0.08	4.34	2.28	1.9	33	1	0.006	39
	14.7	0.15	8.19	4.31	1.9	31	-	0.012	37
	18.1	0.18	10.08	5.30	1.9	34	1.3	0.014	40
	21.5	0.21	11.97	6.30	1.9	32	1.2	0.018	35
B	11.4	0.66	5.91	2.35	2.5	18	3.1	0.21	15
	27.6	1.39	14.07	5.87	2.4	18	3.8	0.55	17
C	27.7	1.24	14.3	6.40	2.2	22	4.0	0.55	22
	27.5	4.05	12.8	6.10	2.1	23	4.0	0.62	25
D	28	3.78	16.6	2.79	7.3	13	3.4	0.1	7
	28	1.52	13.3	3.63	3.7	22	2.6	0.2	14
	28	2.00	12.1	5.58	2.2	32	2.1	0.3	27
	28	1.38	9.96	5.37	1.9	37	2.1	0.3	31
	28	1.52	10.8	10.8	1.0	48	9.5	0.2	52

a. Water soluble non-acidic products (mainly alcohols)
b. Water soluble acidic products
Operating Modes:
A - Near differential (high fresh feed flow, no recycle, low % conversion)
B - Recycle/fresh feed ratio 1.26; H_2 + CO conversion 60% (on fresh feed basis)
C - Recycle/fresh feed ratio 1.1; H_2 + CO conversion 50%
D - Recycle/fresh feed ratio about 2. High conversion levels

For the results shown in Table 19, the same catalyst was used in all cases. In series A and B the fresh feed gas composition was the same throughout. The linear gas velocities were the same in all cases.

From Table 19, sets A and B, it can be seen that for the LTFT process when the H_2/CO ratio of the feed gas is kept constant, increasing the total pressure from about 0.8 to about 2.8 MPa has little or no effect on the hard wax selectivity. Separate experiments were carried out at 2.6, 4.0 and 6.0 MPa, once again at fixed H_2/CO feed gas ratios, and confirmed that even at these higher pressures the wax selectivities remained constant. (In all the test runs the residence times of the gases in the reactor were kept constant by suitable adjustments of the gas flows.) From the correlation shown in Fig. 2 it follows that the total pressure has no effect on the overall hydrocarbon carbon number distribution.

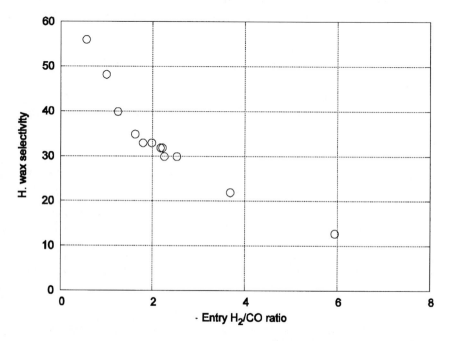

Figure 12 LTFT selectivity

From Table 19, sets A and B and the detail in set D, it appears that the wax selectivity is controlled by the H_2/CO ratio of the gas entering the reactor. In another separate set of runs the reactor entry partial pressure of H_2 was varied from 0.6 to 2.3 MPa, that of CO from 0.25 to 1.14 and that of CO_2 from 0.01 to 0.33 MPa. The correlation between the H_2/CO ratio at the reactor entrance and the hard wax selectivity is shown in Fig. 12 [42]. It is therefore clear that all that is required to control the hydrocarbon selectivity distribution is to control the H_2/CO ratio of the total feed gas to the reactor. As would be expected the alkene content of the hydrocarbon cuts decreases as the H_2/CO ratio increases (see diesel alkene % in Table 19). With some exceptions it appears that as the

H_2/CO ratio increased the alcohol selectivity increased. Over the entire pressure range studied, from 0.8 to 6.0 MPa, the acids appear to increase with increasing total pressure and the CO_2 partial pressure appears to be a significant parameter.

In general agreement with the above results, earlier studies at the US Bureau of Mines [43] found that, over iron lathe turnings as catalyst, the selectivity shifted to lower molecular mass products and the alkene content of the products decreased as the H_2/CO ratio of the feed gas was increased.

Table 20
Reaction over fluidized bed promoted fused iron catalyst[e], 320 °C [3]

a	Reactor entry pressure (bar abs.)					H_2/CO ratio	Selectivity % C atom			$C_2H_4/$ C_2H_6 ratio	Activity[b]
	Total	P_{CO2}	P_{CO}	P_{H2}	P_{H20}		CH_4	NAC[c]	Acids[d]		
A	9.4	1.1	0.90	3.86	0.04	4.3	26	1.4	0.20	0.24	48
	12.9	1.45	1.17	5.17	0.04	5.2	19	1.7	0.29	0.26	68
	14.5	1.52	1.31	5.66	0.04	5.3	17	1.9	0.34	0.25	80
	18.0	1.86	1.58	7.31	0.04	5.8	15	2.3	0.47	0.45	94
	21.5	2.00	1.93	9.66	0.04	5.9	11	3.5	0.81	0.67	120
B	10.8	0.85	1.02	4.27	0.04	4.2	18.3	2.0	0.2	0.5	54
	20.8	1.50	1.95	8.53	0.04	4.4	13.8	3.5	0.6	1.2	106
	40.8	2.82	3.58	14.2	0.04	4.0	10.5	4.1	1.1	1.2	207
	60.8	3.79	5.26	24.6	0.04	4.7	6.8	6.8	1.7	1.3	315
	75.8	4.55	6.70	30.7	0.04	4.6	5.5	7.1	1.9	0.9	375
C	18.0	1.03	1.65	7.24	0.04	4.4	19	1.5	0.2	0.28	98
		2.20	1.31	7.45	0.04	5.7	15	2.3	0.47	0.45	96
		3.10	1.31	7.17	0.04	5.5	9	2.6	0.63	0.34	98
		5.10	1.03	8.07	0.04	7.8	8	3.3	1.3	1.0	97
D	20.8	0.67	1.79	9.84	0.04	5.5	12	3.5	0.6	1.3	107
		0.92	1.80	9.62	0.72	5.3	12	3.2	0.7	1.6	105
		1.20	1.83	10.0	1.40	5.5	12	3.2	1.0	2.0	98
		1.57	1.84	10.3	2.79	5.6	11	5.9	3.2	2.0	83

a. Sets A and C are for medium basicity and B and D for high basicity catalysts
b. Moles $(CO + CO_2)$ converted per hour
c. Water soluble non-acidic products
d. Water soluble acidic products
e. The linear gas velocities were the same in all cases

Table 20 shows the results for four sets of experiments carried out with fused iron catalysts in fluidized bed reactors at about 320 °C (HTFT process). In sets A and B the total pressures were increased. The recycle ratios (ratio of fresh feed flow to recycled tail gas flow, the latter after knock-out of the water and oils) were kept constant and the total feed flow was increased in proportion to the increase in total pressure. The gas linear velocity, and hence the residence time inside the reactor, was thus maintained constant. In set C additional CO_2 was added to the feed gas and in set D additional water was added. Contrary to the findings for the LTFT operations (see Table 19 and Fig. 12) there was no correlation between the total feed H_2/CO ratio and the hydrocarbon selectivity as represented by the methane selectivity (see Fig. 1 for the correlation between methane and the various hydrocarbon selectivities). For set C, where additional CO_2 was added to the feed gas, the methane selectivity in fact decreased despite the increase in the H_2/CO ratio. Whereas for the LTFT process the total pressure had no apparent effect on the hydrocarbon selectivity, for HTFT it clearly shifted to higher molecular mass products, i.e. to lower methane selectivities, as the total pressure was increased (see sets A and B in Table 20).

Note again that the methane selectivity decreased with increasing pressure despite the tendency of the total feed H_2/CO ratio to increase. So again it is clear that the simple ratio of hydrogen to carbon monoxide does not control the overall hydrocarbon selectivity. In sets A, B and C it can be seen that as the CO_2 partial pressure increased the methane selectivity decreased and thus it appears that the latter pressure is also somehow involved in determining the hydrocarbon selectivity.

As observed for the LTFT operation, the acid selectivity increased with increasing pressure and again it appears that the CO_2 partial pressure is involved. The non-acid water soluble chemicals (NACs , mainly alcohols and ketones) also increased with pressure. Likewise the alkene to alkane ratios (as depicted by the ethene to ethane ratios in Table 20) in general also increased with pressure and again the CO_2 pressure appears to be involved. In general agreement with all of the above, earlier work by Anderson [44] found that over fused iron catalysts increasing the pressure resulted in an increase in the average molecular mass of the products as well as an increase in the yield of oxygenated products.

Table 21 shows the results for a series of runs carried out with a 'low basicity' fused catalyst in the HTFT operation. The partial pressures are those of the total feed gas to the reactor before any reactions have occurred. Three different 'selectivity factors' are included in the table. As can be seen the correlation of the methane selectivity with the ratio of $1/P_{CO2}$ is poor. The correlation with the H_2/CO ratio is reasonable, but of the three factors the ratio $P_{H2}^{0.25}/(P_{CO} + 0.7P_{CO2})$ is the best. The latter correlation is shown in Figure 13. This selectivity factor also correlated well with the methane selectivity for 'medium' and for 'high' basicity fused iron catalysts.

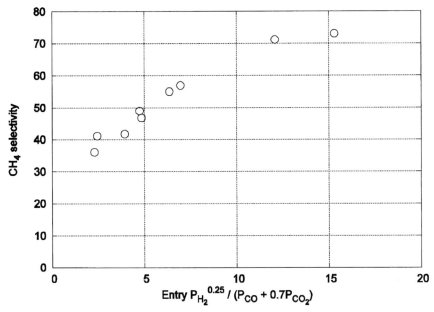

Figure 13 HTFT low basicity

From the previous discussion regarding surface coverage by CO and H_2 it could be expected that besides the partial pressure of CO_2 the partial pressure of H_2O should also influence the selectivity. Using the partial pressures at the stage where half of the conversions had taken place so that the partial pressures of water were significant it was found that the factors:

$$P_{H2}^{0.25}/(P_{CO} + 1.3P_{H2O})$$
$$\text{and} \quad P_{H2}^{0.25}/(P_{CO} + 0.7P_{CO2} + 0.6P_{H2O})$$

also correlated with the selectivity but they were not an obvious improvement on the factor $P_{H2}^{0.25}/(P_{CO} + 0.7P_{CO2})$ at the reactor entrance.

It has, however, been found that when different amounts of catalyst are loaded in a reactor and the feed gas composition and flow are kept constant (which results in different percentage conversions) the selectivities do not change much, despite the change in the gas composition along the reactor length. The value of the selectivity factor used in Fig. 13 does indeed change as the gas moves through the reactor and so an improved factor is required.

Table 21

Low basicity catalyst. Recycle operation, 320°C [3]

CH4 selectivity	PH2 MPa	PCO MPa	PCO2 MPa	$P^{.25}_{H2}/(P_{CO} + .7P_{CO2})$	H2/CO ratio	1/PCO2
36	0.708	0.216	0.261	2.30	3.3	3.8
41	1.81	0.335	0.18	2.52	5.4	5.6
42	1.047	0.242	0.103	3.22	4.3	9.7
47	0.819	0.147	0.068	4.89	5.6	14.7
49	1.09	0.162	0.073	4.79	6.7	13.7
55	0.984	0.116	0.057	6.39	8.5	17.5
57	0.51	0.029	0.131	7.00	17.6	7.6
71	1.23	0.082	0.007	12.12	15.0	14.2
73	1.28	0.068	0.002	15.33	18.8	500

Note: Partial pressures at reactor entrance.

Three different factors were considered, namely,

$$P_{H2}^{0.25}/(P_{CO} + 0.7P_{CO2}),$$
$$P_{H2}^{0.25}/(P_{CO} + 1.3P_{H2O})$$
$$\text{and } P_{H2}^{0.25}/(P_{CO} + 0.7P_{CO2} + 0.6P_{H2O}).$$

Table 22

Comparison of selectivity factors along catalyst bed*[1]

CH₄ Select	Bed Position	Selectivity Factors		
		$P_{H2}^{.25}/(P_{CO}+.7P_{CO2})$	$P_{H2}^{.25}/(P_{CO}+1.3P_{H2O})$	$P_{H2}^{.25}/(P_{CO}+.7P_{CO2}+.6P_{H2O})$
	Entrance	1.12	1.25	1.09
21	Middle	1.51	1.24	1.16
	Exit	2.36	1.13	1.25
	Entrance	0.73	0.86	0.71
12	Middle	1.04	0.70	0.75
	Exit	1.95	0.58	0.79
	Entrance	0.27	0.37	0.27
7	Middle	0.28	0.28	0.27
	Exit	0.28	0.22	0.27

*[1] In all cases the partial pressures are those after the WGS has proceeded to equilibrium. Pressures are in bar abs.

It was found that the values, although somewhat different, of all three factors at the reactor entrance (after WGS equilibrium, but before any FT reaction) correlated equally well with the overall hydrocarbon selectivity. The values of these factors at the entry, half conversion and exit positions in the reactor for three separate FT runs, which had different methane selectivities, are shown in Table 22. As can be seen the factor in which CO, CO_2 and H_2O were included in the denominator, on the whole showed the least amount of variation with catalyst bed depth. This factor may therefore be preferred because it best matches the observation that the selectivity does not change much with extent of conversion in the reactor.

Table 23 and Fig. 14 show the correlation between this factor and the methane selectivity for a catalyst with a 'medium basicity'. Table 24 and Fig.15 show the correlation for a 'high basicity' fused iron catalyst. As can be seen from the latter two figures there appears to be a satisfactory correlation between the 'preferred' selectivity factor and the methane selectivity, and consequently with the overall hydrocarbon selectivity.

From the slopes of the plots of Figs. 13 to 15 it appears that as the basicity of the catalyst increased the gas composition, i.e., the selectivity factor, had a smaller effect on the overall selectivity. In view of the strong influence of the alkali content of the iron catalyst on the selectivity this finding is not surprising. Overall, by suitable manipulation of the gas composition and of the basicity of the iron catalyst the methane selectivity can be varied from about 5 to 80%.

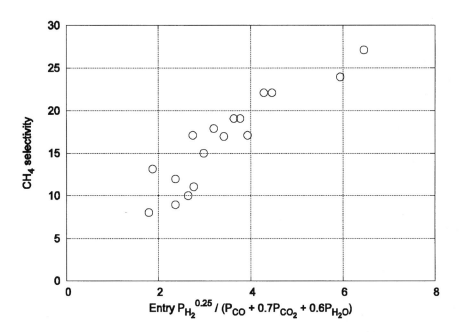

Figure 14 HTFT medium basicity

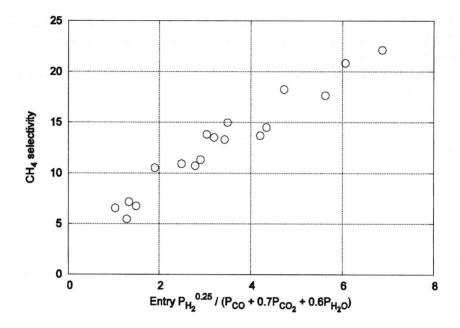

Figure 15 HTFT high basicity

Table 23

Medium basicity catalyst. Recycle operation, 320 °C [3]

CH4 selectivity	PH2 MPa	PCO MPa	PCO2 MPa	PH20 MPa	$P^{-.25}_{H2}/$ $(P_{CO} + .7P_{CO2} + .6P_{H20})$	H2/CO ratio
8	0.752	0.172	0.428	0.075	1.80	4.37
12	0.724	0.248	0.186	0.022	2.38	2.92
15	0.731	0.158	0.186	0.034	2.99	4.63
17	0.566	0.131	0.152	0.026	3.43	4.32
17	0.607	0.2	0.159	0.019	2.73	3.04
18	0.6	0.179	0.124	0.017	3.19	3.35
19	0.71	0.172	0.09	0.015	3.76	4.13
19	0.517	0.117	0.145	0.026	3.63	4.42
22	0.993	0.166	0.069	0.017	4.45	5.98
22	0.738	0.166	0.062	0.011	4.29	4.46
24	0.966	0.138	0.034	0.010	5.92	7.00
27	1.014	0.131	0.028	0.009	6.44	7.74

Notes: Entrance partial pressures after WGS, before FT. For all runs total feed consists of fresh feed plus recycled tail gas (after water and oil knock-out).

In practical commercial operations the water content of the total feed gas to the reactors is low and so in the case of the HTFT process the 'preferred' selectivity factor at the reactor entrance (before any reaction occurs) simplifies to $P_{H2}{}^a/ (P_{CO} + bP_{CO2})$. For the 'high basicity' catalyst the correlation between this simplified factor (at the reactor entrance) and the methane selectivity is shown in Fig. 16. The correlation appears to be as satisfactory as that shown in Fig. 15 and so for practical day to day control of the selectivity of the commercial reactors this simpler factor can be used.

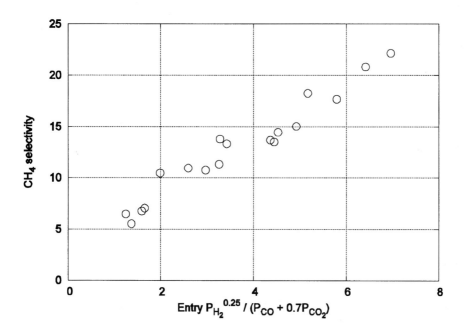

Figure 16 HTFT high basicity

Fig. 17 shows the plot of methane selectivity against the H_2/CO ratio (see Table 24) at the reactor entrance. The plot clearly confirms that, unlike for the LTFT process, the simple ratio cannot be used to predict the selectivity of the HTFT process.

252

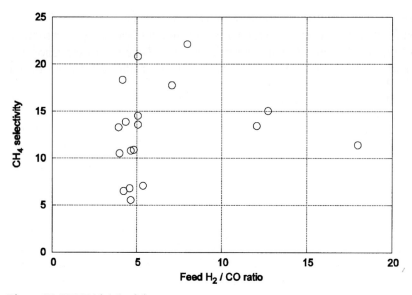

Figure 17 HTFT high basicity

Table 24

High basicity catalyst, recycle operation, 335 °C

CH4 selectivity	PH2 MPa	PCO MPa	PCO2 MPa	PH2O MPa	$P^{.25}_{H2}/$ $(P_{CO} + .7P_{CO2} + .6P_{H2O})$	H2/CO ratio
22.1	1.155	0.145	0.006	0.002	6.89	8.0
20.8	0.49	0.098	0.049	0.008	6.10	5.0
18.3	0.427	0.102	0.085	0.015	4.74	4.2
17.7	1.159	0.164	0.023	0.006	5.65	7.1
15.0[a]	1.322	0.104	0.207	0.094	3.51	12.7
14.5	0.931	0.184	0.051	0.01	4.35	5.1
13.8	0.853	0.195	0.15	0.024	3.06	4.4
13.7	0.964	0.19	0.056	0.01	4.21	5.1
13.5	1.345	0.112	0.234	0.1	3.21	12.0
13.4	1.175	0.296	0.01	0.001	3.43	4.0
11.4	1.148	0.064	0.277	0.159	2.93	17.9
10.9	1.766	0.366	0.117	0.021	2.50	4.8
10.8	1.129	0.241	0.162	0.025	2.79	4.7
10.5	1.422	0.358	0.282	0.035	1.89	4.0
7.1[a]	0.914	0.171	0.672	0.134	1.35	5.3
6.8	2.46	0.526	0.379	0.061	1.51	4.7
6.5[a]	0.766	0.183	0.903	0.124	1.05	4.2
5.5	3.073	0.67	0.455	0.071	1.28	4.6

Notes: Entrance partial pressures after WGS, before FT.
    a. Fresh feed consisted only of CO_2 and H_2.
For all runs total feed consisted of fresh feed plus recycled tail gas (after water and oil knock-out).

For both the commercial LTFT and HTFT processes it has been shown that a satisfactory correlation between the gas composition inside the reactor and the product selectivity can be established. The actual situation inside the individual catalyst particles, however, may be more complex. Under normal FT conditions the catalyst pores are at least partly filled with liquid waxes and other high molecular mass products. The solubility in, and the rates of diffusion of the different gases through, the liquid phase will be different and hence their actual concentration at the catalyst surface sites could be significantly different from those in the gas phase. This does, however, not negate the fact that the gas composition outside the catalyst particles can be used to manipulate the overall product selectivity.

9.3.2 Cobalt catalysts

Studies with cobalt catalysts showed that when the feed gas H_2/CO ratio was lowered the spectrum shifted to higher molecular mass products and the alkene and the alcohol content increased [6]. Tests at Ruhrchemie with a $Co/Ni/ThO_2/Mn/Kieselguhr$ catalyst yielded similar results [6] and later studies with a $Co/ThO_2/Kieselguhr$ confirmed this trend [43, 45]. In general therefore cobalt catalysts react to changes in the H_2/CO ratio in the same way as do iron catalysts in the LTFT process (200 to 240°C). On increasing the operating pressure from 0.1 to 1.5 MPa the wax selectivity increased [9, 46]. In a more recent study with an alumina supported Co catalyst the above pressure effect was confirmed. It was found that as the pressure was increased, at fixed residence time, the wax selectivity increased from virtually zero at 0.2 MPa to about 60% at 4.0 MPa. The light gas selectivity decreased and lined out at about 13% [47]. This effect of pressure over Co catalysts is unlike that observed with Sasol commercial alkali promoted precipitated iron catalysts in the LTFT process which showed the same high wax selectivity as the pressure was increased. As a matter of interest Pichler reported that when the pressure over a ruthenium catalyst was increased up to 100 MPa the wax selectivity increased markedly, virtually into the lower end of polyethylene range [23]. For an alumina supported Ru catalyst it was found that the dominant factor controlling the wax selectivity was the CO partial pressure [8].

When the feed gas consisted of CO_2 and H_2 earlier studies found that over Co, Ni and Fe catalysts methane was the main product whereas at the same temperature and pressure these catalysts produced large amounts of liquid products when CO and H_2 was the feed gas [48-50]. These results were recently confirmed over both silica and alumina supported Co catalysts operating at 210°C and 2.4 MPa [51]. In these experiments the conversion of CO_2 was as high as for the CO runs. A possible reason for the high methane selectivity obtained with CO_2 could be the very poor water gas shift (WGS) activity of Co which would result in very high H_2/CO ratios in the catalyst bed. In an attempt to increase the WGS activity of a silica supported, Pt promoted, Co catalyst, MnO was incorporated into the catalyst but this failed to produce the desired result. Operating at 190°C and 1.0 MPa, the methane selectivity was 95% when feeding CO_2/H_2 while with CO/H_2 the methane selectivity was 10% [52]. Note that when operating with iron based catalysts at the higher temperature of about 330 °C a feed gas consisting of CO_2/H_2 produces a product spectrum similar to normal syngas (see Table 24). The reason for this is of course that alkali promoted Fe at high temperatures is a very active WGS catalyst and so the carbon dioxide is readily converted to CO and thus the normal FT process takes place.

9.4 Production of oxygenated products

From previous discussion it is clear that alkenes probably are primary FT reaction products. Because of the ratios in which alkenes and alcohols are produced it is apparent that the formation of alcohols does not occur via the hydration of the primary alkenes. In fact the opposite, namely the dehydration of alcohols to form alkenes is thermodynamically possible. The probability of the latter occurring is supported by observations in Sasol pilot plant runs carried out under conditions of high alcohol selectivity (about 50% in the oil fraction).

It was found that with time on stream the alcohol selectivity declined while that of the alkenes increased and that of the alkanes remained virtually constant. With time on stream the percent conversion of syngas also declined, i.e. the partial pressure of water vapour inside the catalyst bed declined. This is in keeping with the observed shifts with time on stream, as the rate of alcohol dehydration to alkenes should increase with decreasing water partial pressure. Overall then it appears that alcohols too are primary FT products and it could even be possible that at least some, if not all, of the alkenes are formed by dehydration of these primary alcohols.

Possible reaction sequences resulting to the formation of oxygenated compounds (alhohols, aldehydes and acids) are discussed in Chapter 8. In the Sasol pilot plant runs mentioned above the oil fraction typically contained about 50% alcohols and about 20% alkenes. This is not at all unique to the iron catalysts or to the operating conditions employed in these tests. Previous studies with a cobalt catalyst operating under mild conditions (160 to 175 °C) reported that the oil produced contained 40% alcohols [22]. The same workers found that when the same cobalt catalyst was operated at 200 °C the oil only contained 1% alcohols. Pichler found that with cobalt catalysts only small amounts of oxygenates were produced at atmospheric pressure whereas at 12 bar, 175 °C, 1.0 H_2/CO feed gas to an alkali promoted Co/kieselguhr catalyst the 195 to 250 °C oil fraction contained about 30% alcohols [21].

It should be borne in mind that cobalt is a much more active catalyst than iron carbide for hydrogenation reactions, but despite this, large amounts of alcohols can be formed over cobalt catalysts under mild conditions. This suggests that also for FT over cobalt catalysts alcohols are primary products.

In earlier studies carried out in Germany it was found that operating with iron catalysts at 220 °C produced an oil fraction containing 45% alcohols [53]. The so called 'Synol' process operated with alkali promoted fused iron catalyst at about 20 bar and 190 to 220 °C at high space velocities with a fed gas having a H_2/CO ratio of about one. The 180 to 260 °C oil fraction contained about 65% alcohols, 25% alkenes, the balance being aldehydes, ketones, esters and acids [21]. In other studies it was found that lowering the temperature from 250 to 230 °C resulted in a decrease in the alkene content and a sharp increase in the alcohol content of the FT products [21]. Increasing the pressure from 12 to 50 bar decreased the alkene content but increased the alcohol content of the oil products [21]. Note from the two forgoing findings that there appears to be an inverse relation between alkenes and alcohols as was observed in the observed shifts with time on stream for the Sasol tests mentioned above.

Studies at the U. S. Bureau of mines showed that nitriding alkali promoted iron catalysts increased the FT activity as well as the alcohol selectivity [54]. The overall selectivity, however, shifted to lower molecular mass products. Increasing the operating pressure (up to 10.3 MPa) on nitrided iron catalysts was found to increase both the average molecular mass of the products as well as the alcohol selectivity [55].

Thus overall it can be concluded that the production of alcohols is favoured by operating at relatively low temperatures, high space velocities and high pressures. The high space velocity, i.e. short residence time in the catalyst bed, should minimise the conversion of the primary formed alcohols to other products such as alkanes and alkenes. Regarding the requirement of high pressure it can be presumed that both the CO and H_2 partial pressures need be high as the former will enhance the CO insertion step and the latter will enhance the subsequent hydrogenation to produce the alcohol (e.g. see Fig. 9). The increased alcohol selectivity at lower temperatures could also be related to the concept that CO insertion is the route to alcohol formation. At the lower temperatures the coverage of the surface by CO molecules would be higher because the likelihood of decomposition to C and O atoms would be less.

REFERENCES

[1] K.D. Tillmetz, Chem.-Ing.-Tech. 48 (1978) 1065.

[2] E. von Christoffel, Surjo,I., Baerns, M., Chemiker-Z. 102 (1978).

[3] M.E. Dry, in Catalysis Science and Technology, Chapt. 4, Vol 1, J.R. Anderson, M. Boudart, (eds.), Springer-Verlag, 1981, pp. 159.

[4] H. Pichler, H. Schulz, D. Kuhne, Brennt.-Chem. 49 (1968) 344.

[5] J.H. le Roux, L.J. Dry, Appl. Chem. Biotechnol. 24 (1974).

[6] R.B. Anderson, in Catalysis, Vol. 4, P.H. Emmett, (ed.), Reinhold, 1956

[7] F.R. Craxford, Fuel (London) 26 (1947).

[8] R.C. Everson, E.T. Woodburn, and A.R. Kirk, J. Catal. 53 (1978) 186.

[9] F. Martin, Chem. Fabrik. 12 (1939).

[10] R.J. Madon, E.R. Bucker and W.F. Taylor, in ERDA Report (1977) 8008

[11] E.F.G. Herrington, Chem. Ind. (London) (1946).

[12] R.B. Anderson and R.A. Friedel, J. Amer Chem. Soc. 2 (1950) 2307.

[13] P. J. Flory, JACS, 58, 1877 (1936).

[14] I. Puskas and R.S. Hurlbut, Catalysis Today 84 (2003) 99-109.

[15] M.E. Dry, Applied Catal. A: Gen., 138 (1996) 319.

[16] D.L. King. J. Catal. 51 (1978) 386.

[17] H. Schulz, B.R. Rao, M. Elstner. Erdol Kohle 23 (1970) 651.

[18] L. Caldwell, CSIR (South Africa) report CENG 300, June (1980).

[19] L. Caldwell and. D.S. van Vuuren, Chem. Eng. Sci. 41 (1986) 90.

[20] S.T. Sie and R. Krishna., Applied Catalysis A: Gen. 186 (1999).

[21] H. H. Storch, N. Golumbic and R.B. Anderson, The Fischer-Tropsch and related synthesis, John Wiley, New York, 1951.

[22] D. Gall, E.J. Gibson and C.C. Hall, J. Appl. Chem. (London) 2 (1952) 371.

[23] H. Pichler, Advanc. Catal. vol. 5, Rideal (ed.), N.J. Academic Press, 1952.

[24] H. Schulz, Chemierohstoffe aus Kohle, Thieme Verlag, Stuttgart 1977.

[25] M.A. Vannice, J. Catal. 37 (1975) 449.

[26] E. Iglesia, Applied Catalysis A: Gen. 161 (1997) 59.

[27] D. Schanke, A.M.Hilmen, E. Bergene, K. Kinnari, Rytter, E., Adnanes, E. and Holmen, A., Energy to Fuels 10 (1996) 867.

[28] A. M. Hilmen, D. Schanke, K.F.Hanssen, A. Holmen, Applied Catalysis A: Gen. 186 (1999) 169.

[29] P.J. van Berge, J. van de Loosdrecht, J.L. Visage, WO 01/96014, Sasol Technology (2000).

[30] P.J. van Berge, J. van de Loosdrecht, S. Barradas, A.M. van der Kraan, in Syngas to Fuels and Chemicals Symposium, Div. Petr. Chem. USA (1999) 84.

[31] R.L. Espinoza, J.L. Visage, P.J. van Berge, F.H. Bolder, European Pat. Appl. No. EP 0736326A1 (1996).

[32] E. Iglesia, S.L. Soled, R.A.Fiato, G.H. Via, J. Catal. 143 (1993) 345.

[33] J.G. Goodwin, A. Kogelbauer, R.Oukari, J. Catal. 125 (1996) 160.

[34] S. Eri, J.G. Goodwin, G. Marcelin, T. Riis. U.S. Pat. 4,801,573 (1989). Assigned to Norske Stats.

[35] A. Holmen, D. Schanke, S.Vada, E.A. Blekkan, A.M. Hilmen, and A.Hoff, J. Catal. 156 (1995) 85.

[36] G.J. Haddad, B. Chen and J.G. Goodwin, J. Catal. 161 (1996) 274.

[37] A. Feller, M. Claeys and E. van Steen, J. Catal. 185 (1999) 120.

[38] H. Hammer, D. Bittner, Erdol u Kohle, Erdgas Petrochemie 31 (1978) 369.

[39] H.S. Seelig, H.I.Week, D.J. Voss, and J. Zisson, , in ACS Meeting, Sept. 1952, Atlantic City, USA.

[40] M.E. Dry, T.Shingles, L.J. Boshoff and G.J.Oosthuizen, J. Catal. 15 (1969) 190.

[41] H. Kolbel, W.K. Muller, E. Schottle, and H. Hammer, Angew. Chem. 2 (1963) 554.

[42] M.E. Dry, Encyclopedia of Catalysis, John Wiley and Sons, New York, 2003.

[43] A. Forney, W.P. Heynes, J.J. Elliott, and A.C. Zarochak, ACS Div. Fuel 20 (1975) 3.

[44] RB. Anderson, B. Seligman, J.F. Schulz, R. Kelly and M.A Elliott., Ind. Eng. Chem. 44 (1952) 391.

[45] C. Kibby, R. Panell, and T. Kobilinsky, ACS Div. Petr. Chem. 29 (1984) 1113.

[46] F. Fischer and H. Pichler, Brennst. Chem. 20 (1939) 41.

[47] P.J. van Berge, Natural Gas Conversion IV, Studies in Surface Science and Catalysis 107 (1997) 207.

[48] H. Koch, H. Kuster, Brennst. Chem. 14 (1933) 245.

[49] F. Fischer and H. Pichler, Brennst. Chem. 14 (1933) 306.

[50] W.W. Russel and G.H. Miller, J.A.C.S. 72 (1950) 2446.

[51] J. Zhang, G. Jacobs, D.E. Sparks, M.E. Dry, and B.H.Davis, Catalysis Today 71 (2002) 411.

[52] T. Riedel, M. Claeys, H. Schulz, G. Schaub, S. Nam, K. Jun, M. Choi, G. Kishan and K. Lee, Applied Catalysis A: Gen. 186 (1999) 201.

[53] H.H. Storch, R.B. Anderson, L.J. Hofer, C.O. Hawk,H.C. Anderson, N. Golumbic, U.S. Dept. of Interior, Bureau of Mines technical report no. 709 (1948).

[54] R.B. Anderson, J.F. Schulz, B. Seligman, W.K. Hall, and H.H. Storch, J. Amer Chem. Soc. 72 (1950) 3205.

[55] S. Friedman and M.D. Schlesinger, U.S. Bur. of Mines Bull. (1964) 614.

[56] E.W. Kuipers. I.H. Vinkenburg and H. Oosterbeek. J. Catal.152 (1995) 137.

[57] E.W. Kuipers, C. Scheper, J.H. Wilson, I.H. Vinkenburg and H. Oosterbeek. J.Catal.158 (1996) 288.

Studies in Surface Science and Catalysis 152
A. Steynberg and M. Dry (Editors)

Chapter 4

Synthesis gas production for FT synthesis

**K. Aasberg-Petersen[a], T.S. Christensen[a], I. Dybkjær[a], J. Sehested[a],
M. Østberg[a], R. M. Coertzen[b], M. J. Keyser[b] and A.P. Steynberg[b]**

[a]Haldor Topsøe A/S
Nymøllevej 55, Lyngby, Denmark

[b]Sasol Technology R&D,
P.O. Box 1, Sasolburg, 1947, South Africa

1. INTRODUCTION

The technology used to prepare the synthesis gas used for Fischer-Tropsch synthesis can be separated into two main categories, gasification and reforming. The use of these terms is not unambiguous, see Section 3.1. Gasification is the term normally used to describe processes for conversion of solid or heavy liquid feedstock to synthesis gas, while reforming is used for conversion of gaseous or light liquid feedstock to synthesis gas. Some technologies, particularly high temperature partial oxidation, can be used for a whole range of feeds and the literature may then refer to 'gasification' to include methane reforming. In a recent text book on the topic of gasification by Higman and van der Burgt [1], the term 'gasification' includes what is classified here as adiabatic oxidative reforming. The reader is advised to consult this book [1] for both more detailed description of gasification technologies and a broader description of applications that do not involve Fischer-Tropsch (FT) synthesis.

Future FT synthesis units may include the co-production of electrical power, so an understanding of the issues relating to power generation is also relevant to the application of FT technology. Even without power export, FT technology applications require large scale power generation for in-house use.

The most common feeds used to prepare synthesis gas for FT synthesis are coal, which is rich in carbon, and natural gas, which is rich in methane. The term 'carbonaceous' feed is commonly used and this implies any carbon containing feed material. Other feedstock examples are coal bed methane, heavy oils, bitumen and petroleum coke (petcoke). In general, the feed will become more desirable as the hydrogen content increases. The reasons for this are explained in Chapter 5.

It is usually better to convert carbonaceous liquids directly to refined products, for example by some type of hydroprocessing technology, rather than by conversion via synthesis gas. Gaseous hydrocarbons are already valuable products if they can be piped to nearby consumers. Only remotely located gas is considered for large-scale conversion to liquid fuels. However, smaller scale applications for the production of high value chemicals are common wherever natural gas is found. Coal conversion is more expensive than natural gas conversion but may be worthwhile if the coal price is low enough and if both electricity and higher value hydrocarbon products are co-produced with liquid fuels on a large scale.

2. SYNTHESIS GAS PREPARATION VIA REFORMING

2.1 Introduction

The synthesis gas preparation section is an important part of the entire GTL complex. It is the most expensive of the three process sections (synthesis gas preparation, FT synthesis, and product work-up), and it is responsible for the largest part of the energy conversion in the plant. The design of the synthesis gas preparation unit is, therefore, critical for the economics of a GTL project. However, as explained in Chapter 5, the design of the synthesis gas section depends critically on factors outside the section itself, ultimately of course on the feedstock characteristics and the desired product slate. The product slate will determine the choice of type of FT synthesis (HTFT, LTFT, iron or cobalt-based catalyst, type of synthesis reactor, etc.) and the operating conditions in the chosen synthesis unit, and this will in turn determine the desired characteristics of the synthesis gas. The feedstock characteristics will then, together with the desired product gas, dictate the choice of synthesis gas technology to be applied. When the feedstock is natural gas or similar, the choice will be between various forms of reforming and gasification technologies and combinations thereof.

Without going into details it is noted that with respect to desirable synthesis gas characteristics, mainly two situations exist:

A. The FT synthesis catalyst has, in addition to its activity for the FT synthesis reaction, also activity for the shift reaction. CO_2 is thus a reactant, and the desired synthesis gas composition is similar to that of methanol synthesis gas. It may be characterised by the so-called stoichiometric number (SN) or module $M = (H_2 - CO_2)/(CO + CO_2)$ or by the so-called Ribblett ratio $R = H_2/(2CO + 3CO_2)$. For stoichiometric gas for production of hydrocarbon products with H/C = 2.0 (or of methanol), $M = 2R = 2.0$.

B. The FT synthesis catalyst has no activity for the shift reaction. CO_2 is essentially inert, and the synthesis gas composition is best characterised by the H_2/CO ratio, which should ideally be somewhere between 2.0 and 2.2 depending on the overall H/C ratio in the product. However, the H_2/CO ratio actually required depends on catalyst characteristics and may differ from the ideal value.

In both situations, the ideal synthesis gas has a low content of inerts, mainly CH_4 and N_2, and a high CO/CO_2 ratio. Some claim that a high content of inerts, e.g. N_2 from synthesis gas preparation processes using air, is beneficial, since it assists in controlling the temperatures in the FT synthesis reactors. However, this argument is false. By proper design of the synthesis reactors, satisfactory temperature control can be achieved without dilution with inerts. Furthermore, a high content of inerts in the synthesis gas excludes internal and external recycle of tail gas, and this leads to undue restrictions in the optimisation of overall process schemes, especially when energy export from the GTL complex is undesirable.

In order to identify the optimum overall layout of the synthesis gas section, extensive integration studies involving intimate knowledge about synthesis gas technology and technology for FT synthesis and product work-up are required. The result of such studies will in most cases be that the synthesis gas section will have to process not only fresh feedstock, but also considerable amounts of gas recycled within the complex, with properties which are determined by the design of the other sections of the plant. This means that changes in e.g. operating parameters in one section will have an influence on the conditions for design of other sections, and this of course significantly complicates the studies.

It is not the purpose of the following sections to discuss the specific integrated solutions, which may result from studies for specific situations. This will be discussed in Chapter 5. Rather, this chapter will present technologies for conversion of natural gas to synthesis gas by various forms of reforming. Furthermore, the possible process schemes obtainable by combinations of the technologies will be described.

The technologies to be discussed are:

- Final feed gas purification
- Adiabatic prereforming
- Fired tubular and heat exchange steam reforming
- Adiabatic, oxidative reforming
- Other technologies such as CPO (Catalytic Partial Oxidation) and CMR (Ceramic Membrane Reforming).

An impressive amount of literature is available on the various types reforming. Examples discussing several different technologies and containing further references may be found in [2 – 9].

2.2 Feedstock purification

2.2.1 Feed gas characteristics and purification requirements

The feedstock for a GTL facility may be natural gas or associated gas. In both cases the primary feedstock may be treated in gas treating units to remove and recover higher hydrocarbons (NGL or LPG) and, when relevant, by gas sweetening to reduce high concentrations of CO_2 and/or S compounds. Gas treating technology is considered to be outside the scope of this section. Typically, the normal, treated feed will be lean with a high CH_4 content. However, the upstream treating units may fail, and it will in most cases be necessary to design the synthesis gas preparation unit in such a way that it can absorb significant changes in feed composition without adverse effects in downstream units.

Typical feed gas specifications are shown in Table 1.

Table 1
Typical feed gas specifications

	Natural Gas		Associated Gas	
	Lean	Heavy	Lean	Heavy
N_2, vol%	3.97	3.66	0.83	0.79
CO_2, vol%	-	-	1.61	1.50
CH_4, vol%	95.70	87.86	89.64	84.84
C_2H_6, vol%	0.33	5.26	7.27	6.64
C_{3+}, vol%	-	3.22	0.65	6.23
Max. total S, vol ppm	20	20	4	4
Hydrogen sulphide, vol ppm (typical)	4	4	3	3
COS, vol ppm (typical)	2	2	n.a.	n.a.
Mercaptans, vol ppm (typical)	14	14	1	1

The most important impurities to be considered in the feedstock purification unit are H_2S and other S compounds, since these compounds are poisons for downstream catalysts, both in the synthesis gas preparation section and in the FT synthesis section. In most actual cases details about type and concentration of individual S compounds will not be available or, if they are given, they will often be unreliable and prone to variations over time. However, it must be expected that organic S compounds such as COS, mercaptans, and also heavier compounds such as sulphides, di-sulphides and thiophenes may be present in low concentrations - perhaps too low to allow detection, but still high enough to be harmful to catalysts.

Other impurities, e.g. solids, moisture, and certain trace components such as Hg may be present in the raw feedstock. Removal of such impurities is considered part of gas treating. N_2 and CO_2 are often present in minor quantity. N_2 will generally be an inert with no detrimental effects other than the resulting dilution of the synthesis gas. Trace amounts of N compounds such as NH_3 and HCN may be formed in the reactors in the synthesis gas preparation section and may have to be removed from the synthesis gas before it is passed to the FT synthesis section. The effects of a content of CO_2 are significant and must be taken into account as described in the following paragraphs. Halogens such as organic Cl compounds or HCl are generally not present in typical feeds to GTL plants. If they are present they must be removed to a very low level since they may otherwise cause problems (catalyst poisoning and corrosion) in downstream units. Oxygenates may also be present in the feed, e.g. methanol added to the natural gas to avoid hydrate formation. Oxygenates may also be present in recycle gases originating from other sections of the plant.

There is no absolute value for the acceptable concentration of S compounds in the purified gas. However, for certain types of downstream catalysts, both in the synthesis gas preparation section and in the FT synthesis section, very low concentrations, preferably single digit ppb, are necessary to ensure acceptable lifetime. It may be noted that it is not a trivial task to determine S concentrations at these levels. First of all sampling is difficult, because most surfaces may adsorb and later desorb S compounds depending on the gas phase concentrations. Special materials and procedures are required to obtain a representative sample of the gas. Surveys of available analytical methods for analysis of total S concentration and concentration of individual S compounds in gases may be found in [10, 11]. With the methods described, the detection limit is about 10 ppb for individual S compounds and down to 2 ppb for total S content.

The main problem in final gas purification is thus to remove essentially all S compounds - type and concentration uncertain and variable – from the feed to a concentration in the purified synthesis gas that is preferably below the detection limit, using standard analysis equipment, of a few ppb.

2.2.2 Principles of gas desulphurisation

The typical process concept for desulphurisation of natural gas and similar feedstock is a two-step process based on hydrogenation of organic S compounds (and Cl compounds if they are present) and subsequent adsorption/absorption of H_2S (and HCl). This process concept has been used industrially for decades and is well documented in the literature, see e.g. [12 – 14]. A typical layout for desulphurisation of gas with a relatively high S concentration is shown in Fig. 1.

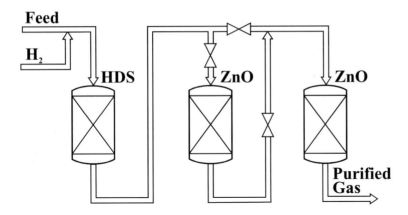

Figure 1 Typical process flow diagram for desulphurisation of natural gas

The feedstock is mixed with a small amount of hydrogen or hydrogen-rich gas, preheated to 350-400°C, and passed to a first reactor containing a hydrogenation catalyst, typically based on cobalt and molybdenum (CoMo) or nickel and molybdenum (NiMo). After the hydrogenation reactor the gas passes to two S absorbers in series, both typically containing zinc oxide (ZnO), which absorbs the H_2S formed in the hydrogenation reactor. In special cases, e.g. when the feed contains high concentrations of CO_2 and very low concentrations of S compounds, a special absorption material may be required downstream the ZnO. If organic Cl compounds are present in the feed, HCl will be formed in the hydrogenator. The HCl can react with ZnO and form $ZnCl_2$. At normal operating conditions in the S absorber, $ZnCl_2$ will sublimate and deposit on downstream catalysts or heat transfer surfaces. Therefore, HCl should be removed by a special absorbent, normally an activated alumina, placed in a vessel between the hydrogenator and the first S absorber.

Variations of the process concept shown in Fig. 1 may be used. In the simplest case, only one vessel is used with a top layer of hydrogenation catalyst and a bottom layer of ZnO. In other cases, the hydrogenation reactor is followed by only one S absorber. In special cases, a last layer of an absorption material with high chemisorption capacity may be installed in the bottom of the last S absorber or in a separate vessel after the last S absorber. In the layout shown in Fig. 1, valves and piping are installed so that the first S absorber can be isolated for change of ZnO as needed while the plant is running. In an alternative layout with two S absorbers, valves and piping are installed so that the vessel with fresh ZnO will always be the last of the two absorbers ('Swing operation'). This arrangement is not well-suited for cases with stringent requirements for purification, since some sulphur will always be absorbed on piping etc. upstream the absorber, and part of this sulphur will, after a 'swing', be passed to downstream catalysts.

Other processes such as cold absorption of S compounds on iron oxide or molecular sieves have been used to desulphurise natural gas before processing in e.g. ammonia and hydrogen plants, see e.g. [15, 16]. However, these processes can not provide the very high purity required in GTL applications; they will not be considered in the following text.

2.2.3 Reactions in the hydrogenator

The conversion of organic S-compounds over the hydrogenation catalyst depends on hydrogenolysis (addition of hydrogen over the S-C bond) by reactions such as:

$$R\text{-}SH + H_2 = RH + H_2S \qquad (1)$$
$$R\text{-}S\text{-}R_1 + 2H_2 = RH + R_1H + H_2S \qquad (2)$$
$$R\text{-}S\text{-}S\text{-}R_1 + 3H_2 = RH + R_1H + 2H_2S \qquad (3)$$
$$C_4H_8S \text{ (tetrahydrothiophene)} + 2H_2 = C_4H_{10} + H_2S \qquad (4)$$
$$C_4H_4S \text{ (thiophene)} + 4H_2 = C_4H_{10} + H_2S \qquad (5)$$
$$R\text{-}Cl + H_2 = RH + HCl \qquad (6)$$

All these reactions have very large equilibrium constants in the temperature range relevant for desulphurisation reactors, see [13]. This means that full conversion is achievable for all types of organic S compounds, if sufficient H_2 is present. If no or too little H_2 is present, the S compounds may react by thermal decomposition forming olefins and H_2S. For some compounds this may happen at temperatures prevailing in the preheaters upstream the hydrogenation reactor. This is undesirable, and H_2 should therefore preferably be added before preheating. It may be added that the same tendency can be seen for dehydrogenation of hydrocarbons. As an example, Fig. 2 shows the equilibrium concentration of butene in n-butane as function of temperature with and without the presence of H_2. It is seen that significant concentrations of olefins may be formed at high temperature in H_2-free gas.

The hydrogenolysis reactions are generally first order with respect to the S compound and between zero and first order with respect to H_2 [13] The reaction rate differs significantly for the different types of S compounds with sulphides ($R\text{-}S\text{-}R_1$) and thiophenes being the most difficult to convert followed by di-sulphides and mercaptans.

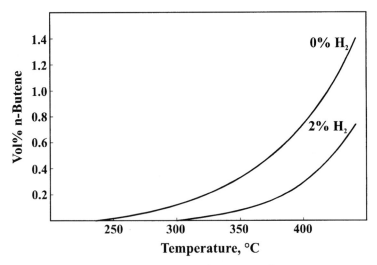

Figure 2 Equilibrium concentration of butene in n-butane at 30 bar g

Oxygenates such as methanol may react with H_2S in the hydrogenator to form organic S-compounds such as mercaptans and sulphides:

$$CH_3OH + H_2S = CH_3SH + H_2O \qquad (7)$$
$$2CH_3SH = (CH_3)_2S + H_2S \qquad (8)$$

If the temperature is too low, the rate of conversion of the organic S compounds may be too low, and some mercaptans or sulphides may pass unconverted through the hydrogenator. At sufficiently high temperature, practically complete conversion to H_2S (and COS) will be ensured.

Carbon oxides and carbonyl sulphide (COS) interact with H_2 and steam according to the following reactions:

$$COS + H_2O = CO_2 + H_2S \qquad (9)$$
$$CO + H_2O = CO_2 + H_2 \qquad (10)$$

Both these reactions will generally be at equilibrium after the hydrogenator. The equilibrium constants are shown in Fig 3.

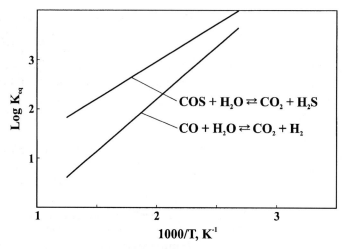

Figure 3 Equilibrium constants for COS hydrolysis and the shift reaction

In total a potentially rather complicated situation exists. However, as indicated above it may be expected that all the reactions (1) – (10) are active in the hydrogenator and reach a close approach to equilibrium. Calculations show that if this is the case, then the concentration of organic S compounds other than COS will be below 1 ppb at all temperatures below about 450°C. The equilibrium concentration of COS may be quite high, especially at high temperature and high CO_2 concentration.

2.2.4 Hydrogenation catalysts

The catalysts used for hydrogenolysis of the organic S compounds are based on either cobalt molybdate (CoMo) or nickel molybdate (NiMo). The catalysts are prepared by impregnation on high surface area carriers, usually alumina. The catalysts supplied by leading catalyst suppliers such as Haldor Topsøe, Südchemie and Johnson Matthey are all of this type, although differences obviously exist in formulation and shape.

The active phase in the operating catalyst is the so-called Co-Mo-S or Ni-Mo-S phases. The Co-Mo-S phase - and the corresponding Ni-Mo-S phase - is not a well-defined compound. Rather "it should be regarded as a family of structures with a wide range of Co concentrations, ranging from pure MoS_2 up to essentially full coverage of the MoS_2 edges by Co" (cited from p. 32 in [17], an authoritative review of hydrotreating catalysts and catalysis).

The hydrogenation catalysts are manufactured and supplied normally in the oxide state and must be converted to the sulphided state to gain full activity. This sulphidation will normally take place by exposing the catalyst as delivered to the normal operating conditions in the plant, i.e. the sulphur for the sulphidation is supplied by the feed and at the concentration at which it is available.

The degree of sulphiding and thereby the catalyst activity depends on the amount of sulphur present in the feedstock used for sulphiding and during normal operation. At high concentrations of sulphur in the feedstock, the sulphiding is rapid and complete and the activity is at a high and constant level. When the sulphur concentration in the natural gas is low, the amount of the active Co-Mo-S phase in the catalyst will decrease and the catalyst activity is likewise lower. The sulphidation will also take a long time, in extreme cases several months. In such cases the sulphidation may be accelerated and enhanced by addition of an easily decomposed S-compound (usually di-methyl-di-sulphide, DMDS) to the feed until full sulphidation has been achieved.

If the S content in the feed is constant and low, the stable S concentration in the catalyst and thus the activity will be low. The reduced activity is normally not a problem as long as the low sulphur content in the feedstock is constant. However, if the organic sulphur content suddenly increases, the catalyst may not convert these compounds sufficiently for a short period (days) due to insufficient activity. After a few days of operation at high S concentration the hydrogenation catalyst will be fully sulphided and the slip of organic sulphur goes down. In cases where such variations are foreseen, problems may be avoided by addition of a sulphur compound e.g. DMDS to the feed in periods with very low S concentration.

2.2.5 Reactions in the sulphur absorber

After the hydrogenation reactor the gas will, as explained above, mainly contain S in the form of H_2S. Any Cl will be present as HCl. If CO_2 is present in the hydrocarbon feed, significant amounts (several hundred ppb) of COS may also be present. Other S compounds may be present only in low ppb concentrations in the product gas from a well functioning hydrogenator.

In the absorption vessel, H_2S reacts with ZnO according to:

$$ZnO + H_2S = ZnS + H_2O \qquad (11)$$

The equilibrium constant for this reaction is shown in Fig.4.

In addition to the bulk phase reaction with H_2S the ZnO also has some activity for reaction (9), COS hydrolysis, and for reaction (10), the shift reaction. Absorption of H_2S will cause the COS hydrolysis (9) to proceed to full conversion, and COS will thus be completely removed by ZnO operating at the proper temperature. Finally, a certain capacity for chemisorption of H_2S must be taken into account when considering the performance of ZnO in the absorption vessel.

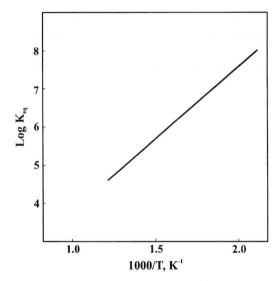

Figure 4 Equilibrium constant for reaction (11)

If the feedstock contains carbon dioxide, the reverse of the shift reaction (10) will cause the steam content in the gas to increase considerably. The increased steam content will have an impact on the equilibrium for absorption of H_2S on the ZnO, reaction (11). Fig. 5 shows, as an example, the equilibrium content of H_2S as function of temperature over ZnO at a CO_2 content in the feedstock of 5% and varying H_2 concentration.

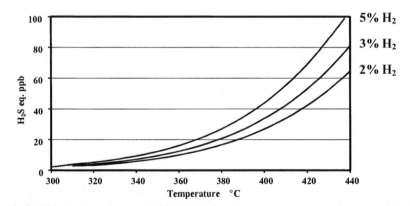

Figure 5 Equilibrium H_2S, 5% CO_2 in natural gas

Fig. 6 shows, for the case with 3% H_2 in the feed, the concentrations of CO, CO_2 and H_2 in the equilibrated gas. It is seen that significant amounts of CO may be formed. The possible formation of carbon in downstream equipment due to this presence of CO must be taken into account in the design, see [18].

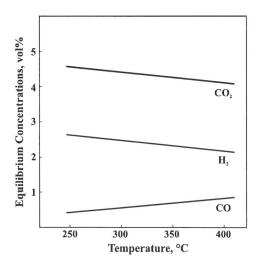

Figure 6   Equilibrium concentrations of CO, CO_2, and H_2; 3% H_2 and 5% CO_2 in the feed

From the above it is obvious that there are two ways to reduce the equilibrium level of hydrogen sulphide and CO over ZnO in cases where CO_2 is present in the feed:

• By reducing the temperature in the HDS section
• By reducing the H_2 recycle

It is noted that the temperature has a greater impact on the equilibrium sulphur content on ZnO than the amount of hydrogen recycle.

However, as previously discussed, both the hydrogen recycle and the temperature play a vital role with regard to the performance of the HDS section. If the hydrogen recycle is reduced, the reaction rate on the hydrogenation catalyst is decreased, and there is a risk that organic sulphur starts to leak. If the temperature is decreased, the hydrogenation reaction rate is again reduced, and the sulphur absorption efficiency of the ZnO becomes lower. Furthermore, the risk for leakage of organic S compounds from the hydrogenator increases at low temperature, and the ability of ZnO to absorb these compounds decreases with decreasing temperature.

The sulphur uptake in a zinc oxide reactor ideally consists of various zones as illustrated in the Fig. 7.

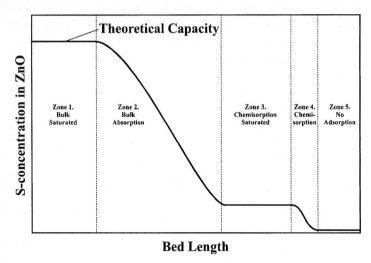

Figure 7 Ideal sulphur profile in a zinc oxide bed

Fig. 7 shows the situation at one particular time. The fronts will gradually move through the ZnO bed towards the outlet, and eventually breakthrough will occur.

Five distinct zones may be identified:

- **Zone 1:** Bulk saturated. The zinc oxide in this zone is fully saturated with sulphur. The gas phase concentration is constant and equal to the feed gas concentration.
- **Zone 2:** The bulk absorption front. The zinc oxide has capacity for absorbing the sulphur. The sulphur is transported through the catalyst pellets by means of solid diffusion and pore diffusion until full saturation is achieved. The gas phase concentration drops to the bulk equilibrium level as determined by reaction (10) and (11), as described above.
- **Zone 3:** Chemisorption saturated. The surface of the zinc oxide is covered with sulphur. The gas phase concentration is constant at the bulk equilibrium level.
- **Zone 4:** The chemisorption front. In cases with low concentration of H_2S and high concentration of CO_2, zone 4, the chemisorption front will develop. The H_2S, which escapes zone 2 due to equilibrium according to reaction (11), will be chemisorbed on the fresh catalyst. The gas phase concentration drops to a very low level. In theory, H_2S and COS are removed to sub ppb levels.
- **Zone 5:** Fresh ZnO. No reactions occur.

If the H_2S concentration in the feed is 'high', and when CO_2 is absent or present in low concentration only, the bulk absorption front will move faster than the chemisorption front, which will then not be visible. There will be only one absorption front, and the gas phase concentration will drop directly from the inlet concentration to the outlet concentration which will in the ideal case correspond to the unmeasurably low chemisorption equilibrium. In such cases the bulk absorption determines the design of the absorption vessel, and the ideal absorption material has the highest possible absorption capacity per volume.

However, in cases where low (single digit ppm or lower) concentrations of H_2S (+ COS) and high CO_2 concentrations (several vol% are not uncommon in natural gas) are present in the feed, calculations may show that the chemisorption front moves faster than the bulk absorption front at normal operating conditions and with normal quality ZnO absorption material. In such cases the operating temperature may be reduced to the lowest acceptable level (to decrease the equilibrium concentration of H_2S according to reaction (11)), and a ZnO with highest possible chemisorption capacity (highest possible surface area per volume) may be selected. If this is not enough, a special absorption material with high chemisorption capacity, e.g. based on Cu, may be installed downstream the ZnO to ensure efficient removal of H_2S.

H_2S appears to absorb on ZnO according to a 'core-shell-model' or 'shrinking-core-model', see e.g. [19]. However, new research, (see e.g. Ref. [20]), as well as industrial feedback indicates that this simple model can not adequately describe all situations. Deviations from expected profiles are sometimes seen, especially at low S concentration and high CO_2 concentration in the feed.

The shrinking-core-model assumes that H_2S (and COS) is absorbed as soon as it meets a clean ZnO surface, and that it is subsequently removed from the surface by solid state diffusion. This means that a partly saturated particle will consist of a 'shell' of ZnS surrounding a 'core' of ZnO. ZnS has a higher molar volume than ZnO, and the conversion of ZnO to ZnS (reaction (11)) will therefore cause an expansion of the solid phase. If the initial porosity of the unconverted ZnO is too low, the volume expansion may lead to closure of the pores so that the unconverted 'core' becomes inaccessible, and full conversion in zone 1 becomes impossible.

2.2.6 S Absorption materials

As mentioned in the preceding paragraphs ZnO is the universal S absorption material in modern desulphurisation units. Leading catalyst suppliers such as Haldor Topsøe, Südchemie, and Johnson Matthey each have their own products with specific properties. ZnO will normally be supplied in the form of extruded cylindrical pellets consisting of almost 100% pure ZnO. In order to ensure highest possible absorption capacity per volume installed absorption material, the highest possible bulk density is normally desired. However, as mentioned above, a certain porosity is required to ensure proper functioning of the material, and this limits the achievable bulk density.

In certain situations it may be desirable to optimize not the bulk absorption capacity, but the chemisorption capacity. In such cases the bulk density and consequently the S-content at full saturation will be lower. It may also, in certain applications, be advantageous to add promoters to the ZnO, e.g. alumina to enhance the ability to absorb COS directly (see e.g. [21]) or Cu or similar metals to increase the chemisorption capacity and/or to add hydrogenation activity to the material.

2.3 Steam reforming

Steam reforming of hydrocarbons is the dominating process for the manufacture of hydrogen especially for refineries. The typical feedstock ranges from natural gas and LPG to liquid fuels including naphtha and in some cases kerosene. Steam reforming is often used in combinations with various oxygen or air-blown partial oxidation processes for production of synthesis gas for ammonia, methanol and various petrochemical products. In recent years steam reforming is also seen as an option for converting the primary feed into a gas suitable for a fuel cell. In this case, the choice of the steam reforming technology depends on the type of fuel cell.

Steam reforming by itself is not the preferred technology for production of synthesis gas for the large-scale GTL applications. Large-scale steam reformers have a poor economy of scale compared to processes based on partial oxidation and air separation. Furthermore, large heat input is required and steam reforming produce gases with an H_2/CO ratio well above the desired value of about 2. Processes based on adiabatic oxidative reforming are therefore in general preferred for the Syngas Generation Unit (SGU) in a GTL plant.

In spite of the above steam reforming plays a role in optimised GTL plants. The steam reforming reactions also take place in adiabatic oxidative reforming reactors. Stand-alone steam reformers may also be used in an optimised GTL plant in the form of adiabatic prereforming or heat exchange reforming. The latter may also in some cases be used for revamp of existing facilities to boost the capacity. Finally, tubular reforming provides a means for pure hydrogen production for example for the hydrodesulphurisation or hydrocracking units in an optimised GTL facility or for adjustments of the composition of hydrogen lean synthesis gas produced by partial oxidation.

Steam reforming is carried out in several different types of reactors in a GTL facility. Each of these is optimised for a specific application. The main types of reactors include:

- Adiabatic prereformers
- Tubular or primary steam reformers
- Various types of heat exchange reformers

The fundamentals of the steam reforming reactions are described in the following Section 2.3.1. This is followed by a description of the characteristics of the above mentioned three types of reactors in Section 2.3.2-4. In Section 2.3.5 the catalysts used in the reactors are described along with the reaction mechanisms and the typical reasons for catalyst deactivation.

2.3.1 The reactions
Steam reforming converts hydrocarbons into hydrogen and carbon oxides. The key reactions involved are given in Table 2 along with the enthalpy of reaction and an expression for the equilibrium constants.

Table 2
Key reactions in steam reforming

Reaction	Std. Enthalpy of Reaction $(-\Delta H^\circ_{298}$, kJ/mol)	Equilibrium Constant $\ln K_p = A + B/T$ [*)	
		A	B
1. $CH_4 + H_2O \rightleftharpoons CO + 3H_2$	-206	30.420	-27,106
2. $CH_4 + CO_2 \rightleftharpoons 2CO + 2H_2$	-247	34.218	-31,266
3. $CO + H_2O \rightleftharpoons CO_2 + H_2$	41	-3.798	4160
4. $C_nH_m + nH_2O \rightarrow nCO + (n+\frac{m}{2})H_2$	-1175[**)]		

*) Standard state: 298K and 1 bar; **) For n-C_7H_{16}

Reaction (1) in Table 2 is the steam reforming of methane. The reaction results in gas expansion, and high temperature and low pressure are needed to give high equilibrium conversions. The conversion of methane at various process conditions is depicted in Fig. 8. High values of the steam-to-methane ratio in the feed are required to give high conversions especially at elevated pressure.

Steam can be partially substituted by carbon dioxide to perform CO_2 reforming (reaction (2) in Table 2). This reduces the H_2/CO ratio in the product gas, which in some cases may be more economical, especially if a source of low-cost carbon dioxide is available. Steam (and CO_2) reforming is always accompanied by the water gas shift reaction (reaction (3) in Table 2) which is generally fast and may be considered in equilibrium at most conditions.

The methane steam reforming reaction is strongly endothermic as illustrated by the high negative value of the standard enthalpy of reaction. The heat required to convert a 1:2 mixture of methane and steam from 600°C to equilibrium at 900°C is 214 kJ/mole CH$_4$ at 30 bar.

Reaction (4) in Table 2 is the steam reforming of higher hydrocarbons (hydrocarbons with two or more carbon atoms). This is also a highly endothermic reaction, which may be considered irreversible at all conditions of practical interest.

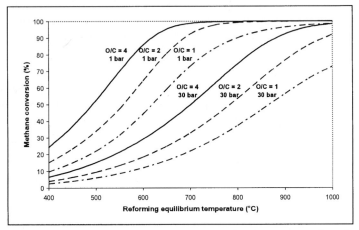

Figure 8   Steam reforming and methane conversion (O/C: ratio of steam to methane in the feed gas).

2.3.2 Adiabatic prereforming

Adiabatic prereforming has during the last 20 years become an integrated step in modern synthesis gas production. Adiabatic gasification of naphtha was already known during the first part of the 1960's for production of town gas and substitute natural gas, see [22-24]. These applications are mainly of historical interest. Today adiabatic prereforming is used widely in the chemical process industry. Adiabatic prereforming is well established in the ammonia and methanol industries, for hydrogen production in refineries, and for the production of synthesis gas for a variety of other chemicals [22].

The adiabatic prereformer converts higher hydrocarbons in the feedstock into a mixture of methane, steam, carbon oxides and hydrogen according to the reactions in Table 2.

All higher hydrocarbons are quantitatively converted by reaction (4) assuming sufficient catalyst activity [22]. This is accompanied by the equilibration of the exothermic shift (3) and methanation (the reverse of methane steam reforming (1)) reactions.

2.3.2.1 Reactor characteristics and operating conditions

The operating conditions in an adiabatic prereformer depend on the type of feedstock and the application. The typical inlet temperature is between 350°C and 550°C. The selection of the operating conditions is in many cases dictated by the limits of carbon formation on the catalyst. For a given feedstock and pressure, the adiabatic prereformer must be operated within a certain temperature window. The formation of a whisker type of carbon will occur above the upper temperature limit. Operation below the lower temperature limit may result either in a polymeric type of carbon formation (gum) or lack of sufficient catalyst activity. The formation of carbon on the catalyst is further discussed in Sections 2.3.2.2 and 2.3.5.3.

The overall prereforming process is often exothermic for heavy feedstock such as naphtha. In Fig. 9 the temperature profile for a prereformer with naphtha feed in a hydrogen plant is shown. For lighter feedstock such as LPG and natural gas the overall reaction may be endothermic, thermoneutral, or exothermic [22, 25]. In Fig. 10 the temperature profile with natural gas feed at high steam-to-carbon (H_2O/C) ratio in an ammonia plant is illustrated.

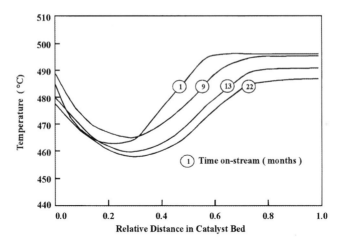

Figure 9 Temperature profile of an adiabatic prereformer with naphtha feed [22]. H_2O/C=3.5. P=26 bars. The movement of the temperature profile with time is due to catalyst poisoning by sulphur (see Section 2.3.5.2).

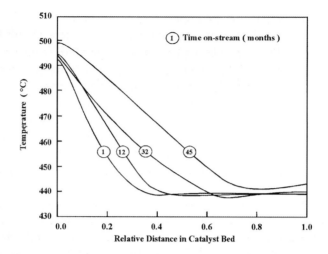

Figure 10 Temperature profile of an adiabatic prereformer with natural gas feed in a 1600 MTPD ammonia plant [22]. H_2O/C=2.8 P=35 bars. The movement of the temperature profile with time is due to catalyst poisoning by sulphur (see Section 2.3.5.2).

The reactor is an adiabatic vessel with specially designed catalysts typically based on nickel. The low operating temperature requires a catalyst with high surface area to obtain sufficient activity and resistance to poisoning especially by sulphur. The optimal shape of the catalyst particle depends on the specific application and on the plant capacity. In many cases catalyst particles of a cylindrical shape in a size of 3-5 mm are used [22]. This particle provides a large surface area for access of the gas into the pore system. The pressure drop over the prereformer is often low (<0.4 bars) for small or medium-scale plants even with such particles giving low void. For large-scale plants, a shape-optimised catalyst will be an advantage and particles in the form of rings or large cylinders with axial holes are usually the preferred choice for minimum pressure drop and high activity [22].

Deactivation of the prereformer catalyst may take place during operation. The cause is typically sulphur but sintering and other poisons may also play a role as discussed in Section 2.3.5. The deactivation of the catalyst can be observed as a progressive movement of the temperature profile as illustrated in Fig. 9 and Fig. 10. The resistance to deactivation is an important aspect in the design of adiabatic prereformers.

The assessment of the performance of an adiabatic prereformer during operation is used to determine the actual rate of deactivation and the optimal time for changing the catalyst [25]. This can be done by monitoring a number of parameters. It is important to follow the content of higher hydrocarbons as an increase in concentration may indicate loss of activity. The approach to equilibrium of the methane steam reforming reaction at the reactor exit is also a parameter, which can be used to monitor the performance. The approach to equilibrium is expressed by a temperature difference defined as:

$$T_R = T(\text{exit catalyst}) - T(Q_R), \quad Q_R = \frac{P_{CO} P_{H_2}^3}{P_{CH_4} P_{H_2O}} \tag{5}$$

in which $T(Q_R)$ is the equilibrium temperature corresponding to an equilibrium constant equal to the reaction quotient Q_R.

The approach to equilibrium and the content of higher hydrocarbons in the prereformer exit are generally close to zero and constant throughout the operation period of the prereformer. In many cases a graphical deactivation plot is used to assess the performance of the prereformer [25]. The deactivation plot shows the length of the reaction front as a function of operation time. The method is illustrated in Fig. 11.

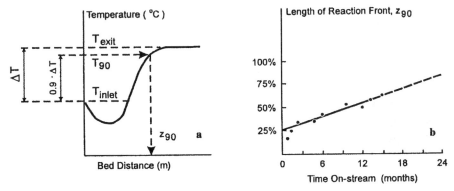

Figure 11 Graphical deactivation plot for performance prediction [25].
a) Estimation of length of reaction front, z_{90} from temperature profile.
b) Deactivation plot.

The temperature difference between the outlet and the inlet is calculated. The axial position (z_{90}) at which 90% of the temperature difference has been obtained is plotted versus time. A steep slope indicates a high rate of deactivation. The inverse slope of the deactivation plot is known as the resistance number defined as the amount of feed required to deactivate 1 g of catalyst. A large resistance number indicates slow deactivation.

2.3.2.2 Adiabatic prereformers in GTL applications

Adiabatic prereforming is a key element in an optimised design of the synthesis gas generation unit (SGU) in a GTL plant. In Fig. 12 the typical installation of an adiabatic prereformer in an SGU based on autothermal reforming in a GTL plant is illustrated [26]. The natural gas is mixed with steam and a minor amount of hydrogen. The resultant mixture is fed to the adiabatic prereformer after desulphurisation and preheating to the desired reactor inlet temperature. The effluent from the prereforming step is further preheated, mixed with a CO_2-rich gas and fed to the autothermal reformer [26].

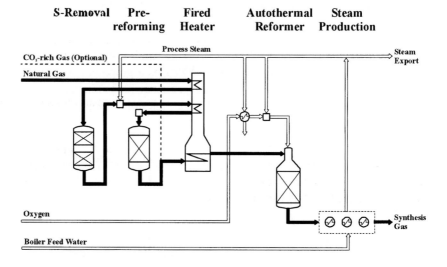

Figure 12   Typical layout of the syngas generation unit with adiabatic prereformer in a GTL plant [26].

The removal of the higher hydrocarbons from the natural gas in a GTL plant has several advantages. A higher feed temperature to the autothermal reformer can be applied without the risk of thermal cracking in the preheater coil. A higher feed temperature reduces the oxygen consumption in a GTL plant (per unit of Fischer-Tropsch product) and improves the carbon efficiency. A temperature profile for an adiabatic prereformer operating at low H_2O/C ratio is given in Fig. 13 [26].

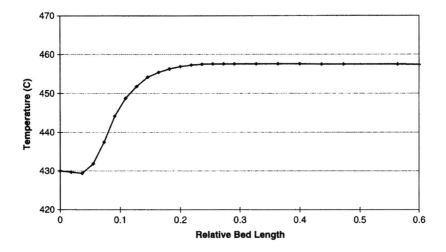

Figure 13 Temperature profile in adiabatic prereformer during pilot plant operation.
$H_2O/C = 0.4$ [26].

A GTL plant is operated at a low H_2O/C ratio for optimal process economics. Operation at low H_2O/C ratio involves the risk of formation of carbon on the catalyst in the prereformer. Carbon may form either from methane or higher hydrocarbons as illustrated below:

$$CH_4 \leftrightarrow C + 2H_2 \tag{6}$$
$$C_nH_m \rightarrow nC + \tfrac{1}{2}mH_2 \tag{7}$$

The selection of the catalyst and the operating conditions of an adiabatic prereformer in a GTL plant are often dictated by the limits of the above reactions. The limits of reaction (6) may in principle be determined from thermodynamics. Carbon will form if the gas shows affinity for carbon formation after establishment of chemical equilibrium of the methane steam reforming and shift reactions [22]. The risk of carbon formation by reaction (6) is most pronounced in the reaction zone where the temperature is highest. In the evaluation of the potential for carbon formation the correct equilibrium constant should be used as described in Section 2.3.5.3.

Carbon formation from higher hydrocarbons is an irreversible reaction that can only take place in the first part of the reactor with the highest concentration of C_{2+} compounds. Carbon may form even if thermodynamics predict no affinity at equilibrium. The criterion for carbon formation can be described as a kinetic competition between the carbon forming and steam reforming reactions. A thorough kinetic analysis, both with fresh catalyst, and towards end-of-run at each point in the reactor, is required to accurately evaluate this criterion. In general the limits for carbon formation from reaction (6) are approached with reduced ratio of steam to higher hydrocarbons and with increased temperature [22, 27].

The knowledge of the carbon limits is imperative for optimal design. Examples of pilot plant experiments at low pressure performed to gather information about these limits are given in Table 3. Post-test analysis indicated that the carbon formation could be ascribed to reactions (5) and (6) [22].

Table 3
Adiabatic preforming at low H_2O/C ratio [22]

Experiment	A	B	C	D
H_2O/C	0.40	0.25	0.13	0.25
Inlet temperature, °C	455	395	400	430
Pressure, Mpa	0.8	1.0	1.0	0.9
Carbon formation	No	No	Yes	Minor

2.3.2.3 Modelling of adiabatic preformers

The use of mathematical models is an invaluable tool in the design and optimisation of adiabatic preformers. The chemical conversion versus time can be determined by combining reaction kinetics, pore diffusion, pressure drop, and effects of catalyst deactivation and poisoning [22].

No radial concentration gradient exists in a preformer due to its adiabatic nature. A one-dimensional axial model is thus sufficient to simulate the temperature and concentration profile. Simulation of (pre)reformers may be carried out by heterogeneous and pseudo-homogeneous models, see sec. 2.3.3.2. The heterogeneous model is based on intrinsic kinetics. This model takes into account the pore diffusion inside the catalyst particle and the transport restrictions across the film on the external surface. Accurate representations of the conditions inside the catalyst particle may be obtained. However, these models are mostly used for development of novel catalysts and catalytic systems and for detailed investigations of deactivation phenomena [22]. For design purposes, pseudo-homogenous models are often used.

The preformers often operate in the diffusion-controlled regime, which validates the use of pseudo-homogenous models. The pseudo-homogeneous model does not take into account the difference in temperature and concentration between the catalyst particle and the bulk gas phase. The transport restrictions are implicitly taken into account by the use of effective reaction rate expressions.

The pseudo-homogeneous model may be expressed as given below (for one reaction, [22]):

$$-u_s \frac{dC}{dZ} = r_v \tag{8}$$

$$u_s p_g c_p \frac{dT}{dz} = (-\Delta H) r_v \tag{9}$$

$$-\frac{d_p}{d_z} = \frac{2 f p_g U_s^2}{g d_{p,v}} \tag{10}$$

The parameters of significance for the model equations (8-10) are:

u_s = superficial velocity
r_v = effective reaction rate, volume basis
p_g = gas density
c_p = heat capacity
ΔH = heat of reaction
f = friction factor
g = acceleration of gravity
$d_{p,v}$ = equivalent particle diameter
P = pressure
T = temperature
C: molar concentration
Z: axial distance

In Fig. 14 [22] calculated and experimental temperature and conversion profiles are given for a pilot plant experiment at low H_2O/C ratio. The good agreement illustrates the validity of the use of pseudo-homogeneous models. The simulation of the concentration and temperature profiles may be coupled with detailed models for catalyst deactivation and carbon formation.

The conversion profile in an adiabatic preformer is largely determined by the kinetics of the steam reforming of the higher hydrocarbons [22]. The rate of reaction can typically be represented by a power law kinetic expression. An example for steam reforming of ethane is given in the following [27]:

$$r = 2.2 \times 10^5 \exp(-9100/T) \, p_{C_2H_6}^{0.54} \, p_{H_2O}^{-0.33} \, p_{H_2}^{0.2} \tag{11}$$

The numerical value of the reaction order with respect to steam decreases with increasing temperature. For heavier components than ethane the reaction order with respect to hydrocarbon may decrease [27]. The reaction orders may depend considerably on the type of catalyst.

282

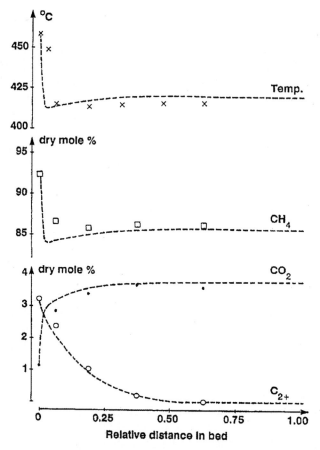

Figure 14 Temperature and conversion profiles for prereforming of natural gas,
H₂O/C ratio = 0.40, P = 8 bar. Measured data: x - temperature;
□ - CH₄ concentration, • - CO2 concentration; o - C₂₊ concentration. Calculated
profiles are shown as dashed lines. [22].

2.3.3 Design of steam reformers

Steam reforming is, in industrial practice, mainly carried out in reactors referred to as steam reformers, which are essentially fired heaters with catalyst-filled tubes placed in the radiant part of the heater. The process may also be carried out in reactors referred to as heat exchange reformers. These are essentially heat exchangers with catalyst-filled tubes. Heat exchange reformer design is discussed in Section 2.3.4. Abundant literature is available on steam reforming and design of steam reformers. Examples are [4, 27–34]. Reference to older literature may be found in [3].

Steam reforming is a mature process [27]. The first industrial steam reformer was installed at Baton Rouge in Louisiana, USA, by Standard Oil of New Jersey and commissioned in 1930. Six years later a steam reformer was commissioned at ICI, Billingham, UK. During the fifties, metallurgical developments made it possible to design reformers for operation at elevated pressures. This improved the energy efficiency of the overall process, because higher pressure facilitates the heat recovery and results in savings in compression energy in ammonia and methanol plants. In 1962, two tubular reformers operating at around 15 atm and using "high molecular weight hydrocarbone" as feed were commissioned by ICI. Less than five years later, a Topsøe reformer was operating at 40 atm.

Already at this time, the basic types of reformers (see below) had been developed and industrialised [28], and all later developments have, in spite of major improvements in performance and cost, not basically changed the designs.

The tubular reformer is an energy converter [27], since most of the energy input for many processes is added via the reformer with the hydrocarbon feed and the fuel (often the same hydrocarbon). The energy is transferred into hot, steam-containing synthesis gas, and hot flue gas. The synthesis gas is subsequently processed further and converted into products by mainly exothermic reactions, releasing more heat. The latent heat of the process streams and of the flue gas must be recovered to achieve high energy efficiency, and this requires tight integration of the reformer with other parts of the process.

2.3.3.1 Mechanical design

The main elements in the design of a steam reformer furnace are:

- Tube and burner arrangement
- Inlet and outlet systems
- Tube design
- Burner characteristics

Tube and Burner Arrangement

Tubes and burners shall ideally be arranged in the furnace box in such a way that adjustment and control of temperature and heat flux along the tube length can be obtained.

Reformers are designed with a variety of tube and burner arrangements. Basically there are four types of reformers as illustrated in Fig. 15 [4, 27, 29, 30]:

284

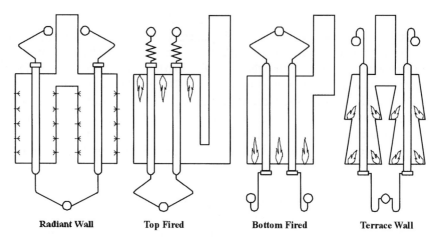

| Radiant Wall | Top Fired | Bottom Fired | Terrace Wall |

Figure 15 Tube and burner arrangements [4]

The radiant wall or **side fired** reformer contains tubes mounted in a single row along the centreline of the furnace. In larger installations, two such furnaces are erected side by side with common inlet and outlet systems and with common fuel supply, flue gas duct, and waste heat section. Burners are mounted in several levels in the furnace walls, and the flames are directed backwards towards the walls. The tubes are heated by radiation from the furnace walls and the flue gas and to a minor extent by convection. The flue gas leaves the furnace at the top so that the flow of process gas and flue gas is counter-current. The most important supplier of reforming technology based on the side fired design is Haldor Topsøe [3, 4, 27-34]. A cross section of a typical side fired reformer for CO production (high outlet temperature), (Topsøe design) is shown in Fig. 16 [4].

The **top fired** reformer features of a furnace box with several rows of tubes. Burners are mounted in the furnace ceiling between the tube rows and between the tubes and the furnace wall. From the burners, long flames are directed downwards, and the tubes are heated by radiation from the flames and the hot flue gas and by convection. The flue gas leaves the furnace box at the bottom, so that the flow of process gas and flue gas is co-current. The top fired reformer design is used by e.g. Howe Baker [35,36], KBR [28], Lurgi [37,38], Technip/KTI [39], and Uhde [40].

A photo of a large reformer (for H_2 production) is shown in Fig. 17 [34].

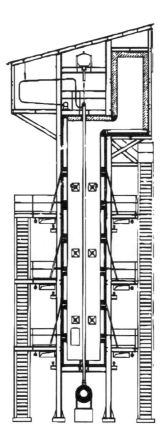

Figure 16 Cross section of side fired reformer

The **bottom fired** type has easy access to the burners and gives an almost constant heat flux profile along the length of the tube. Since the tubes are hot in the bottom a substantial margin is required on the tube design temperature limiting the outlet temperature. The bottom fired design is today considered outdated. It was used in the past by e.g. Exxon and Chemico, and it may still be seen in small-scale plants from e.g. Caloric [41].

The **terrace wall fired** type reformer is a modification of the bottom fired type, having slightly lower tube wall temperatures. Problems can arise at the 'pinch point' in the middle of the furnace where the tubes are subject to both radiation from the burners and to enhanced convection from the flue gas. The terrace wall fired design is used by Foster Wheeler [42].

Figure 17 Photo of Topsøe Reformer [34]

Inlet and Outlet Systems

Inlet and outlet system shall be designed to allow thermal expansion of the hot parts without any risk of excessive stresses.

Gas Inlet

The feed gas to the reformer is normally distributed to the individual reformer tubes from an inlet header through so-called 'pigtails'. The design will depend on the type of tube support and on the temperature of the feed gas. Typical designs are shown in Fig. 18 [4, 29].

For reformers designed for traditional inlet temperatures the inlet hairpin is connected to the top of the reformer tube. The hairpin is designed to absorb the thermal expansion of the catalyst tube and the thermal expansion of the inlet header. Due to the flexibility of the hairpin it can easily be swung aside to give free access to the catalyst tubes for loading or unloading of the catalyst.

In cases where preheat of the gas inlet temperature is high, the inlet hairpin may be connected to the side of the reformer tube. By means of insulation inside the top of the reformer tube this design will bring down the temperature of the flange assembly to a level where tightness can be ensured. This type of reformer will normally have conditions for the reformer tubes where spring support arrangements are needed in order to avoid bending of tubes.

Gas Outlet

The gas outlet system should in all cases be flexible enough to withstand the thermal expansion and contraction during start-up and shutdown as well as during normal operation. The outlet system connecting the catalyst tubes to the manifolds is critical in most reformer designs. A considerable number of failures in manifolds/pigtails have been reported, and therefore a thorough analysis of stress in these areas is important.

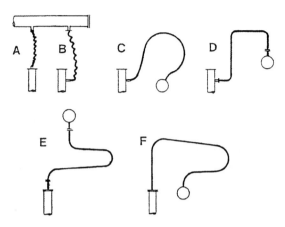

Figure 18 Typical design of gas inlet systems

Two main types of outlet systems are illustrated in Fig. 19 [4, 29].

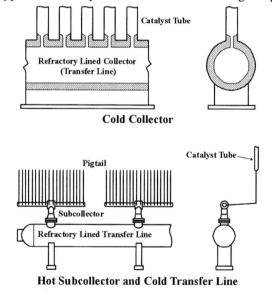

Figure 19 Typical designs of gas outlet systems [4]

In the **cold collector** system the reformed gas is introduced direct from the individual tubes to the transfer line which is located outside the furnace. In this system, special devices are required for connecting the hot parts to the cold collector. Different versions of this concept are used by Haldor Topsøe [4], Lurgi [37, 38], and Uhde [40].

In the **subcollector** system, groups of catalyst tubes are connected through pigtails or hairpins to hot subcollectors, and the subcollectors are joined to the main collector. In relatively small reformers, the pigtails may be connected direct to the transfer line. The pigtails are designed to relax thermal stress. This concept is dominating in reformers designed for 'low' outlet temperatures (below about 900°C).

Tube Design

A steam reformer may contain up to about 1000 high alloy steel tubes filled with catalyst. The outer diameter is typically 100 - 180 mm, the tube wall thickness is 8 - 20 mm, and the heated length may be 10 - 14 m. The traditional materials used for reformer tubes were HK 40 (25 Cr 20 Ni) or IN 519 (24 Cr 24 Ni Nb) [29, 43, 44]. However, the demand for better creep rupture properties led to the development of new alloys with greater strength. The development in the strength of reformer tube materials is illustrated in Fig. 20 [44].

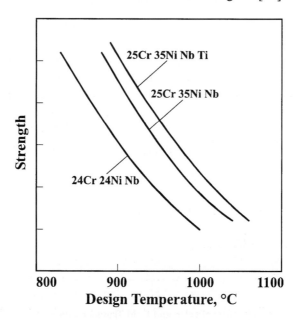

Figure 20 Creep rupture strength for different reformer tube materials [44]

Today, the preferred tube materials are the so-called 'microalloys', typically 25 Cr 35 Ni Nb Ti [43]. The other types of materials will only be used in cases where the strength of the microalloys cannot be used fully, as there is a lower limit for the wall thickness of the tubes.

Reformer tubes are normally designed according to the 'remaining life assessment' technique, API-530 [45], for an average lifetime before creep rupture of 100,000 hours. The main parameters in the design are the design pressure; the design temperature and the creep rupture strength of the material used. However, the determination of these parameters can be ambiguous, and each reforming technology licensor applies in-house procedures to determine the parameters and to introduce necessary design margins.

The calculation of the design temperature is demanding since it requires detailed understanding of the heat transfer. This includes heat transfer to the individual tube by radiation from the furnace internals including furnace walls and neighbouring tubes as well as heat transfer by convection from gas to tube wall, by conduction through the tube wall, and by convection from the inner tube wall to the catalyst and the reacting gas. Secondly, an understanding of reaction kinetics, catalyst ageing, and heat and mass transfer (radial and axial) in the catalyst bed, etc. is required. The interplay between catalyst, reacting gas and reformer tube is also essential for the prediction of the limits for carbon formation.

Burner Characteristics

Reformer burners are a speciality product supplied by a few companies only. For descriptions, reference is made to [46-49]. Originally, burner developments were mainly directed towards control of the radiation heat transfer by control of flame shape and combustion intensity. Lately, environmental issues (NO_x control) have become a major concern, and this has to some extent changed the combustion characteristics. CFD modelling of the furnace side including correct simulation of the combustion is important in the analysis of the effect of this development on reformer design and performance.

2.3.3.2 Modelling of the reformer

Simulation of Furnace Chamber

Tubular steam reforming is a complex interaction of heat transfer and coupled chemical reactions [27]. The heat released by the burners is transferred via radiation and convection to the reformer tubes. The heat passes through the tube walls by conduction and is transferred to the catalyst bed by convection and radiation. At the same time, a network of chemical reactions creates temperature and concentration gradients in the radial direction of the tube and around and within the porous catalyst particles.

An ideal model should be able to simulate the reformer performance on the basis of the individual burner duties, the feed stream characteristics, the properties of the catalyst and the reformer geometry.

Early simulations of the process gas side in tubular reformers were generally uncoupled from the furnace box by assuming an outer tube wall temperature profile or a heat flux profile. These profiles were established or checked by feedback from measurements in industrial plants and monotube pilot plants. It should be pointed out, however, that measurements of tube wall temperatures are difficult [50]. Pyrometric methods involve complex corrections because of reflections from furnace walls and flames. The correction is largest at the coldest position of the tube at the reformer inlet, where reaction conditions at the same time are most complex. Thermocouples welded into the tube wall give more exact information but their life may be limited. Shadowing effects in the tube row cause another uncertainty. The extent of this distortion increases with decreasing tube pitch.

Tube Side

One-dimensional pseudo-homogeneous models are adequate for studying reformers at non-critical conditions and for simulation of the over-all performance. They are, however, insufficient for reformers operating close to carbon limits. For such cases a more detailed analysis of the local phenomena in the reformer is required.

Radial temperature and concentration profiles are included in two-dimensional pseudo-homogeneous models, whereas the gradients in and around the catalyst pellets are neglected.

Such models are generally plug flow reactor models with detailed kinetic schemes considering 2D axi-symmetric radial temperature and concentration gradients within the tube. Heat transfer is calculated as an effective radial conductivity within the bed and a film heat transfer coefficient at the tube wall. The main parameters are the reaction kinetics and parameters in the heat transfer and the pressure drop equations. Such data are proprietary parameters, and generally quite difficult and costly to establish. However, it must be remembered that the usefulness of even the most sophisticated models is not better than the accuracy by which the relevant parameters are known.

Haldor Topsøe's proprietary, in-house process model is described in Refs. [30, 51]. The parameters in this model were determined by experiments in a full-size monotube reformer PDU (Fig. 21) [4, 51] and validated against a large amount of industrial data.

Figure 21 Tubular reformer Process Demonstration Unit (PDU) [4]

An application of this model is shown in Fig. 22 [51], which shows a comparison between calculated and measured axial catalyst bed temperatures at measured outer tube wall temperatures.

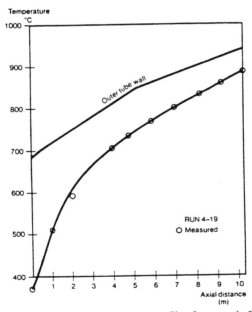

Figure 22 Measured and calculated axial temperature profiles from run in PDU [51]

The data are from an experiment carried out on the monotube PDU at a low steam to carbon ratio of 1.18 but also with a low average heat flux of 50,500 kcal/m^2/h on the inner tube surface. It is seen that there is a good agreement with the measured temperature data. Similar agreement has been obtained in simulations of a large number of data sets, and the mathematical model is therefore used in the design of both heat exchange type reformers and tubular reformers, without further adjustments of the parameters.

Reaction Kinetics

The properties of steam reforming catalysts and fundamental aspects of their behavior in the reactor (reaction mechanism and mechanism of carbon formation and S poisoning) are discussed in Section 2.3.5. Detailed discussions of the reaction kinetics of the reforming reactions may be found in [27, 33]. In the following text only some main points are highlighted.

The activity of a reforming catalyst depends mainly on the nickel surface area. The nickel crystals will sinter quickly at high temperature. This may be partly prevented by a stable micropore system because the nickel particles may hardly grow larger than the pore diameter of the support.

The activated chemisorption of methane is usually considered to be the rate-determining step in steam reforming. Likewise, there is general agreement on first order kinetics with respect to methane, but the impact of steam and hydrogen on the rate is complex. The kinetics depends on the temperature range and the catalyst composition. The rate of the shift reaction is much faster and it can be assumed to be at equilibrium in all positions of the reformer.

For normal steam reforming catalysts, the utilization of the activity, i.e. the effectiveness factor, is far less than 10% because of transport restrictions. The mass transport restrictions are related mainly to pore diffusion, whereas heat transfer restrictions are located in the gas film. The strong endothermic reaction results in a temperature drop of ca. 5-10°C over the gas film. This means that the activity is roughly proportional to the external surface area, and that there is in the normal case a surplus of activity in the reformer tube. The surplus of activity can be used in more advanced reformer designs with only thin layers of active catalyst material e.g. for honeycomb catalysts having low pressure drop or even in designs with the catalyst supported directly on the heat transfer surfaces.

Physical Properties

The catalyst support material must be stable under the process conditions and under the conditions prevailing during start-up and shutdown of the plant. In particular, the conditions during upsets may become critical. Degradation of catalyst may cause partial or total blockage of some tubes, resulting in the development of 'hot spots', 'hot bands' or totally hot tubes. Carbon formation may cause similar problems.

The catalyst particle size and shape should be optimized to achieve maximum activity and maximum heat transfer while minimizing the pressure drop. The high mass velocities in steam reforming plants (40,000 – 70,000 kg $m^{-2}h^{-1}$) necessitate a relatively large catalyst particle size to ensure a low pressure drop across the catalyst bed; but the particle size is limited by the requirement for effective packing. The pressure drop depends strongly on the void fraction of the packed bed and decreases with particle size.

Also, the particle size has a certain impact on the heat transfer coefficient. The optimum choice is a catalyst pellet with high void fraction and high external surface area. Fig. 23 shows an optimized catalyst shape, the 7-hole catalyst.

Figure 23 Steam reforming catalyst. Topsøe R-67 R-7H.

To summarize, the effective catalyst activity is a complicated function of the particle size and shape and the operating conditions. A high intrinsic catalyst activity is still important to control tube wall temperatures, in particular in those parts of the reformer having a high heat flux.

Heat Flux and Activity

The acceptable maximum heat flux in reformer tubes is dictated by mechanical and metallurgical constraints rather than by the catalyst activity. Present catalysts could work at average heat fluxes twice the present maximum if the heat flux profile could be controlled [34]. However, deactivation of the catalyst or unsuitable flux profile may result in overheating of tubes.

Most commercial catalysts have a surplus of catalyst activity. This means that apart from the inlet of the reformer tube, the gas composition is, at all positions in the tube, not far from equilibrium at the average catalyst temperature. In other words, the slope of the temperature profile is the 'driving force' for the reaction, and the reaction proceeds as fast as the required heat can be supplied through the tube wall.

The slope may become very high in the upper part of top fired reformers or in the bottom of bottom fired or terrace wall fired reformers, or if the catalyst looses activity, e.g. by poisoning or aging. In such cases the reaction rate may not be able to follow the rate of heat input, and the temperatures of the gas and the tube wall and also the approach to equilibrium will increase.

It is evident from the above that lack of catalyst activity will cause high temperatures both in the gas and in the tube material. The advantage of high catalyst activity (and the disadvantage of an unsuitable flux profile) is illustrated in Fig. 24 [30] which shows calculated tube wall temperatures in a top fired reformer for various relative catalyst activities in the upper half of the reformer tube. It is seen that loss of activity in the upper part of the tube may lead to significant overheating of the tube.

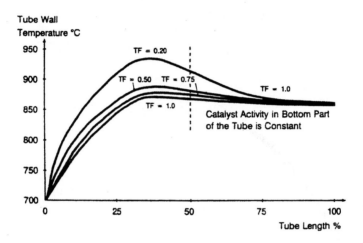

Figure 24 Tube wall temperatures in top fired reformer at various catalyst activity levels (TF) in the upper 50% of the tube.

Modelling by CFD

CFD (Computational Fluid Dynamics) is becoming a preferred tool for modelling and simulation of steam reformers. Results obtained by simulation of top fired furnaces have been reported in e.g. [53, 54] and, of a terrace fired furnace in [55] and of a side fired furnace in [56].

It is emphasised that these descriptions at best show status in 2002. Developments in CFD modelling are rapid, and more comprehensive and accurate models are being elaborated.

At Haldor Topsøe numerous experiments have been performed on a full size PDU. Some of the experiments have been performed in order to provide data for the verification of computational models, including a CFD model. The

PDU contains a single full size catalyst-filled tube located in the centre of the furnace. The furnace has 5 burner rows.

Qualitatively the behaviour of the PDU resembles that of industrial reformers. However, the temperature distribution in the furnace is more homogeneous in the PDU than in industrial reformers. This is due to the smaller number of tubes per unit volume. Furthermore, there are no shadowing effects due to the presence of other tubes.

Fig. 25 compares the calculated outer tube wall temperature to the measured temperature in one run in the PDU. The outer tube wall temperature agrees well with the measured temperature. The small deviation (less than 10 degrees) is well within the measurement accuracy.

Fig. 26 shows for the same run the outer tube wall heat flux calculated by the CFD model and the in-house process model. It is seen that the furnace model accurately predicts the transferred duty, but that the predicted heat flux profiles deviate slightly.

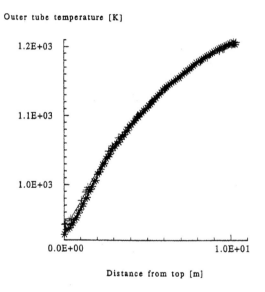

Figure 25 Calculated and measured outer tube wall temperature Measured Outer Tube Wall Temperature [56]

296

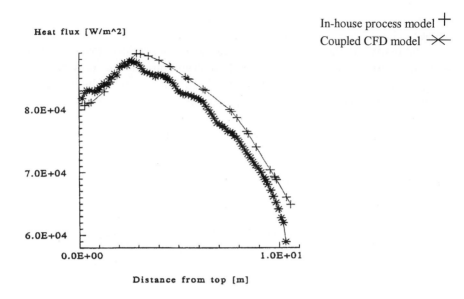

Figure 26 Heat flux calculated by CFD and by in-house process model [56]

2.3.4 Design of heat exchange reformers

Basically, a heat exchange reformer is a steam reformer where the heat required for the reaction is supplied predominantly by convective heat exchange. The heat can be supplied from a flue gas or a process gas - or in principle by any other available hot gas. In the following, the various types of heat exchange reformers are discussed on the basis of source of heat and mechanical concept.

When the heat and mass balance on the process (catalyst) side only is considered, there is no difference between heat exchange reforming and fired tubular reforming, where the heat transfer is predominantly by radiation. This means that all process schemes using heat exchange reforming will have alternatives where the function of the heat exchange reformer is performed in a fired reformer. The process schemes differ 'only', in the amount of latent heat in flue gas and/or process gas and in the way this heat is utilised.

Models for design and simulation of heat exchange reformers are combinations of models for steam reformer catalyst tubes (as described in Section 2.3.3.2) and models for convective heat transfer (as used in design and simulation of normal gas/gas heat exchangers).

2.3.4.1 Types of heat exchange reformers

Three different concepts for heat exchange reformer design have been commercialised by various companies. The three concepts are illustrated in Fig. 27.

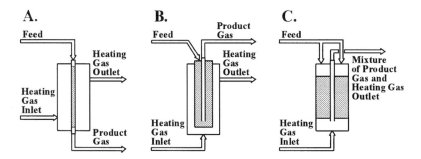

Figure 27 A: Concept with 'straight-through' tubes;
B: Concept with bayonet tubes;
C: Concept with mixing of heating gas and product gas before heat exchange.

Type A and B in Fig. 27 can be used with all types of heating gas, whereas type C can only be used when the desired product gas is a mixture of the heating gas and the product gas from the catalyst in the heat exchange reformer.

Flue Gas Heated Heat Exchange Reformers

These heat exchange reformers are stand-alone reformers, and their function is similar to normal fired reformers. Two designs by Haldor Topsøe A/S, HER [57, 58] and HTCR [59, 60] are examples of this category. As seen in Fig. 28, the HER consists of a number of concentric cylinder shells around a centrally placed burner, while the HTCR as shown in Fig. 29 features a bundle of bayonet tubes and a burner in a separate chamber.

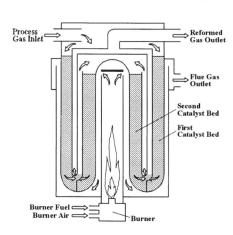

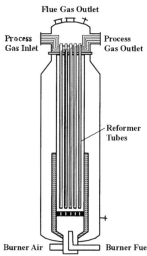

Figure 28 Topsøe Heat Exchange
Reformer (HER) [57]

Figure 29 Topsøe Convection
Reformer (HTCR) [59]

Also others, e.g. KTI [61] have developed heat exchange reformers with bayonet tubes. (It may be argued that reformer concepts with bayonet tubes are partly gas heated reformers (see below) since the process gas is cooled by heat exchange with the catalyst bed, thus providing part of the heat required for the reforming reaction. However, bayonet tubes and similar concepts are in this context only considered as special reformer tube designs).

BP is promoting a so-called 'Compact Reformer' in co-operation with Kvaerner [62]. It is according to the definition above somewhat doubtful whether this is actually a Heat Exchange Reformer. One piece of equipment contains a reformer feed heating/flue gas cooling heat exchanger, a combustion chamber, and a synthesis gas cooling/air and fuel heating heat exchanger. In the combustion chamber, reformer tubes are embedded in "a sea of flames", apparently created by controlled injection of preheated fuel into preheated combustion air. Flame impingement on tubes is mentioned as taking place by design.

Chiyoda have developed a reformer design based on regenerative burners developed by NFK, the so-called HICOT technology [63]. Although this design is not strictly speaking a heat exchange reforming unit, it competes for the same market as heat exchange reformers.

The potential application for large scale GTL and methanol production of the above described types of heat exchange reformers may be doubtful because stand alone steam reforming seems unsuitable for such applications.

Heat Exchange Reformers Heated by Process Gas

Reformers heated by process gas are normally called Gas Heated Reformers. They may be classified in two types depending on the process concept, see Fig. 27. One type, which may be referred to as GHR or 'Two-in, two-out' (both type A and B in Fig. 27 are of this type), can in principle be used in both series and parallel arrangements. (See below under Process Concepts). The other type (type C in Fig. 27), which may be called GHHER or 'Two-in, one-out', can only be used in the parallel arrangement.

GHR or 'two-in, two-out'

Synetix (now Johnson Matthey) have commercialised two types of heat exchange reformers. The first, referred to as GHR, uses bayonet tubes. It was used in the so-called Leading Concept Ammonia (LCA) process, where the GHR is used in combination with an air-blown secondary reformer. Two NH_3 plants based on this concept were started at Severnside in England in 1988 [64], and a copy of one of the Severnside ammonia plants has later been built for Mississippi Chemicals in Yazoo City, USA. The GHR design has later been used in combination with an O_2-blown secondary reformer in the so-called

Leading Concept Methanol (LCM) process, which was used in a 200 MTPD Methanol Research Plant for BHP in Australia, started in 1994.

Synetix have commercialised a second generation heat exchange reformer called Advanced Gas-Heated Reformer (AGHR) [65]. This new design replaced the original GHR design in the methanol research plant in Australia in 1998. The AGHR uses 'straight-through' tubes with two tube sheets and with a special sliding seal on each tube at the hot end.

Toyo claim in a paper describing a process for large-scale production of DME [66] that they have commercialised technologies for autothermal reforming and heat exchange reforming, and a sketch is shown of a so-called TAF-X Reactor. No details are given of the commercial experience. The reactor is a 'two-in, two-out type' heat exchange reformer with two tube sheets at the top and a floating head at the bottom. There are bellows shown at the top of each tube and at the bottom between the floating head and the outlet flange.

GIAP, the State Institute for Nitrogen Industry of the former Soviet Union, developed a gas heated reformer design called the TANDEM reformer [67]. The design was implemented in an ammonia plant based on O_2-blown autothermal reforming of natural gas in Grodno, Belarus. The TANDEM reactor is a "two-in, two-out" design, and the process concept in Grodno was a 'two-step reforming with GHR', see Section 2.3.4.2.

GHHER or 'two-in, one-out'

In a GHHER, the heating gas and the reformed gas from the catalyst bed in the heat exchange reformer are mixed in the equipment and leave through one exit after having supplied heat to the reforming reaction. The principle is illustrated in Fig. 27C. Haldor Topsøe's GHHER [68], is of this type. The GHHER technology has been successfully demonstrated in a large industrial installation in co-operation with Sasol in South Africa, where it has operated at demanding conditions in parallel with an O_2-blown autothermal reformer [68].

Air Products are promoting the 'Enhanced Heat Transfer Reformer' (EHTR) [69]. The concept is promoted in co-operation with Technip/KTI and seems to be what KTI has earlier referred to as 'post-reforming'. The EHTR has been commercialised as revamp of a H_2 plant, operating in combination with the existing fired tubular reformer.

M. W. Kellogg (now KBR) has commercialised the Kellogg Reformer Exchanger System (KRES) technology. KRES is based on a 'two-in, one-out' heat exchange reformer design operating in combination with an air or O_2-blown autothermal reformer. It has been used for revamp of ammonia plants. The KRES reactor has a bundle of open ended catalyst tubes and baffles on the shell side. Reactor design, process concept, and operating experience have been described in some detail in e.g. Ref. [70]. The possible application in production of MeOH syngas is discussed in Ref. [71].

Uhde have developed and commercialised a so-called Combined Autothermal Reformer (CAR [72]). The first - and only - industrial size application, a demonstration facility in Chemco s.p. Strazske in Czechoslavia, was started in 1990. CAR is a combination of a 'two-in, one-out' type heat exchange reformer with a POX reactor.

2.3.4.2 Process concepts

Heat exchange reformers heated by process gas are of course always installed in combination with another reformer, which may be a fired tubular reformer or an air or O_2-blown secondary or autothermal reformer. Evidently, a significant number of possible combinations exist. If there is more than one feedstock, as e.g. in GTL plants, where recycled tail gas from the synthesis may be used as additional feed to adjust the gas composition, the number of possible process concepts increases further. The use of a prereformer may also be considered, also increasing the number of possible process concepts. In the following, an attempt is made to describe systematically the possible combinations of heat exchange reformers with either a fired tubular reformer or a secondary or autothermal reformer. Only cases with one feed are considered. This feed will most often be natural gas. The cases may be divided into two main types, series and parallel arrangements.

Series Arrangements

In series arrangements, all the process feed gas passes through first a heat exchange reformer and then through a second reformer, and the product gas from the second reformer supplies heat to the heat exchange reformer. The second reformer in the series arrangement may be a fired tubular reformer (Fig. 30). This process concept has been referred to as 'Gas Heated Prereforming'.

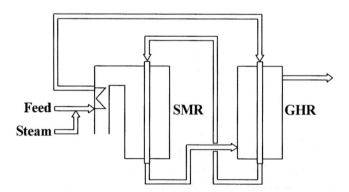

Figure 30 'Gas Heated Prereforming': Fired tubular reforming (SMR) and Gas Heated Reformer (GHR) in series arrangement

Alternatively, the second reformer may be an air or O_2-blown secondary reformer (Fig. 31). This concept, which is often referred to as GHR, is equivalent to two-step reforming and could be called 'two-step reforming with GHR'. (In two-step reforming, a fired tubular reformer is operated in a similar way in series with an air-fired secondary reformer for production of NH_3 synthesis gas or with an O_2-blown secondary reformer for production of synthesis gas for methanol or FT synthesis).

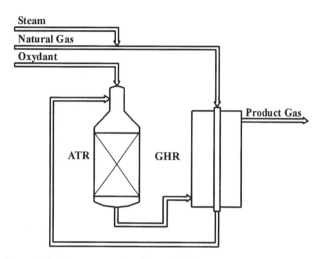

Figure 31 'Two-step reforming with GHR': Autothermal reformer (ATR) and Gas Heated Reformer (GHR) in series arrangement.

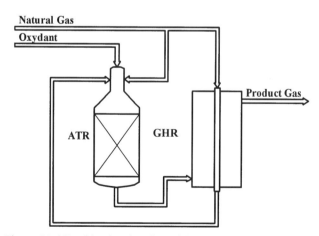

Figure 32 'Combined Reforming with GHR':
Autothermal reformer (ATR) and Gas Heated Reformer (GHR) in series arrangement with partial by-pass of feed over the GHR.

There is an alternative where only part of the feed passes through the GHR, while the rest is bypassed direct to the secondary reformer (Fig. 32). This could be called 'Combined Reforming with GHR'.

The operating conditions (e.g. S/C ratio) may in these concepts be limited by the steam reforming in the GHR, whereas the final gas composition will be determined by the exit conditions from the secondary reforming.

Parallel Arrangements

For obvious reasons, the 'two-in, one-out' concept can only be used in parallel arrangements, i.e. process arrangements, where the feed gas is split into two streams. One goes direct to a conventional reformer, while the other goes to a gas heated reformer heated by the outlet gas from the conventional reformer or by the mixed outlet gases from the two reformers. In parallel arrangements, either the GHR or 'two-in, two-out' design or the GHHER or 'two-in, one-out' design can be used. With a GHR, it is in principle possible to produce two different product gases, whereas the GHHER for obvious reasons allows only production of one product gas, the mixture of product gases from the two reformers. The heat exchange reformer may be coupled with a tubular reformer or with an air or O_2-blown autothermal reformer. Operating conditions (S/C ratio) in the two reformers may be different; final gas composition is determined by the exit conditions from the two reforming catalyst beds. The four possible schemes are shown in Fig. 33-36.

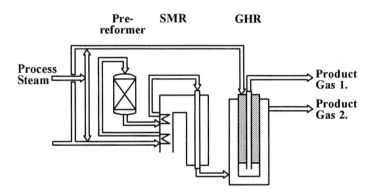

Figure 33 Tubular reformer (SMR) and Gas Heated Reformer ('Two-in, two-out', GHR) in parallel arrangement

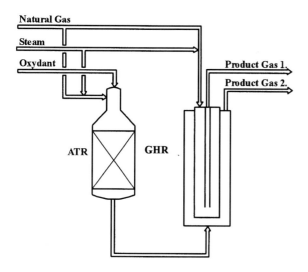

Figure 34 Autothermal reformer (ATR) and Gas Heated Reformer
('Two-in, two-out', GHR) in parallel arrangement

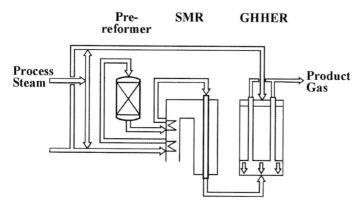

Figure 35 Tubular Reformer (SMR) and Gas Heated Reformer ('two-in, one-out', GHHER)
in parallel arrangement

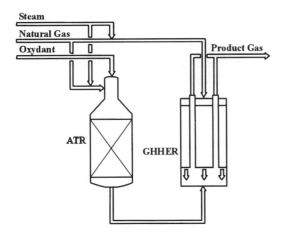

Figure 36 Autothermal Reformer (ATR) and Gas Heated Reformer
('two-in, one-out', GHHER) in parallel arrangement

2.3.4.3 Metal dusting

In all process concepts using heat exchangers heated by process gas, the problem of avoiding metal dusting corrosion on the heat transfer surfaces is a challenge. Metal dusting is a carburisation phenomenon, which may occur on metal surfaces at specific conditions, especially characterised by temperatures in the range 400 – 800°C and contact with a gas with high carbon activity. Metal dusting is well known in the process industry, but unfortunately it is not (yet) completely understood. It results in loss of material, in some cases as 'metal dust', a mixture of metal, carbides, and/or carbon. In severe cases the material wastage can be very fast, leading to catastrophic failure of equipment.

Very significant efforts have been undertaken, especially in recent years, in order to clarify the mechanism and causes of metal dusting and to identify means to cope with the challenge. Reviews of the status of knowledge with comprehensive literature surveys are given in Refs. [73] and [74].

On alloys protected by oxide layers, e.g. stainless steels and Ni-based alloys, metal dusting involves breakdown of the protective oxide layer, transfer of carbon into the base alloy, formation of internal carbides, and disintegration of the matrix. The metal particles generated by the disintegration of the Fe- or Fe-Ni-matrix act as catalysts for carbon formation resulting in coke growing from the corrosion pit [75]. The mechanism on Fe and Fe-Ni-based alloys seems to be different causing different visual appearance of the corroded metal and different types of corrosion products. Full understanding and agreement concerning the mechanism and the kinetics of the attack has not been achieved [74, 75]. The role of gas composition, temperature, and pressure is understood in some detail, but the knowledge is not sufficiently detailed to allow reliable prediction of the risk of metal dusting on different alloys at different conditions and of the progress of possible attack. Studies of metal dusting are therefore empirical, and experimental work is difficult because of the demanding conditions required, and because significant time lag is sometimes seen before an attack takes place. Also the surface condition of the metal and its history may have an influence on the susceptibility for attack. Efforts to improve the understanding are ongoing, see Refs. [73, 74] and for more recent studies see Refs. [76-78].

Metal dusting can be prevented by addition of sulphur to the process gas [74], by proper choice of construction materials [79], or by application of protective coatings [80], especially by Al_2O_3 by so-called alonising [81] or aluminising [82]. Equipment can be designed to avoid 'dangerous' temperatures, special surface treatments including pre-treatment with air or steam at high temperature may be advantageous, while other treatments such as traditional pickling should if possible be avoided. But in spite of these possibilities, metal dusting will remain a challenge in the design of processes and equipment until the ultimate materials or protection methods are developed, and a complete understanding of the phenomena is acquired.

2.3.5 Steam reforming catalysts

The optimal steam reforming catalyst depends to some extent on the application. In an adiabatic prereformer a high activity and a high surface area are desirable to increase the resistance to deactivation and poisoning. The activity is not as important in primary reformers and heat exchange reformers as the reactor design and volume is typically determined by heat transfer and mechanical criteria. In these cases the catalyst activity is normally much higher than needed. Instead the catalysts are designed for maximum heat transfer and minimum pressure drop. However, in all cases the catalyst should ensure equilibrium conversion throughout its lifetime and sufficient resistance to carbon formation and poisoning.

In the following text the typical properties and types of steam reforming catalysts are described. This is followed by sections on sulphur poisoning and carbon formation. Finally, in Section 2.3.5.4 various fundamental aspects and mechanisms of steam reforming catalysis are outlined.

2.3.5.1 The steam reforming catalysts

Most steam reforming catalysts are based on nickel as the active material. Cobalt and noble metals also catalyse the steam reforming reaction but are generally too expensive to find widespread use, although both Ruthenium and Rhodium have higher activity per metal area than nickel [27, 83]. A number of different carriers including alumina, magnesium-aluminium spinel, and zirconia are employed. In Fig. 37 a steam reforming catalyst is shown on a nano-scale.

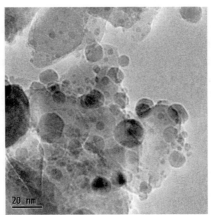

Figure 37 Nickel supported on a MgAl₂O₄ spinel carrier. Recorded at 550°C and 7 mbar of hydrogen using the *insitu* electron microscope at Haldor Topsøe A/S.

The intrinsic activity of the catalyst is proportional to its nickel surface area. The catalyst consists of a huge number of small nickel particles on the ceramic carrier. The active surface area may be calculated from Eq. (1) when the average nickel particle diameter, d_{Nti} and the Ni-loading X_{Ni} (g/m₃) are known:

$$A_{Ni}(m^2 g^{-1}) = \frac{6800 X_{Ni}}{d_{Ni}(\text{Å})} \tag{1}$$

Eq. (1) applies to spherical nickel particles. A_{Ni} is the nickel surface area in $m^2 g^{-1}$.

The catalyst should be optimized for the given application as mentioned above. A key criterion for designing a catalyst for a heated reformer is to maximize the heat transfer at low pressure drop. A high heat transfer coefficient may reduce the tube wall temperature and thereby minimize the required wall thickness. The high heat transfer is not needed for a prereformer catalyst due to the adiabatic nature of this reactor. In all cases the catalyst should be optimized for low pressure drop and high strength.

2.3.5.2 Deactivation by sintering and poisoning

Catalysts in a reforming reactor may lose activity for a number of reasons. In all cases the design of the reactor must be made to take into account the effect of progressive deactivation during the catalysts lifetime.

Sintering is loss of surface area of the active species of the catalyst. The mechanism for sintering is migration and coalescence of nickel particles on the carrier surface [84]. Sintering is a complex process influenced by several parameters including chemical environment, catalyst structure and composition, and support morphology. Factors that enhance sintering include high temperature and high steam partial pressure [84].

Steam reforming catalysts are susceptible to sulphur poisoning. At steam reforming conditions, all sulphur compounds are converted into hydrogen sulphide, which is chemisorbed on the nickel surface:

$$H_2S + Ni_{surface} \rightleftarrows Ni_{surface}\text{-}S + H_2 \tag{2}$$

Sulphur forms a well-defined structure with a S:Ni stoichiometry of approximately 0.5 corresponding to a sulphur uptake of 440 mg S per m^2 of Ni surface. The maximum sulphur uptake on a reformer catalyst is proportional to the nickel surface area and inversely proportional to the extent of sintering. Formation of nickel sulphide only takes place at much higher sulphur levels than normally experienced in a reformer.

The fraction of the nickel surface area covered with sulphur at equilibrium (the sulphur coverage, θ_s) depends on temperature and the P_{H2S}/P_{H2} ratio [22, 27, 30, 85]:

$$\theta_s = 1.45 - 9.53 \cdot 10^{-5} \cdot T + 4.17 \cdot 10^{-5} \cdot T \cdot \ln P_{H2S}/P_{H2} \tag{3}$$

This expression is not valid for θ_s close to zero and close to one.

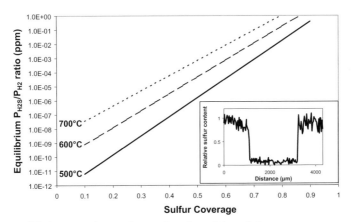

Figure 38 Equilibrium P_{H2S}/P_{H2} ratio as a function of the sulphur coverage. The inserted figure gives the sulphur profile in a strongly sulphur poisoned catalyst pellet

In Fig. 38, P_{H2S}/P_{H2} the ratio at equilibrium is plotted as a function of the sulphur coverage for three different temperatures. The equilibrium partial pressure of H_2S is extremely low at the lower temperature. This means that any sulphur in the feed will be quantitatively withheld in an adiabatic prereformer, thus protecting downstream catalysts against sulphur poisoning.

The poisoning by sulphur takes place as shell poisoning due to pore diffusion restrictions as indicated in the insert of Fig. 38. The average coverage of the particle will be much lower than in the shell and it may take years before the chemisorption front has moved to the centre of the particle [22].

Sulphur is a severe poison for any reforming catalyst [86]. The intrinsic activity of a catalyst decreases rapidly with the sulphur coverage as expressed below:

$$R_{sp} (\theta_s) = R^o{}_{sp} \ (1 - \theta_s)^3 \tag{4}$$

Where $R_{sp} (\theta_s)$ and $R^o{}_{sp}$ are the specific reaction rates for sulphur poisoned and sulphur free catalyst, respectively.

Other poisons include silica and alkali metals. Silica may substantially reduce the activity of the catalyst by acting as a pore mouth poison [25]. The alkali metals reduce the reaction rates in some cases by orders of magnitude.

2.3.5.3 Carbon formation

Steam reforming involves the risk of carbon formation. Three types of carbon are in general considered in steam reforming reactors as shown in Table 4 and Fig. 39 (see also Section 2.3.2.2).

Table 4
Overview of the routes to carbon formation in a reformer

	Critical Parameters	Characteristics	Effects
Whisker carbon	Low H_2O/C Low H_2O/C_nH_m High T	Dissociation of hydro-carbons at the nickel surface and formation of a carbon whisker at the backside of the nickel crystal.	No immediate deactiva-tion but mechanical dis-integration of the cata-lyst and increased ΔP
Pyrolytic carbon	High T Low activity (sulphur poisoning) High partial pressure of C_nH_m	Non-catalytic cracking of higher hydrocarbons to form carbon on catalyst and tube	Carbon formation in tube and on catalyst. Formation of "hot bands" and increased ΔP
Encapsulating carbon	Low T Heavy feed Low H_2O/C Low H_2O/C_nH_m	Encapsulating of the nickel particles by a CH_x film	Deactivation of the catalyst.

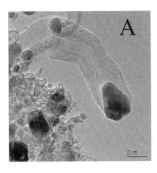 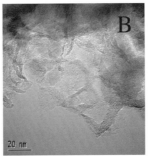 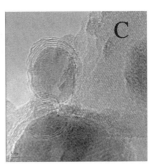

Figure 39 Electron microscopy pictures of whisker carbon (a), pyrolytic carbon on a ceramic
carrier b), and encapsulating carbon (c). C_nH_m: Higher hydrocarbons
(hydrocarbons with two or more carbon atoms)

Whisker carbon is the most destructive form of carbon formed in steam reforming over nickel catalysts. It may be formed from higher hydrocarbons or from methane in equilibrated gas if the overall H_2O/C ratio is too low as described in Section 2.3.2.2. Carbon whiskers grow by the reaction of hydrocarbons at one side of the nickel particle and nucleation of carbon as a whisker on the other side of the nickel particle as shown schematically in Fig. 40.

The carbon whisker has a higher energy than graphite [87]. This means that operation at conditions at which thermodynamics predict formation of graphite may be feasible without carbon formation of the catalyst. The carbon limit also depends upon the crystal size of the nickel particle. Smaller nickel crystals are more resistant towards carbon formation as demonstrated by the data in Fig. 41.

The temperature of the onset of whisker carbon formation was approximately 100°C higher for the catalyst with small nickel crystals (around 7 nm) than for that with large crystals (around 100 nm).

Pyrolytic carbon results from the exposure of higher hydrocarbons to high temperatures. Reddish zones known as 'hot bands' on the walls of tubular reformers are in many cases the result of pyrolytic carbon formed by cracking of higher hydrocarbons often if the catalyst in the top part is deactivated due to sulphur poisoning. The temperature of the tubes at the point of carbon formation increases because pyrolytic carbon thermally isolates the tubes and encapsulates the catalyst pellets resulting in no activity and no consumption of the supplied heat.

Encapsulating carbon (gum) may be formed in reforming of heavy feeds with a high content of aromatic compounds. The rate of gum formation is enhanced by low temperatures and high final boiling point of the hydrocarbon mixture. Encapsulating carbon is a thin CH_x film, which covers the nickel particles and leads to deactivation of the catalyst bed.

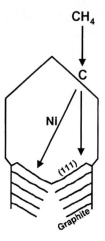

Figure 40 Schematic illustration of the process by which carbon whiskers are formed at the nickel particle during steam reforming.

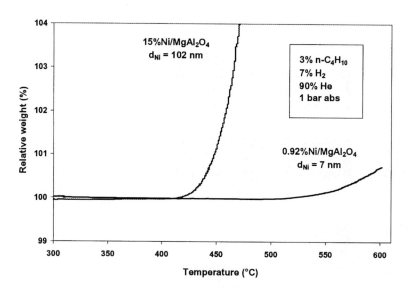

Figure 41 Relative weight increase as a function of the temperature for two catalysts with different nickel crystal size but the same nickel surface area and reforming activity

2.3.5.4 Fundamental aspects of the steam reforming reaction

Fundamental understanding of the reforming process is very important as a guide and as inspiration for new directions in catalyst developments. For example, insight into the mechanism of carbon prevention by promotor addition may lead to new and better promoters. In the following text, a consistent picture of the reforming reaction at the atomic level is given. The description has emerged from DFT (Density Functional Theory) calculations and *insitu* electron microscopy [88-90].

The full potential energy diagrams of the reforming reaction on two different nickel surfaces, Ni(111) and Ni(211) were calculated, and the structures involved in the first four reaction steps for both surfaces are shown in Figs. 42 and 43.

The energetics of the full reforming reaction are shown in Fig. 44. This figure shows the energies of the intermediates on the nickel surface and activation barriers separating the intermediates along the reaction path. Steps are much more reactive than the close packed surface. However, all intermediates are also much more strongly bound at steps than on terraces resulting in more free active sites at terraces. There are therefore (at least) two different reaction channels, one with a low activation barrier, which is associated with steps, and another associated with terraces.

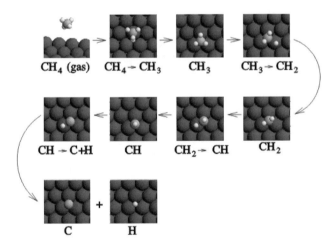

Figure 42 Calculated structures of intermediates and transition states for all elementary processes in the transformation of methane into adsorbed carbon and hydrogen on a Ni(111) surface [89].

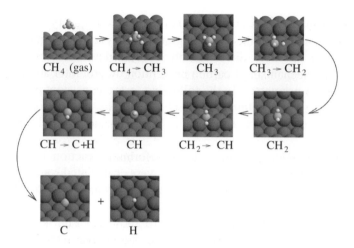

Figure 43 Calculated structures of intermediates and transition states for all elementary
processes in the methane activation reactions on the step sites of a Ni(211) surface
[89].

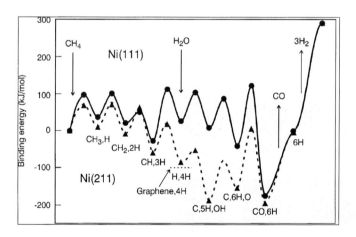

Figure 44 Calculated energies along the reaction path for steam reforming on the Ni(111) and
Ni(211) surfaces. The corresponding structures are shown in Figs. 42 and 43. All
energies are given relative to a situation in which all reactants are in the gas phase
far from the clean surface [89].

It is clear from Fig. 44 that adsorbed atomic carbon is much more stable at
the steps than at the terraces. Consequently, steps are better nucleation sites for
carbon than terraces. When carbon atoms cover step sites, a graphene layer

(single graphite layer) can grow from the step, as illustrated in Fig. 45. In Fig. 44, the energy of a graphene layer on a Ni(111) surface is included as a dashed line. Because this line is below the energy of the most stable form of adsorbed carbon on the surface (at the step), there is a driving force for graphene formation. After a graphene island is nucleated, the growth may continue by transport of carbon atoms to the island. The graphene may eventually cover the whole crystallite. New layers may nucleate at the step by pushing the already formed layer(s) further out. It is therefore assumed that carbon formation is initiated by formation of graphene islands on a nickel surface.

Recently, carbon formation on a supported nickel catalyst was observed by *insitu* electron microscopy. Fig. 46 shows the initial formation of carbon; the formation of the first graphene layers and how the nickel particle is pushed away from the carrier surface. There is an extra small energy gain associated with the formation of graphite-like layers associated with the weak interlayer interaction.

The carbon atoms are so much more stable at a step than on the flat (111) surface that it is energetically advantageous to form new steps as observed in Fig. 46. New nucleation sites can therefore be formed spontaneously during the steam reforming process, and the system has the freedom to rearrange the nickel particles to best form graphene layers.

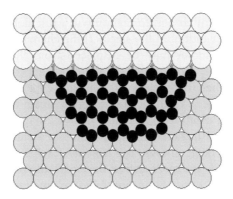

Figure 45 Illustration of a graphene island nucleated from a step on a Ni(111) surface.

These arguments suggest that the availability of step sites is important both for a high turnover rate and for graphite formation. This raises the question of where promoters such as potassium are located on the surface during the catalytic reaction. DFT calculations show potassium was most stable as -K-O-K-O- rows along the steps. Thus, the major carbon-preventing effect of this promoter may be to block the steps and hence remove the nucleation sites for graphite formation. Addition of the promoters should decrease the activity of the catalyst and the decrease should be determined by the promoter coverage at the steps.

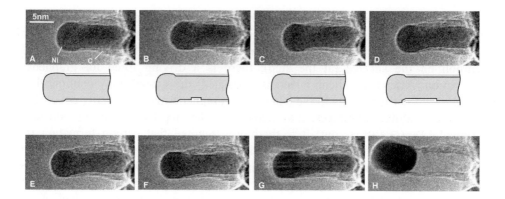

Figure 46 Image sequence of the initial carbon formation on a Ni/MgAl₂O₄ catalyst. Drawings are included to guide the eye in locating the positions of mono-atomic Ni step edges at the C-Ni interface. The images are acquired in situ with $CH_4:H_2 = 1:1$ at a total pressure of 2.1mbar with the sample heated to 536°C.

The promoters need not cover all step sites to prevent nucleation of carbon, because a graphene island of a finite size is needed for it to be stable. The addition of less than an almost complete step coverage will lead to a fraction of the steps being completely covered while the rest are free and open for synthesis. The inhibition of carbon formation will work as long as the promoter is present in sufficient quantities at the steps. If the chemical potential of carbon is high enough, the promoter may be displaced from the step and graphite formation will start. The promotion will therefore appear as a shift in the tolerance towards higher hydrocarbons. Kinetic experiments indicate that promoters may also work via enhanced absorption steam and spill over to the active metal surface on catalysts containing "active" magnesia and alkali [27]. This was reflected in the negative reaction orders of steam obtained for these catalysts. It has also been shown by isotopic exchange experiments that magnesia-based catalysts are more active for steam dissociation than alumina supported catalysts. This observation possibly explains the huge difference in the resistance towards carbon formation between these two supports [89].

2.4 Adiabatic oxidative reforming

In adiabatic oxidative reforming the heat for the reforming reactions is supplied by internal combustion. This is in contrast to fired tubular reforming (Section 2.3.3) and heat exchange reforming (Section 2.3.4), where the heat is supplied by heat exchange from an external source. Initially it may be worth considering: 'What are the benefits of using adiabatic oxidative reforming rather than steam reforming processes such as fired tubular reforming?'

The product gas from a reforming process may be referred to as raw synthesis gas. It consists of a mixture of H_2, CO, CO_2, H_2O, and CH_4 at equilibrium with respect to the steam methane reforming reaction and the carbon monoxide shift reaction. It will further contain any inerts such as N_2 and Ar, which have been introduced to the process.

In steam reforming, where the hydrocarbon feed is reacted with steam alone, the composition of the raw synthesis gas is governed by the steam reforming reaction and the shift reaction only. When an oxidant is used to supply heat for the reforming reactions by internal combustion, additional reactions are introduced, see Section 2.5. The overall reaction is adiabatic, meaning that there is no exchange of heat with the surroundings except a very limited heat loss, and the composition of the raw synthesis gas can be predicted by a heat and mass balance over the reactor. It should be noted that the combustion reactions are all irreversible. At the conditions applied (high temperature and addition of an under-stoichiometric amount of oxidant), all oxygen will be consumed, because this is the limiting reactant.

The raw synthesis gas is often characterised by the ratio between H_2 and carbon oxides, specifically by the so-called stoichiometric number (SN) or 'module' $M = (H_2 - CO_2)/(CO + CO_2)$ or by the Ribblett Ratio $R = H_2/(2CO + 3 CO_2)$, see Section 2.1.

With CH_4 as the reactant, steam reforming alone will produce a gas with a module of 3.0 (or a Ribblett Ratio of 1.5). This makes it possible, by full recycle of CO_2 from the raw synthesis gas back to the steam reformer, to produce hydrogen and carbon monoxide in a ratio of 3. This concept is often used in so-called HYCO plants, where both H_2 and CO are products [91].]. Alternatively, CO may be converted to CO_2 utilising the water gas shift reaction, producing a 4 to 1 mixture of H_2 and CO_2. By removal of CO_2 (and other undesired components in e.g. a PSA unit), pure H_2 may then be produced. This is the dominating process in H_2 production [60].

Adiabatic oxidative reforming will produce raw synthesis gas with different composition. Fig. 47 [92] shows as an illustration the values of the module M and the H_2/CO ratios, which can be obtained by autothermal reforming of CH_4 using O_2 as the oxidant.

It is seen that the use of an oxidant has brought the gas composition closer to desired values, which are a value of M close to 2.0 for methanol, DME and high temperature FT synthesis, and a H_2/CO ratio close to 2.0 for low temperature FT synthesis with Co catalyst. Means for final adjustment of the synthesis gas composition will be described in the following text:

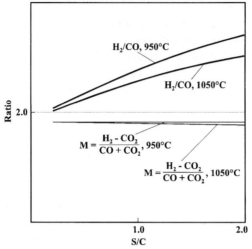

Figure 47 M and H_2/CO in raw gas from ATR

There are of course also other advantages (and disadvantages) related to the use of an oxidant as additional reactant in reforming reactions. The main advantages are related to economics of scale – much larger single stream units are possible with adiabatic oxidative reforming than with steam reforming – and to size of equipment – adiabatic oxidative reformers are very compact units compared to fired reformers. Furthermore, reformer tube materials limit the outlet temperature from fired reformers to a maximum of about 950°C, while the adiabatic oxidative reforming processes easily exceed 1000°C. This makes higher conversion of the feed possible, even at low steam to carbon ratio.

The main disadvantage of adiabatic oxidative reforming, especially with O_2 as oxidant, is that it requires an oxygen source. Oxygen plants are expensive, and the associated investments constitute a major part of the total investments in a synthesis gas generation unit. However, the advantages usually outweigh this disadvantage, especially for the very large capacities required for FT synthesis units and for modern 'mega'-methanol plants.

2.4.1 Process concepts

The process concepts for adiabatic oxidative reforming may be split into three categories considering the type of chemical reactions taking place in the reactor:

- homogeneous reactions
- heterogeneous reactions
- combination of homogeneous and heterogeneous reaction

Furthermore, oxidative adiabatic reforming processes may be characterised by the type of feed. If the feed comes directly from a desulphurisation unit or from a prereformer, and the reactions are carried out homogeneously without the aid of a reforming catalyst, then the oxidative adiabatic reformer is referred to as a gasification or partial oxidation unit. If the reactions are carried out heterogeneously on one or several catalysts, they are referred to as catalytic partial oxidation (CPO). If they are initiated by homogeneous reactions, e.g. in a burner, and completed by heterogeneous catalysis, then the reactor is called an autothermal reformer (ATR). If the feed has been partly reformed in a fired tubular reformer, the ATR reactor is most often called a secondary reformer. Secondary reformers are almost always based on a combination of homogeneous reactions (combustion) followed by catalytic conversion.

Finally, the processes based on adiabatic oxidative reforming may be characterised according to type of oxidant. Here, the split is between processes based on the use of air or enriched air in one group and the use of oxygen in another group.

There is also the possibility to integrate the process layouts with a heat exchange reformer, where the heat of the gas leaving the adiabatic oxidative reformer is used to perform reforming of the feed for the oxidative reformer or of a separate feed stream. Such process schemes are described in Section 2.3.4.2.

Descriptions given of the various process concepts and their principal characteristics and applications are given in the following. A survey is shown in Table 5.

Table 5
Survey of process concepts and characteristics

	Secondary Reformer (Air)	Secondary Reformer (O_2)	Autothermal Reformer (O_2)	Autothermal Reformer (Air)	PDU (O_2)	PDU (Air)
Burner/Mixer Type	Burner	Burner	Burner	Burner	Mixer	Mixer
Hydrocarbon Feed	Process gas [1]	Process gas [1]	Natural gas [2]	Natural gas	NG/diesel [3*]	NG/diesel [3*]
CH_4 content, dry mole%	10-15	10-40	85-100 [3]	85-100 [3]	85-100 [3*]	85-100 [3*]
C_nH_m content, dry mole%	None	None	1-10 [3]	1-10 [3]	1-10 [3*]	1-10 [3*]
Oxidant Feed	Air	Oxygen/steam	Oxygen/steam	Air	Oxygen/ Enriched air	Air
Process Gas Temp. °C	770-830	750-810	350-650	350-650	<200	<AIT [4]
H_2O/C ratio, mole/mole	2.5-3.6	1.5-2.5	0.4-3.5	2-4	0-2	<2.0 [5]
Feed Ratio						
Gas/oxidant, mole/mole	3-4	7-12	1-3	0.5-1.0	1-4	1-5
Flame Temp. °C						
Peak	~2000	~2500	2500-3500	2200-2500	-	-
Adiabatic [6]	~1200	1200-1500	1300-2100	1200-1500	-	-
Catalyst Temp. °C						
Exit	970-1020	950-1020	850-1100	1000-1100	750-1300	750-1200
Typical products	Ammonia synthesis gas	Methanol synthesis gas Hydrogen	Hydrogen/ Carbon monoxide Syngas for FT synt.		Syngas for FT synt.	Fuel Cells Syngas for FT synt.

1) Partly converted process gas from primary reformer
2) Naphtha and LPG feed can also be converted directly in an autothermal reformer
3) For natural gas. In case of naphtha and LPG feed, approx 100% C_nH_m and no methane contents
3*) For natural gas. In case of diesel, 100% and no methane contents
4) Preheat depends on the auto-ignition temperature of the fuel
5) High H_2O/C ratios quench the reactions
6) For the reaction: $CH_4 + 3/2 O_2 \rightarrow CO + 2H_2O + Q$. Higher hydrocarbons react by similar reaction scheme.

2.4.1.1 Processes based on homogeneous reactions.
gasification or partial oxidation (POX)

Production of the synthesis gas based on homogeneous reactions alone is referred to as gasification or partial oxidation (POX). The oxidant and the hydrocarbon feedstock is mixed in a reactor, where they are allowed to react homogeneously at very high temperatures typically 1300-1400°C. The resulting gas is quenched or cooled by steam production, and carbonaceous by-products like soot are removed by washing. This formation of carbonaceous by-products has an influence on the carbon efficiency.

As mentioned in Section 2.1, gasification is a versatile process which can convert a wide range of feedstocks to synthesis gas. Especially the entrained flow gasification technologies have been adapted to operation on liquid or gaseous feed. Both Shell and Texaco have supplied technology for natural gas conversion by gasification for decades, and lately Lurgi is also promoting a multi-purpose gasification process (MPG) which is available in a version adapted for operation on natural gas feed. The information available in the open literature about these technologies is scarce. However, some information can be found as follows: The Shell process: Refs. [93-95]; The Texaco process: Refs. [96-98]; The Lurgi process: Refs. [99,100].

When operating on gas, the gasifiers will most often be in the boiler mode, where the product gas from the gasifier is cooled by steam production in a boiler, which is specially designed to tolerate the aggressive process conditions and the presence of soot. The exit temperature from the gasifier is high, typically 1300-1400°C, to minimise soot formation and ensure close to complete conversion of the feedstock.

Typical operating conditions for a gas gasification unit are shown in Table 6. The feedstock is assumed to be light natural gas; the product gas is in equilibrium with respect to the steam reforming reaction, but equilibrated with respect to the shift reaction at a lower temperature, about 1125°C - the reaction runs homogeneously backwards at the highest temperatures in the cooling unit. The H_2/CO ratio in the product gas is slightly below 2.0; lower values can be achieved by recycle of CO_2, while steam additions (beyond small amounts added for burner cooling) can not be tolerated due to increased soot formation. In GTL applications tail gas (CO_2-rich) may be recycled to the gasifier to improve carbon yield. If this is done, the gas production in the POX unit must be supplemented by gas production in a steam reformer as in the Shell Middle Distillate Synthesis process [93-95]. In the Shell Bintulu GTL plant, each gasifier has a capacity corresponding to about 3000 b/d FT products. However, Shell claim that the gasification units may be scaled up to about 8000 b/d for new projects [95]. Similar capacities are quoted for the Lurgi MPG process.

Table 6
Typical operating conditions for a POX unit operating on natural gas feedstock

Oxygen/NG Ratio, mol/mol	0.55 – 0.65
H_2O/C Ratio, mol/mol	0 – 0.15
Exit Pressure, kg/cm^{2}g	25 – 40 [1]
Exit Temperature, °C	1300 – 1400
Exit Gas Characteristics	
H_2/CO Ratio	1.6 – 1.9
CO_2/CO Ratio	0.05 – 0.1
CH_4 Content, mol%	0.1

[1] Higher pressure is feasible, but probably not relevant for GTL.

2.4.1.2 Processes based on heterogeneous reactions.
catalytic partial oxidation (CPO)

The production of synthesis gas based only on heterogeneous catalytic reactions is normally referred to as CPO (catalytic partial oxidation). The oxidant and the hydrocarbon feedstock are premixed in a mixer before the feed enters the catalytic bed. The catalyst bed consisted, in older designs, of an ignition catalyst followed by a reforming catalyst. In some modern versions, only one catalyst is employed, and ultra short residence time is used. The risk of auto-ignition limits the preheat temperature, which results in a higher consumption of oxidant. The process is reported to be able to handle a wide range of feedstocks including heavy aromatic-containing hydrocarbons like diesel. However, for GTL applications the feed will of course be natural gas.

CPO with air as oxidant has been promoted for GTL applications by Syntroleum [101]. However, as for other adiabatic oxidative reforming processes, the use of air has also in this case severe disadvantages, and the process has not been used in practice. Conoco is developing a process apparently based on CPO with oxygen as the oxidant [102, 103]. The process has been developed to the demonstration stage, but only scarce information has been made available in the open literature. CPO is further described in Section 2.6.1.

2.4.1.3 Processes based on combined homogeneous and heterogeneous reactions.
autothermal reforming (ATR) and secondary reforming

Processes in which the reactions are initiated homogeneously, e.g. in a burner, and completed heterogeneously in a catalyst bed are called autothermal reforming (ATR) or secondary reforming.

ATR

In ATR, the feed goes directly from desulphurisation or from a prereformer to the adiabatic oxidative reformer. ATR is used mainly with O_2 as oxidant, although it has in the past been used with air or enriched air for production of NH_3 synthesis gas [3]. (In order to make a stoichiometric NH_3 synthesis gas $(H_2/N_2 = 3.0)$ by ATR with enriched air, $30 - 40$ vol% O_2 is required in the oxidant).

As shown in Fig. 47, it is not possible to produce synthesis gas for production of methanol, DME or FT synthesis directly by autothermal reforming of pure feed. It is necessary either to adjust the composition of the gas by removal of carbon oxides or by addition of H_2 (for methanol synthesis gas and similar) or by addition of carbon-rich gas (for low temperature FT synthesis). The carbon-rich gas may be recovered or imported CO_2 or carbon-rich tail gas from the synthesis. However, the advantages of ATR are such that it has, especially after development of processes operating at steam to carbon ratios below 1.0, become the preferred route to synthesis gas for these syntheses, especially for very large capacities [8, 26, 104-106].

Two concepts for large-scale production of methanol with synthesis gas production by ATR are shown in Fig. 48 and 49 [104].

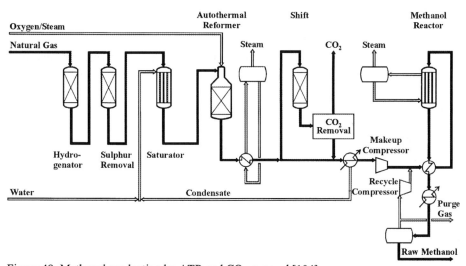

Figure 48 Methanol production by ATR and CO_2 removal [104].

The two process concepts differ mainly in the way the composition of the raw synthesis gas is adjusted in order to match the requirements of the methanol synthesis. As mentioned above and shown in Fig. 47, the raw synthesis gas from an ATR is lean in H_2. Typical values of the module M are 1.7-1.8, whereas a value slightly above 2.0 is preferred for methanol (and HTFT) synthesis.

322

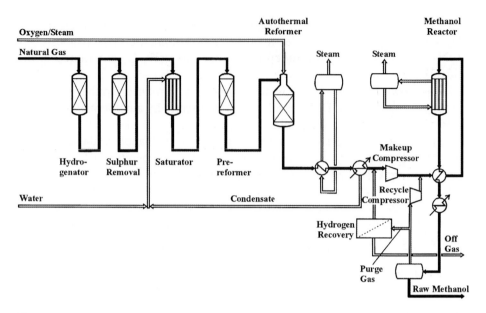

Figure 49 Methanol production by ATR and H_2 recovery and recycle [104]

In Fig. 48, the module is adjusted by removal of CO_2. The natural gas feedstock is desulphurised and saturated with steam to a steam to carbon ratio of 0.6 – 1.0. The mixed feed is passed to a prereformer and, after preheating to a high temperature, to the ATR, which operates at an outlet temperature of typically 1050°C and an outlet pressure of 30 – 40 bar g. Higher outlet pressure is possible [106], but not advantageous at very low steam to carbon ratio due to the resulting increase in CH_4 leakage in the synthesis gas.

In Fig. 49, the gas composition is adjusted by addition of hydrogen. The hydrogen is recovered from the tail gas from the methanol (or HTFT) synthesis loop. Apart from this the process scheme and the operating conditions are similar to the scheme and conditions in Fig. 48.

Production of synthesis gas for low temperature FT synthesis by ATR is illustrated in Fig. 50 [26].

This process has after careful analysis of alternatives been chosen by Sasol and others as the most economically attractive process for this purpose. As a consequence it will be employed in the large GTL units under realisation in Qatar and Nigeria and in future GTL plants considered by Sasol and their partners [107].

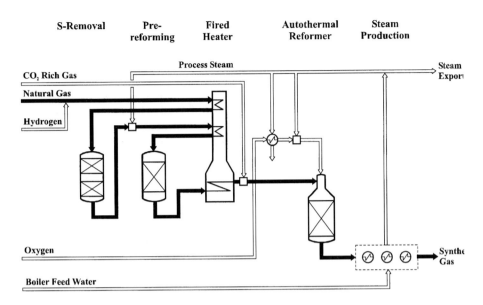

Figure 50 Typical process flow diagram for synthesis gas production for GTL

The process scheme is quite simple. The natural gas feedstock is desulphurised, and process steam is added to a steam to carbon ratio of 0.6. The mixture of feed and steam is passed to a preformer and further, after mixing with carbon rich tail gas from the FT synthesis and preheating, to the ATR. Here it reacts with O_2 to form a synthesis gas with the desired composition. Heat is recovered from the synthesis gas by production of high pressure steam, and the gas is cooled for removal of H_2O, before it is passed without compression to the synthesis section.

Although simple in principle, the process offers significant challenges. One is the size of equipment, piping, valves, etc. For production of 34,000 bbl/d of FT products, about 1.5 million Nm^3/h of synthesis gas will be produced in just two ATR reactors. For future projects, even larger reactors are contemplated. Another challenge is the risk for metal dusting corrosion in downstream equipment, see Section 2.3.4.3.

Design of the ATR reactor and the mixer burner is treated in Section 2.5.

ATR may, as fired tubular reforming, be combined with heat exchange reformers in a number of process schemes. Such schemes are described in Section 2.3.4.2.

Secondary reforming

Air-blown secondary reforming is the dominating process for manufacture of synthesis gas for NH_3 production from natural gas and naphtha [3]. A typical process scheme is shown in Fig. 51 [108].

The process concept is used by all important suppliers of technology for NH_3 production. Natural gas is desulphurised, mixed with process steam, and passed to a fired tubular reformer, the 'primary reformer'. The product gas from the primary reformer is reacted with air in the secondary reformer to produce the raw synthesis gas, which is processed further by shift conversion, removal of CO_2, and methanation to give the final synthesis gas, a 3 to 1 mixture of H_2 and N_2 with small amounts of inerts, mainly CH_4 and Ar. The amount of air added to the secondary reformer is adjusted to give the correct ratio of H_2 and N_2 in the synthesis gas. Variations of the process scheme exist, see [3].

The secondary reformer is a refractory lined vessel with a mixer/burner (normally a multi-nozzle design), a combustion chamber, where homogeneous reactions take place, and a bed of Ni-based reforming catalyst, where the shift- and reforming reaction are equilibrated by heterogeneous reaction on the catalyst.

In a modern NH_3 plant, the steam to carbon ratio at the inlet of the primary reformer is in the range $2.5 - 3.5$, the pressure at the outlet of the secondary reformer is $25 - 35$ bar g, and the outlet temperatures from the primary and secondary reformers are 750-850°C and $950 - 1050$°C, respectively.

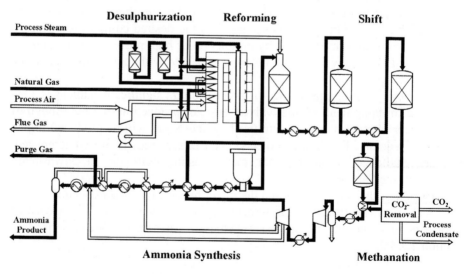

Figure 51 Complete ammonia plant [108]

Use of enriched air as oxidant in the secondary reformer has been suggested as a revamp option [109] in combination with a fired reformer.

Combinations with heat exchange reformers are used in Johnson Matthey's LCA process and in KBR's KRES process. These processes are described in Section 2.3.4.2. Both concepts with air and enriched air are described.

When N_2 is an undesired constituent in the synthesis gas, O_2 is may be used as the oxidant in the secondary reformer. This is the case in production of synthesis gas for methanol, dimethylether (DME), and high temperature FT synthesis.

A process scheme used by Haldor Topsøe for production of methanol synthesis gas by so-called 'two-step reforming', using prereforming, fired tubular reforming, and O_2-blown secondary reforming is illustrated in Fig. 52 [104]. This process scheme was used in a large methanol plant in Norway [110]. The natural gas feed is desulphurised, and process steam is added in a saturator. The mixture of feed and steam is passed to a prereformer, a primary reformer, and an O_2-blown secondary reformer. The steam to carbon ratio at the inlet to the prereformer is 1.5 – 2.0, the outlet pressure from the secondary reformer is about 35 bar g, and the outlet temperatures from the three reformers are about 450°C, 750 – 800°C, and 1000 – 1050°C, respectively.

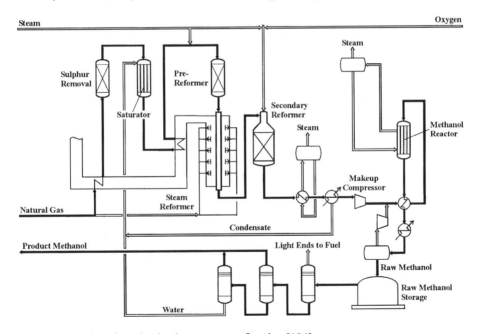

Figure 52 Methanol production by two-step reforming [104].

The design of the secondary reformer is very similar to the design used in air-blown processes. However, the operating conditions in the secondary reformer are more severe than in the air fired concept, and multi-nozzle burner design can not be used. Instead, a design similar to the design of burners for autothermal reformers (see Section 2.5) is used.

In a variation of the concept, the natural gas feed is split into two parts. One is added to the primary reformer, whereas the other part is sent directly to the secondary reformer. This concept, originally suggested by Foster Wheeler, is normally referred to as 'combined reforming'. It is mainly used by Lurgi [111] in their process for production of methanol synthesis gas. It has also been used in the high temperature FT synthesis plant owned and operated by PetroSA (Mossgas) [112].

The advantage of using O_2 instead of air as oxidant is obvious. The presence of N_2 as an inert in the final synthesis gas is avoided. Use of air as oxidant has been suggested in production of synthesis gas for methanol [113] and FT synthesis [92,114]. However, this is not economically feasible [115], since the presence of large amounts of N_2 – about 50 vol% in the dry synthesis gas – makes recycle concepts in the synthesis gas impossible leading to low overall efficiency. Furthermore, compression of the large amounts of air required consumes more power than required for production and compression of O_2 in O_2-blown concepts.

2.5 Autothermal reforming

The autothermal reforming (ATR) process has been used to produce hydrogen- and carbon monoxide-rich synthesis gas for decades. In the 1950's and 1960's autothermal reformers were used to produce synthesis gas for ammonia production and methanol [116, 117]. In ammonia plants, hydrogen production was maximised by operating at high steam to carbon ratios ranging from 2.5 to 3.5 on molar basis, while in the methanol units carbon dioxide removal adjusted the synthesis gas composition. In the early 1990's, the technology was improved and operation at much lower steam to carbon ratios was achieved [118].

In production of CO-rich synthesis gas for production of pure CO and H_2 and CO mixtures as feed for methanol or Fischer Tropsch synthesis, operation at low steam to carbon ratio feed ratios are beneficial. Operation at H_2O/C ratio of 0.6 has been demonstrated in pilot scale [118, 132, 155] and industrial scale [119].

2.5.1 Chemical reactions

ATR is a combined combustion and catalytic process carried out in an adiabatic reactor. A mixture of natural gas and steam is partially converted by pressurized combustion at fuel-rich conditions and the final conversion of hydrocarbons into an equilibrated synthesis gas is made over a fixed catalyst bed. The overall chemical reactions taking place in the ATR reactor can be described by the below reactions:

Combustion $CH_4 + {}^{3}/_{2} O_2 \rightarrow CO + 2 H_2O$ $- \Delta H^{0}_{298} = +519$ kJ/mole (1)

Reforming $CH_4 + H_2O \rightleftarrows CO + 3 H_2$ $- \Delta H^{0}_{298} = -206$ kJ/mole (2)

Shift $CO + H_2O \rightleftarrows CO_2 + H_2$ $- \Delta H^{0}_{298} = +41$ kJ/mole (3)

The ATR reactor is described in section 2.5.2. It can be divided into three zones;

- Combustion zone
- Thermal zone
- Catalytic zone

The *combustion zone* is a turbulent diffusion flame, where hydrocarbon molecules and oxygen are gradually mixed and combusted. The combustion reactions are exothermic and very fast, and from a global point of view it can be assumed progressing as 'mixed is burnt'. Combustion in an ATR is sub-stoichiometric with an overall oxygen to hydrocarbon ratio of 0.55 to 0.6, but when simplified as a one-step model the flame zone can be described as a single reaction (1) of CH_4 to CO and H_2O with an O_2/CH_4 ratio of 1.5. Although simplified in a single reaction (1) above, combustion of hydrocarbons consists of a large number of homogenous radical reactions. Combustion chemistry is described in more details in Section 2.5.1.1. The local stoichiometry in the flame zone will vary from very fuel-lean to very fuel-rich. The core of the flame will have parts that are close to stoichiometric, and therefore the centre of the flame will be very hot as the adiabatic flame temperature for the stoichiometric mixture exceeds 3000°C.

In the *thermal zone* above the catalyst bed, further conversion occurs by homogeneous gas-phase-reactions. These reactions are slower reactions like CO oxidation and pyrolysis reactions involving higher hydrocarbons. The main overall reactions in the thermal zone are the homogeneous gas-phase steam methane reforming (2) and shift reaction (3). Reactions between N_2 and hydrocarbon radicals, which may form by-products like HCN and NH_3 are also proceeding in the thermal zone. The N-chemistry in relation to sub-stoichiometric combustion is described in Section 2.5.1.4.

In the *catalytic zone* the final conversion of hydrocarbons takes place through heterogeneous catalytic reactions including steam methane reforming

(2) and shift reaction (3). The catalytic zone is described in more detail in Sections 2.5.2.3 and 2.5.2.4.

Fuel-rich combustion in partial oxidation processes involves the risk of incomplete combustion. Methane combustion at fuel-rich conditions is mainly proceeding through reaction steps with C_2-radicals as intermediates, which may react to soot precursors like poly-aromatic hydrocarbons (PAH) and further to soot particles [120]. ATR operation is soot-free at normal circumstances, and this is achieved through a careful burner design, by addition of small amounts of steam and finally by catalytic conversion of soot precursors, and therefore the exit gas contains no other hydrocarbons than methane. Soot formation chemistry is described in Section 3. Soot formation is unwanted and. It would reduce the carbon efficiency of the process, and soot particles would need to be removed from the synthesis gas.

2.5.1.1 Combustion chemistry

The gas phase combustion reactions between hydrocarbons (i.e. methane) and oxygen proceeds through radical reactions. The reaction mechanism consists of a large number of species (molecules and radicals) and reactions. Even for a simple hydrocarbon like methane a precise description includes several hundred reactions. The combustion chemistry of the lighter aliphatic hydrocarbons present in natural gas has been studied intensively during the recent years [120, 121]. Fig. 53 shows the simplified reaction path for high temperature methane combustion.

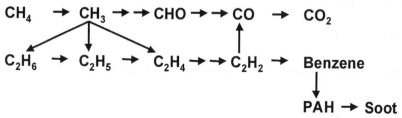

Figure 53 Simplified reaction path for C_1 and C_2 chemistry CH_4 combustion [118].

The combustion chemistry is controlled by radicals; the primary radicals are the hydrogen radical (H), the hydroxyl radical (OH) and the oxygen radical (O). A large number of hydrocarbon radicals exist as well, some of them are included in Fig. 53 (CH_3, C_2H_5, and CHO). The hydrocarbon radicals are generally produced as combustion intermediates primarily by reaction between a molecular hydrocarbon and a radical, but it may also be formed from a hydrocarbon molecule alone. Examples of these types of reactions are given below by the reaction between methane and the hydroxyl radical (4), and the reaction between ethane and a third body molecule (M) (5) forming methyl radicals:

$$CH_4 + OH \rightarrow CH_3 + H_2O \tag{4}$$
$$C_2H_6 + M \rightarrow CH_3 + CH_3 + M \tag{5}$$

In combustion of hydrocarbons, the C_1 chemistry is the basis for the reactions leading to oxidation to CO and CO_2, while the C_2 chemistry leads to formation of unsaturated hydrocarbons, aromatics and soot precursors. Even in a methane flame with methane as the only hydrocarbon source, ethane, ethylene and acetylene will be formed especially in the fuel rich parts of the flame.

The chemistry of hydrocarbon combustion can be modelled as a large number of elementary reactions described in detailed kinetic models. This will be further described in Section 2.5.3.2.

2.5.1.2 Ignition

The establishment of a flame, normally referred to as ignition, is a special part of the radical combustion chemistry. The ignition reactions involve chain branching and chain termination reactions. At T>1000 K the radicals H, O, OH become responsible for the chain branching reactions. Once the ignition is established the propagation reactions proceed by radical attacks on the fuel [122, 123].

The ignition is often controlled by two competing reactions involving the oxygen molecule leading either to radical branching (ignition) or radical termination. The branching reaction is:

$$O_2 + H \rightarrow OH + O \tag{6}$$

At lower temperature, this reaction may lead to formation of the peroxide radical instead:

$$O_2 + H (+ M) \rightarrow HO_2 (+ M) \tag{7}$$

The peroxide radical is less reactive than other radicals meaning that it has lower rates when reacting with hydrocarbons, and it will lead to radical termination through reactions with another hydrogen radical:

$$HO_2 + H \rightarrow H_2 + O_2 \tag{8}$$

The net reaction of (7) and (8) is a recombination of two hydrogen radicals to form a hydrogen molecule, (i.e. radical termination), while (6) reaction leads to radical branching. Furthermore, the oxygen radical is very reactive and is likely to react in a new branching reaction:

$$H_2 + O \rightarrow OH + H \tag{9}$$

Once the flame is ignited, the process condition for the ATR process with temperatures in the combustion chamber above the auto-ignition temperature of the process gas ensures the flame stability.

To understand more about ignition behaviour the general overview given by Westbrook [123] is recommended.

Although the ATR process is a non-sooting technology, there is a potential for soot formation in the combustion chamber at abnormal operating conditions and during start-up transients. To avoid soot formation, it is necessary to have detailed knowledge of the mechanisms for soot formation and limits for on-set of soot formation.

Soot may be formed as a by-product in all combustion processes involving hydrocarbons [124-126]. One of the best known flames is that of a candlelight with the yellowish flame. The yellowish flame is due to soot formation in the fuel rich part of the diffusion flame. Close to the wick, the colour of the flame is a darker bluish. This is where the fuel oxidation is initiated, but with insufficient air to allow for complete combustion. Therefore, pyrolysis reactions are allowed to occur resulting in fuel decomposition and formation of aromatic species subsequently leading to soot particles. In candlelights, the soot is oxidised as air is mixed with the fuel rich gas leaving the overall flame non-sooting. However, if the wick becomes large allowing more fuel to evaporate or if the flame is interacting with other flows e.g. crosswind, the temperature in the diffusion layer may drop, quenching the soot particles oxidation and leaving a black plume of soot.

In methane combustion, reaction steps involved in formation of hydrocarbon radicals, aromatic species and formation of soot particles are schematically illustrated in Fig. 54 [126, 127]. Once aromatic species are formed, there is a potential for growth of the aromatic structure forming poly-aromatic hydrocarbons (PAH), and subsequently PAH clusters will lead to inception of soot particles.

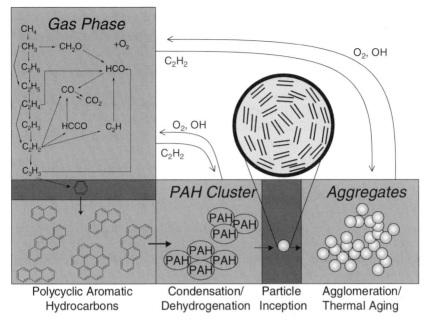

Figure 54 Schematic representation of reaction steps involved in the soot formation in methane combustion, taken from [128].

Oxidation of PAH and soot particles are competing reactions to soot particle formation. Although soot is formed in the flame, it may have time to be oxidised making the overall flame non-sooting as observed in most candlelights. Limitations in the reaction pathway to soot are illustrated in Fig. 54 as dark zone; Benzene formation and particle inceptions. The first step is the formation of the initial aromatic species allowing the PAH chemistry to initiate. The second step is the inception of the initial soot particle. While the gas phase chemistry leading to the formation of the aromatic species is well understood [124-127], the chemistry leading to formation of PAH clusters and soot particle inception cannot presently be described in detail [128].

2.5.1.3 N-chemistry

The feed streams for an ATR also contain some inerts; the inerts are primarily argon from the air separation unit and nitrogen from the hydrocarbon source. Nitrogen is not completely inert in combustion processes. Two reaction mechanisms can break the nitrogen-nitrogen bond in the nitrogen molecule forming nitrogen oxides and hydrogen cyanide. The dominant mechanisms are the thermal NO_x and the prompt NO_x reactions, respectively, and main reaction steps are given below as reaction (10) and (11).

$$N_2 + O = NO + N \tag{10}$$
$$N_2 + CH_i = H_iCN + N \tag{11}$$

While the thermal NO_x reaction takes place on the fuel lean side of the flame at elevated temperatures (above 1200°C), the prompt NO_x reaction takes place in fuel rich areas. In the prompt NO_x mechanism, therefore nitrogen reacts with a hydrocarbon to form amines or cyano compounds, which may react further to NO_x. The index i denotes 1, 2 or 3 meaning that nitrogen will react with all C_1 fuel radicals. In traditional combustion systems, e.g. furnaces and gas turbines, which operate with a fuel lean flame, the H_iCN species will be further oxidized to NO_x. The ATR flame is fuel rich and the fired nitrogen species will remain in reduced form as H_iCN or NH_i.

The fixed nitrogen species formed in the flame zone will all be reduced in the hydrogen and fuel rich post flame zone [129-131], entering the catalytic bed mainly as ammonia or hydrogen cyanide. Nitrogen species will further be converted in the catalyst bed, and in the exit synthesis gas the nitrogen containing molecules will be either molecular nitrogen N_2, ammonia or hydrogen cyanide. The concentration of the latter two will be in the ppm range. The concentration will depend on the amount of nitrogen in the natural gas and in the oxygen. In air-blown processes more ammonia and hydrogen cyanide will be formed than in oxygen-blown processes.

2.5.2 ATR process and reactor design

The ATR reactor design consists of a burner, a combustion chamber, a fixed catalyst bed section, a refractory lining, and a reactor pressure shell as illustrated in Fig. 55. The key elements of the ATR technology are the burner and the catalyst:

- The *burner* provides proper mixing of the feed streams, and the fuel-rich combustion is taking place as a turbulent diffusion flame. Intensive mixing is essential in order to avoid soot formation.
- The *catalyst* equilibrates the synthesis gas and destroys soot precursors. The catalyst particle size and shape is optimized to achieve high activity and low pressure drop in order to obtain a compact reactor design.

A careful design of the process burner and combustion chamber is required in order to avoid excessive temperatures and formation of soot particles. Further, the detail design and construction of the whole ATR reactor including refractory and catalyst bed is of utmost importance for ensuring safe design and operation of the syngas unit. Predictions and design are facilitated by reactor models based on use of computational fluid dynamics (CFD) and on chemical kinetics.

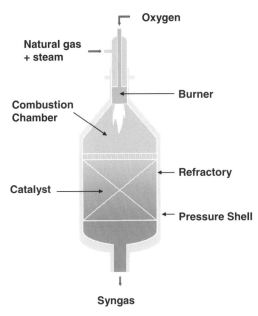

Figure 55 Illustration of an ATR reactor [taken from 132].

2.5.2.1 The ATR burner and combustion chamber

The ATR process burner is a key element of the technology. The burner provides the mixing between the hydrocarbon feed and the oxidant in a turbulent diffusion flame. The flame core may exceed 3000°C. It is essential to minimise heat transfer by thermal radiation or hot gas recirculation to the burner parts.

In the design of the burner and the combustion chamber the following reaction-engineering aspects must be considered in order to ensure optimal reactor performance, safe operation and satisfactory equipment lifetime:

- Effective mixing at the burner nozzles
- Low-metal temperature of the burner
- Soot-free combustion
- Homogeneous gas and temperature distribution at the catalyst bed entrance
- Protection of the refractory from the hot flame core.

Recirculation of the reacted gas from the thermal zone to the burner can protect the refractory and the burner from the hot flame core and gases from the combustion zone. An efficient external re-circulation will enhance the position of the flame core along the centreline of the combustion chamber and protect the refractory from the hot flame core [134]. The temperature of the gas circulating along the walls and into the catalyst bed is reduced to the range 1100-1400°C by the endothermic reactions proceeding in the thermal zone.

An optimised re-circulation will also provide a homogeneous distribution of gas and temperature at the entrance to the catalyst bed. Inhomogeneity will result in larger distance to equilibrium, and the methane concentration in the exit gas is increased. The more even the distribution of gas to the catalyst bed, the better the utilization of the catalyst activity. An even flow and temperature distribution is obtained by proper design of the geometry of the combustion chamber.

The flow velocities in the burner nozzles can be selected within large ranges. Highly turbulent mixing intensity of the diffusion flame is obtained with high velocities in the nozzle gaps. For applications with oxygen or enriched air as oxidant, the flame speed will be faster than for similar air flames [133]. The position of an oxygen flame will be very close to the burner nozzles especially at highly turbulent mixing intensities. A turbulent diffusion flame is in steady-state seen over a certain time period, but an inherent feature of the turbulent flame is that the flame is dynamic and changes position within short time frames.

A process burner concept, the CTS burner [134], was developed in 1991 by Haldor Topsøe. CTS burners have been in successful operation since 1992 in both ATR reactors and oxygen-blown secondary reformers for production of a variety of products ranging from methanol, hydrogen, and ammonia to pure CO and Fischer-Tropsch products. A new and optimized version of the CTS burner has been developed with special focus on operation at low S/C-ratios with high flame temperatures. The flow patterns and mechanical design of the burner nozzles have been optimized by use of tools like CFD (computational fluid dynamics), stress-analysis with Finite Element Analysis (FEA) and real-environment testing in pilot scale and in full size demonstrations.

Burners in secondary reformers and autothermal reformers operation are exposed to high operating temperatures. Industrial operation has from time to time faced problems ranging from burner wear without serious process implications to severe failure. However, safe operation and long burner lifetime has been obtained by proper design of the burner [134, 156], whereby severe consequences to equipment and production loss have been avoided. And example of a severe failure of an oxygen-blown secondary reformer refractory and reactor shell following a burner related incident is described by Mossgas [135]. Burners for high temperature reformers can be designed with a focus on mechanical and thermal integrity in combination with the combustion chamber design [134]. CFD (Computational Fluid Dynamics) can be used to predict the flow pattern and avoid unwanted behaviour.

2.5.2.2 Vessel and refractory

The reactor vessel is lined on the inside with refractory. The refractory insulates the steel wall of the pressure vessel from the high temperature reaction environment. The refractory is commonly constructed of several layers with different materials and insulation properties. Efficient refractory design ensures that reasonable low mechanical temperature design temperatures can be applied for the pressure shell. Typically the temperature of the reactor wall is reduced to 100-200°C at normal operation.

In air-blown secondary reformers it is common practise today to use a design with two refractory layers. In older designs a design with only one layer was applied, but such a design was sensitive to cracks in the refractory layers, which resulted in gas flow and transfer of heat to the shell and thereby in hot spots on the pressure shell [136]. In oxygen-blown secondary reformers and ATR reactors the operating conditions are more severe including a higher operating temperature in the combustion zone. In ATR reactors, a refractory design with three layers of refractory is commonly used today. The inner layer has high thermal resistance and stability and is typically a high density alumina brick layer. The installation of the refractory lining is important and involves skilled craftsmen.

Transport of hot gasses from the high temperature combustion chamber and the catalyst bed through the refractory layers to the reactor wall is not happening with a good refractory design and installation. But it is a potential risk and may lead to increased temperatures at the reactor walls, which could develop into so-called 'hot spots''', where the design temperature of the vessel is approached or exceeded. The risk of gas bypass through the refractory is especially high in the combustion chamber where the temperature is highest, and at the catalyst bed where the catalyst bed creates a pressure drop which may force the gas through weak zones of the refractory.

The outside of the reactor vessel is either non-insulated and exposed to the atmosphere or covered in a water jacket. In non-insulated designs a thermal sensitive paint is applied on the reactor surface. Such paint allows visual monitoring and inspections for colour changes indication increased temperature in case of hot spot development. With a water jacket it is more difficult to detect hot spots, but this can be done to a certain degree by monitoring the increased evaporation rate from the water jacket.

2.5.2.3 Catalyst bed

The hydrocarbons are only partly converted in the combustion chamber, and the gas leaving the combustion chamber contains a significant amount of methane and traces of other hydrocarbons. In the catalytic zone the final conversion of methane and other hydrocarbons takes place. The reactions are heterogeneous steam reforming reactions (see Section 2.5.1) over the fixed catalyst bed. The methane steam reforming reaction is endothermic, and the temperature will decrease from typically 1200-1300°C at the inlet to the catalyst bed to approx 1000°C at the exit of the catalyst bed. The catalyst bed operates adiabatically with only a small heat loss to the surroundings. The gas composition is ideally in total equilibrium when the catalyst has sufficient activity. But the gas distribution to the catalyst bed and especially the temperature and homogeneity of the inlet gas composition will affect the average product gas composition.

A layer of protecting tiles is often placed on top of the catalyst bed to protect it from the very intense turbulent flow in the combustion chamber. The radiation from the flame and the circulation velocities in the combustion chamber require that the tiles have a high thermal stability and can resist the thermal shocks during start-up and trips.

The requirements for the catalyst include;

- high thermal stability
- sufficient activity to reach equilibrium
- low pressure drop to avoid bypass of gas through the refractory

The preferred catalyst for autothermal reforming is a nickel based reforming catalyst. The catalyst is exposed to high operating temperatures and the nickel metal is subject to a high degree of sintering resulting in a low intrinsic catalyst activity.

It is also a requirement for the support for the nickel catalysts that it has a high thermal stability to achieve sufficient strength at the high operating temperatures. In ATR and secondary reformers carriers of both alumina (α-Al_2O_3) and magnesium alumina spinel ($MgAl_2O_4$) are used. Spinel has higher melting point and in general higher thermal strength and stability than the alumina based catalysts.

The shape of the catalyst pellet is an important design parameter for the catalyst bed. The pressure drop should be kept low in order to reduce the risk of bypass around the catalyst bed through the refractory. Bypass would have the consequence that flow through the refractory layers transports heat into the refractory layers, which could increase temperature on the pressure shell and result in an increased risk of local hot spots., see section 2.5.2.2. A shape optimised catalyst e.g. seven-axial holes with low pressure drop should be selected [134]. In shape optimization, a pellet shape with high catalyst pellet activity can be selected. The optimal loading of the catalytic fixed bed may consist of several layers of different types of catalysts.

2.5.2.4 Catalyst deactivation in ATR operation

Generally an ATR catalyst is robust with a high thermal stability and not prone to deactivation. However, some deactivation may occur, mainly due to:

- sintering
- sulphur poisoning
- fouling by ruby deposition leading to ΔP increase

Sintering proceeds as in all steam reforming catalysts, see Section 2.3.5.2. However, the activity of the catalyst in ATR service is rapidly reduced due to the high operating temperatures, and after this initial sintering only minor further deactivation is expected due to sintering.

The main poison for the catalyst in ATR operation is sulphur. Sulphur is a well-known poison on the nickel based steam reforming catalyst, as described in Section 2.3.5.2. Sulphur originates from impurities in ppb levels in the feed streams that remains in the hydrocarbon feed stream after feed purification or in the process steam. The sulphur is adsorbed on the nickel surface as nickel sulphide and thereby deactivates the catalyst [28].

The adsorption is described by sulphur adsorption isotherm (cfr. eqs (2) and (3) in Section 2.3.5.2). The sulphur coverage at lower temperatures is close to 90-100%. However, only a partial coverage of the nickel surface is found at the elevated catalyst temperatures of the ATR. At 1000°C the sulphur coverage can be calculated to approx 30%, when the inlet gas to the catalyst bed contains a low sulphur concentration. Thus, desulphurization is normally not necessary for a stand alone ATR unit. In GTL plants sulphur is removed upstream of the ATR in order to protect the prereforming catalyst and this then also protects the downstream FT synthesis catalyst.

In high temperature ATR and secondary reformers, it is common to observe a deposition of white and pink crystals on the catalyst outer pellet surface. The crystals are mixtures of alumina and chromium-alumina spinel. The latter is also known as the ruby with a purple colour. Ruby formation is not a poisoning as such, but it reduces run time between shutdowns, because pressure drop over the catalyst increases and may increase the risk of formation of hot spots on the reactor wall. Ruby formation and deposition are well known to industry, but the knowledge about the mechanism of ruby formation is very empirical. A case history is described in Ref. [137] supported by some more theoretical viewpoints. The primary process leading to fouling seems to be transport of 'rubies' from the refractory and deposition in the catalytic bed. The 'ruby' formation is caused by case story describes the "ruby formation" as evaporation of aluminium species, probably AlOOH, from the high-alumina bricks in the refractory. When the gas is cooled in the catalytic bed because of the reforming reaction, AlOOH will condense and together with impurities of chromium and iron deposit as rubies. The chromium and the iron come from construction materials upstream of the combustion chamber and the ATR reactor. Often ruby deposition is seen in a narrow section in the upper part of the catalyst bed. In such cases, only 'skimming' of the layer with rubies may be sufficient to solve pressure drop problems.

2.5.3 Modelling

Modelling of an ATR reactor may be done in several steps with focus on the different reactions zones and different levels of details.

The modelling approaches described in the following sections include:

- Computational Fluid Dynamics (CFD)
- Chemical kinetic modeling
- Combined models (chemical kinetics and flow)
- Catalytic fixed-bed modeling.

The combustion in the ATR combustion chamber is a complex interaction of combustion chemistry and fluid dynamics. Complete modelling of this combustion process includes detailed modelling of the gas phase radical reactions, flow equations (e.g. Navier Stokes), radiation and turbulence models, all implemented into CFD. However, due to the complexity of such models and the fact that the turbulent flow and the chemical reactions have different time scales, the mathematical problems are so complex that even with supercomputer access, long computing times are required. Therefore, work has been carried out on de-coupled models, where the variations in the flow pattern are solved separately from the chemistry.

2.5.3.1 Computational fluid dynamics

Computational fluid dynamics (CFD) is a powerful tool in design and performance prediction of the ATR burner and combustion chamber. CFD is used to calculate the flow pattern of the ATR combustion chamber and of the flame [134, 138].

Computational Fluid Dynamics (CFD) software is a general purpose fluid flow simulation software which makes it possible to analyse turbulent and laminar flow problems in arbitrary complex geometries. Use of CFD simulation software is widespread today for design of advanced apparatus like diesel engines, gas turbines and slurry bed reactors. All applications are characterized by interaction between complex flow fields and chemical reaction. Various software packages are available among which Fluent, CFX-5 from ANSYS CFX and STAR-CD from CD Adapco Group are the most common.

The first step in each CFD calculation is division of the reaction space into a three-dimensional grid. Each grid cell is a discrete computational cell in which the equations for fluid flow and chemical reaction are solved. A typical grid may contain from 10.000 to 100.000 cells. The next steps are selection of models for thermodynamic properties, chemical reaction kinetics, turbulence, and thermal radiation, as well as definition of boundary conditions.

The chemical models implemented into CFD are usually relative simple, often only a few chemical reactions in order to maintain reasonable calculation times and at the same time describe the chemical conversion in sufficient details. For combustion processes such chemical models is a compromise as the combustion proceeds through numerous radical reactions as described in Section 2.5.1.1. In some applications, the 'mixed-is-burnt' approximation is an acceptable approach in order to de-couple the fluid flow from the chemical reaction in the flame, as most of the combustion reactions are very fast. But in order to describe thermal expansion and slower reactions the use of more detailed chemical models is advantageous.

The choice of a turbulence model is essential, and in some cases the simple k-ε model is too simple an approach, so it is necessary to include a higher order turbulence model, like a Reynolds stress model [138, 139].

Besides the fluid flow pattern and temperature distribution in the combustion chamber, the CFD simulation also gives information about the temperature on the burner nozzle parts and on the refractory surfaces.

An example of a CFD simulation results for an ATR burner and combustion chamber is shown in Fig. 56. The simulation is made with CFX-4 [140] using a 2D model assuming rotational symmetry in the combustion chamber. The calculation shows that a flow pattern in the combustion chamber with a centrally positioned flame is obtained. A good re-circulation is obtained, which keeps the refractory cool, provides sufficient cooling of burner parts, and ensures a homogeneous composition and flow at the inlet to the catalytic bed [134].

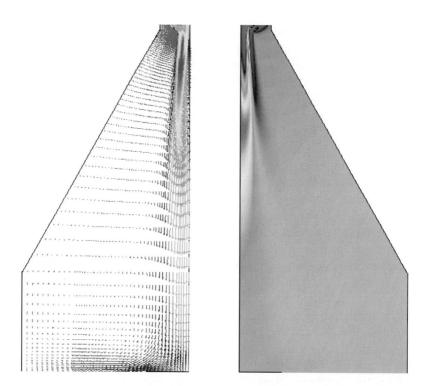

Figure 56 Example of CFD modelling results for the combustion chamber in an ATR reactor.
The left part show velocity vectors and the right part show temperatures. [134]

2.5.3.2 Chemical kinetic modelling

Chemical kinetic modelling is based on a detailed kinetic gas phase model in a simplified reactor, either a plug flow reactor or a perfectly stirred reactor. It is solved using e.g. CHEMKIN II [141].

Reaction schemes with the combustion chemistry of methane and ethane are presently well described for ambient pressure and at stoichiometric conditions. Detailed kinetic models describing the chemistry based on elementary reactions are available in the literature and some can be downloaded from the Internet [121]. However, to adopt a model, it should carefully be considered what needs to be simulated and how the selected model can be verified for reaction conditions corresponding to those to be modelled. Since most of the present kinetic models have been developed based on ambient pressure conditions, adoption for high-pressure conditions, as is the case for an ATR reactor, should be done with caution.

Detailed chemical kinetic simulations with the CHEMKIN II software package [141] can be used for the simulation of the combustion chamber of the ATR reactor. Reaction path analyses [142] can be used to deduce reaction mechanisms and preferred reaction channels. Pathways in soot formation in methane oxidation at fuel-rich conditions were studied at atmospheric pressure [142], and a channel producing benzene was identified. The mechanism describes that benzene subsequently reacts to phenanthrene and pyrene, which could lead to inception of soot particles.

The chemistry and flow interaction in the combustion chamber of an ATR reactor introduce a lot of complexities regarding modelling. Detailed kinetic models have been developed to describe the combustion of hydrocarbons [120, 121]. However, most detailed kinetic models have focussed primarily on fuel lean combustion at ambient pressure. This is far from the conditions in the combustion chamber of an ATR reactor, where the pressure is much above atmospheric and the overall chemistry very fuel rich. Nevertheless, the detailed kinetic models have been adopted for this purpose and used to model the conversion in the combustion chamber with surprisingly good results [132] as further described below.

Ignition may be modelled using detailed kinetic models as well, but the combustion models described above are typically optimised for higher temperatures and may not provide reliable results for auto-ignition calculations. However, there are more appropriate models available e.g. from studies of spark engine knocking [143, 144], although these studies are made for air as oxidant. Modelling of the ignition has the purpose to predict the necessary temperature at start-up for the feeds to auto-ignite. Normally, the start-up conditions are characterized by lower temperature and possibly also lower pressure than normal operation, and they are therefore more critical with respect to flame ignition and stability.

2.5.3.3 Combined modelling of flow and chemistry

The complex interaction between fluid flow and chemistry in the turbulent diffusion flame of the ATR combustion chamber makes it difficult to apply the simplified models in both the detailed chemical kinetic simulations (detailed chemistry, simple reactor geometry) and the CFD simulation (complex reactor geometry, detailed fluid flow and simplified chemistry). Two solutions have been suggested. A reactor network consisting of plug-flow reactors and perfectly stirred reactors has been developed [132] as a means of including more detailed information about fluid flow into the detailed chemical kinetic simulations. The model has been further developed into CFD post processing tool [128, 148], in which detailed chemistry for the combustion reactions and for the PAH and soot chemistry is calculated in all cells of the CFD-grid.

A reactor flow network model consisting of a combination of plug flow reactors and perfectly stirred reactors was defined [145] using information from the CFD simulation about flow pattern and residence time in the ATR reactor at given conditions. Such a model has been tested with two different detailed kinetic models and compared with experimental measurements of the gas composition in the combustion chamber at the ATR process demonstration unit. The GRI-Mechanism, Version 2.11 (279 reactions with 49 species) [146], has been compared with a mechanism developed by Miller and co-workers (301 reactions with 65 species) [147]. Both consist of reactions including H, C, O, N and Ar species up to C_2 hydrocarbons, but in the mechanism by Miller and co-workers, the selected reactions and species leading to formation of benzene have also been included. Each kinetic model is used together with the reactor model to describe the chemistry in the combustion chamber and to predict the composition of the gas entering the catalyst bed. The calculations have been performed with the CHEMKIN-II software packages [141]. The results of the experimental measurements and modelling taken from [132] are listed in Table 7.

Table 7
Comparison between experimental measured concentrations and chemical kinetic modelling results for an ATR combustion chamber [132]

	H_2	H_2O	CO	CO_2	CH_4	C_2H_6	C_2H_4	C_2H_2	N_2
Experimental	48.8	16.0	24.9	3.51	6.46	0.0151	0.0672	0.0168	0.190
Calculated:									
GRI-MECH Ver. 2.11	48.7	16.1	24.9	3.46	6.62	0.0025	0.0137	0.0020	0.190
Miller and co-workers	48.7	16.1	24.9	3.47	6.61	0.0027	0.0148	0.0025	0.190

This type of modelling can provide detailed information about the chemistry occurring in the combustion chamber, but it is at the expense of loss of information about the complicated flow pattern and the influence of turbulence on the chemistry especially in the flame.

In order to keep the detailed kinetic models and get further details of the flow in the combustion chamber, a chemical kinetic post-processor for CFD calculations has been suggested [128, 148]. In this case the grid structure used to solve the CFD modelling is adapted as a reactor network consisting of perfectly or partially stirred reactors. The turbulence intensity may be included in the post processor and the reaction rates can therefore be determined either by turbulence or by the finite reaction rates calculated by the detailed kinetic model. This will especially improve the modelling of the turbulent diffusion flame, where turbulent mixing is controlling the reaction between hydrocarbon feed and oxidant.

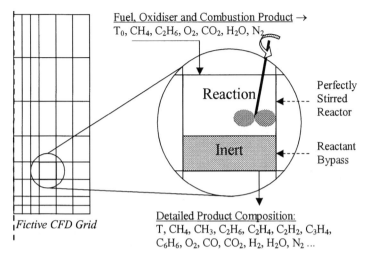

Fuel, Oxidiser and Combustion Product →
T_0, CH_4, C_2H_6, O_2, CO_2, H_2O, N_2

Reaction

Perfectly
Stirred
Reactor

Inert

Reactant
Bypass

Fictive CFD Grid

Detailed Product Composition:
T, CH_4, CH_3, C_2H_6, C_2H_4, C_2H_2, C_3H_4,
C_6H_6, O_2, CO, CO_2, H_2, H_2O, N_2 ...

Figure 57 Schematic illustration of the partially stirred reactor approach used to model individual cells in the detailed chemical kinetic post-processing of the post-processing procedure taken from [148].

2.5.3.4 Fixed bed simulation of the catalytic bed

Modelling of the catalyst bed of the ATR reactor is in itself a challenge. Simulation of the catalyst bed includes radial distribution of gas and temperature, gas diffusion in the gas film and catalyst pore system, heterogeneous catalytic reactions, and pressure drop. The catalyst bed is simulated by fixed bed reactor models using input from the CFD simulation with respect to the gas and temperature distribution profiles into the fixed bed. Although the pressure drop is the simplest part, it may nevertheless be the most important, since a catalyst with too high pressure drop will result in gas bypass through the refractory leading to hot-spots on the pressure shell.

Fixed-bed reactor models are divided into two groups: pseudo-homogeneous and heterogeneous models [149]. The catalyst in an ATR is exposed to very high temperatures and thus a high degree of sintering. The catalyst performance is controlled by a combination of both film diffusion and pore diffusion, and relatively successful simulations were achieved with both a heterogeneous model [150] and with a pseudo-homogeneous model [151]. A key parameter in determining the necessary catalyst volume is the gas distribution at the entrance to the catalyst layer. In air blown secondary reformers, it is well-known that burners of poor design and damaged burners may cause an increased average methane leakage from the catalyst bed. This is a result of inefficient mixing in the combustion chamber leading to uneven flow and temperature distribution at the entrance to the catalyst bed [134]. Even with an optimal burner design, some gradients in temperature and concentrations are observed along the radius of the reactor as a result of the mixing efficiency. Variations in the O/C and H/C atomic ratios along the inlet to the catalyst bed will result in different adiabatic equilibrium states at the exit of the catalyst bed, which may be seen as an increased overall approach to equilibrium which is caused more by mixing and gas distribution than by catalyst activity.

2.5.4 Autothermal reforming (ATR) in GTL applications

In the mid-1990s, Autothermal Reforming (ATR) was identified as being an attractive process for conversion of natural gas to liquid fuels (GTL) [152]. ATR has been selected as synthesis gas technology for a number of large-scale GTL plants for the production of synthetic fuels by Fischer-Tropsch (FT) synthesis (30-35,000 bbl/d liquid products) [153, 154]. The preferred lay-out of the syngas production section is a combination of adiabatic prereforming and ATR, see fig. 50. The lay-out results in large flexibility for variations in natural gas and in FT recycle-gas composition as well as reduced oxygen consumption per unit synthesis gas produced. Operation at low steam-to-carbon ratio (H_2O/C) is one of the key parameters to achieve good economic performance for GTL applications.

The H_2O/C ratio of the autothermal reforming process was previously in the high range $1.5 - 2.5$ [118]. ATR operating at low H_2O/C ratios was developed by Haldor Topsøe through the 1990's [92, 118, 132, 155]. The breakthrough for the use of ATR technology for application in Fischer-Tropsch process flow schemes has been enabled by soot-free operation at low H_2O/C ratios, of 0.6 and below, and by the development of new burners ensuring safe operation and high on-stream factors [134]. Application of ATR technology in mega-size plants involves aspects such as scale-up of equipment design, maximum size of equipment and capacity for single-line plants. In GTL applications, a synthesis gas with H_2/CO ratio $\cong 2.0$ is often needed. ATR can produce this gas by recycling a gas rich in carbon oxides from the FT synthesis unit for H_2/CO adjustment. The recycle gas with FT applications is rarely pure CO_2 but rather contains CO and small amounts of hydrocarbons of various types, e.g. paraffins and olefins.

Operation at low H_2O/C ratios challenges the limitations of the technology; soot formation must be avoided, and the process burner must have a sufficient lifetime without excessive burner wear and degradation at the high operating temperatures.

2.5.4.1 Industrial operation at low H_2O/C ratio

ATR operation at low H_2O/C around 0.6 has become the state-of-the-art synthesis gas technology for Fischer-Tropsch applications. An industrial unit based on ATR technology for production of CO-rich synthesis gas has been in operation in Europe at $H_2O/C=0.6$ since 2002. Two units with a combined production of synthesis gas of 430.000 Nm^3/hr (H_2+CO) will start at Sasolburg, South Africa, in 2004. The synthesis gas will be used for production of chemicals and waxes by FT synthesis (see Chapter 5, section 3.1). Large dedicated GTL plants will start up in 2005 in Qatar and in Nigeria in 2007 (107). These plants will both feature 2 synthesis gas units based on ATR at S/C=0.6 with a combined production of about 1.5 million Nm^3/hr synthesis gas.

The first operation at industrial scale was a demonstration test in an ATR unit operating with approx. 20,000 Nm3/hr hydrocarbon feed gas [119]. This unit is one of 16 autothermal reformers owned and operated by Sasol at their site in Secunda, South Africa. These reforming units were originally revamped in 1997 to use burner technology from Haldor Topsøe [156]. One of the units was modified specifically for the demonstration run at H_2O/C ratio of 0.6 in stead of the normal high ratio of 1.5 - 2.0. During the demonstration run which lasted 5 weeks, satisfactory and stable operation at the low H_2O/C ratio was demonstrated without soot formation. The demonstration test pushed the limits for application of the ATR technology.

2.5.4.2 Development and demonstration at PDU-scale

The ATR process for operation at low H_2O/C ratios was developed through the 1990's. Operating parameters were studied in a process demonstration unit (PDU) [118, 155] built at the pilot plant facilities of Haldor Topsoe, Inc., Houston, Texas in 1990 (Fig,. 58). The process demonstration unit was built using the same design principles as industrial units. The layout of the PDU-unit is shown schematically in Fig. 59.

The ATR plant is operated with natural gas feed at a capacity corresponding to typically 100 - 150 Nm^3/hr and produces synthesis gas (CO + H_2) in the range from 250 to 450 Nm^3/hr. The feed streams consist of natural gas, steam, pure oxygen ($O_2 > 99.5\%$), hydrogen and, optionally, carbon dioxide and liquid hydrocarbons. At normal operating conditions, no soot formation is taking place, but when testing operating conditions close to and beyond the limit for onset of soot formation, minor amounts of soot can be formed and captured in the condensate. The condensate is investigated through both gravimetric and spectrophotometric methods.

Figure 58 ATR Process Demonstration Unit, Houston, Texas [132]

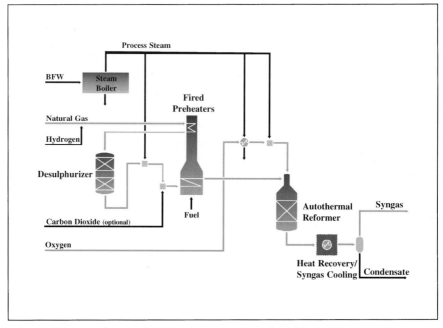

Figure 59 Process layout of ATR Process Demonstration Unit

Soot-free operation and satisfactory burner performance was demonstrated at a H_2O/C ratio of 0.6 in the early 1990's [118]. Similar results at H_2O/C ratios down to 0.2 were obtained in later demonstrations [155], see Section 2.5.4.3.

2.5.4.3 ATR process performance tests related to GTL applications

Numerous tests with varying operating parameters were performed in the ATR PDU in order to establish safe design background at the very low H_2O/C ratio of 0.6 and below. This included the influence of variations in H_2O/C ratio, temperatures, and pressure as well as the effect of feed composition such as content of higher hydrocarbons, CO_2, and H_2 on the limits for soot-free operation.

Limits for soot formation were studied intensively in a pilot-scale the ATR PDU reactor representing real-environment operation. Various operating conditions were tested, and the limits for onset of formation of soot particles in the exit condensate were identified. Tests included preformed natural gas, variations in feed gas composition with different levels of C_{2+} in natural gas, and variations in recycle gas composition, e.g. CO_2 and hydrocarbons. Tests with preformed natural gas showed a larger margin to the onset of soot formation than with a natural gas containing higher hydrocarbons at similar operating conditions. Even though the preformed natural gas implies an improvement with regards to the risk of soot formation in the ATR, it is still prone to form soot at certain operating conditions, which made it necessary to establish the limits and extend the models to include a wider range of feedstocks [157].

Explorative tests as well as tests of longer duration with steam-to-carbon ratios in the range of 0.2-0.6 are described in Ref. [155]. The tests were made with the purpose of reducing the H_2O/C ratio as much as possible without reaching the range of soot formation. Results from various pilot programs are shown in Fig. 60 [118, 155] and Table 8.

All data represent operating conditions without soot formation, but they do not represent the limits of the technology. The ATR can produce synthesis gas within a wide range of H_2/CO-ratio, when the H_2O/C ratio, CO_2-recycle and exit temperatures are optimized.

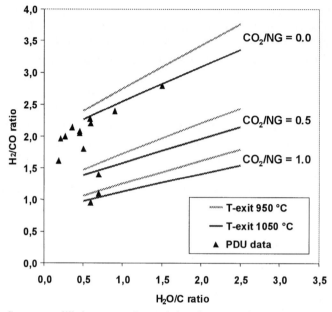

Figure 60 Syngas equilibrium; experimental data from soot-free ATR-operation [155]

Table 8 ATR PDU Runs [118, 155].
ATR pilot plant demonstration runs

TEST		A	B	C	D	E
Feed Ratios[1]	*(mole/mole)*					
H_2O/C		0,59	0,21	0.51	0.60	0.36
CO_2/C		0,01	0,01	0.19	0.01	0.01
O_2/C [2]		0,62	0,59	0.62	0.58	0.57
Product Gas						
Temperature	(°C)	1065	1065	1025	1020	1022
Pressure	(bar)	24,5	24,5	27.5	28.5	28.5
H_2/CO	(mole/mole)	2,24	1,96	1.80	2.30	2.15
CO/CO_2	(mole/mole)	5,05	9,93	4.44	4.54	6.78
CH_4-leakage	(dry mole %)	0,48	1,15	0.92	1.22	1.66

Notes:
1) Mole per mole of hydrocarbon C-atoms
2) The O_2/C-ratio is approximately 5% higher than truly adiabatic reactors with same exit temperature.

2.6 Other technologies

Most or all of the technologies described in the previous section are in commercial use today for production of synthesis gas. However, substantial efforts to develop new technologies are undertaken due to the fact that the most capital intensive part of a GTL plant is the Syngas Generation Unit (SGU). The focus of many of these developments is to reduce or eliminate the use of oxygen and/or reduce the size of the primary reactor in the SGU. In the following text the production of synthesis gas by Catalytic Partial Oxidation (CPO) and Oxygen Membrane Reforming (OMR) will be briefly described.

2.6.1 Catalytic Partial Oxidation

The principle of CPO is illustrated in Fig. 61. The hydrocarbon feed and the oxidant are mixed in an inlet zone upstream the catalyst bed. In the catalytic section, the mixture reacts by heterogeneous reactions (partial and total combustion along with the methane steam reforming and shift reactions). The catalyst is normally based on noble metals and the space velocity is in many cases very high. Catalysts in the form of pellets, monoliths, and foams have been used to perform the reaction.

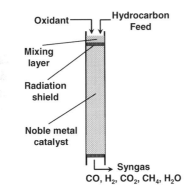

Figure 61 CPO Principle

The methane steam reforming and shift reactions are typically at or close to equilibrium at the reactor exit. It has been claimed [158] that methane reacts by partial oxidation according to reaction (1) below and that methane conversions above those corresponding to thermodynamic equilibrium of the methane steam reforming reaction can be obtained.

$$CH_4 + \tfrac{1}{2}O_2 \rightarrow CO + 2H_2 \qquad (1)$$

However, in practice the products of reaction (1) will further oxidise and fundamental studies have shown that partial oxidation is only kinetically favoured at temperatures above 900°C [159]. Gas compositions indicating higher conversions than thermodynamic equilibrium most likely reflect the temperature of the catalyst [160].

350

CPO differs from ATR especially by the fact that no burner is used. Instead all the chemical reactions take place in the catalytic zone. Total combustion takes place to some extent in the first part of the catalyst layers making the catalyst very hot in this region. Laboratory measurements have indicated that this temperature may be higher than 1100°C [161]. In order to avoid overheating of the gas upstream of the catalyst a thermal shield is often employed as indicated in Fig. 1. It should be noted that the gas temperature remains comparatively low as compared to the catalyst surface temperature in the inlet zone [161].

CPO has been investigated extensively for many years. Before 1992 most studies were carried out at moderate or low space velocities at a residence time of 1s or above [161]. However, during the later years CPO has been carried out at least in the laboratory at very short contact times between 0.1 and 10 ms in some cases without preheating the feedstock and with no steam addition. Additional information regarding research, mainly of a fundamental nature, can be found for example in a series of papers by L.D. Schmidt and co-workers [159, 162-166].

Both air and oxygen may in principle be used as oxidant in a CPO reactor. An example of the use of oxygen in a CPO reactor in the laboratory is given in Fig. 62, illustrating that stable conversion may be obtained at elevated pressure. Experiments with CPO and air as oxidant have been conducted at the Topsøe pilot plant in Houston, Texas. Selected results at various pressures with and without steam in the feed are presented in Fig. 63. In all cases, the methane conversion corresponds closely to the equilibrium of the methane steam reforming reaction.

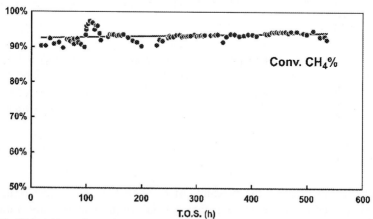

Figure 62 CPO with methane feed and oxygen as oxidant [161]. P=1.5 Mpa. H_2O/CH_4=0.26; CO_2/CH_4=0.11;O_2/CH_4=0.56

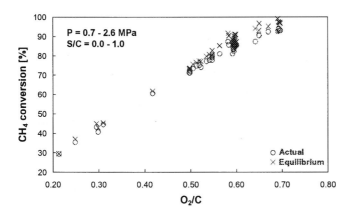

Figure 63 Methane conversion in CPO pilot plant experiments with air as oxidant [167]

Table 9
Auto-ignition temperatures for natural gas in air [168]

Pressure (bar)	Auto-ignition Temp., (°C)
1	465
4	313
20	267
40	259

Table 10
Relative oxygen and natural gas consumption for CPO- and ATR-based GTL front-ends for production of hydrogen and carbon monoxide [26]

Reactor	S/C ratio	Hydrocarbon feed temperature, reactor inlet (°C)	Oxygen consumption (relative)	Natural gas consumption (relative)
ATR	0.6	650	100	100
CPO	0.6	200	121	109
ATR	0.3	650	97	102
CPO	0.3	200	114	109

An adiabatic pre-reformer is located upstream the ATR. CO_2 is introduced before the partial oxidation reactor at 200 °C in an amount to give $H_2/CO=2.00$.
Pressure: 25 bar abs. Oxygen temperature: 200 °C. Exit temperature: 1050 °C.

The presence of a flammable mixture in the inlet zone upstream the catalyst may in some cases make use of CPO at high inlet temperatures difficult especially at elevated pressures. In Table 9 the auto-ignition temperature is given for natural gas in air as a function of pressure. The auto-ignition temperatures are lower with oxygen as oxidant. For safety reasons, the inlet feed temperatures of the hydrocarbon feedstock and oxidant must be kept low. This increases the oxygen consumption as shown in Table 10. Higher oxygen consumption increases the ASU investment and the level of inerts in the synthesis gas.

2.6.2 Oxygen Membrane Reforming

The principle of Oxygen Membrane Reforming (OMR) is indicated in Fig. 64.

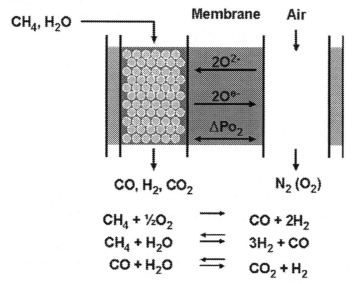

Figure 64 Principle of oxygen membrane reforming

Very significant efforts are undertaken to develop OMR, see e.g. [169] for a recent overview. Air is introduced at one side of a membrane through which oxygen in the form of ions is transported selectively to the other side of the ceramic membrane. At the other side of the membrane the oxygen reacts with the hydrocarbon feedstock to produce synthesis gas. The concept simultaneously avoids the capital cost of the Air Separation Unit and a high content of inert nitrogen in the synthesis gas. The catalyst on the synthesis side of the membrane may be in the form of pellets or directly attached to the membrane itself.

The membrane itself is made out of a ceramic material often in the form of a perovskite or a brownmillerite. The driving force across the membrane is proportional to the logarithm of the ratio of the partial pressures of oxygen on the two sides. Hence, in principle air may be introduced at ambient pressure to supply oxygen to the other side at elevated pressure because the oxygen partial pressure on the process side is extremely low. The temperature should probably be above 750-800°C for sufficient oxygen flux.

The membrane material must enable a high flux probably in the range of more than 10 Nm^3 O_2/m^2 hr. The membrane should also withstand reducing gas on one side and air on the other. Various types of composite membranes have been proposed and research is also undertaken to place a thin membrane on a stronger porous support.

A key challenge in the development of OMR is the absolute pressure difference across the membrane. It may render the process non-economical if air must be compressed to ensure similar pressures on the two sides of the membrane. It sets great demands on the mechanical integrity of the membrane if ambient pressure air is used. In any case the process may be best suited for small or medium scale applications as the scaling factor of the membrane unit will be close to unity.

3. SYNTHESIS GAS PREPARATION VIA GASIFICATION

3.1 Introduction

Gasification involves the reaction of a source of carbon, possibly associated with hydrogen, with a source of hydrogen (usually steam) and/or oxygen to yield a gas containing hydrogen, carbon monoxide, carbon dioxide and methane. Proportions of these component gases depend on the ratio of the reactants used and on the reaction conditions [170]. It is a versatile process which can be used to convert a variety of solid or liquid carbonaceous feedstocks into synthesis gas (syngas).

Once the feedstock has been converted to a gaseous state, undesirable substances such as sulphur compounds and entrained solid particles may be removed from the gas by a range of techniques.

Coal gasification yields a wide variety of useful products for residential, utility and industrial markets [171]. Clean syngas, essentially a mixture of carbon monoxide (CO) and hydrogen (H_2), can be converted into gaseous fuels, liquid fuels, chemicals, electric power or a combination of these products.

The first companies to convert coal to combustible gas through pyrolysis were chartered in 1912, while the first true gasifier, a Lurgi moving bed gasifier, was in operation by 1887. During the 1930's, the first commercial coal gasification plants were constructed, followed by town gas applications in the

1940's. In the 1950's, chemical process industries started applying gasification for hydrogen production. Sasol is now by far the largest single user of gasification technology. Sasol's first coal gasification plant was commissioned in 1954. Studies to generate electric power in a more efficient manner in an Integrated Gasification Combined Cycle (IGCC) followed in the 1970's, and are still being actively pursued. At present, there is about 11 GW of gasification capacity in operation around the world in the chemicals industry or synthetic transport fuels production [170] and a little over 40 GW total syngas production capacity [1]. Gasification can be said to be a commercially proven, mature technology, and furthermore, proven in combination with various downstream technologies [170].

Users of gasification technologies are seldom willing to experiment with new equipment, preferring proven technology [172], since gasification often forms a vital link in extensive flow schemes and involves rather extreme reaction conditions. Selecting proven gasification technology minimises the risk associated with the high capital investment cost for new installations. Therefore most of the current research, development and demonstration efforts are aimed towards technical and economic improvements for established technologies [171]. Improvements in efficiency, reliability and capital and operating cost of gasification are all means to the same end – cheap and 'clean' coal conversion processes.

There is a growing realisation that the world will be dependent on fossil fuels for the foreseeable future to meet ever-increasing energy requirements. Dwindling supplies and rising prices of other fossil fuels like oil and gas are increasing the attractiveness of the vast remaining coal reserves. A higher energy demand in several countries has resulted in a re-focus on indigenous coal resources and available technologies for the conversion of coal into useful products [173]. At the same time increasing environmental pressure is challenging increased coal utilisation. As gasification is the cleanest coal technology available, there is a strong drive for its further growth.

3.2 Possible applications for gasification processes

There are many possible uses for coal. Gasification is a prominent coal conversion technology, yielding the most diverse group of products (see Fig. 1). Besides coal, a wide variety of other, usually low value, carbonaceous feedstocks can be gasified to produce fuels and chemicals. These feedstocks include renewable energy sources e.g. biomass, wastes e.g. paper pulp and municipal solid wastes (MSW) and refinery by-products e.g. petcoke and vacuum residues.

Synthesis gas production through gasification was growing at a rate of about 10% per annum in 1983 [174]. From about 1985 to 1995 there was a period of constant capacity before the growth in capacity resumed [171]. In the USA, the September 11-events in 2001 and the more recent political conflict with Iraq brought about a renewed urgency concerning energy security and independence. In this regard, the utilisation of the abundant coal resources in places such as China, India, South Africa, Australia and the USA through clean technologies gives new impetus to the interest in gasification technology. Many of the coal-fired power stations in the USA and elsewhere are over thirty years old and operating with typical efficiencies of 30-35%. Replacement with cleaner and more efficient gasification based power generation at efficiencies of about 46% would extremely beneficial to the environment.

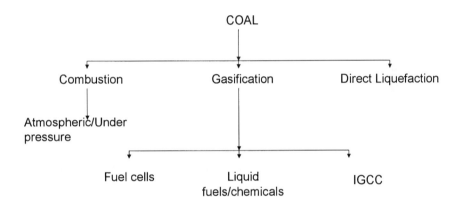

Figure 65 Different options for coal conversion and power generation. [175]

After gas cooling and purification, pure synthesis gas is then ready for conversion into the numerous possible end products. The product range depends on the hydrogen to carbon ratio of the clean synthesis gas. Products include, among others, methanol, acetic acid, oxo-chemicals, liquid fuels and chemicals via the Fischer-Tropsch process and electric power [176]. The synthesis gas can also be converted into synthetic natural gas (SNG) for use in town gas and heating fuel applications.

3.3 Integrated Gasification Combined Cycle (IGCC)

The combustion of coal to generate power is the historical backbone of the electric utility industry. This is due to the availability, low price and security of supply of coal compared to other fuels, and the relative simplicity of the combustion process. Tightening environmental requirements result in substantial increases in both capital and operating cost for conventional coal-fired combustion plants, while adversely affecting plant efficiency and reliability. A combination of gasification with a combined cycle is the only coal-based technology that can approach the environmental performance of natural gas-fired systems [177].

The IGCC plant design essentially involves two stages of combustion with gas cleaning between the two stages [178]. A number of variations of IGCC plant designs exist, primarily due to differences in the selected gasifier technology [179] and subsequent raw gas cleaning methods. Integration of gasification and combined cycle facilities is achieved mainly through the plant steam system. IGCC technology has been proven for a variety of fuels, particularly heavy oils, heavy oil residues, petcokes, and bituminous coals [180].

The first stage of the IGCC system employs gasifiers where partial oxidation of a solid or liquid fuel occurs to produce a combustible gas. The produced gas is then cleaned to remove particulates, sulphur compounds and other undesirables, before being fed to a gas turbine combustor to complete the combustion. The combustion turbine is normally followed by a turbine driven by steam generated from the hot tail gas of the combustion turbine [181].

IGCC systems can achieve high thermal efficiencies, which correspond to lower CO_2 production. Low emissions of acid rain precursors (NO_x and SO_x) are achieved [181]. The IGCC process offers several additional advantages over conventional coal-based power generation technology, including [182]:

- Decreased coal consumption due to increased generating efficiency
- Reduced land area and cooling water requirements
- Modular construction which allows economic increments of capacity to match load growth
- Potential for design standardisation due to modular construction, which should reduce required engineering effort, construction time and delays due to permitting complications for future plants
- Greater fuel flexibility.

It is generally accepted that IGCC is clean and efficient. Regrettably, it is also expensive and not entirely commercially proven. There are, however, clear factors favouring the selection of IGCC for future applications [177]:

- An absence of cheap natural gas, and not enough proven (local) gas reserves

- Tight emission limits

- Higher coal prices, demanding high efficiency

- The opportunity for co-gasification of wastes and biomass.

High capital and operating costs remain the greatest challenge to IGCC implementation. Even though IGCC is at least 10 to 12% more efficient than a conventional coal-fired power plant [182], the capital cost may be about 20% more than for a pulverised coal combustion (PCC) boiler plant [183]. At present, operational costs are considerably more for IGCC than for PCC boiler plants, with operating costs for IGCC estimated at $ 7.00 - $ 9.00/MWh compared to $ 3.00 - $ 5.00/MWh for a conventional PCC plant. Other hindering factors, which have to be addressed, can be summarised as follows [184]:

- Concerns regarding long-term reliability and operability
- Uncertain long term operating and maintenance costs
- Few deeply capitalized and credit worthy technology suppliers
- Few qualified and experienced operators
- Uncertainties regarding competitive economics
- Money lenders are strongly disposed to highly demonstrated technologies
- Environmental and utility regulators are more comfortable with demonstrated technologies
- Poor availability [177]
- Poor operational flexibility [177].

Apart from the barriers listed above, another major reason why IGCC has yet to impact significantly on the power generation landscape is that most of the recent coal-fired capacity development occurred in countries such as India and China, where there is a particularly strong emphasis on reliability and cost, both still challenges to IGCC [177]. However, with the developed world once again moving towards coal utilisation, IGCC technology has now reached the commercialisation stage in the USA and EU with a number of plants already in operation [185]. Coal-based IGCC plants that are or were in operation up to the year 2000 are listed in Table 1.

Although IGCC technology was originally conceived as a clean method of using coal for electricity production, it has been proposed as a solution to the growing problem of surplus heavy fuel oil. For example, the ISAB (Industria Siciliana Asfalti e Bitumi) complex in Sicily, which started up in 1998, is one of the largest IGCC installations in the world, and produces hydrogen and electric power through gasification of high-sulphur heavy liquid residues from an adjacent refinery [186]. The complex, which uses Texaco gasifiers, can produce 540 MW.

Serious operational problems have been encountered at various IGCC installations. One of the most crucial problems is that gas turbines are not always able to meet the demands set by the IGCC plant. A number of failures have been reported in hot gas cleanup systems, e.g. breakage of ceramic candle filters and stress corrosion cracking in heat exchangers [181]. Availability has proven to be a major challenge, but is improving with greater operating experience. For example, Shell reported an availability of 86%, including 4%

planned outages, at the Pernis refinery in the Netherlands in 2001 [172]; an installation that was retrofitted for IGCC, commissioned in 1997 and runs on heavy residues.

Table 1
Commercial IGCC plants across the world [181, 187]

PROJECT	CAPACITY	OPERATION	REMARKS
Coolwater Plant, Barstow,California, USA	125 MW	Operated 1984-88	Texaco Gasifier (1000TPD)
Plaquemine Plant, Louisiana,USA	160 MW	In operation since April,1987	Dow (Destec) Gasifier (2200TPD)
Demkolec Buggenum Plant, Netherlands	253 MW	Started operation in 1993.	Shell Gasifier – Initial problems encountered in Gas Clean-up System. Now operating with good availability.
PSI Energy, Wabash River Plant, USA	262 MW	Commissioned November,1995	Destec Gasifier
Tampa Electric Polk Power Plant, USA	260 MW	Commissioned Sep. 1996	Texaco Gasifier
Sierra Pacific Pinon Pine Plant, USA	100 MW	Commissioned 1998	KRW Gasifier
ELCOGAS, Puertollano, Spain	335 MW	Commissioned 1998	Prenflow, Krupp Uhde gasifier commissioned in 1998
Schwarze Pumpe, Germany	40 MW	Commissioned September,1996	Noell KRC, 7 fixed bed gasifiers. Power/methanol production.

A key development area is the integration of emission control systems for simplification and cost reduction. The concept of integrated environmental control has several dimensions [179]: (1) interactions between control methods for air, water and solid waste emissions, so that a reduction in one type of discharge does not unduly increase others, (2) the integrated use of pre-combustion, combustion and post-combustion emission control methods, and (3) the development of new processes for combined pollutant removal in lieu of separate processes for individual pollutants.

Texaco was once the world leader in IGCC gasification technology but ChevronTexaco sold this business recently (2004) to General Electric (GE). GE is the leading supplier of gas turbines to the power industry. Several installations use Texaco gasification technology for IGCC, but only one plant uses coal as feedstock, the Tampa Electric plant in the USA, which co-feeds coal and petroleum coke.

Although the integration of systems complicates plant operation and control, IGCC technology has matured to the point where it is cleaner and more energy efficient than conventional combustion/steam station power generation plants. Further efficiency improvements are envisaged with more research and better integration. Studies are in progress to investigate the combination of IGCC technology with Fischer-Tropsch synthesis for the combined production of liquid hydrocarbons and electricity (see Chapter 5).

3.4 Characteristics of coal important for gasification

Coal is a polymeric substance with no repeating monomers [171]. It can be defined as a macromolecular gel and the concepts and techniques of polymer science may be applied in the study of coal. The main constituents of coal are hydrocarbon chains, mineral matter and moisture, while other components, e.g. halogens, are present in small concentrations.

The composition and structure of coal influences the applications that a specific coal is suitable for. Coals are classified by rank, i.e. according to the degree of metamorphism in the series from lignite to anthracite [171]. Time and pressure gradually transform the initial biomass from lignite to sub-bituminous coal to bituminous coal and then to anthracite. For gasification, coal of three ranks are used; sub-bituminous, bituminous and lignite [188]. Lignite is also called brown coal, particularly in Europe and Australia. Also important for gasification are the coal caking properties, water content and the ash properties.

Chemical and physical properties of coal show a large variation from source to source. Because these properties relate directly to gasifier behaviour, detailed analyses of coal are essential to predict gasification performance when a specific coal source or mixtures of coal sources are to be gasified.

Some of the tests that are usually conducted on coal sources to determine suitability for gasification purposes include the following:

- Proximate and ultimate analysis
- CO_2 gasification reactivity
- Particle size distribution
- Thermal fragmentation
- Caking properties
- Ash melting properties and ash viscosity
- Ash composition

Other analyses that may also be relevant include mechanical fragmentation, Fischer Assay to determine tar production, total sulphur content, determination of the heating value of the coal, maceral composition and rank. The importance of the different analyses depends on the specific gasifier technology. For example, a Fischer Assay is more important in a fixed bed gasifier where tars and oils are produced as by-products, while ash viscosity is important in a slagging gasifier since ash will be extracted in a liquid state.

ASTM or ISO standards exist for most of the typical analyses done to classify a specific feedstock. Additionally, gasification technology suppliers generally have a suite of in-house developed feedstock characterisation techniques to ensure the compatibility of a particular feedstock with their gasifiers. Most of that information, however, is not available in the open literature.

The starting point for an understanding of the composition of the coal is the proximate and ultimate analysis. The proximate analysis determines the moisture, volatile matter, fixed carbon and ash in the coal. The methods for performing these analyses have been standardized by all the major standards institutions (e.g. ASTM, ISO, DIN, BS, and others). The standards differ in some details so it is important to specify the method used.

For the ultimate analysis the percentage of carbon, hydrogen, oxygen, sulphur and nitrogen are determined for the hydrocarbonaceous part of the coal (i.e. excluding non-volatile the mineral matter and water). Higman and van der Burgt [1] may be consulted for guidance on the significance of the various standard coal analyses.

3.5 Gasification chemistry

Gasification is the conversion of coal or other solid or liquid carbonaceous feedstock (hereafter, only coal will be mentioned for simplification) to a combustible gas via pyrolysis and heterogeneous reactions [183]. The conversion is usually accomplished by partial combustion of coal in an air/steam or oxygen/steam mixture [189].

Upon entering the gasification vessel, coal undergoes several transitions as it moves through different temperature zones in the moving bed gasifier:

- Drying, to evaporate moisture
- Devolatilisation, or pyrolysis, producing gas, vaporised tars and oils and solid char residue
- Gasification, or partial oxidation of the solid char and pyrolysis tars and gases
- Combustion

The mineral matter remaining from coal then exits the gasifier as ash.

For the moving bed type gasifiers, the different stages listed above occur in distinct zones. For other types of gasifiers no separate zones can be identified e.g. entrained flow gasifiers which operate at constant high temperatures and fluidized bed gasifiers which are well-mixed.

Upon initial exposure to heat in the gasifier, coal loses its surface moisture. Since the drying process consumes energy, the feed moisture content has an influence on the thermal efficiency of the gasification process.

Pyrolysis can be defined as the "decomposition of organic substances by heat" [190]. These decomposition reactions commence at 350 to 400°C, depending on the coal properties, most specifically the rank of the coal [191]. Pyrolysis is a complex process, involving cracking, hydrogenation and free radical reaction mechanisms. As a host of feedstock and process parameters have an influence on the pyrolysis reactions, it is problematic to predict the eventual yield and composition of the pyrolysis products and difficult to fully understand all the process steps.

Pyrolysis products are destroyed in most types of gasifiers, but not in the Sasol-Lurgi fixed bed dry bottom gasifier, where tars, pitches and phenols are valuable co-products from gasification. It is widely recognised that the initial pyrolysis step in coal conversion processes has a noteworthy effect on the yield and distribution of end products and emissions.

A few examples of pyrolysis reactions follow [192]:

$$Ar\ CH_2CH_2CH_2Ar \rightarrow ArCH_2CH_2\bullet + \bullet CH_2Ar$$
(1)

$$ArCH_2CH_2\bullet + \bullet Ar \rightarrow \begin{cases} ArCH_2CH_3 + HAr \\ \\ ArAr + CH_2{=}CH_2 \end{cases}$$
(2)

$$ArCOOH \rightarrow ArH + CO_2$$
(3)

where Ar represents an aromatic ring compound.

After drying and pyrolysis, gasification (or reduction) and combustion (or oxidation) reactions occur in the gasifier, with the exothermic combustion reactions supplying energy to the endothermic gasification reactions. The gasification process takes place in a variety of parallel and/or consecutive reactions. Presented in simplest form, the process can be regarded as the reaction of carbon with the gasifying agent [170].

In an oxygen and steam fed gasifier the reactions can essentially be summarised as [170, 184, 193]:

$C + H_2O \rightarrow CO + H_2$ $\Delta H = 119$ kJ/mol ...(4) , steam/carbon reaction and
$C + \frac{1}{2} O_2 \rightarrow CO$ $\Delta H = -123$ kJ/mol ...(5) , partial oxidation.

In addition to partial oxidation, total combustion can also occur. Partial combustion predominates at high temperatures, while total combustion predominates at lower temperatures [194].

Water-gas shift (WGS) and methanation reactions proceed to near chemical equilibrium in the case of most gasifiers:

$CO + H_2O \rightarrow \quad CO_2 + H_2$ $\Delta H = -40$ kJ/mol ... (6), WGS and
$CO + 3H_2 \quad \rightarrow CH_4 + 2 H_2O$ $\Delta H = -206$ kJ/mol ...(7), methanation.

The WGS reaction can be used to manipulate the H_2/CO ratio of the final syngas to a certain extent.

Methane formation is favoured at high pressures (> 70 bar) and low temperatures (760-920°C) [175]. Methane content is usually higher than equilibrium would predict because methane is also formed during devolatilisation.

The rates and degree of conversion for the various reactions involved in gasification are functions of temperature, pressure and the nature of the coal being gasified. At higher operating temperatures, the conversion of carbon to CO and hydrogen increases, while the production of methane, water and CO_2 decreases [188].

Overall carbon conversion cannot be predicted from thermodynamic principles, as this is often dictated by slow reaction kinetics and diffusion limitations, which is a function of the specific feedstock properties (e.g. ash content, rank, maceral composition, etc).

Sulphur in the feedstock reacts to produce hydrogen sulphide, carbonyl sulphide, and smaller amounts of other sulphur compounds, while nitrogen in the feedstock is converted to ammonia, hydrogen cyanide, other nitrogen compounds and molecular nitrogen. The presence of coal-derived nitrogen in the product gas is one reason why it is not essential to use ultra pure oxygen for for coal gasification. The percentage of the nitrogen in the coal that is converted into nitrogen molecules in the synthesis gas will depend on the type of nitrogen compoundsin the coal. It is widely accepted that SO_x and NO_x compounds do not exit the gasifier, except perhaps in trace amounts in lower temperature gasifiers, because a reducing atmosphere exists in the vessel. As a result sulphur and nitrogen is reduced to H_2S, NH_3 etc. [170].

3.6 Classification of gasifiers

A wide variety of gasifier technologies exist. More than a 100 different gasification technologies can be found in the literature. These technologies are in various stages of development, ranging from laboratory to commercial scale.

Gasifier designs differ primarily in the type of reactor bed selected for coal gasification, which is directly related to the method of contact between the solid and gas phases during gasification [195]. Most gasifier technologies can be classified into one of the following three generic groups according to the type of gasifier bed used [174]:

- Fixed bed or moving bed, also known as 'descending bed' gasifiers [196], of which the Sasol-Lurgi gasifier is a prominent example. Coal flows counter-current with the steam and oxidant and the highest temperatures are reached towards the bottom of the gasifier.
- Fluidized bed gasifiers, such as the Kellogg-Rust-Westinghouse (KRW) gasifier. The coal, steam and oxidant are well-mixed, leading to a more uniform temperature distribution in the gasifier.
- Entrained flow gasifiers, like the Texaco and Shell gasifiers. Entrained flow gasifiers are plug flow reactors in which the coal and reactants move co-currently through the reactor. These gasifiers essentially operate using a flame with high and uniform temperatures throughout.

Other gasifier types have been developed based on e.g. rotary kilns, molten baths or in-situ (underground) gasification, but there are no gasifiers of these types near to commercial application [191].

Gasifiers are also classified according to how carbonaceous material, steam and the oxidising agent are injected in relation to each other, e.g. counter-current or co-current.

Another, more theoretical method classifies gasifiers according to their operating temperatures as either high-temperature or low-temperature gasifiers [188]. Low-temperature gasifiers are thermally more efficient, produce more methane and also produce tars. These processes have low throughput. High-temperature processes avoid tar production and have higher throughput. However to avoid low thermal efficiencies it is necessary to recover energy in a downstream boiler and/or power turbines, increasing operational complexity. Gasifiers operating at lower temperatures release the mineral content of the feedstock as an ash, whilst those operating at higher temperatures produce a molten slag which can be water-quenched to produce a hard glassy slag that is generally non-leachable [170].

It is important to note that there is not one particular gasification technology which will be the most economic and practical selection in every instance. Rather, the optimum technology selection is dependent on the specific process application. The selection of a gasification process cannot be done without considering the supporting processes, as each gasification technology has its own unique set of characteristics which have separate challenges such as gas cleanup, by-product disposal, etc. [188]. Some of the important factors that have to be taken into account when selecting a gasification technology include the following [197, 198]:

- Feedstock characteristics
- Quality requirements for clean gas
- Quality of waste products
- Operating characteristics
- Site-specific conditions
- Environmental legislation

Of the above factors, feedstock, i.e. coal, may be the least flexible factor for economical, geographical and political reasons and it is usually necessary to select the gasification technology according to the feedstock that has to be processed [191]. Some of the important operating parameters and characteristics, to illustrate the differences between types of gasifiers, are included in Table 2.

The composition and maximum outlet temperature of the syngas which is produced by the gasifier will depend on the feedstock utilized, the steam to oxygen ratio and the temperature and pressure at which the gasifier is operated, all directly related to the type of gasifier selected [170]. Important characteristics of the raw gas include the methane and ammonia content, whether there are tars and oils present (which influence the gas cooling and cleaning sections), and most importantly, the H_2/CO ratio obtained.

Gasifier design determines the vessel's ability to handle a particular feedstock. For instance, a very high ash coal cannot be processed in a slagging gasifier, as the energy required to slag the ash would be excessive, impacting negatively on the economy of the process, while caking coal cannot be processed in a moving bed type gasifier. Some gasifiers can handle feedstock in liquid form better than others.

Other factors that influence the choice of gasifier include the ability to cope with a specific size particle, e.g. moving bed gasifiers cannot handle fine coal well, as well as the steam and oxidant requirements.

Table 2
Operating parameters for generic types of gasifiers [199, 200]

	FIXED BED	FLUIDIZED BED	ENTRAINED BED
Preferred feedstock	Lignite, reactive bituminous coal, wastes	Lignite, bituminous coal, cokes, biomass, wastes	Lignite, reactive bituminous coal, petcokes
Coal feed size (mm)	6 – 75	< 6	< 0.1
Ash content	No limit, for slagging type < 25% preferred	No limitation	< 25% preferred
Exit gas temperature (°C)	420 – 650	920 – 1050	±1200
Ash conditions	Dry/slagging	Dry/agglomerating	Slagging
Key distinguishing characteristic [192]	Hydrocarbon liquids in the raw gas	Large char recycle	Large amount of sensible heat energy in the hot raw gas
Key technical issue[192]	Utilisation of coal fines and hydrocarbon liquids	Increased carbon conversion	Raw gas cooling
Well-known technology supplier	Sasol-Lurgi	Kellogg Rust Westinghouse (KRW)	Texaco

Common gasifying agents used in industrial gasifiers include a mixture of steam and air or oxygen, with the amount of oxygen being generally one-fifth to one-third the amount theoretically required for complete combustion [191]. The choice between air and oxygen as the oxidant depends on several factors, including the reactivity of the feedstock, the type of gasifier used and the application(s) the produced gas will be used for. Oxygen-blown gasification delivers syngas with a heating value of 10 to 16 MJ/kg, or about one fourth the heating value of natural gas, compared to a heating value of less than 5 MJ/kg for air blown gasification [197], since the syngas is then diluted with nitrogen [175]. The nitrogen from the air also results in greater volumes of gas to be treated downstream.

Depending on the gasification application three types of syngas, with different calorific values, can be produced [191]:

- Low heating value gas, with a calorific value (CV) of 3.8 – 7.6 MJ/m^3. The applications include fuel gas for gas turbines and smelting and reduction of iron ore. It cannot be used as a natural gas replacement or for chemical synthesis due to its high nitrogen content, since this type of gas is obtained from certain air-blown gasifiers.
- Medium heating value gas, with a CV of 10.5 – 16 MJ/m^3. The usage of this type of gas includes fuel gas for gas turbines, a replacement for natural

gas and for chemical synthesis. Depending on the H_2/CO ratio, this type of gas can be used in Fischer-Tropsch synthesis, methanation, methanol and ammonia production and H_2 production for refinery, fuel cell and other applications.

- High heating value gas with a calorific value of more than 21 MJ/m^3. This type of gas is mostly used as a substitute for natural gas.

The average compositions of the exit gas for some gasifiers are reported in Table 3.

Table 3
Typical gasifier raw gas outlet gas compositions (dry basis].

	Sasol-Lurgi dry bottom (fixed bed) [196]	KRW (fluidized bed) [196]	Texaco (entrained flow) [201]
H_2	38-41	31	26-36
CO	21-26	44	31-47
CO_2	26-29	18	6-16
CH_4	8-10	6	<0.3
H_2/CO ratio	1.7-2.0	0.7	0.7-0.9
Heating value (MJ/Nm^3)	12 - 14	10 – 12	10 - 11

Selection of the method of cooling the hot raw gas depends on the composition of this gas (e.g. is there tars and oils present or not) and the intended application. Cooling can either be done by direct contact with the cooling medium (quench) or via an indirect cooling method [191, 202, 203]. While the quench approach is somewhat less efficient, it is also less expensive and more suitable for feeds with a high metal content. The indirect, or heat-exchange approach offers higher overall efficiency but requires a greater capital investment, and there is a higher risk involved [186].

With direct cooling, the hot raw gas is quenched with liquid or cooled recycle gas. The direct liquid-quench method is often used, especially in systems where entrained particles must be removed from the raw gas before further processing. The water quench approach is especially suitable where the hydrogen content of the syngas needs to be increased by the shift reaction, for example when modifying the syngas for ammonia production. Some synthesis gas shifting to increase hydrogen content may also be needed for methanol and HTFT synthesis. For these intermediate H_2/CO ratio requirements, a combination of quench and waste-heat boiler cooling is possible [204].

If a water quench is used in power generation applications then energy efficiency can be improved by operating the gasifier at high pressure and recovering energy from the water vapour when it passes through an expansion turbine together with the solids free syngas. Texaco and others have proposed the use of a quench IGCC design in which the gasification is conducted at a higher pressure of about 70 bar [205]. Additional power can then be generated by expansion of the clean fuel gas to about 20 bar followed by gas reheat and firing in a gas turbine. The gas cooling by an expander can also be used to provide much of the refrigeration demand of a physical absorption desulphurization system [206].

Syngas expansion is being used in two of the heavy oil IGCC plants at Falconera and Priolo in Italy. This approach is especially suitable for the integration of IGCC power generation with FT synthesis. In this case expansion down to about 30 bar would be followed by further clean-up and then FT synthesis to provide FT tail gas to a gas turbine at 20 bar. The steam produced by the FT reactor cooling system is integrated into the IGCC steam system. This approach is discussed in more detail in Chapter 5.

For the Lurgi moving bed dry bottom gasifier design, as well as for the direct quench version of the Texaco gasifier, raw gas is quenched with recycled aqueous liquor (also known as 'gas liquor'), resulting in the condensation of oils and tars and removal of particulate matter. Thereafter, the liquor is separated into aqueous and hydrocarbon fractions for further treatment. Quenched raw gas is subsequently passed through a waste heat boiler for further cooling, but only low quality steam can be generated as most of the sensible heat is lost during the quench, thus giving lower thermal efficiencies.

For indirect cooling, a high temperature syngas cooler is used [191]. High pressure steam is produced from the recovery of sensible heat. The oxygen-blown high temperature coal gasification processes with indirect cooling have significant steam raising capability in the raw gas cooling section. This allows for the development of clean fuel gas plants with combined steam turbine power generation which efficiently produce large quantities of electrical power [203].

High temperature gasifiers may use dry coal feed systems or coal-water slurry feed systems. Dry-coal fed gasifiers have the advantage over coal-water slurry fed gasifiers that they can operate with a minimal amount of steam and avoid the oxygen combustion required to provide the energy needed for the latent heat to vaporize the water. This implies in practice that they have a 20 – 25% lower oxygen consumption than coal-water fed gasifiers. However, steam generation facilities need high quality boiler feed water while the water quality is a minor issue for coal-water slurries.

The big advantage of coal-water slurry fed gasifiers is the more elegant method of pressurizing the coal. Lock hoppers as used in dry-coal fed gasifiers are costly and bulky equipment with complex valve systems that have to provide a gas-tight block in a dusty atmosphere [1]. Pumping a coal-water slurry is not a simple matter either, but it is definitely less complex. The practical pressure limit for dry-pulverized coal lock hoppers is about 50 bar, wheras for coal water slurry pumps the pressure could, in principle, be as high as 200 bar [1]. Typical commercial operating pressures for high pressure slurry fed gasifiers are between 60 and 100 bar.

In order to compensate for virtually all drawbacks of using water slurry the slurry can be preheated [1]. This provides the following advantages:

- The water has to be heated less in the reactor, and the heat of evaporation becomes lower at higher temperatures.
- Atomization becomes better with hotter liquid containing feedstocks, increasing the accessibility of the coal for gaseous reactants, especially when the feedstock slurry is preheated such that the carrier flashes upon introduction into the gasifier. Also, the reduction in surface tension of the carrier liquid at higher temperature enhances atomization.
- More reactor space becomes available for the gasification per se, which will increase the carbon conversion.
- The oxygen consumption will decrease and the cold gas efficiency will increase.
- The water will expand resulting in lower water requirement to maintain good slurry conditions.

If the water is heated close to its critical point then the above effects will be most pronounced and the heat of evaporation is then minimal. Furthemore, the coal in the slurry is also preheated to above 300°C, a feature that is not possible using dry-coal feeding because the coal particles will become sticky.

With a suitable preheat system, the efficiency of a slurry fed IGCC system is almost equal to that of a dry-coal fed system (49 versus 50%) [1].

A brief overview of the three generic gasification technologies will be presented in the next three sections.

3.6.1 Fixed bed gasifiers

Fixed bed gasifiers, have been in use since the beginning of the gasification industry. Somewhat confusingly, these gasifiers are sometimes also referred to as moving bed gasifiers. The coal moves downwards through the gasifier but the various zones within the gasifier are 'fixed' in location. These gasifiers have maintained a strong position in the broad range of reactor types used for coal gasification [189].

Carbonaceous feedstock enters the gasifier at the top periodically during operation, resulting in an essentially full gasifier at all times during normal operation. As coal has to enter the gasifier during operation, a coal lock hopper is used to feed the coal or other carbonaceous material into the pressurised gasification reactor, after being pressurised itself [185, 201].

The feedstock is consumed as it moves downward through the bed and ash is removed at the bottom. A rotating grate with ash lock is employed for dry ash removal systems, which is absent in the case of slagging fixed bed gasifiers. This grate also serves as agent distributor.

To remove entrained solid particles and condense heavy hydrocarbons (tars and oils) in the hot raw gas leaving the gasifier at the top, a wash cooler is attached to the gasifier, where the gas is scrubbed with recycled gas liquor (water produced when raw gas is cooled). A large amount of the sensible heat in the raw gas is degraded by the scrubbing step which substantially decreases the temperature of the scrubbed gas entering the downstream heat recovery systems.

Fixed bed gasifiers are suitable for solid feedstocks, usually in the form of coarse coal particles. Coarse particles ensure good bed permeability and help to avoid excessive pressure drop and channel burning, both of which can lead to unstable gas outlet temperatures and gas composition. As coarse coal particles are required, this greatly simplifies feedstock preparation steps leading to reduced gasification cost.

Run-of-mine (ROM) coal is typically screened at about 6 mm, resulting in a fine coal fraction which can be routed to steam plants or pulverised coal combustion units for electricity generation. The only additional coal preparation required in some instances is the proper blending of multiple coal sources, or the homogenisation of single or multiple sources, which have highly varying properties, to ensure stable gasifier operation.

Pressure drop and entrained solids can limit gas throughput in certain circumstances. While feedstock particle size distribution (PSD) can be managed correctly to ensure stable gasifier operation [183], mechanical and thermal fragmentation during gasification could make it difficult to control the PSD at all times during the process. These fragmentation phenomena are determined by the feedstock characteristics, and high moisture content in coal has been shown to be largely responsible for particle size reduction due to thermal fragmentation when coal is exposed to the high gasifier temperatures [207, 208].

Co-current fixed bed gasifiers have not been able to succeed commercially yet, although the concept has been widely studied. Commercial fixed bed gasifiers are all counter-current. Coal flows counter-current to the steam and oxidant feed. Cold coal comes into contact with hot ascending product gases. As the coal moves downward under gravity in the reactor, it passes through several zones which are defined according to the reactions and temperatures occurring at different heights in the fuel bed [209]. These are the drying, devolatilisation, reduction and combustion zones. These zones are somewhat better defined than is the case in other gasification technologies. The distinct temperature zones aid the overall conversion of coal to gas [195]. In the reduction zone mainly endothermic reactions occur and in the combustion zone exothermic oxidation reactions take place to supply energy to the endothermic gasification reactions.

The 'cold gas efficiency' of a gasifier is defined as the chemical energy from the coal that is contained in the raw gas exiting the gasifier. Fixed bed gasifiers have high cold gas efficiencies. This is due to the heat exchange achieved in the counter-current operation which allows the gas and solid product streams to exit at relatively low temperatures, with the minimum of energy contained as sensible heat in the product gases [171]. Other advantages of the fixed bed type gasifier include high heating value syngas due to its high methane content and low oxygen consumption [175].

Tars and oils, formed in the pyrolysis zone at the top of the gasifier, exit the vessel with the raw gas. Due to the presence of these compounds in the raw gas, gas cleanup is more complex than for gasifiers in which condensable hydrocarbons are destroyed during the gasification process. To avoid deposition of these substances in downstream equipment, the raw gas must be quenched with water or gas liquor, resulting in reduced raw gas heating values. However, these tars and oils can be utilised as feedstocks for the production valuable products like coke and creosotes for wood preservation.

The biggest challenge for fixed bed gasifiers is their limited ability to handle fine coal, necessitating the disposal of the portion of the run-of-mine coal that is too fine to be used for feed.

A shortcoming is the strong possibility of excessive coal segregation inside the fixed bed gasification vessel, aggravated by broad particle size distributions. Segregation can lead to difficulty in maintaining uniform gas flow through the coal bed [195], which in turn will lead to unstable gasifier operation. This complexity makes scale-up to very large units difficult, so that smaller units will be duplicated rather than risking scale-up, even though larger units would provide better economic performance. Multiple units result in high maintenance cost, but are advantageous in that very high plant availability can be achieved, so that the downstream plant does not suffer due to loss of syngas production, a phenomenon which generally occurs in plants with a few large gasifiers.

Another disadvantage of fixed bed gasification is the relatively high steam consumption. A large excess of steam is injected into the gasifier to reduce the bottom temperature, protecting the vulnerable mechanical ash grate from extreme temperatures [171, 189]. This results in lower overall efficiency, but the water-gas shift reaction is pushed towards higher hydrogen production, so that syngas with an H_2/CO ratio directly acceptable for all Fischer-Tropsch processes is produced. For large scale applications such as the Sasol plants in South

Africa, separate steam generation plants are required to supply most of the high-pressure steam demand.

There are only three commercial fixed bed gasification processes [191]. Two of these were originally developed by Lurgi, these being the Sasol-Lurgi fixed bed dry bottom gasifier (dry ash gasifier) and the British Gas Lurgi (BGL) slagging gasifier. The other fixed bed gasifier, also a dry ash gasifier, was developed by BHEL (Bharat Heavy Electricals Ltd).

For the dry ash versions, temperatures at the bottom of the bed are controlled below the ash fusion temperature to allow coal ash to be removed as a solid. As a result of the low operating temperatures that must be maintained, these gasifiers are suited to reactive coals like lignites [191]. For the BGL, bottom temperatures are high enough to allow the ash to melt and to be removed as a molten slag.

There are currently 152 Lurgi dry bottom gasifiers operational throughout the world, most notably 97 which are operated by Sasol in South Africa, producing syngas which is converted to Fischer-Tropsch liquids (fuels and chemicals). Another significant installation is at the North Dakota Gasification Company in the USA, where 14 gasifiers are used to produce synthetic natural gas (SNG).

For IGCC applications, there are presently three installations which gasify their feedstock in fixed bed gasifiers. These are at Schwarze Pumpe, where 8 fixed bed gasifiers are used, one of which is a BGL slagging gasifier; in Scotland at the FIFE Energy plant where 1 BGL slagging gasifier is in operation [184] and in the Czech Republic, where 26 Lurgi dry bottom gasifiers have been in use since 1996 to produce electricity and steam. In Luenen, Germany, 5 Lurgi dry bottom gasifiers were in operation from 1972 to 1977, but the plant was decommissioned because it was not economical.

The two main types of fixed bed gasifiers, these being the Sasol-Lurgi dry bottom gasifier and the BGL slagging gasifier will be described in more detail in the following two sections.

3.6.1.1 Sasol-Lurgi fixed bed dry bottom gasifier

The Sasol-Lurgi fixed bed dry bottom gasifier was originally developed and marketed by Lurgi. In South Africa, 97 of these gasifiers are operated by Sasol. In 2002 the Sasol-Lurgi joint venture was established to further develop and market the fixed bed dry bottom gasification technology, with Sasol and Lurgi each holding a 50% share in the new company.

The Sasol-Lurgi gasifier is a pressurised fixed bed gasifier, and a schematic drawing is shown in Fig. 2.

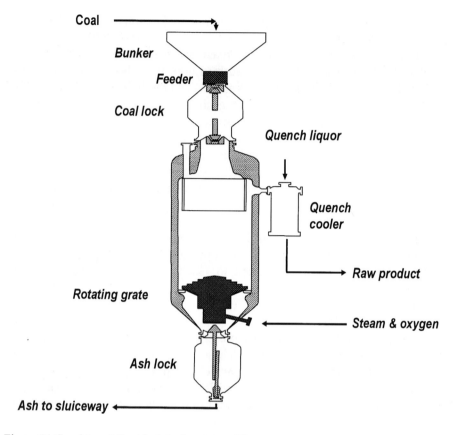

Figure 66 Sasol-Lurgi fixed bed dry bottom gasifier [191]

Lump coal, sized to a typical particle size distribution of 6 mm to 75 mm, is fed into the top of the gasifier through a lock hopper. The hopper is filled, sealed and pressurised and the coal is then released in the reactor. The process is continuous, so that the gasifier is almost filled with coal at all times. Steam and oxygen enter at the bottom and react with the coal as the gases move up through the bed.

Due to the counter-current mode of operation, hot ash exchanges heat with cold incoming agent (steam and oxygen), while hot raw gas exchanges heat with cold incoming coal at the top. As a result, the ash and raw gas leave the gasifier at relatively low temperatures compared to other types of gasifiers, which improves the thermal efficiency and lowers the steam consumption. The Sasol-Lurgi dry ash gasifier consumes the lowest amount of oxygen per unit coal feed due to the high thermal efficiency. A water jacket cools the gasifier vessel and part of the steam for gasifier consumption is generated in the jacket.

A typical temperature profile for a Lurgi dry bottom gasifier is presented in Fig. 3. It can be seen that the temperature of the agent (steam/oxygen mixture) and the hot ash is around 300 °C, but can vary depending on the position of the ash bed, which is determined by the rate of ash extraction. The agent inlet temperature can be used to control the ash temperature. The raw gas outlet temperature is typically in the region of 500-600 °C, and depends on process parameters as well as coal properties (e.g. moisture content).

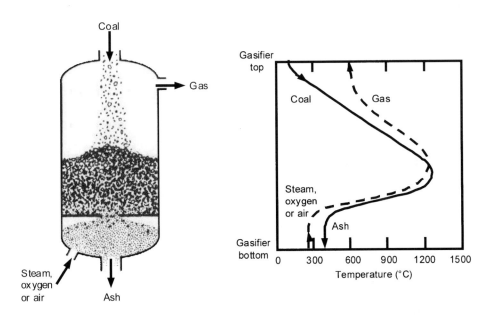

Figure 67 Lurgi dry bottom gasifier and typical temperature profile [174]

The cold gas efficiency of these gasifiers is typically about 80%, and can rise to 89% if liquid hydrocarbon products are included in the calculation.

Ash is removed at the bottom of the gasifier by a rotating grate and lock hopper. Coal with high ash content can be gasified without severe losses in thermal efficiency, as the ash is not extracted in a molten state. The Sasol-Lurgi gasifiers have been shown to be capable of handling coal with an ash content as low as 6% to as high as 35% [173]. However, it is important to maintain the ash content as constant as possible in order to achieve stable gasifier operation, especially at high load conditions.

The H_2/CO ratio of the syngas produced is in the range of 1.70 to 2.0, which makes it suitable for methanol synthesis and for Fischer-Tropsch synthesis (see Chapter 3) without the need for water-gas shift conversion to adjust the ratio.

By-products like tars, pitches, oils and chemicals like ammonia are present in the exiting raw gas, and the raw gas is quenched to separate these products from the syngas. Approximately 85% of the tar is condensed in the crude gas quench, and the remainder is condensed as oils and naphtha in downstream gas purification units [195]. The treatment of these by-products is an added cost to the technology, and is regarded as a disadvantage in instances where no market exists for the by-products. These by-products can be recycled to the gasifiers and converted to additional synthesis gas. However, with very large scale applications such as the Sasol plants, large enough quantities of by-products are produce to achieve good economics of scale for the upgrading of the by-products to valuable products. Examples from the Sasol operations include the production of high quality anode coke from pitch, as well as creosotes, phenols and other valuable aromatic compounds from the tars. If the syngas is converted to transportation fuels, the aromatic by-products from the gasification process can be used as octane enhancers, since synthetic fuels contain substantially lower amounts of aromatics than crude oil derived fuels.

At the Secunda petrochemical complex, Sasol operates the largest coal to synthesis gas conversion process in the world. Low-grade sub-bituminous coal with relatively high ash content is upgraded through gasification and gas cleaning to purified synthesis gas, which is used as feed to a Fischer-Tropsch conversion process for the production of fuels and chemicals.

The first 10 fixed bed dry bottom gasifiers which were installed by Sasol in 1954, had a coal throughput of approximately 28 tons/h and produced about 23500 m^3_n/h of dry raw synthesis gas [173]. Two scaled-up versions were developed. The original design was first scaled up to a throughput of 54 tons/h of coal (producing 45500 m^3_n/h of dry raw gas). Most of the fixed bed gasifiers in operation today are of this type, with Sasol operating 83 of these gasifiers. A further scale-up to 75 tons/h of coal followed (producing 63000 m^3_n/h of dry raw gas), and one of these gasifiers is commercially operated by Sasol. The further optimisation of the Sasol-Lurgi gasification process is ongoing, and it is believed that further gains is throughput, efficiency and utility consumption are achievable [175]. In 2003 approximately 52 million tons of coal was mined by Sasol's coal mining division, of which about 40 million tons was consumed by two Sasol synthetic fuels plants at Secunda. About 6 million tons was consumed by Sasol's smaller chemical producing plant at Sasolburg, and about 3 million tons was exported. Approximately 70% of the total coal consumption is for gasification to synthesis gas by Sasol-Lurgi fixed bed dry bottom gasifiers, while the remaining 30%, a finer coal fraction not suitable for gasification, is combusted to produce steam and electric power required by the process. The total pure synthesis gas production from Sasol's 97 fixed bed dry bottom gasifiers is approximately 3.6×10^6 m^3_n/h.

In addition to coal, Lurgi gasifiers can handle a variety of different feedstocks, e.g. at Schwarze Pumpe seven Lurgi fixed bed gasifiers are employed to treat solid wastes, such as plastics, sewage sludge, rubber, contaminated wood, paint residues and household wastes [210]. However, a feedstock preparation step such as briquetting or pelletising may be required to ensure the desired feedstock particle size for optimum gasifier operation.

3.6.1.2 British Gas Lurgi (BGL) slagging gasifier

The BGL gasifier is a dry feed pressurised fixed bed slagging gasifier. It was developed by British Gas, London and Lurgi for the gasification of coals and cokes, and is primarily a methane and fuel gas producer [188]. It is the only gasifier in which gasification of large particles and vitrification of inorganic matter can be carried out under pressure in the same reactor [210]. The BGL can handle up to 40 tons per hour of coal feedstock.

The gasifier is provided with a motor-driven coal distributor/mixer to stir and evenly distribute the incoming coal mixture (coarse coal, fines, briquettes and flux). Oxygen and steam are introduced to the gasifier vessel through sidewall-mounted tuyeres at the elevation where combustion and slag formation occur (at the bottom of the vessel).

The coal mixture descends through several reaction zones, similar to that of the Sasol-Lurgi fixed bed dry bottom gasifier. Below the gasification zone, any remaining carbon is oxidised and the mineral matter in the coal is liquefied, forming slag. The slag is withdrawn through a slag tap at the bottom of the gasifier, quenched with water and removed by a lock hopper.

The crude gas exiting the gasifier contains tars and oils from the carbonisation zone in the gasifier. The gas is scrubbed with recycled gas liquor to remove the condensable hydrocarbons. These are then separated by gravity from the liquor. The tar and oil by-products, as well as any solids which are entrained from the gasifier, can be recycled to the top of the gasifier, or re-injected trough the tuyeres at the bottom of the gasifier. A limited amount of coal fines can also be fed through the tuyeres. The cold gas efficiency of the BGL is about 88% [173] and It has a nominal dry gas capacity of 21 000 to 24 000 Nm^3/hr, depending on the degree of tar-oil recycle.

High ash content, combined with a high ash melting point, will result in addition of a fluxing agent, so that the BGL could be costly to operate. The tapping temperature (1400°C) is lower than for most entrained flow gasifiers, and for some feedstocks, high amounts of fluxing agent have to be added to comply with this operating limit. The optimum slag viscosity for a BGL is approximately < 5 Pa.s, in comparison to < 15 Pa.s in entrained flow gasifiers [191].

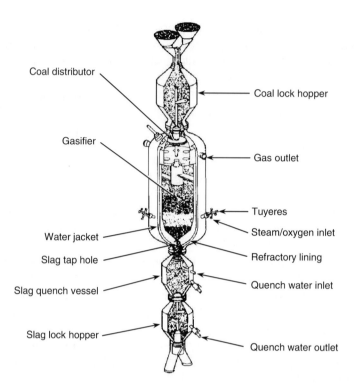

Coal distributor

Coal lock hopper

Gasifier

Gas outlet

Tuyeres

Steam/oxygen inlet

Water jacket

Slag tap hole

Refractory lining

Slag quench vessel

Quench water inlet

Slag lock hopper

Quench water outlet

Figure 68 BGL gasifier [174, see also 177]

3.6.2 Fluidized bed gasifiers

Fluidized bed gasification with oxygen and steam was first done in a gasifier developed by Winkler [171]. These gasifiers are completely mixed, partially cocurrent reactors.

A fluidized bed gasifier consists of a vertical, cylindrical, refractory-lined vessel with cyclones above the fluidized bed that separate entrained solids from the outlet gas; bottom ash cooling and if required, dry fly ash removal and a wet gas scrubbing system [199].

Solid, crushed dry fuels are used as feed. Reactive coals like lignites and brown coals are favoured as feedstock, as the relatively low operating temperatures make it difficult to maintain high carbon conversion without very high solid recycle rates.

It is usually not desirable to have coal particles agglomerating, as it causes uneven fluidisation in dry ash fluidized bed gasifiers. Therefore, it is necessary to process coals that have a higher ash fusion temperature than the operating temperature of the gasifier. Mineral matter is a major constituent of the bed and therefore coal ash characteristics can have a large impact on the operation of the gasifier [191]. Highly varying ash characteristics are problematic, and the feedstock should be well-mixed to maintain the same characteristics as far as possible. Coals with a low swelling index are also preferable to avoid agglomeration.

The presence of pyrite and sodium silicates formed during gasification is believed to be among the factors that cause agglomeration in fluidised bed systems. Therefore, very careful control of the gasifier operating temperature is required when processing coals with high alkali content. However, from an environmental point of view, fluidized bed gasifiers are tolerant to coals with high sulphur content as up to 90% of sulphur in the coal feed can be retained by adding a sorbent, like calcium oxide.

Dry coal particles of less than 6 mm are introduced at the bottom of the gasifier and then fluidised using oxidant and steam [174]. Heat exchange between the cool steam/oxidant feed and the hot ash discharged increases the thermal efficiency. Upon entering the gasifier, the solids undergo rapid heating. Coal particles are suspended in the gas flow and feed particles are mixed with particles already undergoing gasification reactions [191], so that no clear reaction zones can be distinguished. Because the coal is well-mixed with steam and oxidant more uniform and moderate temperature distributions are achieved than for other types of gasifiers. Unreacted entrained particles are recovered by cyclones for recycle to the vessel. The recycle of solids efficiently transfers heat to the bed material.

Residence time for the solids is typically in the order of 10 to 100 seconds. Ash can be discharged in dry form or as agglomerated particles, as for the KRW gasifier [191].

Fluidized bed gasifiers are usually operated at near atmospheric pressure, and therefore well suited for smaller capacities (approximately 20 tons per hour of coal) [200]. If the gasifier is operated at elevated pressures, higher gasifier capacities can be achieved. Operating temperatures are relatively low at around 1000°C. At the bottom of the gasifier, where oxidising conditions prevail, the temperatures must be lower than the ash fusion temperature of the coal, unless the gasifier has an agglomerating ash removal system.

No pyrolysis by-products are produced, as they are combusted to provide energy for endothermic gasification reactions. The raw gas exiting the vessel is essentially free of hydrocarbons heavier than methane.

Fluidized bed gasifiers use a moderate amount of steam and oxygen during gasification. Due to relatively low operating temperatures, there is little thermal stress on the equipment. A typical temperature profile for a fluidized bed gasifier is presented in Fig. 5.

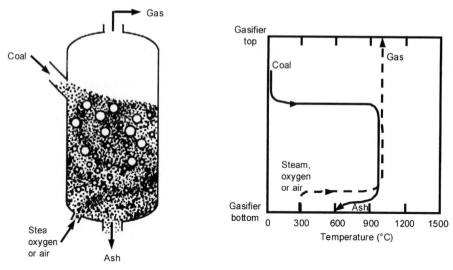

Figure 69 Fluidized bed gasifier and typical temperature profile [174].

At present, 21 fluidised bed gasifiers are operational, most notably 8 GTI U-GAS gasifiers in China, applied to produce fuel- and town gas.

The two main contenders for fluidized bed gasifier technology are the High Temperature Winkler (HTW) and the Kellogg Rust Westinghouse (KRW) gasifiers. KRW is now known as KRW Energy Systems. These technologies have not been applied extensively on commercial scale due to various reasons, including low capacity throughput, high operating cost, low carbon conversion, large char recycle and ash agglomeration problems. From a recent survey conducted by the Gasification Technologies Council it is evident that fluidized bed technology is not a major contender for future applications [178].

Perhaps the most promising fluidized bed technology is the transport gasifier developed by Kellogg, Brown and Root Inc., which is currently being tested at the Power Systems Development Facility in Wilsonville, AL [203]. This type of gasifier, as well as the HTW and KRW gasifiers will be discussed in more detail in the following sections.

3.6.2.1 KRW fluidized bed gasifier

Coal, via a lock hopper, and oxidant (air) enter the gasifier at the bottom of the vessel via a burner, ensuring thorough mixing of coal and oxidant.

Upon entering the vessel, coal releases volatile matter almost instantaneously. The volatiles burn rapidly and supply heat to the endothermic gasification reactions. The resultant gas forms large bubbles that rise up the centre of the gasifier, causing char and sorbent (if added) in the bed to move down the sides of the reactor and back into the central jet [199].

The gasifier bed is maintained in a fluidized form by injecting controlled amounts of steam, air and recycled gas through nozzles in the combustion zone. Steam reacts with the char in the bed, converting it to fuel gas. If a sorbent like limestone was added to the coal upon feeding, the calcined form (CaO) now reacts with H_2S released from the coal during gasification to form CaS, removing a portion of the sulphur gases from the raw gas before it exits the vessel.

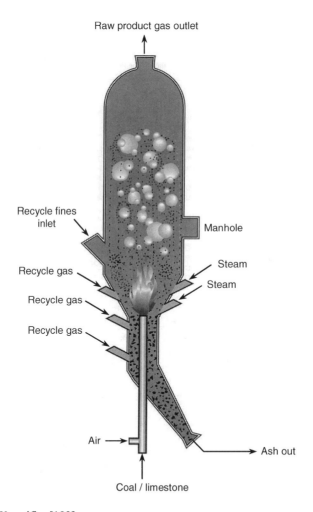

Figure 70 KRW gasifier [192]

To remove ash, particles are recycled through the flame of the burner at the bottom of the gasifier, where low-melting mineral matter partially melts, causing particles to agglomerate [199]. The ash agglomerating zone, which is hotter than the main fluidized bed, makes the process distinctly different and more complex than conventional fluidized bed gasifiers such as the HTW (High Temperature Winkler) [174], where dry ash removal is done. The agglomeration allows for the efficient recycle of even small particles in the feed. The velocity of gases in the reactor is selected to maintain most of the particles in the bed and smaller agglomerated particles return to the bed for further reaction.

Some of the finer particles are carried out of the gasifier, but are recaptured in a high efficiency cyclone and returned to the conical section of the gasifier. Eventually most of the smaller particles agglomerate as they become richer in ash and gravitate to the bottom of the gasifier. Since the ash and spent sorbent particles are substantially denser than the coal feed, they settle to the bottom of the gasifier where they are cooled and classified by a counter-flowing stream of recycled gas.

The DOE and its predecessor agencies have supported development of KRW Energy Systems' fluidized bed gasifier technology since 1972 when the design of a process development unit was first initiated under a contract with Westinghouse Electric Corporation [192]. This unit was completed in 1975 at Westinghouse's Waltz Mill Facility in Pennsylvania. The addition of limestone and dolomite to the gasifier for in-bed sulphur removal was successfully demonstrated during the 1980's. During this same period, numerous coal types were tested as feedstock to the KRW gasifier. Sasol in South Africa collaborated with Westinghouse to construct a demonstration unit in South Africa, but due to economic and other considerations the project was aborted.

3.6.2.2 The High Temperature Winkler (HTW) gasifier

The HTW process is a modified version of the Winkler fluidized bed gasification process, which was first developed in the 1920's.

The HTW is a refractory-lined fluidized bed gasifier with water jacket [174, 180]. Feed coal is crushed and dried before it is pneumatically conveyed to feed bins. Coal in the receiving bin is then dropped via a gravity pipe into the fluidised bed of ash, semi-coke and coal particles [190].

The gasifier is fluidized from the bottom with either air or oxygen and steam. These reactants are injected at two levels in the fluidized bed to maximise carbon conversion [177] and reduce methane and other hydrocarbon yields [174]. The gasifier is operated at 750 to 800 °C [174]. Operating pressure can vary from 1 to 3 MPa, depending on the use of the syngas produced [191].

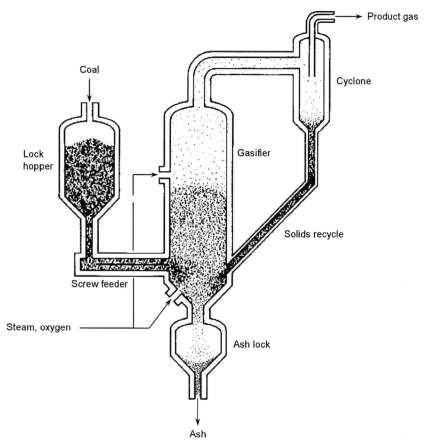

Figure 71 HTW gasifier [174, see also 177]

A portion of ash and char is entrained in the reactor effluent. The raw gas passes through a cyclone to remove particulates and is then cooled [191]. Solids that are recovered in this way are recycled to the gasifier and dry ash is removed at the bottom of the gasifier via a discharge screw.

The choice of feedstock is restricted to high reactivity feedstocks like reactive coals, chars derived from lignite carbonisation, peat and wood. Other feedstock requirements are an ash melting point of above 1200°C and non-caking tendencies, as the formation of agglomerates would disrupt the uniformity of the fluidized bed.

The cold gas efficiency (chemical energy from the coal that is contained in the raw gas exiting the gasifier) of the HTW is approximately 82%, and if the energy content of steam generated during gas cooling is added to the fuel gas energy content, and the energy for coal drying and steam feed to the gasifier is deducted, the efficiency increases to 85%.

The HTW gasifier was first developed by Rheibraun in Germany to gasify lignites for the production of a reducing gas for iron ore [191], and was a natural outflow from the low temperature process previously used. It was successfully applied for the synthesis of methanol from lignites at Berrenrath in Germany between 1986 and 1997. It was shut down at the end of 1997, because the process was no longer economically viable. Another plant utilising HTW gasification technology has been operating in Finland from 1988, producing ammonia from peat. A plant to study HTW technology with the aim of utilising it for IGCC in future plants was constructed at Wesseling in Germany in 1989. This 140 t coal/day plant provided operational data required to design a potential 300 MW power plant (KoBra), but the plant was never built. There is, however, an ongoing project to replace 26 Lurgi moving bed gasifiers at Vresova (Czech Republic) with two HTW units.

3.6.2.3 The transport gasifier

The transport reactor is an advanced circulating fluidised bed reactor that can function as a combustor or as a gasifier. Ground coal, sorbent (for sulphur capture), air and recycled solids are mixed in the mixing zone and injected into the reactor. Small particles, together with multiple passes of the coal/char through the gasification zone result in high carbon conversion at reasonable temperatures.

Gas with entrained solids moves up from the mixing zone into the riser, and from there into the disengager (see Fig. 8). The disengager removes a major portion of the solids from the gas by gravity separation. The separated solids flow downward through the aerated vertical standpipe. The gas stream then flows to the primary cyclone where additional particles are removed from the gas stream. These solids flow downward through a dipleg into the standpipe. The solids flow from the standpipe through the J-leg back into the mixing zone.

Gasification in the transport reactor is conducted at temperatures of 870°C to 1000°C and at a pressure of up to 17 bar(g).

The design is capable of very high throughputs and could lead to economically attractive IGCC fluidised bed designs. Further development and research is taking place at the Power Systems Development Facility (PSDF), which is specifically focused on reducing capital cost and increasing efficiency of an IGCC plant [178]. The transport reactor was designed to allow for scale-up of fluidised bed gasification technology to commercial applications, and a 99.9% combustion efficiency, 99% sulphur removal, very low particulate emissions (below 1 ppm) and reliable system operation are among its successes.

In September 1999 the transport reactor was converted to gasification mode. Its viability as a gasifier was confirmed by the results that included: (1) production of synthesis gas with a heating value high enough to fire a combustion turbine, (2) stable operation and pressure drop in the HTHP filter, (3) stable operation of the transport reactor, and (4) control of oxygen content during start-ups and process upsets.

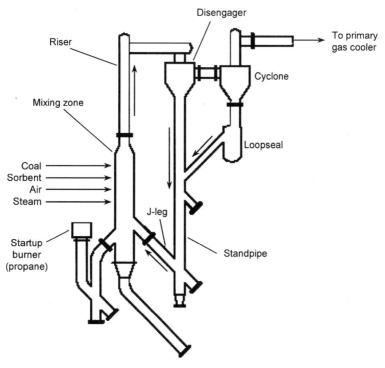

Figure 72 Transport reactor [191]

3.6.3 Entrained flow gasifiers

The entrained flow type of gasifier is the most widely used design [170]. These gasifiers are co-current, plug flow reactors. Finely ground feed, in the case of solids, or liquid feed is introduced into the gasifier along with the oxidant and steam or liquid water as moderator. Coal and gases flow co-currently at high speed. As a result, very short gas residence times are experienced, and the coal has to be pulverised to ensure high carbon conversion. All entrained flow gasifiers are of the ash slagging type.

A strong driver for the application of this technology is the wide range of feedstocks that can be accommodated [191]; including organic wastes, refinery by-products, solids and liquids, as well as fine coal, making this type of gasifier the most versatile. Entrained flow gasifiers operate at high temperatures (above ash slagging temperatures) to ensure high carbon conversion and syngas free of tars and phenols [191]. The fact that no hydrocarbons are produced which require clean-up and processing, makes it a desirable technology for many applications [174]. Texaco has marketed its entrained gasification technology very successfully as add-ons to existing refineries so that heavy liquid and solid by-products can be processed. The synthesis gas can be applied elsewhere in the refinery for clean heating fuel or to supplement the hydrogen demand, and there are also applications where electric power is generated from the synthesis gas.

Entrained flow gasifiers can accept almost any type of coal as long as the coal can be pulverised to very fine particles. However, there are certain feed requirements linked to each type of entrained gasifier design, e.g. a minimum ash content is required for gasifiers with slag self-coating walls [174]. High ash coals are usually not economical to gasify in this type of reactor, as the energy required to slag the ash could raise the oxygen consumption to unacceptable levels. If sulphur and halogens are present in considerable quantities, the cooling, cleaning and ash tapping equipment may have to be constructed from specialised materials. Other coal requirements, which varies slightly depending on the gasifier design, are an ash fusion temperature of below 1400°C, a slag viscosity of < 15 Pa.s at 1400°C, little or no flux requirement (< 3 wt%), a critical temperature viscosity and a SiO_2/Al_2O_3 ratio of about 2, which minimises limestone addition and avoids slag crystallisation at tapping temperatures [191].

Control of the fuel/oxidant ratio is more critical in entrained flow gasifiers as they have a smaller heat capacity and no inventory of process feedstock. Therefore, operating entrained flow gasifiers at varied load can prove difficult [191].

No reaction zones can be distinguished in the entrained flow gasifier. Gasification reactions typically occur at temperatures in excess of 1200°C. The high operating temperatures ensure high carbon conversion. Bed temperatures are above the ash fusion temperature, and hence the ash forms a slag. The raw gas usually exits the vessel above 1000°C [174].

The high operating temperatures have an impact on burners and refractory life and require the use of expensive materials of construction [191]. In Fig. 9, a typical temperature profile for an entrained flow gasifier is shown.

The outlet gas from the gasifier is free of hydrocarbons heavier than methane. Methane production is low due to the unfavourable thermodynamic equilibrium for methane formation at high temperatures. Likewise, the production of CO in preference to CO_2 is favourable due to the effects of low steam addition and high operating temperature on the water-gas shift equilibrium. A large amount of sensible heat is present in the raw gas leaving the gasifier.

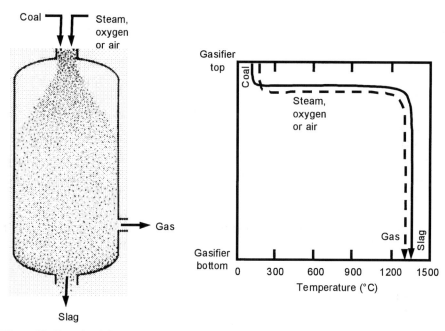

Figure 73 Entrained flow gasifier and typical temperature profile [174].

To maintain high thermal efficiency, the sensible heat has to be recovered from the raw gas, which is complicated by the possibility of molten ash deposition on heat transfer surfaces [195]. To limit the impact of molten ash, sophisticated high temperature heat exchangers have to be used to cool the syngas below the ash softening temperature in order to avoid fouling and control corrosion problems [191].

The entrained flow gasifier offers many advantages. These include the ability to handle a wide variety of coals, the elimination of tar and oil formation (not necessarily positive in all scenarios) and water-condensate and solid residue which can be easily disposed of [171].

One of the main disadvantages arises from the low inventory of coal in suspension. Due to cocurrent flow, the temperature drops from the inlet to the outlet while the carbon content and reactivity decreases. It is thus difficult to obtain high carbon conversion, as a very large reactor would be required to compensate for long residence times. To keep the reactor volume within bounds, high temperatures are used to speed up the reaction rates, and a certain amount of unconverted coal is acceptable [195].

Some operational difficulties are experienced with entrained flow gasifiers. These include material limitations regarding high operating temperatures, limited refractory life and difficult slag control. Gasification of coal with high ash content can be problematic, since the ash has to be melted and therefore high ash content feedstock impacts negatively on the thermal efficiency and oxidant requirement. High ash fusion feedstock could be problematic, and a suitable flux may be required to achieve ash melting at acceptable temperatures in order to protect gasifier linings. Operation at atmospheric pressure is also a handicap with some of the entrained flow gasification technologies, as the produced syngas has to be pressurised for most applications. However, most of the entrained flow gasification technologies have been adapted to operate under pressure.

Most commercial gasifiers in the world are entrained flow gasifiers. There are more than 240 entrained flow gasifiers operational in the world, and 50 more in the planning phase. Of the operational gasifiers, more than 50% of the gasifiers are Texaco gasifiers and 36% are Shell gasifiers. It is also the type of gasifier most favoured for IGCC applications at present, specifically Texaco gasifiers, which is the most popular in these installations (used in 7 IGCC projects up to 2001) [174].

3.6.3.1 Texaco gasifier

Texaco is the world's leading licensor of gasification technologies. The first gasification plant utilising Texaco gasification technology was commissioned in 1956, and used oil as feedstock. The first coal-based gasification plant of this nature was started up in 1978. There are currently 72 commercial facilities that employ Texaco gasification technology, of which 60 are operating and 12 are under construction or in the engineering phase. A total of 125 gasifiers are installed in these facilities [176], with a combined nominal capacity of 154 200 000 m3n/day. The technology is currently owned and marketed by ChevronTexaco after the merge between the Texaco and Chevron companies.

The Texaco gasifier is a cylindrical pressure vessel. The upper gasification section is refractory-lined, while the lower part extends into a water reservoir in the case of the direct quench type. The reservoir acts as a slag quencher and has a restricted orifice through which slag from the gasification section exits with the raw gas [188].

These gasifiers are single-stage, downward firing units. The oxidant can be pure oxygen, air or oxygen-enriched air [201], and enters the vessel via a specially designed burner. If solids like coke or coal are used as feedstocks, it is introduced as a pumpable slurry in water. A solid concentration in the range of 60-70% is preferred.

Slurryability is an important coal property to take into account. As the water requirement for producing a pumpable slurry increases, oxygen demand of the gasifier increase due to the larger quantity of water that has to be evaporated, leading to lower thermal efficiencies and poor operating economics [174]. Slurryability and grindability are interrelated as size distribution affects the slurry properties of the coal and the conversion in the gasifier, i.e. if a coarse grind size distribution is used, a high solid concentration slurry can be produced but the larger coal particles will not gasify as well as the smaller particles [191].

The Texaco gasifier is operated at temperatures of 1260°C to 1430°C and 41 bar is a typical operating pressure [188]. Coal residence time is a few seconds, and the ash is removed as a molten slag, which is inert and can be safely disposed of by landfilling. Due to the gasification conditions, the raw gas is rich in hydrogen and carbon monoxide, with a higher concentration of CO than H_2. The hydrogen content of the produced gas can be increased by further processing through a shift converter, if required.

Entrained flow slagging gasifiers have a relatively high oxygen requirement due to high operating temperatures and oxygen consumption for Texaco gasifiers is even higher than that for other entrained flow gasifiers which operate with a dry feed, due to the energy required to vaporise water in the slurry feed [174]. However, since the coal is injected as a coal-water slurry, no additional steam is required, unless the temperature in the reactor is above the desirable temperature, in which case a moderator, usually steam, is injected to control the temperature [188].

One of the disadvantages of the Texaco gasifier is a limited ability to economically handle low-rank coals because of the water/slurry feed system, high temperature operation which can lead to high maintenance costs and high-temperature gas cooling equipment that has to be utilised. The wet feeding system does however offer the advantage of being safe due to the elimination of dust and the associated explosion hazard [201]. Also, the water need not be processed to the high purity required for steam-fed gasifiers.

The operating temperature (1260°C-1430°C) of the gasifiers has to be high enough for the coal mineral matter to melt and flow freely down to the bottom of the gasifier. The ASTM ash fusion temperature method is used to determine the minimum gasifier operating temperature for successful tapping operation [191]. A maximum of 12% ash (on a weight basis) is recommended for coals to be processed in a Texaco gasifier.

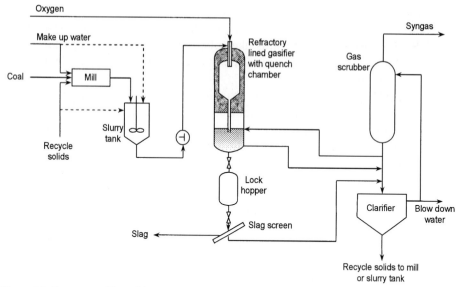

Figure 74 Texaco gasifier with quench cooling[1]

The Texaco gasifier has a typical cold gas efficiency of 77%, with the quench cooling approach that is usually used to minimize the capital cost and increase gasifier availability. If high-pressure is steam generated during cooling of the raw gas then the efficiency can be as high as 95% [174].

To cool the syngas, a direct or indirect cooling method can be employed. In the quench (direct) variant, raw gas from the gasifier is shock-cooled with water (see Fig. 10). Crude syngas leaves the bottom of the gasifier through a quench tube, which is submerged in a pool of water. In passing through the water, crude gas is cooled to the saturation temperature of the water and is cleaned of slag and soot particles. The cooled, saturated syngas then exits the vessel through a duct on the side wall for further cooling and cleaning or direct use [177]. Most Texaco gasifiers currently in operation use the quench variant. The major advantage is that it is cheaper and more reliable, but it is less thermally efficient.

In the full heat recovery (indirect) variant, the raw gas leaves the gasifier section and is cooled in a radiant syngas cooler from 1400°C to approximately 700°C. The recovered heat is used to generate high-pressure steam. Molten slag flows down the cooler and is quenched in a bath at the bottom, from where it is removed through lock hoppers. The partly-cooled syngas leaves the bottom of the gasifier and is then cooled further in convective coolers before being cleaned for use in the downstream application [177].

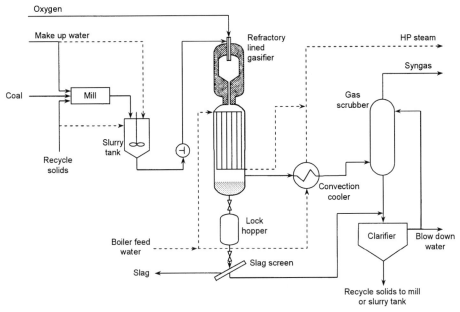

Figure 75 Texaco gasifier with radiant cooling [1].

3.6.3.2 Shell gasifier

The Shell gasifier is a dry feed pressurised entrained slagging gasifier
[199]. Feed coal is pulverised and dried with the same type of equipment used
for conventional pulverised coal boilers. The coal is then pressurised in lock
hoppers and fed into the gasifier with a transport gas by dense-phase conveying.
The oxidant is preheated to minimize oxygen consumption and mixed with
steam as moderator prior to feeding to the burner. The coal reacts with oxygen at
temperatures in excess of 1370°C to produce principally hydrogen and carbon
monoxide with very little CO_2. Operation at elevated temperatures eliminates the
production of hydrocarbon gases and liquids. The high temperature gasification
process converts the ash into molten slag, which runs down the refractory-lined
wall of the gasifier into a water bath, where it solidifies and is moved through a
lock hopper as a slurry [174]. Some of the molten slag collects on the cooled
walls of the gasifier to form a solidified protective coating. The raw gas leaving
the gasifier at 1600°C contains small quantities of unburned carbon and about
half of the molten ash. To make the ash less sticky, preventing it from fouling
surfaces, the hot gas leaving the reactor is partially cooled by quenching with
cooled recycle product gas to approximately 900°C. Further cooling takes place
in the waste heat recovery unit, which consists of radiant, superheating,

convection, and economising sections where high pressure superheated steam is generated before particle removal.

Cooled syngas is filtered through ceramic filters. About 50% of the cooled syngas is then recycled to the top of the gasifier to act as quenching medium for the exiting gas. The remainder is washed to remove halides and ammonia and then passed on for desulphurisation [177].

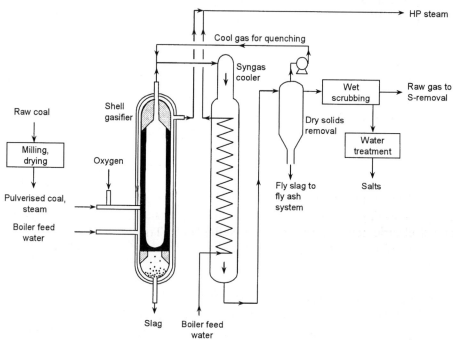

Figure 76 Shell gasifier [218, 219]

Shell claims that this type of gasifier has high flexibility in terms of coal rank and ash, sulphur and moisture content [191]. Any coal that can be milled to the right size and pneumatically transported can be gasified in the Shell entrained flow gasifier, although some adjustments are necessary in the case of specific coals. For example, bituminous coals require steam injection and oxygen/coal (dry, ash free) ratios from 0.85 to 1.05 for producing a syngas with an H_2/CO ratio of 0.4 to 0.45 and 1-2.5% CO_2. Sub-bituminous coals on the other hand do not require steam injection and can be operated with oxygen/coal ratios of between 0.8 and 0.9.

The ash content of a coal has an impact on the performance of the gasifier in terms of efficiency, as slag forms part of the insulation of the wall of the gasifier and so prevents excessive heat loss during gasification [191]. An ash content of more than 8% but lower than 15% is recommended, as increased ash content increases the overall cost of the operating system. However, coals with an ash content of up to 40% can be processed. The ash content can also have an impact on the performance of the slag tap and the slag handling system. Elements like Cl, F, Pb, K, Na and P need to be taken into account as they vaporise during gasification and then condense in the syngas cooling systems, sticking to ash particles and causing fouling.

Large variations in the coal sulphur content have the largest impact on the gasification process, followed by variation in ash content and moisture content [191].

The Shell gasifier has a cold gas efficiency of 80%, which is higher than that of the Texaco entrained flow gasifier due to reduced oxygen consumption and lower CO_2 generation. If the energy content of the high pressure steam generation in the gasifier jacket and the gas coolers is added to that of the fuel gas and if the energy for coal drying and steam feed to the gasifier is deducted, the actual efficiency would increase to 94% [174].

IGCC using the Shell gasification technology has been demonstrated over the last few years at the Buggenum facility in the Netherlands. The single Shell gasifier has a throughput of approximately 23 kg/s coal [211], which translates to 2070 tons/day of coal. The plant is currently owned by Nuon, which is the largest electric power supplier in the Netherlands. Trials are under way to co-feed renewable energy sources such as agricultural waste and chicken litter [212].

3.6.3.3 Lurgi Multi-purpose gasifier (MPG)

The MPG developed from a well-proven partial oxidation technology that was acquired by Lurgi to offer all production and treating processes for gas production from a single source [213]. The MPG was developed from Lurgi's fixed bed coal gasification technology and a demonstration unit has been in operation at Schwarze Pumpe since 1968. The MPG can process an extreme variety of hydrocarbon-containing feeds ranging from natural gas, coal tars and oils, heavy refinery residues, asphalts, etc. It is even possible to feed several streams of non-mixable substances simultaneously.

The feed enters the refractory-lined vessel through a top-mounted robust burner, which is able to process fluids of high viscosity, emulsions and slurries containing particles of several mm in diameter [211]. Oxygen is mixed with steam before being fed to the burner. The steam acts as temperature moderator. The process is non-catalytic and is carried out at temperatures of 1200°C to 1450°C and pressures of 30 to 75 bar.

Depending on the feed composition, the oxidant and the actual gasification temperature, the raw syngas contains H_2, CO, CO_2, etc. in various concentrations. The H_2/CO ratio of the obtained syngas is in the range of 0.5 to 0.8. Typical characteristics of the raw gas exiting the MPG, compared to the Sasol-Lurgi dry bottom fixed bed gasifier, are listed in the table below.

Table 4
Comparison of raw gas composition and temperature for the Sasol-Lurgi gasifier and Lurgi MPG [211]

		Sasol-Lurgi gasifier	MPG
H_2	% vol.	39	41
CO	% vol.	18	53
CO_2	% vol.	31	4
CH_4	% vol.	11	0.15
$N_2 + Ar$	% vol.	1	1.85
Raw gas temp.	°C	300 - 600	1200 / 220
H_2 / CO		2.2	0.77

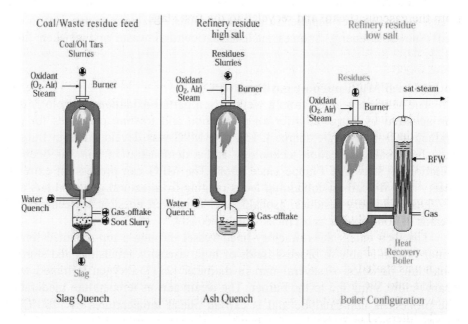

Figure 77 Different modes of operation for a MPG [211]

Soot and ash are removed from the raw syngas by means of wash water [213]. The resulting slurry is routed to a treatment step. After separation from the solid particles, the greater part of this water is recycled to the washing units.

To provide maximum feedstock and product flexibility, two variants for the syngas cooling section are offered for the MPG. The MPG gas cooling can either be a quench configuration or boiler configuration. Application of a waste heat boiler utilizes a major portion of waste heat to generate high-pressure steam. However, if the feedstock contains high levels of precipitation-prone contaminants which could block boiler tubes of the waste heat boiler, the quench mode is preferred. This option offers highest flexibility at lowest investment cost traded off against lower energy efficiency.

3.6.3.4 E-Gas two-stage gasifier

The E-Gas Technology, formerly Destec, process is an entrained flow, two-stage oxygen-blown gasifier. Coal is slurried, which is a more elegant way of pressuring coal than lock-hoppering [1], combined with pure oxygen, and injected into the first stage of the gasifier.

Two-stage gasification involves a first high-temperature slagging stage to which only part of the reactants are fed and a second non-slagging stage where the hot gas from the first stage drives the endothermic reactions. Unreacted carbon and dry ash which exits the second stage with the raw gas is removed from the gaseous stream and recycled to the first stage, so that almost all ash is removed as a slag. The two-stage gasifier has a higher thermal efficiency than single-stage gasifiers as the sensible heat in the raw gas is reduced and has a lower oxygen requirement than single-stage gasifiers. However, for a two-stage slurry-feed gasifier, such as the E-Gas gasifier, the disadvantage of a higher steam consumption, which is the case for dry-feed two-stage gasifiers, is not applicable, since the amount of steam is dictated by the coal:water ratio in the slurry, which is not affected by the staging arrangement.

In the case of the E-Gas gasifier, coal slurry undergoes partial oxidation at temperatures high enough to bring the coal ash above melting point in the first stage. Fluid ash falls through a tap hole at the bottom of the first stage into a water quench, forming an inert vitreous slag. The hot gas flows to the second stage, where additional coal slurry is injected.

The E-Gas gasifier is the only two-stage process with an operating commercial-scale demonstration plant, at the Wabash River site in Indiana, which was started up in 1996. Until 2000, sub-bituminous coal-water slurry was injected into the hot gases from the first stage. Particulate matter (char) was removed from the exiting gas in a particulate-removal unit featuring metal candle filters, after which it was injected with oxygen and/or steam into the first slagging stage, operating at a temperature of approximately 1400°C. The result was that, although sub-bituminous coal is used, the slagging part of the gasifier saw a feed upgraded by a dry char stream that required relatively little oxygen to be gasified.

Until the year 2000, 2550 tpd of coal was processed at the Wabash complex. Thereafter, the feedstock was switched to petcoke, and currently, 2000 tpd of petcoke is processed, with a 262 MWe net output. In fact, the E-Gas gasifier offers the highest petcoke throughput possibility of any gasifier at 2700 t per vessel per day [214].

The Wabash River Coal Gasification Repowering project, an IGCC complex, is claimed to be the cleanest coal-fired power plant in the world, operating at emissions well below its 1993 permit. The E-Gas technology has been operated in this IGCC application for over 15 years, and provides a highly efficient, environmentally superior and competitive based-load power alternative. IGCC achieves low SO_x and NO_x emissions, near zero particulate emissions and more than 90% of mercury in the feedstock can be removed. Coal ash from the E-gas process is an inert saleable construction material.

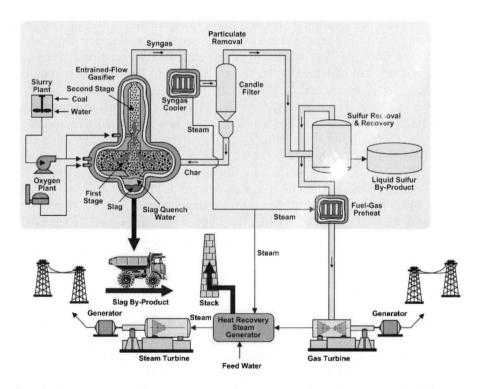

Figure 78 Wabash River Coal Gasification Repowering plant [215]

Numerous problems were experienced during the first years of operation. These included (1) ash deposition on the walls of the second stage of the gasifier and downstream piping, (2) particulate breakthrough in the hot gas filter system, (3) chloride and metals poisoning of the COS catalyst, (4) cracking problems with the gas turbine combustion liners and tube leaks in the heat recovery steam generator. However, these problems were all solved by the fourth year of operation.

The Wabash plant has been used to test certain improvements to the E-Gas gasification technology and integration into IGCC. ConocoPhillips claims that the technology will have a reduced capital cost, improved availability, GT-ASU integration and cyclone/hot gas filter hybrid system for gas cleaning from particulates to offer in future. The immediate future for E-Gas Technology appears to lie with applications where low-cost feedstocks can be used and co-production options are desired, e.g. the co-production of steam, fuels/chemicals and electricity [215].

The Wabash River Coal Gasification plant was owned by Dynegy at the beginning of the re-powering project. At the end of demonstration in December 1999, Global Energy Inc. purchased Dynegy's gasification assets and technology, and marketed the technology under the name E-Gas Technology. ConocoPhillips purchased the E-Gas technology in 2003, since the gasification technology was identified as a means to solve potential disposal problems of wastes from coking operations, and to meet refinery hydrogen needs. E-Gas Technology is a proven clean and efficient commercial process for converting coal or petroleum coke into a hydrogen-rich syngas, ideally suited for refining, power and chemicals applications.

ConocoPhillips was formed in 2002 with the merger of Conono Inc. and Phillips Petroleum Company. ConocoPhillips is the world's number one petcoke producer and also licenses coker technology [216].

3.7 Combinations of Sasol-Lurgi and entrained flow gasifiers

Sasol-Lurgi dry bottom fixed bed gasifiers have several advantages:

- The H_2/CO ratio of the syngas produced is in the range of 1.70 to 2.0, which makes it ideally suited to applications for high temperature Fischer-Tropsch (HTFT) synthesis without the need for water-gas shift conversion to adjust the H_2/CO ratio.
- High ash coals can be used as feedstock without major economic implications.
- The proven reliability and availability of these units is extremely high. As several smaller units are used, maintenance is easy and can be done without much impact on the process, by taking one unit off-line at a time.
- Low oxygen consumption and low capital cost for suitable applications.

A strong driver for the application of entrained flow gasifiers is that a wide range of feedstocks can be processed, including organic wastes and refinery by-products. Also, most types of coal can be gasified in entrained flow gasifiers. However, each type of entrained flow gasifier has certain requirements of its feed, depending on the design. There is a minimum ash content required for gasifiers with slag self-coating walls, which have to be covered by slag to function and minimise heat loss through the wall [217]. A maximum ash content is usually also fixed for each type of entrained flow gasifier, as the tolerance to ash content depends on economic and technical factors.

The outlet gas from the gasifier is free of any hydrocarbons heavier than methane, which is present in very low concentrations due to the unfavourable thermodynamic equilibrium for methane formation at high temperatures. Likewise, the production of CO over CO_2 is favoured due to the effects of low steam addition and high operating temperature on the water-gas shift equilibrium. A large amount of sensible heat leaves the gasifier in the raw gas, but it can be recovered in a waste heat boiler with the generation of high-pressure steam. However, this heat recovery is relatively expensive.

To address the challenges related to entrained flow gasifiers and Sasol-Lurgi gasifiers, combinations can be considered. Entrained flow gasifiers could be used to treat tars, oils, aqueous by-products, fine coal, and even other problematic material not specifically related to Sasol-Lurgi gasification such as refinery residues and chemical waste streams. Entrained flow gasifiers could even be used to partially reform methane, reducing or eliminating the separate reformer capacity often required in schemes employing Sasol-Lurgi gasifiers.

In some instances, syngas with a relatively low H_2/CO ratio is required, e.g. for low temperature FT synthesis (LTFT) using iron catalyst. The Sasol-Lurgi fixed bed dry bottom gasifier can usually only be operated within a very narrow H_2/CO band. The gas from an MPG or entrained flow gasification unit could be used to manipulate the H_2/CO ratio of syngas to obtain the optimum composition for the downstream synthesis process.

In summary, employing entrained flow gasifiers together with with Sasol-Lurgi fixed bed dry bottom gasifiers would have the following advantages:

- Simpler flow scheme, since tar and phenosolvan (aqueous product) work-up can be simplified or eliminated.
- The possibility to produce the maximum amount of synthesis gas from coal, as all black products will be converted to syngas as well.
- Better flexibility in terms of feedstock, as Sasol-Lurgi gasification offers the opportunity to process high-ash coals, while entrained flow gasifiers offer the opportunity to process fine coal and hydrocarbon and aqueous wastes.
- Better flexibility in terms of products, as the syngas can be manipulated to suit the downstream synthesis processes.

3.8 Particulate removal

Before the raw syngas from a gasifier can be fed to an expansion turbine, gas turbine or condensing heat exchanger it is necessary to remove entrained solids. Currently this requires a cooling step immediately after the gasifier. This may be a significant source of energy degradation and compares unfavourably with gas reforming that does not require this solids removal step.

In most existing plants the solids are washed out with water in venture scrubbers or wash towers. This scrubbing takes place below the dewpoint of the gas so that the finest solid particles can act as nuclei for condensation, thus ensuring that all solids are removed efficiently [1].

ChevronTexaco and others have proposed the use of a quench IGCC design in which the gasification is conducted at higher pressure of about 70 bar [1]. Additional power can then be generated by expansion of the clean syngas fuel to about 20 bar followed by reheat and firing in the gas turbine. Fuel gas expansion is being used in two of the heavy oil IGCC plants at Falconara and Priolo in Italy.

A significant development has been the use of dry solids removal at gas temperatures up to 500 °C using ceramic candle filters [1].

Dry particulate removal must take place at a temperature that is sufficiently low that alkali compounds can be removed as solids so that they do not form corrosive alkali (hydro-) sulphates. This temperature is about 500 °C, which represents an upper bound for the particulate removal [1]. The lower bound is governed by the ammonium chloride sublimation temperature and lies at about 280 – 300 °C [1]. Within this allowable range of 280 - 500 °C it is preferable to remain close to the upper limit when feeding the syngas to a gas turbine to ensure the highest possible efficiency if this gas is to be fed to a turbine.

REFERENCES

[1] C. Higman and M. van der Burgt, Gasification, Elsevier Science, USA, 2003.
[2] Å. Solbakken, J. Stud. Surf. Sci. Catal. 61 (1991) 447.
[3] I. Dybkjær., in: A. Nielsen (ed.), "Ammonia", Springer-Verlag, Berlin (1995) 199.
[4] I. Dybkjær, Fuel Processing Technology, 42 (1995) 85-107.
[5] S.W. Madsen, Paper presented at the International Symposium on Large Chemical Plants 10, Antwerp, 1998.
[6] K. Aasberg-Petersen, J.B.H. Hansen, T.S. Christensen, I. Dybkjær, P. Seier Petersen, C. Stub Nielsen, S.E.L. Winter-Madsen and J.R. Rostrup-Nielsen, Appl. Catal. A: General, 221 (2001) 379.
[7] J.R. Rostrup-Nielsen, Catal. Today, 71 (2002) 243.
[8] I. Dybkjær, in: Fundamentals of Gas to Liquids, Petroleum Economist (2003) 16.
[9] Nitrogen & Methanol, 266 (2003) 33.
[10] J.W. Evans, Sulphur Measurement Methods and their Applications, http://www.thermo.com/eThermo/CDA/Technology/Technology_Detail/1,121 3, 12002-11418-136,00.html.

398

[11] Trace Sulphur Component Analysis Models 4025/4026/4027/4028/4029
http://www.amelinc.com/HTMLobj-2183/pe4025-4029.pdf.
[12] J.J. Philipson, in: Catalyst Handbook, Wolfe Scientific Books, London, 1970, 46.
[13] J.H. Carnell, in: Twiggs M.V. (ed) Catalyst Handbook, 2nd ed. Wolfe Publishing Ltd.,
London, 1989, 199.
[14] Nitrogen, 72 (1971) 34.
[15] J.Y. Livingston, Hydcar. Proc., 50(1) (1971) 126.
[16] A. Hidalgo-Vivas and B.H. Cooper, in: Vielstich, W. et al. (ed.) Handbook of Fuel
Cells, John Wiley & Sons, Ltd., 2003.
[17] H. Topsøe et al., Hydrotreating Catalysis, Springer-Verlag, Berlin, Heidelberg, 1996.
[18] J. Richardson and R. Drucker, Ammonia Plant Saf., 38 (1998) 21.
[19] P.J.H. Carnell and P.J. Denny, Ammonia Plant Saf., 25 (1985) 99.
[20] H. Fan, C. Li, H. Gno and K. Xie, J. Nat. Gas Chem., 12 (2003) 43.
[21] P.E. Jensen and K. Søndergaard, Ammonia Plant Saf., 24 (1984) 47.
[22] T.S. Christensen, Appl. Cat. A: General, 138 (1996) 285.
[23] H.S. Davies, K.J. Humphries, D. Hebden and D.A. Percy, Inst. Gas Eng. J., 7 (1967)
708.
[24] F. Moseley, R.W. Stephens, K.D. Stewart and J. Wood, J. Catal., 24 (1) (1972) 18.
[25] T.S. Christensen and J. Rostrup-Nielsen, in: Deactivation and Testing of Hydrocarbon
Processing Catalysts, ACS Symposium Series 634, ACS, Washington, 1996.
[26] K. Aasberg-Petersen, T.S. Christensen, C.S. Nielsen and I. Dybkjær, Fuel Processing
Tech., 83 (2003) 253.
[27] J.R. Rostrup-Nielsen, in: J.R. Anderson and M. Boudart (Eds.), Catalysis, Science and
Technology, Springer-Verlag, Berlin, 1984.
[28] A.V. Slack and G.R. James, Ammonia, Part 1, Marcel Decker, Inc., New York, 1973,
145.
[29] T. Kawai, K. Takermura and M.B. Zaghloud, Paper presented at International Plant
Engr. Conf., Bombay, 1984.
[30] J.R. Rostrup-Nielsen, I. Dybkjær and L.J. Christiansen, in: NATO ASI Chemical
Reactor Technology for Environmentally Safe Reactors and Products, H.J. de Lasa et al.
(eds.), Kluwer, Dordrecht, 1992, p.249.
[31] Nitrogen, 214 (1995) 38.
[32] J.R. Rostrup-Nielsen and L.J. Christiansen, in: Tubular Steam Reforming. Chemical
Reaction and Reactor Design Tominaga, H. and Tamaki, M. (eds.), John Wiley,
Chichester, 1997, chap. 5.2.
[33] J.R. Rostrup-Nielsen, J. Sehested and J.K. Nørskov, Adv. Catal., 47 (2002) 65.
[34] T. Rostrup-Nielsen, Hydcar. Eng., 7 (8) (2002) 51.
[35] D.L. King and C.E. Bochow, Jr., Hydcar. Proc., 79 (5) (2000) 39.
[36] K. Hoitsma and P. Snelgrove, World Ref., 12 (July/Aug 2002) 24.
[37] F. Hohmann, Hydcar. Proc., 75 (3) (1996) 71.
[38] U. Herrlett and U. Wolf, Hydcar. Eng., 7 (8) (2002) 55.
[39] T. Johansen et al., Hydcar. Proc., 71 (8) (1992) 119.
[40] www.uhde.biz/competence/fertilisers/steam reforming.
[41] R.E. Stoll and F.V. Linde, Hydcar. Proc., 79 (12) (2000) 42.
[42] J.D. Fleshman et al., NPRA 1999 Annual Meeting, San Antonio, Texas, USA (1999).
[43] T. Mohri, K. Takemura and T. Shibasaki, Ammonia Plant Saf., 33 (1993) 86.
[44] K. Hosoya, N. Shiratori, K. Satoh and K. Yamamoto, PTQ, Autumn (2001) 115.
[45] API Recommended Practice 530, 4th edition, 1996.

[46] www.callidus.com.
[47] www.hamworthy-combustion.com.
[48] www.johnzink.com.
[49] B.R. Fisher, Ammonia Plant Saf., 41 (2001) 303.
[50] B.J. Cromarty and S.C. Beedle, Ammonia Plant Saf., 33 (1993) 63.
[51] J.R. Rostrup-Nielsen, L.J. Christiansen and J.H.B. Hansen, Appl. Cat., 43 (1988) 287.
[52] L. Storgaard, Nitrogen '91 Conference, Copenhagen (1991).
[53] D. Barnett and D. Wu, Ammonia Plant Saf., 41 (2001) 9.
[54] B. Cotton and B. Fisher, Ammonia Plant Saf., 42 (2002) 106.
[55] V. Mehrota, B. Rosendall, A. Heath and J. Berkoe, PVP-Vol. 448-2, Computational Technologies for fluid/Thermal/Structural/Chemical Systems with Industrial Applications - 2002, Vol. II, ASME 2002, PV 2002-1581, p.119.
[56] P. Nielsen and L.J. Christiansen, Proceedings of: 4th International Symposium on Computational Technologies for Fluid/Thermal Chemical Systems with Industrial Applications, August 4-8, Vancouver, B.C., Canada.
[57] H. Stahl, J. Rostrup-Nielsen and N.R. Udengaard, in: Fuel cell seminar 1985. Tucson, Arizona (1985) 83.
[58] N.R. Udengaard, L.J. Christiansen and W.A. Summers, Endurance testing of a high-efficiency steam reformer for fuel cell power plants. EPRI AP-6071, Project 2192-1. Electric Power Research Institute, California (1988).
[59] I. Dybkjær, J.N. Gøl, D. Cieutat and R. Eyguessier, NPRA Annual Meeting, San Antonio, USA, paper No. AM-97-18, March 16-18 (1997).
[60] I. Dybkjær and S.W. Madsen, Int. J. Hydcar. Eng., 3 (1) (1997/98) 56.
[61] F. Giacobbe and O. Loiacono, Stud. Surf. Sci. Cat., 119 (1998) 937.
[62] J. Font Freide, T. Gamlin and M. Ashley, Hydcar. Proc., 82 (2) (2003) 53.
[63] T. Mohri, T. Yoshioka, Y. Hozuma, T. Shiozaki, T. Hasegawa, S. Mochida and S. Maruyama, Ammonia Plant Saf., 43 (2002) 115.
[64] K.J. Elkins, A.J. Gow, D. Kitchen and A. Pinto, Proceedings No. 319, The Fertiliser Society, London, 1992.
[65] P.W. Farnell, Ammonia Plant Saf., 40 (2000) 173.
[66] T. Mii and K. Hirotani, Economic Evaluation of a Jumbo DME Plant, presented to WPC Asia Regional Meeting, Shanghai, Sep. 17-20 (2001).
[67] Nitrogen, 179 (1989) 16.
[68] S.W. Madsen and I. Dybkjær, Novel Revamp Solutions for Increased Hydrogen Demands, presented to ERTC, London, Nov. 17-19 (2003).
[69] S. Ratan and C.F. Vales, Hydcar. Proc., 81 (3) (2002) 57.
[70] A. Malhotra and L. Hackemesser, Ammonia Plant Saf., 42 (2002) 223.
[71] R.V. Schneider III and G. Joshi, PTQ, summer (1997) 85
[72] H.-D. March and N. Thiagarajan, Ammonia Plant Saf., 33 (1993) 108.
[73] R.T. Jones, K.L. Baumert, Corrosion 2001, paper No 01372, NACE (2001).
[74] H.J. Grabke., Materials at High Temperatures, 17 (4) (2000) 483.
[75] C.H. Chum, T.A. Ramanarayaran and J.D. Mumford, Materials and Corrosion, 50 (1999) 634.
[76] M. Maier, J.F. Norton and P. Puschek, Materials at High Temperatures, 17 (2) (2000) 347.
[77] H.J. de Bruyn, E.H. Edwin and S. Brendryen, Corrosion, paper 01883 (2001).
[78] C.H. Toh, P.R. Munroe and D.J. Young, Oxidation of Metals, 58 (1/2) (2002) 1.
[79] B.A. Baker, B.D Smith and S.A. Meloy, Ammonia Plant Saf., 42 (2002) 257.

400

[80] C. Rosado and M. Shütze, Materials and Corrosion, 54 (1) (2003) 831.
[81] G.T. Bayer, Corrosion 2001, paper No 01387, NACE.
[82] A.B. Smith, A. Kempster, A. Lambourne and J.R. Smith, Corrosion 2001, paper No. 01391, NACE (2001).
[83] J.R. Rostrup-Nielsen and J.H.B. Hansen, J. Catal., 144 (1993), 38.
[84] J. Sehested, J. Catal., 217 (2003) 417.
[85] J.R. Rostrup-Nielsen, Steam Reforming Catalysts, Technical Press, Copenhagen, 1974.
[86] J.R. Rostrup-Nielsen, J. Catal., 85 (1984) 31.
[87] I. Alstrup, J.R. Rostrup-Nielsen and S. Røen, Appl. Catal., 1 (1981) 303.
[88] H.S. Bengaard, J.K. Nørskov, J.S. Sehested, B.S. Clausen, L.P. Nielsen, Molenbrock and J.R. Rostrup-Nielsen, J. Catal., 209(2002) 365.
[89] J.R. Rostrup-Nielsen, J. Sehested and J.K. Nørskov, Adv. Catal., 47 (2002) 65.
[90] S. Helveg, C. López-Cartes, J. Sehested, P.L. Hansen, B.S. Clausen, J.R. Rostrup-Nielsen, F. Abild-Pedersen and J.K. Nørskov, Nature, 427 (2004) 426.
[91] R. Vannby, C.S. Nielsen and J.S. Kim, Hydcar. Tech. Int., (1993).
[92] T.S. Christensen, P.S. Christensen, I. Dybkjær, J.H.B. Hansen and I.I. Primdahl, Stud. Surf. Sci. Catal., 119 (1998) 883.
[93] W. De Graaf, F. Schrauwen, Hydcar. Eng., 7 (5) (2002) 55.
[94] The Shell Gasification Process. www.Uhde.biz.
[95] A. Hoek, The Shell Middle Distillate Synthesis Process, CatCon2003, Houston, 5-6 May (2003).
[96] J.E. Philcox and G.W. Fenner, Oil & Gas J., 95 (28) (1997) 41.
[97] T.M. Beaudette et al., Project Success, Hydcar. Eng., 7 (7) (2002) 41.
[98] R. Menon, T. Vakil, M. Hawkins and G. Saunders, Oil & Gas J., 99 (12) (2001) 70.
[99] W. Liebner and D. Ulber, MPG – Lurgi Multi Purpose Gasification: Application in "gas-Gasification", 2000, Gasification Technologies Conference San Francisco, Cal. Oct. 8-11 (2000).
[100] H. Schichtling, "Update on Lurgi Syngas Technologies", Gasification Technologies 2003, San Francisco, Oct. 12-15 (2003).
[101] K. Arcuri, K. Schimelpfenig and J. Laegen, CatCon2003.
[102] Remote Gas Strategies, 6 (1) (2002) 11.
[103] Remote Gas Strategies, 6 (4) (2002) 5.
[104] J. Haugaard, H. Holm-Larsen, Recent Advances in Autothermal Technology – Reducing Production Cost to Prosper in a Depressed Market, Paper presented at the World Methanol Conference, San Diego, Cal., USA, Nov. 29-Dec 1 (1999).
[105] E.L. Sørensen and H. Holm-Larsen, Flexibility in Design of Large-scale Methanol Plants, Paper presented at the World Methanol Conference, Phoenix, Arizona., USA, Nov. (2003).
[106] O. Olsvik and R. Hansen, Stud. Surf. Sci. Catal., 119 (1998) 875.
[107] M. Jay and P. Cook, in: Fundamentals of Gas to Liquids, Petroleum Economist, (2003) 46.
[108] S.E. Nielsen, Ammonia Plant Saf., 42 (2002) 304.
[109] J. Koenig, A.J. Kontopoulos, I. Dybkjær and T. Rostrup-Nielsen, Ammonia Plant Saf., 38 (1998) 206.
[110] A. Gedde-Dahl, H. Holm-Larsen, Ammonia Plant Saf. 39 (1999) 14
[111] www.lurgi-oel.de/english/nbsp/main/info/methanol_combined_reforming. pdf.
[112] H. De Wet, R.O. Minnie and A.J. Davids, Ammonia Plant Saf., 38 (1998) 64.
[113] Hydrocarbon Asia, 9 (1999) 56.

[114] A. Jess, R. Popp and K. Hedden, Appl. Catalysis A: General, 186 (1999) 321.

[115] I. Dybkjær and T.S. Christensen, Stud. Surf Sci. Catal., 136 (2001) 435.

[116] Topsoe-SBA Autothermal Process, The Journal of World Nitrogen, May (1962).

[117] Chem. Eng., 69 (14) (1962) 88.

[118] T.S. Christensen, I.I. Primdahl, Hydcar. Proc., 73 (1994) 39.

[119] W.S. Ernst, S.C. Venables, P.S. Christensen and A.C. Berthelsen, Hydcar. Proc., 79(3) (2000) 100-C.

[120] J. Warnatz, U. Maas and R.W. Dibble, Combustion, I. Glassman (ed.), 3rd edition, Springer-Verlag, Heidelberg, Academic Press, San Diego, 1996.

[121] G.P. Smith, D.M. Golden, M. Frenklach, N.W. Moriarty, B. Eiteneer, M. Goldenberg, C.T. Bowman, R.K. Hanson, S. Song, W.C. Gardiner, Jr., V.V. Lissianski, and Z. Qin, GRI-MECH 3.0, Gas Research Institute (1990), http://www.me.berkeley.edu/gri_mech/.

[122] F.L. Dryer, The Phenomennology of Modeling Combustion Chemistry, in Fossil Fuel Combustion, W. Bartok and A.F. Sarofim (eds.), John Wiley & Sons, 1991.

[123] C.K. Westbrook, Chemical Kinetics of Hydrocarbon Ignition in Practical Combustion Systems, 28th International Symposium on Combustion, The Combustion Institute (2000) 1563.

[124] , B.S. Haynes, Soot and Hydrocarbons in Combustion, in: Fossil Fuel Combustion, W. Bartok and A.F. Sarofim (eds.), John Wiley & Sons, 1991.

[125] H. Bockhorn (ed.), Soot Formation in Combustion - Mechanism and Models, Springer-Verlag, Berlin, 1994.

[126] M.S. Skjøth-Rasmussen, P. Glarborg, M. Østberg, J.T. Johannessen. H. Livbjerg, A.D. Jensen and T.S. Christensen, Formation of polycyclic aromatic hydrocarbons and soot in fuel-rich oxidation of methane in a laminar flow reactor, Combustion and Flame, 136 (2004) 91.

[127] M.S. Skjøth-Rasmussen, P. Glarborg, M. Østberg, M.B. Larsen, S.W. Sørensen, J.E. Johnsson, A.D. Jensen and T.S. Christensen, A Study of Benzene Formation in a Laminar Flow Reactor, Proceedings of the Combustion Institute, 29 (2002) 1329.

[128] M.S. Skjøth-Rasmussen, Modeling of Soot Formation in Autothermal Reforming, Ph.D.-Thesis, Department of Chemical Engineering, Technical University of Denmark (2003).

[129] J.A. Miller and C.T. Bowman, Mechanism and Modelling of Nitrogen Chemistry in Combustion, Progress in Energy and Combustion Science, 15 (1989) 287.

[130] C.T.Bowman, Chemistry of Gaseous Pollutant Formation and Destruction, in: Fossil Fuel Combustion, W. Bartok and A.F. Sarofim (eds.), John Wiley & Sons, 1991.

[131] P.G. Kristensen, Nitrogen Burnout Chemistry, Ph.D.-Thesis, The Technical University of Denmark (1997).

[132] T.S. Christensen, M. Østberg and J.-H. Bak Hansen, Process Demonstration of Autothermal Reforming at Low Steam-to-Carbon Ratios for Production of Synthesis Gas, AIChE Annual Meeting, Reno, Nevada USA, November 4-9 (2001).

[133] C.E. Baukal Jr. (ed.), Oxygen-Enhanced Combustion, CRC Press, Boca Raton, 1998.

[134] T.S. Christensen, I. Dybkjær, L. Hansen and I.I. Primdahl, Ammonia Plant Saf., 35, (1994) 205.

[135] G. Shaw, H. de Wet and H. Hohmann, Ammonia Plant Saf., 35 (1994) 315.

[136] M.B. Sterling and A.J. Moon, Ammonia Plant Saf., 17 (1974) 135.

[137] D. Pasaribu, I.I. Primdahl and C. Speth, Ammonia Plant Saf., 42 (2002) 175.

[138] CFX-4 manual, AEA Technology plc, United Kingdom.

[139] H.K. Versteeg and W. Malalasekera, An Introduction to Computational Fluid Dynamics, The Finite Volume Method, Addison Wesley Longman, Harkow, 1995.

[140] http://www-waterloo.ansys.com/cfx/.

[141] R.J. Kee, F.M. Rupley and J.A. Miller, CHEMKIN-II: A Fortran Chemical Kinetics Package for the Analysis of Gas Phase Chemical Kinetics, Sandia National Laboratories Report SAND89-8009 (1989).

[142] M.S. Rasmussen, P. Glarborg, A. Jensen, T. Johannessen, H. Livbjerg, M. Østberg, and T.S. Christensen, Kinetic Modeling of Rich Combustion based on an Experimental Investigation of Soot Formation in a Laminar Flow Reactor, Proc. First Biennial Meeting, The Scandinavia-Nordic Section of the Combustion Institute, April (2001) 263.

[143] M.J. Pilling (ed.), Low-Temperature Combustion and Autoignition, in Vol. 35, R.G: Compton and G. Hancock (eds.), Comprehensive Chemical Kinetics, Elsevier, 1997.

[144] H.J. Curran, P. Gaffuri, W.J. Pitz, C.K. Westbrook and W.R. Leppard, 26th Symposium International on Combustion, The Combustion Institute (1996) 2669.

[145] Østberg, M. unpublished results.

[146] C.T. Bowman, R.K. Hanson, D.F. Davidson, W.C. Gardiner, V.V. Lissianski, G.P. Smith, D.M. Golden, M. Frenklach and M. Goldenberg, GRI-MECH. Version 2.11, www.me.berkeley.edu/gri_mech/ (1999).

[147] J.-F. Pauwles, J.V. Volponi and J.A. Miller, Combustion Science and Technology, 110-111 (1995) 249.

[148] M.S. Skjøth-Rasmussen, O. Holm-Christensen, M. Østberg, T.S. Christensen, and P. Glarborg, Development of a Procedure to Apply Detailed Chemical Kinetic Mechanisms to CFD Simulations as Post-Processing, Paper no. 20, Proceedings of the European Combustion Meeting, Orléans, France (2003).

[149] G.F. Froment and K.B. Bischoff, Chemical Reactor Analysis and Design, Wiley, New York, 1990.

[150] L.J. Christiansen and J.E. Jarvan, NATO ASI Ser. E, 110 (1986) 35

[151] J.R. Rostrup-Nielsen, L.J. Christiansen and J.H.B. Hansen, Appl.Catal., 43 (1988) 287.

[152] I. Dybkjær and J. Sandholm Hansen, Synthesis gas Production Technology for Conversion of CO2 Rich Hydrocarbon Feedstocks, Indonesian Conference on Natural Gas for Petrochemicals, Jakarta, June 26-27 (1996).

[153] J.R. Rostrup-Nielsen, I. Dybkjær and G.R.G. Coulthard, Synthesis gas for Large-scale GTL-plants, AIChE Spring National Meeting, Houston USA, April 22-26 (2001).

[154] T. Chang, Oil Gas J., 98(51) (2000) 46.

[155] T.S. Christensen, J.H.B. Hansen, P.S. Christensen, I.I. Primdahl and I. Dybkjær, Synthesis gas Preparation by Autothermal Reforming for Conversion of Natural Gas to Liquid Products (GTL), Second Annual Conference "Monetizing Stranded Gas Reserves", San Francisco, December 14-16 (1998).

[156] W.S. Ernst and A.C. Berthelsen, Ammonia Plant Saf., 43 (2003) 172.

[157] T.S. Christensen, M. Østberg, L.S. Nielsen, M.S. Skjøth-Rasmussen, K. Aasberg-Petersen, "Autothermal Reforming – A Preferred Technology in Natural Gas Conversion for GTL Applications, 2003 AIChe Spring Meeting (3rd Topical Conference on Natural Gas Utilization), New Orleans, Louisiana, USA, March 30-April 4, 2003.

[158] V.R. Choudhary, A.M. Rajput and B. Prabhakar, Catal. Lett., 15 (1992) 363.

[159] L.D. Schmidt, Stud. Surf. Sci. Catal., 1 (2001).

[160] J.R. Rostrup-Nielsen, Catal. Today, 18 (1993) 305.

[161] G. Basini, K. Aasberg-Petersen, A. Guarinoni, M. Oestberg, Catal. Today, 64 (2001) 9.

[162] D.A. Hickman and L.D. Schmidt, J. Catal., 138 (1992) 267.

[163] D.A. Hickman and L.D. Schmidt, Science, 259, 5093 (1993) 343.

[164] D.A. Hickman and E.A. Haupfear, L.D. Schmidt, Catal. Lett., 17 (1993) 223.

[165] P.M. Torniainen, X. Chu and L.D. Schmidt, J. Catal., 146 (1994) 1.

[166] A.G. Dietz III and L.D. Schmidt, Catal. Lett., 43 (1995) 15.

[167] K. Aasberg-Petersen, L. Basini, A. Guarinani and M. Oestberg, Presented at the AIChE Spring Meeting, New Orleans, March 10-14 (2002).

[168] Physical Properties of Natural Gas, Groningen N.V., Nederlandse Gusunie, 1988.

[169] J. Shen, V. Venkataraman and D. Gray, in: Fundamentals of Gas to Liquids, Petroleum Economist (2003) 24.

[170] http://66.102.11.104/search?q=cache:ks13z4B1jRkJ:www.environment-agency.gov.uk/commondata/105385/129352+Texaco,+gasifier+%22indirect+cooling%22&hl=en&ie=UTF-8

[171] R.H. Perry and D. Green, Perry's Chemical Engineers' Handbook. McGraw-Hill, Inc, 1984.

[172] J. De Graaf, Overview of Shell Global Solutions' Worldwide Gasification Development. Paper presented at the Gasification Technologies Conference, San Francisco (2002).

[173] H.B.D. Erasmus and J.H. Scholtz, Proven Coal Liquefaction Processes: Sasol-Lurgi Coal Gasification and Sasol Fischer-Tropsch Gas Conversion. China International Hi-tech Symposium and Exhibition on Coal Chemical Industry and Conversion November 6-9, Beijing, PRC, 2002.

[174] D.R. Simbeck, R.L. Dickenson and E.D. Oliver, Coal gasification systems: A guide to status, applications and economics, Prepared for Electric Power Research Instititute by Synthetic Fuels associates, inc. June 1983.

[175] http://www.ept.ntnu.no/fag/sio40ae/innhold/Kullkraft_kompendium_.pdf

[176] N. Richter, Introduction to gasification. Gasification Technologies Public Policy Workshop held in Washington, DC, 2001.

[177] http://www.dti.gov.uk/energy/coal/cfft/cct/pub/tsr008.pdf

[178] http://psdf.southernco.com/overview.html

[179] http://www4.ncsu.edu/~frey/Frey94.pdf

[180] http://chemed.chem.purdue.edu/genchem/topicreview/bp/1organic/coal.html

[181] http://www.crosswinds.net/~nqureshi/igccscan.html

[182] http://www.epa.gov/fedrgstr/EPA-IMPACT/1994/November/Day-14/pr-145.html

[183] M.J. Keyser and J.C. Van Dyk, Full Scale Sasol/Lurgi Fixed Bed Test Gasifier Project: Experimental Design and Test Results. Paper presented at the 17th Annual International Pittsburgh Coal Conference, Pittsburgh, USA, 11-15 September 2000.

[184] D.M. Todd, IGCC Power Block – Lessons Learnt. IChemE Conference 'Gasification in Practice', Assolombarda, Milan, 26-27 February 1997.

[185] http://www.fwc.com/publications/heat/heat_pdf/9806-040.pdf

[186] http://www.et.byu.edu:8080/~tom/cpd/CPD_Summary.pdf

[187] L.F. O'Keefe and K.V. Sturm, Clean Coal Technology Options – A Comparison of IGCC vs. Pulverized Coal Boilers. Paper presented at the Gasification Technologies Conference, San Francisco, California, 2002.

[188] E.C. Mangold, M.A. Muradaz, R.P. Ouellette, O.G. Farah and P.N. Cheremisinoff, Coal Liquefaction and Gasification Technologies. Ann Arbor Science Publishers, 1982.

404

[189] A. Bliek, Mathematical Modeling of a Cocurrent Fixed Bed Coal Gasifier. Dissertatie Drukkerij Wibro, 1984.

[190] W. Jucks and A.G. Sandhoff, Theory of Coal Pyrolysis. The Pennsylvania State College, State College, Penna, 1980, pp. 567-569.

[191] A.-G. Collot, Matching gasifiers to coals. IEA Clean Coal Centre, October 2002.

[192] http://www.lanl.gov/projects/cctc/newsletter/documents/cct_summer96.pdf

[193] http://www.gnest.org/Journal/vol3_no3/androutsopoulos_171_178.pdf

[194] http://www.lib.kth.se/abs98/chen1124.pdf

[195] C.F. Clark, C.V. Fojo, J.P. Henry, J.L. Jones, S.M. Kohan, N. Korens, A.J. Mathias, A.J. Moll, M.A. Moore, W.J. Schumacher and J.G. Witwer, Evaluation of Processes for the Liquefaction and Gasification of Solid Fossil Fuels Volume 1: Coal Mining and Conversion (including lignite). Stanford Research Institute, 1975.

[196] J.H. Slaghuis, Coal Gasification: A study guide for the national diploma in Fuel Technology. Coal Processing III, Part A, 1993.

[197] P. Harasgama, K. Reyser and T. Griffin, The GT13E2 Medium Btu Gas Turbine. IChemE Conference 'Gasification in Practice', Assolombarda, Milan, 26-27 February 1997.

[198] R. Durrfeld, IGCC – treading the path between option and acceptance. IChemE Conference 'Gasification in Practice', Assolombarda, Milan, 26-27 February 1997.

[199] SFA Pacific, Gasification: Worldwide use and acceptance. Report prepared for the U.S. Department of Energy, Office of Fossil Energy, National Energy Technology Laboratory and the Gasification Technologies Council, January 2000.

[200] http://www.cleantechindia.com/eicnew/News/news-coalgasif.htm

[201] Texaco Development Corporation. Texaco gasification process for solid feedstocks. Information brochure, 1993.

[202] I.M. Chellini, P.V. Chaintore, F. Starace and R.D. Braco, api ENERGIA 280 MW IGCC plant: A description of the project from the technical and contractual point of views. IChemE Conference 'Gasification in Practice', Assolombarda, Milan, 26-27 February 1997.

[203] V. Ramanathan, A.D. Rao, S.J. Siddoway, A.C. Simon and S.C. Smelser, Economics of the Tecaxo Gasification Process for Fuel Gas Production. Electric Power Research Institute, Inc. 1982.

[204] H. Jungfer, Synthesis Gas from Refinery Residues. Linde Reports on Science and Technology (40), May 1985.Van Dyk, J.C. Development of an alternative laboratory method to determine thermal framentation of coal sources during pyrolysis in the gasification process. FUEL 80, 2001, pp.245-249.

[205] R.W. Allen and J. Griffiths, The Application of Hot Gas Expanders in Integrated Gasification Combined Cycles. Paper presented at ImechE Conference 'Power Generation and the Environment', London, November 1990.

[206] U. Zwiefelhofer and H.-D. Holtman, Gas Expanders in Process Gas Purification. Paper presented at VDI-GVC Conference, 'Energy Recovery by Gas Expansion in Industrial Use', Magdeburg, March 1996.

[207] J.C. Van Dyk, Development of an alternative laboratory method to determine thermal framentation of coal sources during pyrolysis in the gasification process. FUEL 80, 2001, pp.245-249.

[208] J.C. Van Dyk, Thermal friability of coal sources used by Sasol Chemical Industries (SCI) for gasification – Quantification and statistical evaluation. M.Sc. Thesis, University of the Witwatersrand, 1999.

[209]J.A. DelaMora, J.R. Grisso, H.W. Klumpe, A. Musso, T.R. Roszkowski and B.H. Thompson, Evaluation of the British Gas Corporation/Lurgi Slagging Gasifier in Gasification-Combined-Cycle Power Generation. Electric Power Research Institute, Inc., 1985.

[210]H. Hirschfelder, B. Buttker and G. Steiner, Concept and realisation of the Schwarze Pumpe, FRG 'Waste to Energy and Chemicals Centre'. IChemE Conference 'Gasification in Practice', Assolombarda, Milan, 26-27 February 1997.

[211]http://www.dieter-ulber.de/ichem/Ichem.htm

[212]M. Kanaar, Operations and Performance Update Nuon Power Buggenum. Paper presented at the Gasification Technologies Conference, San Francisco, California, 2002.

[213]http://www.dieter-ulber.de/lurgi-english/mpg_referenz.html

[214]http://www.sws.uiuc.edu/hilites/confinfo/energy/ppt/t7amick.htm

[215]http://www.lanl.gov/projects/cctc/factsheets/wabsh/wabashrdemo.html

[216]http://www.fuelstechnology.com/egas/petcoke.asp

[217]D. Simbeck and H. Johnson, World Gasification Survey: Industry Trends and Developments. Paper presented at Gasification Technologies Conference, San Fransico (2001).

[218]P.C. Richards, J-B. Wijffels and P.L. Zuideveld: Clean and Efficient Power Generation with the Shell Coal Gasification Process, Paper presented at Powe-Gen Eourope '93 Conference, Paris, May 1993.

[219]S.A. Posthuma, E.E. Vlaswinkel and P.L. Zuideveld, Shell Gasifiers in Operation, Gasification Technology in Practice, IChemE Conference, Milan, 1997.

Studies in Surface Science and Catalysis 152
A. Steynberg and M. Dry (Editors)

Chapter 5

Commercial FT Process Applications

M. E. Dry[a] and A. P. Steynberg[b]

[a]Catalysis Research Unit, Department of Chemical Engineering,
University of Cape Town, Rondebosch, 7701, South Africa

[b]Sasol Technology R&D,
P.O. Box 1, Sasolburg, 1947, South Africa

1. FEEDSTOCK AND TECHNOLOGY COMBINATIONS

There are a number of combinations of feedstock, Fischer-Tropsch technology and products that have found commercial application or have been proposed for commercial application. The most important applications are summarised in Table 1.

From Table 1 it is clear that certain technologies compete for application with the same feedstock. Notable competitors are Synthol (HTFT) and cobalt catalyst (LTFT) for natural gas feedstock as well as iron catalyst (LTFT) and Synthol (HTFT) for carbon rich feedstock. While FT technology costs may be an important issue, it is not necessarily the primary decision criteria. The opportunities to add value to the potential project by means of more lucrative co-products is an important consideration [1]. So the desired chemical co-products may narrow down the choice of technology.

Whatever combination of feedstock, technology and products is selected, it will be highly desirable to involve a party experienced at integrating the various technologies used with each other and with the associated utility systems. A further requirement for successful implementation is a highly skilled team to implement the mega-project on time and within budget, often with special challenges associated with a remote location. The track record of the technology supplier may be a vital consideration. Mega-projects are required to reap the benefits from economics of scale. History has shown this to be necessary for energy conversion projects, as is clearly evident by analogy to similar industries such as crude oil refining and electrical power generation.

Table 1

FT applications

FEEDSTOCK	FT TECHNOLOGY	PRODUCTS	APPLICATION
High Ash Coal	Synthol (HTFT)	**Primary Product –** Gasoline **Co-products -** Diesel, kerosene, LPG, ethylene, propylene, 1-hexene, oxygenates, gasifier by-products, pipeline gas	Secunda, South Africa
High Ash Coal	Iron Catalyst (LTFT)	Waxes, paraffins, LPG, ammonia, hydrogen, gasifier by-products	Sasolburg, South Africa
Petroleum Coke	Iron Catalyst (LTFT)	Diesel, naphtha, lubricant base oils, elec. power	ChevronTexaco & Rentech proposals
Low Ash Coal	Synthol (HTFT)	**Primary Products –** Diesel and electrical power **Co-products –** Naphtha, LPG and various commodity chemicals	Proposed for application in China.
Associated Gas	Co Catalyst (LTFT)	**Primary Product –** Diesel **Co-products -** Naphtha, LPG	Escravos, Nigeria
Lean Natural Gas	Iron Catalyst (LTFT)	Waxes, paraffins, LPG, ammonia, hydrogen	Sasolburg, South Africa
Lean Natural Gas	Synthol (HTFT)	**Primary Product –** Gasoline **Co-products –** Diesel, LPG, oxygenates	Mossel Bay, South Africa Brownsville, Texas, USA
Lean Natural Gas	Co Catalyst (LTFT)	Waxes, paraffins, diesel, naphtha, LPG	Bintulu, Malaysia
Lean Natural Gas	Co Catalyst (LTFT)	**Primary Product –** Diesel **Co-products -** Naphtha, LPG, detergent feedstock, lubricant base oils, electrical power	Ras Laffan, Qatar Other proposals: Iran Australia Indonesia
Lean Natural Gas	Synthol (HTFT)	**Primary Product –** Gasoline **Co-products** Diesel, kerosene, LPG, ethylene, propylene, 1-hexene, oxygenates	Proposal for Oman

It almost goes without saying that a new mega-project should not be based on technology that has not been proven on a large scale. Smaller technology companies will need to co-operate with governments and oil majors or multinational chemical companies.

Natural gas is the preferred feedstock for new plants using Fischer-Tropsch technology. GTL (gas-to-liquids) is a popular terminology used to describe these plants. Fischer-Tropsch (FT) based GTL technology brings together technologies derived from gas processing, industrial gas manufacturing, refining, power generation and effluent treatment on a huge scale, with investments exceeding those for new grass roots crude oil refineries. This sets considerable challenges in terms of cost and performance that have only recently been met in a way that allows large scale commercial application.

The experience in South Africa has demonstrated that the use of F-T technology to produce fuels and chemicals from coal can be economically viable if there is government assistance for the initial capital investment. Thus coal based FT must initially be supported by a strategic initiative from the host government but is thereafter competitive on the basis of cash operating costs.

2. ALTERNATIVE ROUTES FOR THE PRODUCTION OF FUELS AND CHEMICALS

Currently the bulk of the world's fuels and commodity chemicals are produced from carbon containing raw materials (fossil resources). Taking the reserves of crude oil as unity, Table 2 gives an approximate comparison of some of the known reserves of such materials. New deposits, particularly of crude oil and of natural gas are regularly being discovered and thus the numbers in Table 2 would require regular updating.

Table 2
World reserves of 'carbon' relative to oil

Source	Reserves, oil equivalent
Crude oil	1.0
Tar sands	0.7
Shale oil	1.2
Natural gas	1.5
Coal	26

At present the world demand for crude oil is about 12 million tons a day. The reserves of crude oil, however, are large (estimated at about 2 trillion i.e. 2×10^{12} barrels or about 240×10^9 metric tons) and could possibly last for fifty or more years. Nevertheless, for both political and economic reasons it is foreseen that alternative sources of carbon containing materials could be utilised to a greater extent in the not too distant future. In the long term, because of the huge amount available, coal could become the main source for the production of liquid fuels and of commodity chemicals.

Extensive research into the recovery of shale oil has been carried out in the past but to date there have been no commercial applications and its future is uncertain. The huge tar sand deposits in Alberta, Canada are being utilised by the companies Syncrude, Suncor and Albion. Significant amounts of 'crude oil' are being produced. The sands contain between eight and twelve percent of bitumen and to produce one ton of oil requires the work up of about fifteen to twenty tons of sand. As in the case for shale oil, the tar sand oil is highly aromatic and so is not suitable for the production of low molecular weight alkenes for the chemical industry. Although the diesel oil fraction after severe hydrotreatment does meet the current specifications, it is unlikely to do so when Cetane numbers above say 50 are required.

The direct conversion of coal to fuels and chemicals is no longer applied to any significant extent and the direct conversion of methane to chemicals such as methanol, ethylene or benzene has met with little success to date. However, the conversion of coal or of methane to sulphur- free synthesis gas (a mixture of CO, H_2 and CO_2) are well established processes. The synthesis gas (syngas) can be used for the generation of power by combined cycle operation or can be converted to environmentally friendly fuels and chemicals via the Fischer-Tropsch process. For environmental and economic reasons methane is currently preferred to coal for syngas production. The capital cost of the plant is lower and the process is more efficient in the case of methane since about 25% of the carbon is converted to CO_2, whereas for coal it is over 50%.

New reserves of natural gas are regularly being discovered, but often these reserves are situated very far from the markets. The laying of long distance pipelines is too expensive and can also be politically very hazardous if the pipelines have to cross international borders. Liquifaction of the methane and transportation in specialised tankers is uneconomical over long distances. A solution to these problems is converting the methane to oil via the Fischer-Tropsch (FT) process and then transporting the oil in normal tankers.

The thermal efficiencies of producing liquid fuels from methane have been given [2] as follows:

Gasoline from the MTG process	58%
FT diesel fuel	63%
Methanol	72%

This published methanol thermal efficiency appears to be slightly optimistic (see, for example, de Jong [3] who published a figure of 67% for the thermal efficiency for methanol production) perhaps because the utility energy requirements are not fully accounted for. The overall thermal efficiencies for methanol production should range from 66 to 72% and for modern FT synthesis from about 60 to 66%. An inherent advantage for methanol production is the very high selectivity of the reaction for the desired product while for the FT processes typically about 10% of the carbon is converted back to methane, ethane and other products which are not recovered. This means that the syngas generation equipment is nearly 10% larger for the equivalent quantity of useful product.

MTG is Mobil's methanol to gasoline process. This process is no longer considered to be competitive with the direct production of hydrocarbons (predominantly diesel) via FT synthesis. Direct use of methanol appears to be the most thermally efficient option; however, its introduction would require the duplication of the fuel distribution networks as well as the required changes in the design of engines. The fuels from FT synthesis do not of course have these drawbacks. The possible higher efficiency of methanol production is also offset by the fact that the energy content per litre of fuel is higher in the case of diesel.

When it is considered that the main aim is the production of a fuel that is easily transported then FT diesel has a number of advantages:

- High energy density
- Low vapour pressure
- Not soluble in water
- Biodegradable
- Compatible with existing engines and fuel infrastructure
- Used in the most efficient currently available engines
- Fast growing practically infinite existing market

No other option can rival these advantageous characteristics. Di-methyl ether (DME), produced by methanol dehydration, has been proposed as a diesel substitute. DME is compatible with efficient compression ignition engines unlike methanol that would be used in less efficient spark ignition engines. However, with the exception of the use in efficient engines, DME shares none of the other advantages listed above for FT diesel.

The key factor determining the viability of the FT process for the production of fuels and chemicals is the price of normal crude oil with which the FT oils have to compete in the open market. For various reasons, particularly political upheavals in the main crude oil producing countries, the price of crude oil can be very unpredictable. Fig. 1 gives an indication of the variation of the price of crude oil per barrel over the period 1970 to 2002. Note the huge rise in the early 1980's and again the sharp rise during 1999. Subsequently the price declined to the mid-$20 level and then rose again to about $30 per barrel. These ongoing and unpredictable large price fluctuations make the decision to put up a FT plant a very hazardous affair if the cash cost of producing synthetic crude is more than about $10 per barrel. However, projects based on low cost remote natural gas can now survive the low cost periods and become lucrative when the oil price is above $20 per barrel.

During the 1930's in South Africa the mining house Anglovaal extracted oil from shale associated with a coal seam. The viable shale deposit was soon depleted and Anglovaal decided to get involved in producing oil from coal. They obtained a licence to build a Fischer-Tropsch plant in 1935 but World War II intervened. After the war all the oil from coal plants that had been operating in Germany were shut down because they could not compete with crude oil in an open market. There was, however, at that time the perception that due to the limited known deposits of crude oil, together with the increasing demand, the price would increase and hence the FT process could become viable. Because of this perception research in FT and the development of improved FT reactors was continued in the USA and in Germany. South Africa possessed no oil deposits but there were huge deposits of coal which could be mined at low cost because of the thick seams at relatively shallow depths. The South African Coal, Oil and Gas Corporation (SASOL) was established with government funding in 1950. At that time the price of crude oil was about $2 per barrel and that of coal about $0.6 per ton. The FT process was considered to be viable for the following reasons: (a) The low-cost coal deposits were near the main market, Johannesburg. (b) The high cost of transporting fuels to Johannesburg, which is about 600 km from the coast at an altitude of 6000 feet. (c) The perception that the crude oil price was going to rise as the world supplies dwindled.

Construction of the first plant at Sasolburg commenced in 1951 but as fate would have it, the huge deposits of crude oil in the Middle East were discovered, before the plant came on steam in 1955. Consequently the price of crude did not rise but stayed in the vicinity of $2 to $3 a barrel till up to about 1970. Needless to say the plant was not a financial success and amongst the reasons for its survival was the exporting of the highly priced FT linear waxes, the installation of new plants to produce ethylene, butadiene and styrene, and ammonia based fertiliser complex and the sale of 500-BTU gas to nearby domestic and industrial consumers.

In 1960 OPEC was formed and in the early 1970's the price of crude started to increase. At this stage Sasol had resolved all its technical and catalyst problems and it was decided to construct another much larger plant at Secunda. Construction started in 1976. Political upheavals in Iran lead to further large increases in the crude oil price and while the second Sasol plant was still being constructed it was decided to erect a third plant, also at Secunda. The second plant came on stream in 1980 and the third plant in 1983 when the oil prices were at their peak at about $35 a barrel. So, contrary to the first Sasol plant the next two were successful from the start. All three plants were coal based.

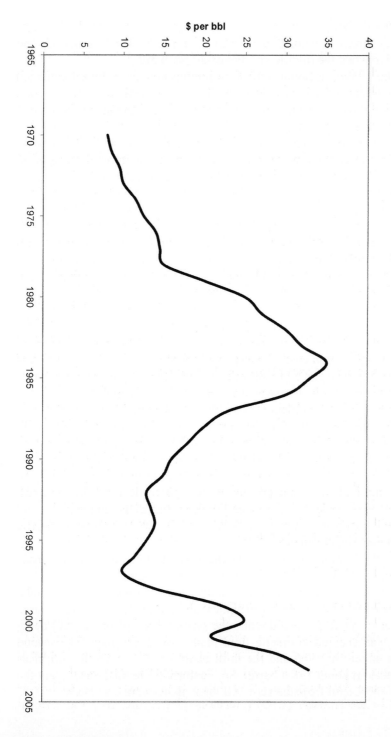

Figure 1 Crude oil price

The price at which fuels derived from FT are sold is the same as those derived from crude oil. There is only a weak link, if any, between the operating cost of a FT plant and the price of crude oil. Because of this if the crude oil price increases the profitability of the FT process rises. If the price of crude drops then of course the reverse holds.

In 1979 Sasol was listed on the Johannesburg stock exchange and within a few years the South African government sold most of its remaining share holding and thus Sasol become a fully public owned company. In 1982 Sasol was listed on the New York NASDAQ stock market and the listing was moved to the New York stock exchange in 2003. Early on it was realised that to improve the profitability of the FT operation it was necessary to not only produce fuels but also to extract the higher value chemicals present in the product streams and also convert other components into higher value chemicals. By 2001, 57% of total sales were from exported chemical products and the Group's overseas operations.

The South African government funded the construction of another separate FT complex, named Mossgas (now PetroSA), at the coastal town Mossel Bay. The plant is based on natural gas obtained from offshore rigs and came on stream in 1992. In the next year the Shell FT plant at Bintulu in Malaysia came on stream. It was also based on offshore methane. At that stage the price of crude oil was in the vicinity of $15 a barrel and hence these plants were not as fortunate as the two large Sasol plants which came on steam when the price of crude was about $35 a barrel. Both these plants are small compared to the Secunda complex and were therefore not able to benefit from economies of scale. The Shell plant initially used four fixed bed reactors to produce 12 500 barrels/day of product. Shell now claims that 8 000 bbls/day can be produced in a single maximum capacity fixed bed reactor. The Mossel Bay plant has three circulating fluidized bed (CFB) Synthol reactors that are each capable of producing 8 000 bbls/day but the plant design called for the ability to produce 80% of the total throughput with two reactors on-line. Thus the nominal FT synthesis capacity is 20 000 bbls/day. Today a single Sasol Advanced Synthol reactor, with a simple fluidized bed design, can produce 20 000 bbls/day.

3. COMMERCIAL FT PLANTS

3.1 Sasol (Sasolburg and Secunda, South Africa)

Up until 2004 the raw syngas for all three Sasol plants was produced from coal. (From 2004 the Sasolburg plant uses methane piped in from Mozambique.) The production of syngas from coal is described in Chapter 4. About 40 million tons of low grade coal is consumed per year by Sasol.

At the present Sasolburg plant (Fig. 2) the pure syngas is fed to the low temperature Fischer-Tropsch (LTFT) reactors. There are five multitubular reactors and one high capacity slurry phase reactor. These reactors are operated at conditions which maximise the production of linear alkanes/alkenes and waxes. Sasol is the world's largest producer of linear paraffinic waxes. These products are hydroprocessed to convert all the alkenes and oxygenated compounds to alkanes. After recovery of the LPG (C_3 and C_4) from the FT tail gas, hydrogen is extracted from part of this gas and used mainly in the synthesis of ammonia. The balance of the tail gas is sold as fuel gas.

At the Secunda plants (Fig. 3) the purified syngas is fed to the high temperature Fischer-Tropsch (HTFT) fluidized bed reactors. These reactors are described in Chapter 2. The operating conditions are aimed at the production of 1-alkenes and gasoline. Iron based catalyst is used (see Chapter 7). The selectivities and the control thereof are discussed in Chapter 3 and Sections 4 and 5 in this Chapter. Since the FT reactions produce a wide spectrum of products ranging from methane to high molecular mass oils the separation and refining of the products is complex. On leaving the FT reactors the products are condensed yielding a gas, oil and an aqueous phase. The aqueous phase contains water soluble alcohols, aldehydes, ketones and acids. The non-acid chemicals are recovered by distillation and the alcohol and ketone cuts are hydroprocessed. N-Crotonaldehyde and n-butanol is produced from acetaldehyde. Acetic and propionic acid is extracted by a liquid-liquid process using a light solvent in packed bed extractors. The effluent water is biologically treated to remove any remaining organic compounds and the purified water is then used as cooling water make-up in the plant.

After the condensation of the oil and water phases the tail gas containing unconverted syngas and gaseous hydrocarbons is cryogenically separated into a hydrogen-rich, a methane-rich and three light hydrocarbon streams. The latter are fractionated and purified to yield high value 1-alkenes. A portion of the H_2 rich gas is fed to a PSA (pressure swing absorber) unit to produce H_2 required in the various hydrotreating operations and the balance is recycled to the FT reactors. The bulk of the CH_4 rich gas is fed to autothermal catalytic reformers and the product syngas is recycled to the FT reactors. The balance of the CH_4 rich gas is sold as fuel gas. Ethylene is piped to polyethylene production units in Sasolburg. Propene is fed to polypropylene, acrylonitrile and acrylic acid plants. Thus all the feed materials required to produce acrylates, namely propylene, acrylic acid, n-butanol and ethanol are all available within Sasol. The remaining C_3 cut together with the C_4 cut is oligomerised over phosphoric acid/kieselguhr catalyst to yield LPG (light petroleum gas), gasoline and diesel fuel. The C_5 to C_8 1-alkenes are extracted purified and sold as co-monomers for the production of polyethylene. The longer chain linear olefins are hydroformulated to produce about 120 000 tons of detergent alcohols per year. The light FT naphtha is hydrotreated and then catalytically reformed (Pt/Al_2O_3 catalyst) to produce gasoline having the required octane number. The diesel fuel cuts, the straight-run FT diesel as well as that produced in the oligomerisation plant, are hydrotreated to convert the alkenes and oxygenated products to alkanes. Currently Sasol produces a total of about 7500×10^3 tons per year of various products and provides about 30% of South Africa's liquid fuel requirement. The quality of FT gasoline and diesel fuels is discussed in Chapter 6.

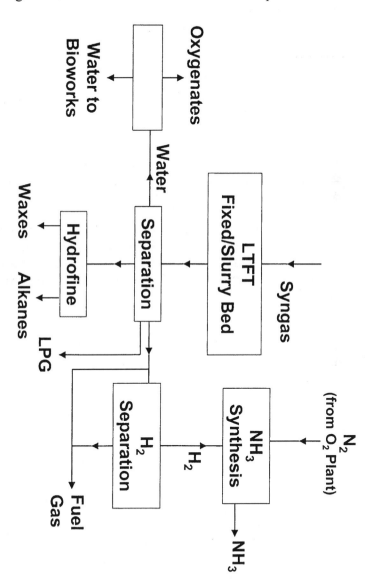

Figure 2 Sasolburg plant block diagram

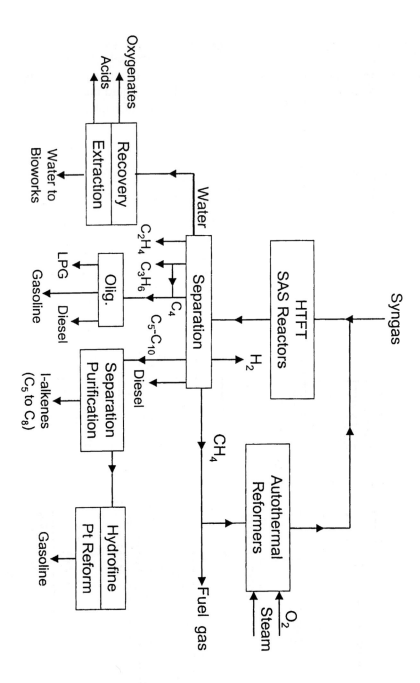

Figure 3 Secunda plant block diagram

3.2 PetroSA/Mossgas (Mossel Bay, South Africa)

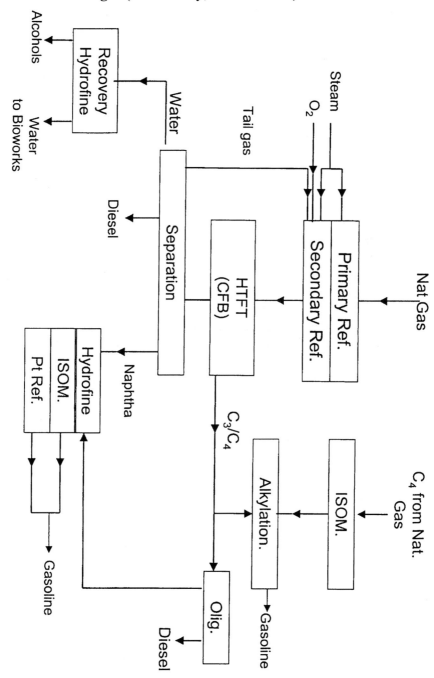

Figure 4 Mossel Bay PetroSA block diagram

Fig. 4 shows the flow diagram of the Mossgas/PetroSA plant. The plant produces mainly gasoline and diesel fuel [4]. Syngas is produced from off-shore natural gas. After condensing out the lighter hydrocarbons present the methane is catalytically reformed in multi-tubular steam reformers followed by autothermal reformers [5, 6]. The synthesis gas is fed to the circulating fluidized bed (CFB) FT reactors. Fused iron based catalyst is used in these reactors.

The alcohols, ketones and aldehydes dissolved in the condensed FT water phase are extracted and the ketones and aldehydes are hydrogenated to alcohols. The FT C_3 and heavier hydrocarbons in the tail gas are recovered in a chilling (refrigeration) unit. The remaining tail gas containing unconverted syngas, CO_2, CH_4, C_2H_4 and C_2H_6 is recycled to the secondary reformers. Butane from the natural gas is isomerized to isobutane (Butamer process) which is then alkylated with FT C_3 and C_4 alkenes to produce high octane gasoline. The balance of the C_3^+ alkenes from the tail gas is oligomerized over a shape selective acid zeolite catalyst to produce diesel fuel and gasoline. All the FT gasoline cuts are hydrotreated, the C_5/C_6 alkanes are isomerized (Penex process) and the C_7^+ alkanes catalytically reformed (Pt) to produce high octane gasoline. All the diesel cuts are also hydrotreated. The total fuel production is approximately 1020×10^3 tons per year.

3.3 Shell SMDS (Bintulu, Malaysia)

The SMDS (Shell Middle Distillate Synthesis) plant is based on off-shore methane [7, 8, 9]. The plant layout is shown in Fig. 5. The syngas is produced by a non-catalytic partial oxidation (POX) process at high pressure and about 1400°C. The reformer carbon efficiency is greater than 95% and the methane slip is about 1%. The H_2/CO ratio of the syngas is approximately 1.7 and since this is below the usage ratio of about 2.1 required for the cobalt based catalyst used in the FT section some additional hydrogen-rich gas is required. The latter gas is provided by catalytic steam reforming of the FT tail gas. This is a low efficiency and high cost operation. This plant also provides the hydrogen for the hydrotreating/hydrocracking operations in the product work-up sections. The FT reactors are multi-tubular reactors similar to Lurgi methanol reactors. Operation is at about 3MPa and 200 to 230°C. Conversions of approximately 80% and C_5^+ selectivities of about 85% are claimed [9]. The objective is high wax production. After condensing out the FT water and liquid oils and waxes the tail gas, presumably containing all the C_1 to C_4 and also some C_5^+ products, is fed to the catalytic steam reformer to generate syngas which is recycled to the FT reactors. There are two modes of product work-up. In the one the waxes are hydrofined to eliminate alkenes and oxygenated compounds and fractionated into different grades of waxes. In the other mode the waxes are hydroisomerized/hydrocracked to yield high quality diesel and kerosene fuels.

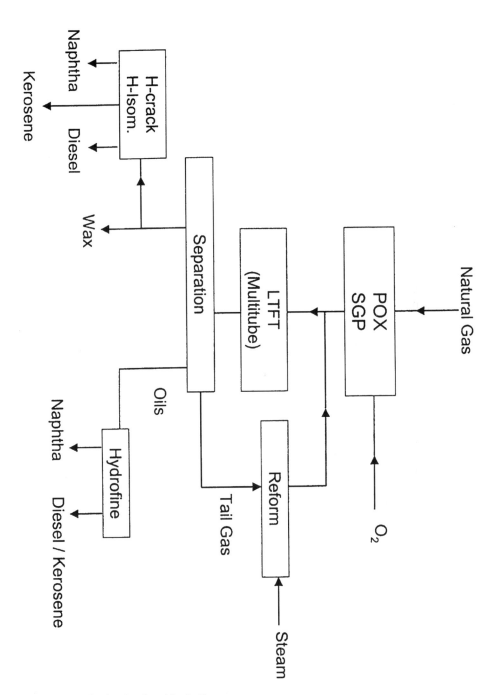

Figure 5 Shell Bintulu plant block diagram

4. INTRODUCTION TO THE GAS LOOP

The application of Fischer-Tropsch Technology involves much more than simply feeding a synthesis gas to a Fischer-Tropsch (FT) reactor. The gases and vapours leaving the FT reactor require further processing and the recycle of some of the gases is usually required for an optimised design. All the commercial applications to date involve the cooling of the FT reactor outlet gases and vapours to condense hydrocarbon products and reaction water which are separated from a tail gas. Some of this tail gas is typically recycled to the FT reactor for reasons explained in Chapter 2. This recycle stream is known as internal recycle. The remaining tail gas undergoes further processing to provide gases that may be recycled to the FT reactor and/or a methane reformer. Typically some of the tail gas is used as fuel and in some cases hydrogen is produced from the tail gas. The gas processing scheme used is known as the gas loop.

An outlet or purge from the gas loop is always necessary to prevent a build up of inert gases (e.g. nitrogen) in the loop. When the outlet is kept to a minimum required for inert gas control, this is known as a closed gas loop design. When large amounts of gas are exported as a fuel product or for processing to make hydrogen, this is referred to as an open gas loop design.

Various gas loop designs are described in the following sections in a way that should facilitate an understanding of the important considerations for the preparation of such designs. It is important to stress at the outset that a computer model is an essential tool to produce an optimum gas loop design.

4.1 Gas loop for HTFT synthesis with a Sasol-Lurgi fixed bed coal gasifier

This type of gas loop has been used in both open and closed loop configurations by Sasol in South Africa. The closed loop configuration is illustrated in Fig. 6 [10].

Given that this gas loop is applied commercially, it is appropriate to discuss some of the design and operational considerations. A volumetric composition for the Pure Gas from the fixed bed gasifiers after clean-up in the Rectisol plant for the early application in Sasolburg was dependant on the operating mode of the gasifier used to produce the desired H_2/CO ratio gas [11]:

Component	High H_2/CO (Volume %)	Medium H_2/CO (Volume %)	Low H_2/CO (Volume %)
N_2	0.8	0.8	0.8
H_2	59.5	55.6	53.6
CO	24.8	28.5	31.5
CO_2	0.9	0.9	0.8
CH_4	14.0	14.2	13.3
H_2/CO	2.4	1.95	1.7

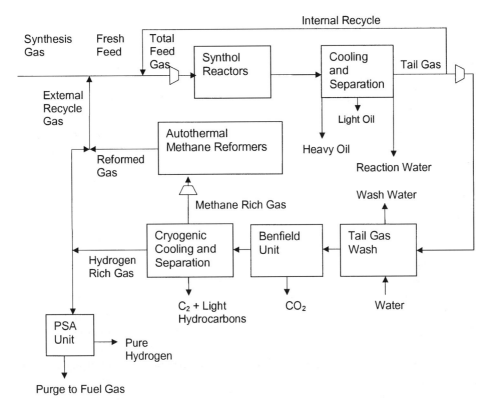

Figure 6 Sasol-Lurgi Gasifier HTFT Gas Loop

Currently the Pure Gas H_2/CO ratio at Secunda is as low as 1.7 (without CO2 recycle) accompanied by a lower methane content of about 12.0%.

Total removal of CO_2 is required in the Benfield unit to avoid freezing up the cryogenic unit (cold box) with solid CO_2.

For the gas loop design shown in Fig. 6, inert gases are purged via the PSA (pressure swing adsorption) unit off gases and some methane rich gas export for use as fuel gas. If there is a bottleneck in any of the units in the gas loop, this is usually handled by sending some methane rich gas to the flare. It is sometimes necessary to flare hydrogen rich gas to avoid an upward spiral of hydrogen in the gas loop.

Note that the gas fed to the Synthol (Fischer-Tropsch) reactors is a mixture of a number of streams i.e. synthesis gas, reformed gas, hydrogen rich gas and internal recycle gas. If the reactants in the total feed gas are not in stoichiometric balance, the reactor performance will suffer due to the build up of the reactant that is in excess. This build up is magnified by the internal recycle. To ensure optimum performance, the Fresh Feed is kept in stoichiometric balance i.e. the synthesis gas molar consumption ratio, $\dfrac{H_2}{2CO+3CO_2}$ is kept at a value of 1.03 (see Chapters 2 and 3).

This is preferably done without resorting to gas flaring which will result in the loss of potential product.

In the case where carbon is in excess, this can be corrected by decreasing the conversion of reactants in the Synthol reactors. This allows more carbon dioxide to proceed to the Benfield unit where it is removed thus correcting the carbon excess. Since the build up of carbon species will cause a decrease in conversion at the Synthol unit this process is self correcting. The undesirable consequence of this self correction is an increase in the value of the carbonisation factor $P_{CO}/P_{H_2}^2$ (See Chapter 3) which causes a more rapid rate of catalyst deterioration. To avoid this, the usual remedy is to decrease the internal recycle rate or to take one of the Synthol reactors out of service. For the CFB type of reactor, a further degree of control was available by decreasing the catalyst flow to the reactor by closing the slide valve. When changing the flow to a Synthol reactor, care must be taken to avoid upsetting the operation of the cyclone separators in these reactors.

In the case where hydrogen is in excess, this can be corrected by increasing the conversion of reactants in the Synthol units if there is scope to achieve this. If not, then the synthesis gas feed rate must be decreased or, more likely, some hydrogen rich gas is purged from the loop. Typically the installed capacity of the highly reliable Sasol Advanced Synthol reactors at Secunda is such that the hydrogen is rarely in excess.

4.1.1 Advanced gas loop design considerations

The following discussion is only of interest to those seeking an advanced insight into design considerations for this particular gas loop.

It is tempting to regard the synthesis gas H_2/CO ratio as a control variable to adjust the composition of the total feed gas. However, studies have shown that it is desirable to operate the Lurgi gasifiers at the highest practical operating temperature, resulting in the lowest, practical H_2/CO ratio. This results in the lowest carbon dioxide content in the gasifier raw gas and the greatest quantity of synthesis gas that can proceed through the Rectisol units. However, to avoid bottlenecks in the downstream gas loop, there must be sufficient Benfield unit CO_2 removal capacity to cope with the removal of excess carbon.

With the autothermal reformer design used at Secunda, the reformed gas is also rich in carbon relative to the stoichiometric requirement. Typically, 80% of the methane converted at the reformers is from methane entering the gas loop in the synthesis gas. (The remainder is produced in the Synthol reactors.) A typical reformed gas composition [12] is as follows:

Component	Volume%
N_2+Ar	5.47
H_2	63.23
CO	16.99
CO_2	13.23
CH_4	1.08

Based on this composition the stoichiometric ratio (Ribblett) is 0.86 which is only slightly higher than the coal derived Pure Gas Ribblett ratio of about 0.82. The hydrogen rich gas from the cryogenic separation (cold box) is required to bring the Fresh Feed Ribblett ratio up to the required value of 1.03.

The use of a lower steam to carbon ratio for the autothermal reformer improves the reformed gas Ribblett ratio in spite of the lower H_2/CO ratio. The following reformed gas composition was reported for steam/carbon (S/C) ratios of 1.5 and 0.6 [13]:

Component	Volume% (0.6 S/C)	Volume% (1.5 S/C)
N_2+Ar	3.8	1.51
H_2	63.66	65.75
CO	25.93	22.67
CO_2	6.21	9.62
CH_4	0.4	0.45
Ribblett ratio	0.90	0.89

The Ribblett ratio increases to 0.9 primarily as a result of the decreased CO_2 content in the reformed gas.

At Secunda, an opportunity has been identified to add gas heated reformers in parallel to the existing autothermal reformers. These gas heated reformers use the heat from the hot reformer outlet gases to drive the endothermic methane-steam reforming reaction. This has the combined effect of increasing the hydrogen content of the reformed gas and the reformer capacity. The extra hydrogen consumes CO_2 in the Synthol reactions before it arrives at the Benfield unit. This means that larger quantities of Pure Gas can be processed in the gas loop.

The design of the utility systems can be improved for future coal conversion plants to allow the production of electrical power for export. The use of a second stage Synthol reactor may be considered because this approach allows the use of lower internal recycle rates and results in a more energy efficient design. A second stage reactor has recently been installed at Secunda.

The Secunda gas loop can be easily modified to export methane rich gas for sale as pipeline gas [10]. The lost reformed gas can be replaced by additional coal derived synthesis gas feed to avoid a decrease in the production rate of liquid hydrocarbon products.

The need to use computer models now requires some discussion. Such models are used to calculate the capacities of the units in the gas loop and the flows and compositions of the streams that connect them. This is an area of specialist proprietary know-how. The reason for this is that the product spectrum produced by the Synthol reactors depends on the composition of the gas fed to these reactors. Specifically, the methane selectivity depends on this composition, so the quantity of reformed gas depends on the composition of the total feed gas while the composition of the total feed gas depends in turn on the quantity of reformed gas. In addition to the requirement to keep the composition of the total feed gas in stoichiometric balance, there is a need to avoid an excessively high rate of carbon formation on the Synthol catalyst. Carbon formation is proportional to $\dfrac{P_{CO}}{P_{H_2}^2}$ so that a hydrogen rich gas is desirable. On the other hand methane production is proportional to $\dfrac{P_{H_2}^{0.5}}{P_{CO} + P_{CO_2}}$ and in this case a hydrogen lean gas is desired. To complicate matters further, the way in which the catalyst responds to the gas composition depends on the level of promoters used in the catalyst and the temperature at which the reactor operates. However, with modern computer software, it is relatively easy to determine the optimum combination once the trends can be accurately predicted.

Having designed and operated laboratory reactors, pilot plants and commercial HTFT reactors over about half a century, Sasol has refined these predictions to be extremely reliable over a very wide range of operating conditions and catalyst formulations.

4.2 Gas loop for HTFT synthesis with high temperature gasification of carbon rich feedstocks

While the Sasol-Lurgi fixed bed gasifier is the technology of choice for the conversion of high ash coal, it may be desirable to use a high temperature entrained flow gasifier for a low ash, carbon rich feedstock. The types of gasifier are discussed in the Chapter 4.

The high temperature gasifiers produce a synthesis gas with a very low ratio of hydrogen to carbon monoxide. This gas is not suitable as direct feedstock to Synthol reactors. The reactant ratio is easily adjusted to the desired stoichiometric ratio by means of a high temperature shift reactor upstream of the carbon dioxide removal unit. The resulting synthesis gas has the desirable characteristic that it contains very little methane. Compared to the design for Sasol-Lurgi gasifiers, it is possible to use only about one fifth of the reformer capacity in the gas loop design and it is possible to obtain about 10% more products from a given size Synthol reactor. The sizes of the tail gas wash, the CO_2 removal unit and the cold box are also significantly decreased. In fact, it may be possible to avoid the need for a CO_2 removal unit.

The basic principles for the gas loop design and operation remain the same except that there is more freedom to adjust the synthesis gas H_2/CO ratio to optimise the gas loop design and operation. This adjustment is achieved by adjusting to the shift reactor operation. The steam required for the shift reactor is conveniently provided by quenching the hot gasifier outlet gases with water.

In spite of these advantages mentioned above the overall energy efficiency is lower when high temperature gasifiers are used rather than the Sasol-Lurgi gasifiers. Also much more oxygen is required from expensive air separation units for the high temperature gasifiers. This is discussed in more detail in Section 6. Only detailed studies for a specific site will reveal the appropriate choice for the gasifier technology.

Fluor Daniel proposed an open loop gas loop design with co-production of electricity using petroleum coke as the feedstock for syngas preparation [14].

While a low ash carbon rich feedstock may lead to a lower cost plant design than a high ash coal feedstock, the plant will still be more expensive than a plant using a natural gas feedstock. This is because of the capital cost required to remove the inevitable production of carbon dioxide for the carbon rich feedstock. Carbon dioxide separation is not essential when using a natural gas feedstock. This is clearly evident from the following overall equations:

$$2\,C + H_2O + {}^1\!/_2O_2 \quad \rightarrow \quad (\text{-}CH_2\text{-}) + CO_2 \tag{1}$$
$$CH_4 + \tfrac{1}{2}O_2 \quad \rightarrow \quad (\text{-}CH_2\text{-}) + H_2O \tag{2}$$

These equations represent stoichiometric limits. The thermodynamic constraints are discussed in Section 6. Clearly the carbon efficiency for a carbon rich feed is much less than the carbon efficiency for methane feed.

A carbon rich feedstock requires much more process steam than a methane rich feedstock which adds further cost to the conversion process. Coal/carbon fed plants require water import whereas methane based plants will export water. Due to the nature of the HTFT process, Eq. (1) requires the addition of water on both sides of this equation which has some negative cost implications but is beneficial from an energy efficiency perspective (See Section 6).

The application of HTFT technology with low H_2/CO ratio synthesis gas is constrained by the need to avoid excessive rates of carbon formation on the catalyst. This rate of carbon formation is proportional to the ratio $\frac{P_{CO}}{P_{H_2}^2}$ in the total feed gas entering the reactor. This requires the feed gas to be rich in hydrogen with the result that water is formed as a product of the FT reaction.

4.3 Gas loop for LTFT synthesis using iron catalysts

Using LTFT iron catalyst technology, carbon rich gases can be tolerated but the reactors and catalysts are more expensive when aiming for the high levels of overall conversion attained with the HTFT technology.

It has been proposed by the US Department of Energy (DOE) supported by contracted studies by MITRE Corporation [15,16] and companies such as Chevron [17], Texaco [18], Air Products [19, 20] and Rentech [21, 22], that LTFT reactor technology can be used together with coal or petroleum coke gasifiers effectively in an open loop design. Chevron first proposed the use of FT synthesis in combination with power generation based on gasification of solid carbonaceous feed. The DOE sponsored studies and the Air Products and Rentech patents followed after the Chevron patent had expired. At the time of the Chevron patent fixed bed gasification was the preferred syngas generation technology. Later proposals tend to prefer Texaco gasification technology.

In 1992 the DOE operated a demonstration scale slurry phase reactor using LTFT iron catalyst [23, 24, 25]. This demonstration had strong industrial backing. In addition to the DOE, this run was sponsored by Air Products, Exxon, Shell, Statoil and UOP. The run was regarded as a success except that the catalyst/wax separation system did not work. The DOE has not yet solved the catalyst/wax separation operation in a way that can be applied on a commercial scale. In 1993, Sasol started commercial wax production with a slurry phase reactor using LTFT iron catalyst. The synthesis gas used with the Sasol reactor is derived from fixed bed gasifiers and has a higher H_2/CO ratio of about 1.9. Sasol solved the catalyst/wax separation problem. The original demonstration of the slurry bubble column design at the Rheinpreussen plant in Germany in the 1950's [26] used a precipitated iron catalyst (similar to that for LTFT applications) at an intermediate temperature and it was claimed that high conversions were obtained when using low H2/CO ratio synthesis gas derived from a high temperature gasifier. Catalyst/wax separation was less important due to the lighter product spectrum.

The claimed slurry phase reactor productivity does not compare favourably with the productivity of the Sasol Advanced Synthol reactor. No commercial closed loop designs have been proposed for slurry phase reactors using iron catalyst with synthesis gas prepared by gasification of carbon rich feedstocks. Closed loop designs for LTFT iron catalyst reactors will therefore not be discussed further. For proposed gas loop designs, based on natural gas feedstock, for this FT technology see US Patent number 5,620,670 by Rentech

Inc. or visit their website at rentech.com. These configurations will not be competitive with applications using supported cobalt catalyst or HTFT technology.

Recently a novel combination of LTFT followed by HTFT technology has been proposed [27]. This will be discussed in more detail when considering the options for the large scale production of commodity chemicals.

4.4 Gas loop for HTFT with natural gas feed

Sasol licensed the Synthol CFB reactor technology to the South African Government to convert natural gas to liquid hydrocarbon fuels at Mossel Bay in South Africa. At the time of writing, this was still the world's largest gas to liquids (GTL) plant. The more recent Sasol Advanced Synthol (SAS) reactor technology would considerably decrease the capital cost and catalyst consumption rate. Also the lower internal recycle used with SAS reactors results in a more energy efficient utility system design. For the same gas loop configuration the process efficiency (of natural gas used to liquid hydrocarbon products produced) would remain at about 63% on a lower heating value (net heating value) basis. The difference would be that the export of large amounts of power (in addition to the same amount of liquid product) would be possible.

Fig. 7 shows a block flow diagram for the Mossel Bay plant gas loop [28]:

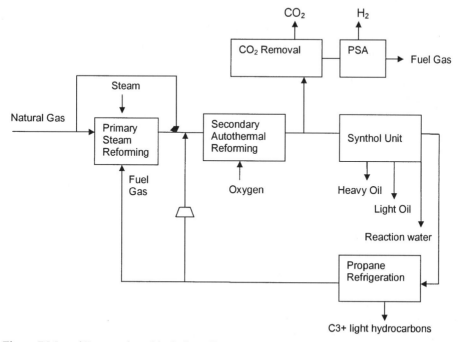

Figure 7 Mossel Bay gas loop block flow diagram

The stoichiometric requirements for HTFT are nearly identical to those for methanol production. For methanol, the term stoichiometric number (SN) is used, where $SN=(H_2-CO)/(CO+CO_2)$. The requirement for methanol is $SN=2.0$ which is the same as Ribblett ratio=1.0.

Any improvement in synthesis gas production technology that applies to synthesis gas for large methanol plants is also likely to apply to synthesis gas for a HTFT plant. One such development is the use of heat exchange reforming or gas heated reforming. This uses the heat available at the outlet of the autothermal reformer to decrease the use of fuel gas as a heat source. However, the capital cost of the gas heated steam reformer is higher than the steam reformer using radiant heat from the burning of fuel gas. It would be desirable from a thermal efficiency perspective to add steam reforming capacity upstream of the autothermal reformer that uses the heat from the autothermal reformer outlet gas and operates at the same low steam to carbon ratio. Steam reforming technology using such a low steam to carbon ratio is not yet commercially available. Unlike methanol plants the Fischer-Tropsch process produces methane and the ability to recycle this methane to the reformer is constrained by the build-up of inerts in the gas loop. For this reason a purge stream is required which can beneficially be used as fuel gas to a primary steam reformer. Thus the choice between a gas heated reformer and a fuel gas fired steam reformer may depend on the level of inerts in the natural gas feed.

Yet another alternative for large scale plants is to only use the highly cost effective autothermal reforming with low steam to carbon ratios below 0.6. This gas will be slightly lean in hydrogen but some of the FT tail gas can be treated to remove carbon dioxide and mixed with the feed gas to provide a gas feed in stoichiometric balance. Given the reduced gas volume to be treated in the carbon dioxide removal unit, this is a low cost gas loop option. C2 hydrocarbons can also be easily recovered downstream of the carbon dioxide removal unit. A slightly sub-stoichiometric feed gas can also be used in an open loop design for the lowest capital cost approach.

The scheme used for the production of hydrogen at Mossel Bay is not always the best approach. Stand alone package hydrogen units are often less expensive than trying to extract hydrogen from synthesis gas. This is partly because extracting hydrogen from the synthesis gas decreases the capacity of the Synthol unit resulting in fewer products to pay for the capital invested. It might be desirable to take a slip stream from the primary reformer product for hydrogen production if this unit is not at its maximum capacity and synthesis gas production is constrained by the maximum size of the air separation unit used to provide oxygen to the secondary reformer. If no primary reformer is used then hydrogen is best recovered from the Synthol tail gas downstream of the tail gas carbon dioxide removal unit described above.

Another factor to consider for hydrogen production from natural gas is the cost of CO_2 removal. When oxygen is used to prepare synthesis gas for conversion to hydrogen, there is a greater quantity of CO_2 to be removed per unit of hydrogen produced, compared to steam reforming. This is why steam reforming is still the technology of choice for the production of relatively small quantities of pure hydrogen. If large quantities of hydrogen are required then the greater benefits obtained by increasing the size of oxygen fired reformers and their higher efficiency may provide savings that exceed the additional cost for CO_2 removal.

There is a reasonably high selectivity to C_2 hydrocarbons with the HTFT process and it may be desirable to recover the C_2's for the production of ethylene. The separation of C_2's requires cryogenic cooling by expansion of the tail gas. To avoid solidification of carbon dioxide, this gas is removed (see the Secunda block diagram) prior to the cryogenic cooling. With modern designs, that do not require the separation of methane and hydrogen, it now seems possible that some C2 hydrocarbons may be recovered without upstream removal of carbon dioxide. All the proven carbon dioxide removal technologies yield carbon dioxide at low pressure, which makes recycle of removed carbon dioxide to the reformer very costly. Also, after expansion to cool the tail gas to low temperatures it becomes costly to recycle the tail gas to the reformer. Recovery of C_2's thus favours an open loop plant design.

For an open loop design it will be desirable to decrease the level of reactants in the tail gas to very low levels. The use of a second reactor stage in series after water knock-out may be justifiable, particularly for very large scale plants with multiple first stage reactors. Compared to the Mossel Bay plant design, the refrigeration unit is replaced by a cryogenic unit and the recycle stream to the autothermal reformers is eliminated. The conversion level in the Synthol unit can be increased to ensure that the Synthol tail gas heating value does not exceed the fuel gas demand for the primary reformer. If no primary reformer is used then a very high conversion is required to avoid excess fuel gas after satisfying the fuel gas demand for the reformer preheater, steam superheating and the other lesser fuel gas consumers.

In the 1950's Carthage Hydrocol Inc. constructed an HTFT plant fed with natural gas in Brownsville, Texas [29]. This plant used partial oxidation reforming to generate the synthesis gas. This would yield a hydrogen deficient synthesis gas which most likely contributed to the problems associated with a high rate of carbon formation on the catalyst used at this plant. No mention is made of recovery of C_3 and lighter products and it appears that an open loop design was used since mention is made of the use of the tail gas as fuel gas in the facility.

4.5 Gas loop for LTFT cobalt catalyst with natural gas feed

Two possible methane reforming technologies are appropriate for large scale gas to liquids (GTL) plants. These are autothermal reforming using a low steam to carbon ratio or partial oxidation reforming [30, 31, 32]. These technologies were discussed in the previous chapter. The use of gas heated reforming [33, 34] together with these technologies is a possible future improvement. In this approach, the hot outlet gases from the oxygen fired primary reformer are used to provide heat to drive the endothermic steam reforming reaction.

A characteristic of the cobalt catalysts used for Fischer-Tropsch synthesis is the negligible activity for the water gas shift reaction. This means that the stoichiometric consumption of reactants depends only on the Fischer-Tropsch reaction. Depending on the methane selectivity, the H_2/CO consumption or usage ratio will be in the range from about 2.06 to 2.16. Using only a natural gas feed, partial oxidation reforming produces a synthesis gas that is below the usage ratio [35] while the use of autothermal reforming or any scheme using gas heated reforming will tend to produce a synthesis gas that has more hydrogen than required for the usage ratio. The synthesis gas may be adjusted to the required H_2/CO ratio by means of a recycle stream and/or gas separation technology [32, 33, 34, 36].

Compared to the HTFT process, the C_2 hydrocarbon selectivity for the LTFT processes is low and it is highly unlikely that recovery of C_2 hydrocarbons will be economically justifiable. This means that the FT tail gas will typically be cooled to a level that allows for the efficient recovery of C_3+ hydrocarbons.

The Fischer-Tropsch reactor will operate most efficiently with a synthesis gas that has a H_2/CO ratio equal to the consumption ratio. However, this does not always lead to the optimum gas loop design. The reason is that the methane selectivity is dependent on the H_2/CO ratio of the gases in the reactor. To quote from an early Shell patent relating to their supported cobalt catalyst [37], "It is observed that when non-converted hydrogen and carbon monoxide is re-circulated over the catalyst bed, it is possible to choose the circumstances in such a way that the catalyst is contacted with a synthesis gas having a substantially lower H2/CO ratio than the feed synthesis gas has. Thus, the selectivity to longer hydrocarbon chains may be improved."

According to Hansen et al [34] there is a substantial improvement in the C5+ selectivity with decreasing feed H_2/CO ratios and this effect becomes more pronounced at higher conversions provided the per pass conversion does not exceed 90% (which is unlikely). A lower C5+ selectivity is inevitably accompanied by higher methane selectivity, which is the main disadvantage. It may be possible to suppress methane formation by using catalytic promoters but in the absence of these promoters, the optimum gas loop design will result in a feed gas to the FT reactor that is below the H_2/CO usage ratio. Gas loop designs are therefore divided into two categories: (a) Those that use synthesis gas at the consumption ratio and (b) those that use synthesis gas below the consumption ratio. In each category, the use of partial oxidation or autothermal reforming is considered below.

(a)   Synthesis gas at the consumption ratio

For POX reforming, a possible gas loop scheme is shown in Fig. 8. It is also possible to operate a steam reformer and a POX reformer in parallel to produce the desired syngas composition. In this case the FT tail gas is simply used as fuel gas.

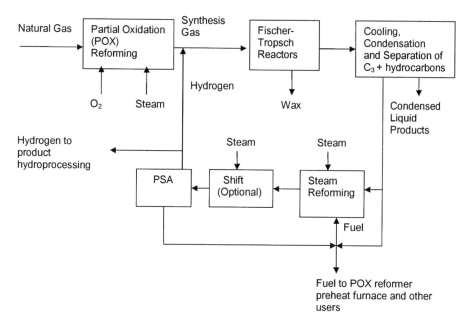

Figure 8 Gas loop for POX reformers

If autothermal reforming is used, the gas loop scheme could be as shown in Fig. 9. Autothermal reforming (ATR) is similar to POX reforming in that there is a burner at the reformer inlet. Even POX reforming requires some steam to be fed to the burner. For the ATR more steam is used and the catalyst at the reactor exit allows a closer approach to equilibrium so the outlet temperature is lower than for POX reforming. For the gas loop scheme in Fig. 9, the flow of external recycle is set to produce synthesis gas with the required H_2/CO ratio. This approach typically results in excess fuel gas availability when compared to the

432

process requirements. This is so even if all the reactants are consumed in the Fischer-Tropsch unit. The external recycle stream consists predominantly of carbon dioxide and methane and it is the carbon dioxide that reacts in the ATR to decrease the synthesis gas H_2/CO ratio. The excess fuel gas situation can be corrected by the addition of steam reforming capacity either by means of a conventional gas fired steam reformer or by using a gas heated (heat exchange) reformer. The former option will obviously increase the fuel gas demand so that the fuel gas balance will be closed with a lower recycle flow than when a gas heated reformer is used. The basic principle is that the hydrogen rich steam reformer can be balanced by the carbon rich external recycle to produce the required Synthesis Gas H_2/CO ratio.

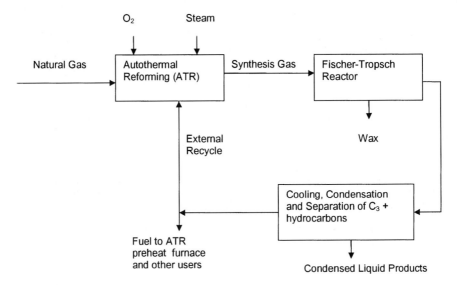

Figure 9 Gas loop for autothermal reformers

The most efficient overall plant design is achieved when the steam reformer capacity is selected to ensure that there is no excess fuel gas available. A small natural gas import may be preferable to ensure that there is never a need to flare excess gas. A variation on this theme has been patented by Statoil in which the FT tail gas is processed in a steam reformer prior to mixing with the synthesis gas from an autothermal reformer [36]. This approach assumes that excess tail gas is available to feed the steam reformer and this is not necessarily the best approach for all cobalt catalysts and FT reactor design concepts. In most cases, internal recycle is best used to avoid the excess tail gas situation.

For both the POX and ATR schemes above, it is highly desirable to avoid excessive amounts of reactants (H_2 + CO) in the FT tail gas. The use of a tandem reactor configuration (2 stages) may be justified to achieve a high overall conversion level. This will depend on the catalyst activity. It is anticipated that catalysts operating at the usage ratio will have a high enough activity that a single reactor stage with tail gas recycle will be able to achieve 96% conversion of the reactants in the synthesis gas.

(b)   Synthesis gas below the usage ratio

If POX reforming is used, a once through design approach is possible with no external recycle. Unfortunately there will be un-reacted carbon monoxide in the tail gas used as fuel gas since hydrogen will be the limiting reactant. An additional disadvantage is that the purge to fuel gas will exceed the fuel gas requirements. Steam reforming capacity can be added to allow the recycle of excess tail gas increasing the synthesis gas H_2/CO ratio above the value of around 1.7 typically obtained with a POX reformer alone [35]. The most efficient way of introducing this steam reforming duty is by means of a gas heated reformer that uses the heat in the POX reformer hot outlet gases to drive the endothermic steam reforming reactions. This technology is not yet commercially proven and is likely to be expensive. The use of simple gas fired tail gas steam reforming is an acceptable alternative.

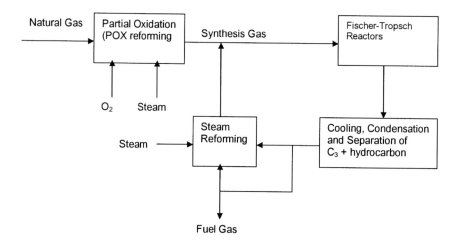

Figure 10 Alternate gas loop for POX reformers

434

Compared to the POX scheme in (a) above, the steam reformer capacity will be larger and there will be a build up of methane and carbon dioxide in the gas feeding the Fischer-Tropsch reactors. If, on the other hand, hydrogen alone is recycled as shown in scheme (a) above, then less hydrogen is used so that the steam reformer will now be smaller but then there is a larger flow of material to the fuel gas system possibly exceeding the fuel gas demand. With the low order kinetics typically obtained with cobalt catalysts the build up of methane and carbon dioxide is not particularly disadvantageous.

Another option for introducing steam reforming capacity is:

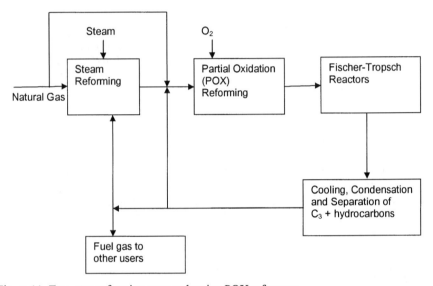

Figure 11 Two-step reforming approach using POX reformers

This second flowsheet is similar to that used for the HTFT process. In this case the bypass stream around the primary reformer will be larger for the LTFT cobalt catalyst process compared to the HTFT process. As a result, the cost of synthesis gas generation should be less for the LTFT process. Further advantages for the LTFT process are the higher selectivity to $C_3 +$ hydrocarbons and the simpler, less costly, processes used to upgrade the primary liquid products. However, the excessive recycle of unreacted reactants and the loss of these reactants to the fuel gas purge may negate these perceived advantages for the LTFT process compared to the HTFT route. The HTFT process is capable of converting carbon dioxide produced in the reforming step into hydrocarbon products via the reverse water gas shift reaction.

In addition to the above two schemes, the steam reformer may be placed in parallel with the POX reformer. This approach is less efficient unless a gas heated reformer is used.

In the case where ATR is used the simple scheme shown in part (a) above and repeated below may be used but now without excess fuel gas.

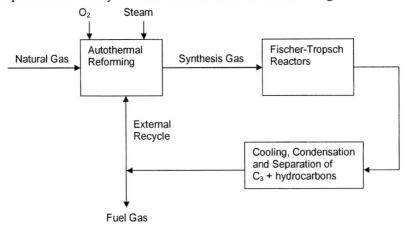

Figure 12 Simple gas loop for autothermal reformers

The disadvantage of this scheme is the loss of reactants via the external recycle and the purge to fuel gas. A 1936 patent [38] describes a method to decrease this loss that could be applied in the following scheme:

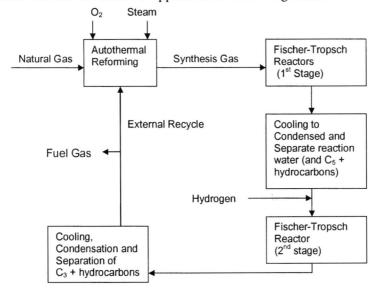

Figure 13 Enhanced gas loop for autothermal reformers

In this scheme, the hydrogen is typically introduced as a hydrogen rich syngas stream. The added hydrogen increases the H_2/CO ratio of the feed to the second stage reactor to close to the usage ratio. The lower H_2/CO ratio gas in the first stage reactor minimises the methane selectivity for the reactants that are converted in this stage. Most of the remaining reactants are consumed in the second stage with a higher selectivity to methane. Thus, to minimize the methane selectivity, the quantity of reactants converted in the first stage should be as high as possible.

Another approach patented by Sasol [39] is to use a single stage FT reactor and employ a separation process such as pressure swing adsorption (PSA) on the FT tail gas to allow H_2 and CO to be recycled directly for the Fischer-Tropsch reactors leaving mainly carbon dioxide, methane and inerts in the fuel gas and external recycle. The drawback of this scheme is the higher compression costs for the external recycle. The amount of external recycle can be decreased by decreasing the steam to carbon ratio used for the autothermal reformer. Both this approach and the tandem reactor approach allow higher plant capacities for a given capacity of air separation unit, typically about 6 to 7% higher.

4.6 Technology targets for gas to liquids (GTL) applications

A paper with the above title was presented at the 2001 Spring National Meeting of the AIChE [33]. The content is reproduced below and placed in context with subsequent developments and other approaches.

Designs and cost estimates have been prepared for feasibility studies regarding the application of the Sasol Slurry Phase Distillate (SPD) Process at Ras Laffan, Qatar and Escravos, Nigeria. These studies show that it is possible to convert natural gas to liquid products with an installed capital cost of less than $ 25 000 per daily barrel. The carbon efficiency for the Escravos process is 75%. The Qatar design has a slightly higher carbon efficiency for approximately the same thermal efficiency. The reason being that the Qatar feed gas is leaner in higher hydrocarbons.

Using the Escravos design as a benchmark, the potential for technology improvements has been examined. For this purpose a stand-alone facility is assumed with no import or export of energy or utilities. The Qatar study, however, has shown that synergies with existing facilities can have a favourable impact on the project economics.

The sources for carbon losses are examined below. The impact on the cost of producing the products is considered and then attention is given to the potential for further capital cost reduction.

4.6.1 Carbon losses

An examination of the reasons for the carbon losses, for the benchmark designs mentioned above, provides useful insights. This analysis is very specific to the starting point benchmark design but the same approach can easily be used for other designs. The carbon losses from the natural gas feed are due to seven clearly identifiable reasons shown in the Fig. 14.

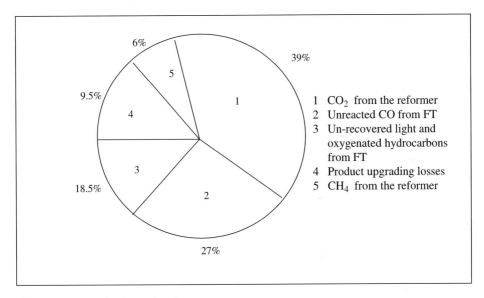

Figure 14 Categories for carbon losses

Carbon losses can be classified into three broad categories. Firstly there are those losses associated with the hydroprocessing of the primary Fischer-Tropsch products together with vent losses resulting from stabilisation of liquid products for storage. These hydrocarbons are mostly recovered for use as fuel gas. There is very little potential to reduce carbon losses from this source although improved hydrocracking catalysts may result in some benefits. This category (item 4 in Fig. 14) accounts for about 9.5% of the carbon losses.

The second category of carbon losses (item 3), light and oxygenated hydrocarbons, can only be significantly decreased by improving the selectivity of the Fischer-Tropsch catalyst. This category includes losses to oxygenated hydrocarbons in the reaction water. Studies and experiments are in progress to investigate the recycle of the oxygenated hydrocarbons to the reformer, which could potentially reduce losses from this source to less than 1%. Most of the oxygenated hydrocarbons are recovered for use as a fuel in the benchmark design. Most of the carbon losses are to light hydrocarbons (mainly methane) that are produced by the Fischer-Tropsch reaction but not recovered as products. These components are all used as fuel gas in the process.

The third category (items 1, 2 and 5) carbon conversion inefficiencies are of most interest. These include losses that are potentially (at least partially) avoidable through process improvements. This includes a 27% carbon loss due to unreacted carbon monoxide which is in the Fischer-Tropsch tail gas stream used as fuel gas. There is also 6% carbon that passes through the reformer as unconverted methane and ends up being used as fuel gas. Most importantly 39% of the carbon loss is due to carbon dioxide generated in the reformer. The total in this category is 72%.

4.6.2 Ways to avoid carbon losses

Focussing now on the third category described in the previous section: Studies have shown that it is not cost effective to aim for a CO conversion in the FT reactor exceeding an upper limit that depends on the syngas composition, even with catalysts that are more active than those currently available. The unreacted carbon monoxide can potentially be recovered from the Fischer-Tropsch tail gas for recycle to the Fischer-Tropsch reactor. In order to achieve this two problems need to be solved:

a) A cost effective separation technology is required to separate the CO and H_2 from the other components in the FT tail gas. Recent developments in PSA technology are showing some promise in this regard.

b) The fuel gas balance must be improved to avoid substituting burning of CO with burning of methane. The fuel gas requirements can be decreased by, for example, more efficient furnace design and energy integration.

Another approach is to adjust the composition of the FT reactor tail gas by adding a hydrogen rich gas and then reacting this hydrogen rich mixture in a second stage FT reactor. This shifts the selectivity in the second stage FT reactor towards lighter products but overall there is an improvement in the carbon efficiency due to the high level of CO consumption. The challenge with this approach is to find the most cost effective method of producing the required hydrogen rich stream.

Potentially the most significant method for improvement of the carbon efficiency is to decrease the amount of CO_2 produced by the reformer. This is discussed in more detail in the following section.

4.6.3 Avoiding reformer CO_2 production

The benchmark GTL design makes use of commercially proven autothermal reforming (ATR) technology with a 0.6 steam to carbon ratio. The process scheme also includes a pre-reformer, which allows the reformer feed gas to be preheated to 650°C. The reformer feed preheating is the main fuel gas consumer followed closely by steam superheating duties.

439

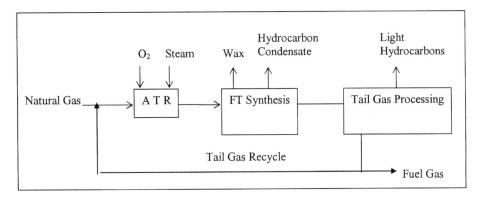

Figure 15 Basic GTL gas loop using ATR reforming

This selected synthesis gas generation process (Fig. 15) requires recycle of Fischer-Tropsch tail gas (containing CO_2) to the reformer feed in order to provide the desired synthesis gas H_2/CO ratio. Studies have shown that the amount of CO_2 generated in the reformer can be decreased by using lower steam to carbon ratios and decreasing the tail gas recycle. This approach has the disadvantages that the reformer methane slip increases and the technology is, as yet, not commercially proven. Using this approach with recycle of excess methane, it is possible to improve the carbon efficiency to about 81%. However, the cost for the methane separation and recycle using available technology cannot be justified. This approach is illustrated in Fig. 16.

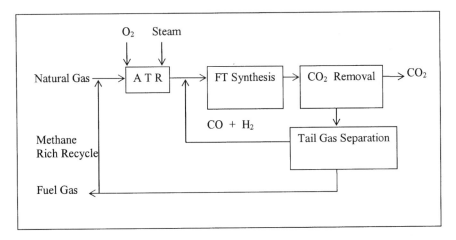

Figure 16 Gas loop with tail gas separation

Note that the above approach also recycles most of the unreacted CO avoiding this carbon loss as discussed previously.

Another approach that can be used to improve the carbon efficiency is to make use of gas heated reforming. The basic concept is to use the heat available in the hot ATR outlet (typically 1050°C) to provide heat for the endothermic steam reforming reaction. This concept requires CO_2 recycle in order to decrease the syngas H_2/CO ratio to meet the requirements for FT synthesis. Gas heated reforming (GHR) can be done in a parallel or a series configuration. The series configuration is illustrated in Fig. 17.

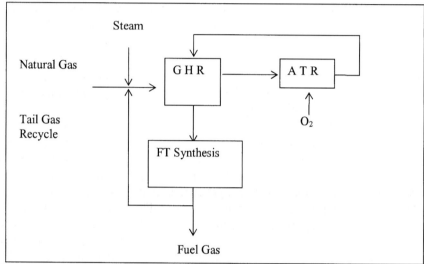

Figure 17 Gas loop configuration using series gas heater reforming

Using this approach, the carbon efficiency can be improved to about 83%. However, exotic materials are required for the gas-heated reformer with a relatively high surface area and the increased reforming capital cost and technology risks are at present difficult to justify. Further combinations of gas heated reforming with ATR together with a tandem FT reactor approach are still being investigated and it is considered possible that a carbon efficiency as high as 85% may be achievable but it is expected that the cost effective carbon efficiency target will most likely be less. The problem is that the carbon efficiency increase is obtained mainly from a decrease in the natural gas feed rate without producing significantly more product. It is difficult to justify the required additional capital expenditure based on saving inexpensive natural gas.

4.6.4 Potential to decrease capital costs

There is no doubt that the usual learning curve will apply and that improved project management and reduced commissioning schedules as well as lower engineering costs will decrease capital costs. Use will be made of the lessons learned during the construction of the first GTL plants. Sasol and Shell have, however, already moved a fair distance along this learning curve with Sasol's existing large scale, coal based, Fischer-Tropsch facilities and the GTL experience from the Mossgas (Synthol) and Bintulu (SMDS) projects.

The greatest potential for cost reduction remains further economy of scale. The benchmark applications for Qatar and Nigeria have capacities of the order of 34 000 bbl/day. They consist of two trains for the oxygen plant, ATR and FT units with single train product upgrading. The maximum single train capacity is set by the maximum practical capacity for the oxygen plant. Today nobody would consider building an oil refinery with such a low capacity. The same drivers apply to GTL plants. Future large GTL facilities will most likely have capacities of the order of 100 000 bbl/day.

Increasing the carbon efficiency generally allows more products to be produced for a given oxygen plant capacity. The FT unit and the ATR unit for the benchmark designs are not at the maximum possible capacities. It is conceivable that a 110 000 bbl/day plant could consist of 6 air separation units (ASU's), 4 ATR units, 4 FT units and 2 product upgrading units.

Based on existing experience with FT reactor diameters as large as 10,7m at the Secunda site, the above capacity would not exceed Sasol's proven experience for the FT units. The oxygen plant and product upgrading capacities would also not exceed proven industrial experience. Further scale-up of the ASU unit capacity is also likely to be achieved. A 6 train design for the major units of ATR, ASU and FT has been found to be a particularly cost effective target.

Further improvements to the Fischer-Tropsch catalyst has the potential to decrease the capital cost by at most $ 1 000 per daily barrel. Operating cost savings could be as much as $ 0.5 per barrel. Considerable progress has been made towards increasing the catalyst life and the catalyst activity as well as decreasing the methane selectivity compared to the assumptions for the benchmark designs.

There is also some scope for cost savings through the development of catalysts specifically tailored for the hydroprocessing of the Fischer-Tropsch primary products. This is likely to result in increased capacities for the product upgrading units.

With all of the above in mind, a capital cost below $ 20 000 per daily barrel is considered to be achievable.

442

4.6.5 Technology targets

Fig. 18 indicates an approximate breakdown of the cost of producing hydrocarbon products with a reasonable return on the capital invested using a realistic price for remote natural gas.

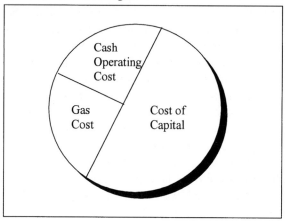

Figure 18 Cost breakdown

This corresponds to a product cost of about $ 20 per barrel. Ignoring any quality premium and taking into account a typical refinery margin, this corresponds to a crude oil price of about $ 16 per barrel.

It can be seen that capital cost is the most important cost component. It is therefore important that attempts to save gas cost by improving the carbon efficiency do not adversely affect the capital cost per unit product. Fig. 19 illustrates the breakdown of capital cost for the most important process units for the benchmark design.

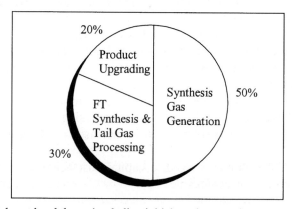

Figure 19 Capital cost breakdown (excluding initial catalyst costs)

Product upgrading includes the cost of producing the hydrogen used. The Tail Gas Processing only accounts for about 4% and it is likely that extra capital invested in this area to improve the carbon efficiency of the process will provide significant economic benefits.

The 50% contribution from synthesis gas generation can be split up as 32% for Air Separation and 18% for Reforming. In addition, a large proportion of the utility capital cost is required to support the air separation unit (ASU). This unit is the main consumer of superheated steam. As mentioned previously, a significant amount of fuel gas is required to superheat the process steam. Thus the ASU is a major consumer of energy in addition to being the most costly unit in the process. Investments in reforming can easily be justified if they result in reduced oxygen consumption. This provides the driver for research into the use of oxygen transfer membranes (OTM) in the reforming step. However, this technology is not yet ready for commercial application.

Considering all the identified options, it is considered feasible for future generation plants to decrease the gas cost by 10% corresponding to a carbon efficiency improvement from 75% to 82.5%.

It is also considered feasible to decrease the benchmark cash operating cost by 10% mainly as a result of decreased catalyst and maintenance costs. (Maintenance costs tend to be proportional to the capital cost.)

As discussed previously, it is expected that a 20% decrease in capital cost is achievable. Summing the above cost element savings, a product cost of $ 17 per barrel, which competes with a crude oil price of about $ 13 per barrel, seems an achievable technology target.

4.7 Expected future plant configurations for natural gas conversion

Without using gas heated reforming, no scheme described above is likely to increase the thermal efficiency much above the 63% obtained with the commercial HTFT process. The choice of technology thus hinges mainly on the following considerations:

(a) overall capital cost
(b) equipment and process reliability
(c) desired products

4.7.1 Overall capital cost

An important factor influencing the overall capital cost is the potential to use the benefits of economics of scale. GTL plants are producing products that compete with products from conventional crude oil refineries and the tendency to build larger refineries to benefit from economy of scale is well known. Due to the capital intensive nature of the GTL business, economy of scale is even more important. On the other hand, the huge amounts of capital required make it difficult to finance a large capacity installation at one time. It is therefore anticipated that plants may be constructed in a phased approach. The minimum capacity will most likely correspond to the capacity of two (current) world scale air separation units producing 34 000 to 36 000 barrels per stream/day of liquid hydrocarbon products. It is conceivable that a single train consisting of one air separation unit, one methane reformer and one FT reactor will, in future, produce this amount of product but this will not be the case for the next plants.

The product rate may be enhanced by the recovery of light hydrocarbons from the natural gas prior to reforming the lean natural gas. Excluding these natural gas liquids and assuming a capital cost of $25 000/barrel, the capital cost is not likely to be less than $900 million. With later additions of further trains with the same design and making use of common utility and other infrastructure is may be possible to approach $21 000/barrel so that the final investment for a 110 000 bbl/day facility would be of the order of $2 300 million. Other variations on the theme may be slightly better, for example starting with three air separation units yielding 51 000 to 54 000 barrels per stream day and doubling capacity for each expansion. Shell has proposed a base plant capacity of about 75 000 barrels per day, most likely using four air separation units.

Air separation unit (ASU) vendors have indicated that within the next decade single unit capacities at sea level of 7000 t/d of oxygen can be considered. This would provide enough oxygen to produce 120 000 bbl/day of product using only three units. With the planned further capacity increases for all the other units it is quite conceivable that a plant with this capacity could cost $20 000/barrel corresponding to a total investment cost of $2 400 million.

Most of the technologies used will scale with the typical power factor of about 0.7, the exception being the steam reformers which scale more linearly with capacity. However, once the maximum train capacities have been reached the main units will tend to scale in a similar fashion to the steam methane reforming units. However, there are still benefits from scaling up common utilities and infrastructure and decreased engineering and construction costs.

Foster Wheeler published a discussion of the capital costs for a generic GTL plant design as follows [40]:

"Capital costs for this GTL process have been calculated using a combination of automated techniques and manual take-offs. In addition, all significant equipment has been cost estimated via actual quotations, and benchmarked against in-house databases from recent projects. This has resulted in an estimated accuracy of ±15%. Obviously any capital cost estimate must take into account the considerable local conditions, as well as the local customised plant design. However for the purposes of this generic plant, the following have been used:

- Generic Middle East location;
- Cost factor equivalent to U.S. Gulf Coast;
- Stick-built construction; Process configuration as per generic plant;
- Coastal location, with access to port facilities.

These result in the following costs:

Equipment, materials and labour	$14 000/bbl
Total constructed cost (excl. Owner's cost)	$17 000/bbl

To put these figures into perspective by way of example, the total plant cost including owner's costs, start-up and commissioning costs and contingency is less than $25 000/bbl.

We can split these costs by process unit and obtain the breakdown shown in Fig 20:

For the purposes of Fig. 20, the air separation unit has been included with the natural gas reformer, as both of these units are required to produce synthesis gas. Together, these comprise the largest portion of the plant capital cost, with the cost of the actual FT Synthesis unit being one of the smaller portions. Utility costs include the entire boiler feed water and start-up systems, but not in the in-process steam generators.

This distribution shows that the costs within the GTL process are widely spread across the entire plant; with no one single entity dominating. This is also a sign of the integrated nature and complexity of the plant, with systems and services widely distributed.

The final critical piece of the project puzzle, and also something which affects the capital costs is the schedule.

A variety of scenarios has been investigated with respect of the plant construction philosophy. While having direct impacts on the capital cost in terms of materials, it also has a significant impact on the project schedule. A concerted effort has been made to shorten the project schedule, while at the same time balancing the need to properly optimise the plant for a particular location and also ensure that the cost does not increase as a result. The generic plant is key to the overall schedule of the optimised project as it acts as a launch pad for early engineering activities.

This results in a competitive and challenging 30-month EPC schedule from award to ready for start-up (RFSU). Achievement of such a schedule requires an intimate knowledge of the characteristics of the facilities involved, as well as their interactions. For a remote site, successful commissioning and start-up requires critical support utility systems to be ready several months prior to the end, to allow for the remainder of the plant to be commissioned; critical for a plant which during normal operations relies on self utility generation."

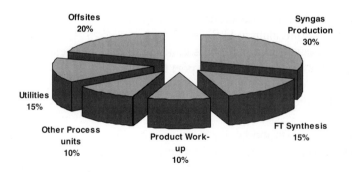

Figure 20 Capital cost breakdown showing the significance of utilities and offsites

There is no published study with similar detail for the HTFT process but some qualitative comparisons can be made. The air separation unit costs will be less since the reformer is more efficient without tail gas recycle so the cost of producing syngas will be less. FT unit capacities and costs should be similar. The HTFT product recovery and upgrading is more complex and costly. On the other hand, the utility systems will be less costly for the HTFT process due to the higher quality steam produced by the FT reactor cooling system. Sasol has published figures [1] which indicate that, at a nominal 50 000 bbl/day plant capacity, the HTFT process producing both fuels and chemicals from natural gas can be competitive with a 30 000 bbl/day LTFT plant producing only diesel, naphtha and LPG. This comparison did not include the benefits that can be obtained by using more advanced reforming technology with HTFT. It is now understood that the capital cost/daily barrel of primary product for both HTFT and LTFT routes are very similar for the current status for these technologies.

Another interesting comparison is with the production of methanol from synthesis gas for which the process of producing synthesis gas is virtually identical. The HTFT technology has more opportunity to reap benefits from economies of scale due to the much larger product market. The HTFT synthesis uses lower operating pressures, lower internal recycle flows, larger capacity reactors and a less expensive catalyst than methanol synthesis. As a result the HTFT synthesis step costs less. The product upgrading is somewhat more expensive but this step is the least costly of the three steps required to convert the feedstock to final products. The latest HTFT technology is capable of generating 70 bar steam in the synthesis reactor cooling system. This is a much better quality steam than that obtained by cooling the methanol synthesis due to the higher reaction temperature used in the Synthol reactor. The HTFT process has the potential to produce significant quantities of power for export at a low cost. This power can be beneficially used in a complex that is designed to co-produce a wide variety of high value chemicals.

One large scale use for methanol is the production of MTBE which is used as a high octane gasoline blending component. Methanol has also been converted into gasoline using a ZSM-5 zeolite catalyst at a commercial facility in New Zealand with technology developed by Mobil (now ExxonMobil) [32, 33, 34]. The similarities with HTFT technology in terms of feedstock and product markets and overall energy efficiencies are clearly evident.

MTBE production has been the preferred fuel product because of its premium value in the market. More recently, the production of MTBE is being phased out in the USA due to environmental concerns with groundwater contamination. The HTFT technology would rely on economies of scale, as well as the value obtained from chemical co-products and electrical power, to compete. The advantages of shared synthesis gas production facilities, utilities and infrastructure for production of methanol together with HTFT products are clear. A detailed study for an actual proposed site would be necessary to confirm that an economically viable application is possible for the latest HTFT technology.

4.7.2 Equipment and process reliability

Whenever a plant consists of a number of units in series, the overall availability becomes a multiple of the availability of each individual unit unless there is intermediate storage. So for a GTL plant with a 98% availability for each of the air separation, methane reforming and FT units, will only have a total plant availability of $0.98 \times 0.98 \times 0.98 = 0.94$ i.e. 94%. If the availability of complex integrated utility systems are also considered it can be appreciated that care must be taken to ensure that each unit is highly reliable and where possible a failure of one unit or one piece of equipment should, as far as possible, not cause the shut-down of the entire plant. Storage facilities for liquid oxygen may be considered to decrease the impact of a temporary air separation unit failure on the total on-stream factor. Actual operating experience with commercial FT plants is invaluable in making design choices that give the best balance between plant cost and plant availability.

Many reasonable decisions can be made, by any reputable engineering contractor with good data on the reliability of the individual equipment items, if the plant owner is aware of the plant availability pitfall and instructs the engineering contractors accordingly. Since the ultimate reliability of the completed plant is not the prime concern of the engineering contractors, input from owners and licensors in the design phase is very important.

Sasol has had good experience with both HTFT and LTFT processes and there is little, if any, difference in the equipment and process reliability.

4.7.3 Desired products

The desired products will be the prime consideration when selecting the FT technology for a GTL project. If the main desired product is diesel and there are opportunities to sell naphtha (for cracking to light olefins), lubricant base oils and detergent industry feedstock, then the technology of choice is the LTFT technology. On the other hand, where the preferred fuel product is gasoline, with some diesel and LPG and there are also opportunities for the sale of ethylene, propylene, alpha-olefins and oxygenated hydrocarbon products, then the HTFT technology is a better choice. HTFT technology can also be configured to produce diesel and naphtha by using oligomerisation technology to convert naphtha boiling range olefins to the diesel boiling range.

Currently, the consumption of gasoline far exceeds the diesel consumption in the USA, the world's largest consumer of motor fuels. The LTFT technology is best suited to serve the European and Asian markets while the HTFT technology would be better suited to the current USA market. It remains to be seen whether the USA will eventually follow the trends in other parts of the globe. It is likely that there will be opportunities for both types of products in the foreseeable future, as replacement for fuels currently derived from crude oil.

5. OPTIONS FOR THE PRODUCTION OF HIGH VALUE HYDROCARBONS

In reviews of the low temperature Fischer-Tropsch (LTFT) technology [41] and high temperature Fischer-Tropsch (HTFT) technology [1] the opportunities to produce high value hydrocarbons in large quantities has already been mentioned. Considering first the Sasol SPD™ process, the following opportunities exist:

- Convert the naphtha to lower olefins by steam cracking.
- Convert the olefins and paraffins to alkylates.
- Recover high value co-monomers (such as C6 and C8) from the product slate
- Recover linear paraffins for sale as solvents or for conversion to olefins which can be converted into a variety of final products (such as detergent alkylates or detergent alcohols).
- Produce lubricant oil base stocks from the primary wax product.

The Sasol SPD™ Process makes use of a supported cobalt catalyst in a slurry phase FT reactor. Another approach is to use a precipitated iron catalyst (also in a slurry phase reactor). The same product options are available for the LTFT iron catalyst. In general, Fe based LTFT technology would offer a higher concentration of olefins in the product than Co based LTFT. This approach has the disadvantage that it is more costly to achieve high reactant conversions. Another disadvantage, for natural gas applications, is that carbon dioxide is produced with the LTFT iron catalyst although this feature may be advantageous for low H_2/CO syngas derived from coal.

The well proven HTFT process can also be considered. Traditionally this process has been applied for the maximum production of gasoline but more and more use is being made of the highly olefinic products to produce higher value chemicals. A fortuitous advantage of extracting the linear $1-C_5$ to $1-C_8$ alkenes (i.e. alpha olefins) from the FT product streams is that this improves the quality of the gasoline pool since the octane numbers of these linear products are low. The HTFT products contain more branched hydrocarbons than the highly linear products produced by LTFT synthesis. This together with some aromatic components makes it possible to consider gasoline production from HTFT products but considerable refining of the primary product is still required. For a new application, the HTFT process can be reconfigured for maximum olefin production and using this approach the process can be simplified to decrease capital cost. Thus higher value products (mainly propylene, ethylene and alpha olefins) can be produced using less capital.

HTFT, on the other hand, is also well suited for the production of diesel by oligomerisation of the highly olefinic C_5 to C_9 cut together with the hydroprocessing of the straight run diesel (See Chapter 6). The remaining C_5 to C_{10} material can be hydroprocessed for sale as petrochemical naphtha. With this approach both the HTFT and LTFT processes produce similar proportions of naphtha product (less than 30%). Again extraction of alpha olefins is desirable since these products have a higher value than diesel and the capital cost for the oligomerisation unit is decreased.

With little or no purification of the raw FT feedstock the longer chain 1-alkenes can be selectively converted to primary alcohols by hydroformylation using homogenous cobalt or rhodium complexes as catalysts. Since the product alcohols have much higher boiling points than those of the hydrocarbon cuts fed, the separation of the alcohols by distillation is a simple operation. The C_{12} to C_{14} alcohols produced are used in the production of biodegradable detergents. The HTFT process also directly produces significant quantities of short chain oxygenated hydrocarbon products e.g. ethanol, methanol and acetone.

5.1 Lower olefins from naphtha

Even for very large GTL plants, there is not sufficient naphtha produced to justify the construction of a world scale naphtha cracker. As a result it is the operators of existing naphtha crackers that will benefit from the high yields to lower olefins using Sasol SPD™ naphtha. Tests have been done to quantify the increased olefin yield [42]. The Sasol SPD™ naphtha used in these tests was a fully hydrogenated product. Its performance was compared with typical yields reported for a light petroleum naphtha.

The ethylene yield attainable with the Sasol SPD™ naphtha is about 8% higher than that of the light petroleum naphtha at relatively low severity. If the comparison were based on the total olefins yield, the SPD™ naphtha would yield about 8% more olefins at the same severity, and up to 12% more when compared with the highest yields observed during the pilot plant tests.

Using industry correlation techniques to estimate the best yields that could be achieved at optimum commercial conditions, it was predicted that for operation aimed at maximum ethylene production, the yield to ethylene would increase by about 10% relative to a light petroleum naphtha.

5.2 Detergent alkylates and paraffin processing

The olefin content and quality of the Sasol SPD process straight run product cut is greater than that obtained with typical paraffin de-hydrogenation processes used to convert paraffins to olefins that are alkylated to linear alkylbenzene (LAB). This makes the production of LAB an obvious candidate for higher value products.

Two options can be considered for the production of LAB. Firstly only the straight run olefins could be alkylated and the remaining stream could be re-combined with the residual hydrocarbon condensate (after the latter has been hydrotreated). These hydrocarbons could then either be fractionated into paraffin products or cut to make naphtha, kerosene and diesel. The second option is to include a process that de-hydrogenates the product paraffins (after the first pass olefin conversion) into olefins for recycle to the alkylation unit. This approach will increase the amount of LAB product and may result in improved economy of scale. The material balances presented in Fig.21 assume that the first processing option is used.

Iron based LTFT product can also be considered for the production of LAB. In this case advantage can be taken from the higher olefin content of the feed from the iron based process.

5.3 Lubricant base oils

It has been reported that a substantial part of the primary products from a GTL process may be converted to lubricant base oils [43]. Although the world lubricant base oil market is not growing rapidly, it is large, estimated at some 40 million tons per annum [9], and a substantial fraction of Fischer-Tropsch base oils will be able to substitute product in the premium market. This is then also an obvious candidate for higher value products.

5.4 Light olefins

For the production of light olefins, the HTFT process offers many advantages. Using HTFT technology (with optimized catalysts) for the production of light olefins the amount of olefin product can be further enhanced by cracking (via fluidized catalytic cracking i.e. FCC) unwanted longer chain molecules to higher value shorter chain molecules. It is believed that an overall mass selectivity of 15-30% towards propylene is viable via such a two step approach. Overall selectivity towards ethylene can be as high as 10-20%. An illustrative mass balance, as shown in Fig. 21, shows an overall selectivity of about 19% towards propylene. This assumption is based on current FT catalyst performance and the results from first round cracking tests. Similarly it assumed that 10% of the hydrocarbons can be recovered as ethylene and 13% as various butenes. About 23% will be diesel boiling range material while about 13% occurs in the gasoline range (excluding C4 components that may eventually end up in the gasoline). About 12% of the product is water soluble oxygenated hydrocarbons. The remainder of the product from the converted synthesis gas consists of ethane, LPG and fuel gas (that may be partly recycled to the syngas generator). The current proposal differs from previous publications, for example Ref. 1, where the possibility of using FCC-type technology to enhance lower olefin yields was not considered. It is anticipated that the modified HTFT technology could in future provide the most cost effective route to short chain olefin production.

5.5 Combined LTFT and HTFT

Besides the higher olefin content already mentioned, the LTFT promoted iron catalyst also has a higher wax yield than cobalt catalysts. As a result of these two differences, the LTFT iron catalyst may offer advantages for the production of various chemicals, i.e. the production of lubricant base oils, and as mentioned earlier, production of olefins and olefin derived products such as detergent alkylates and alcohols. As mentioned previously, several disadvantages are also associated with the use of iron based LTFT. A key disadvantage associated with the LTFT iron catalyst use is that it is expensive to reach high syngas conversions. However, for conversions up to about 50% the capital cost expressed per unit product is similar to that for cobalt catalysts.

An attractive option to consider is a two stage process that uses LTFT iron catalyst for the first stage with HTFT technology as a second stage. This combination will allow for the production of lubricant base oils, olefin derived products such as detergent alkylates and alcohols, light olefins and oxygenated hydrocarbons (such as ethanol and methanol). All these products will be produced with a competitive advantage when compared to conventional processes.

The lubricant base oils, detergent alkylates, co-monomers and other olefin derived products from the first stage (that is the iron based LTFT plant) are high value products. The second stage (HTFT) provides a lower capital cost method to consume the residual reactants; consumes carbon dioxide that is generated in the upstream processes; and provides a way to produce additional commodity chemicals. The result is a high yield of total hydrocarbon products per unit of syngas and a high average product price.

5.6 The simplified HTFT chemical hub concept

The HTFT schemes shown in Figure 21 may be seen as rather complex with difficulties associated with marketing a wide variety of large volume chemical products. In addition the FCC unit is a high cost refinery unit and significant quantities of feed are lost to coke and fuel gas. An alternative is to reconfigure the HTFT refinery to only produce diesel, naphtha, propylene and LPG. This is done by reacting ethylene with butylene to produce propylene by metathesis. The oxygenated products can be dehydrated to produce more olefins. The naphtha cut olefins are oligomerised to be combined with the straight run diesel cut. With this approach a viable footprint plant can be built. Later the alpha olefins and other chemicals can be extracted incrementally from the refinery feed for sale as higher value products when market conditions are favorable. With this approach the initial product would consist of about 40% diesel, 30% naphtha, 26% propylene and 4% LPG on a mass basis. A large proportion of the diesel product can later be diverted to commodity chemical products with this approach.

5.7 Conclusions for the production of high value hydrocarbons

The use of cobalt based FT catalysts is the preferred approach for the production of fuels from natural gas. If it is desired to target large scale commodity chemicals as the main products with fuel as a by-product then iron catalysts could be applied using well proven technology. The HTFT process might become the preferred technology to produce olefins. If it is desired to produce lubricant base oils and LAB then these products can be produced as co-products from the Sasol SPD™ process or by using LTFT iron catalyst in plants integrated with other processes.

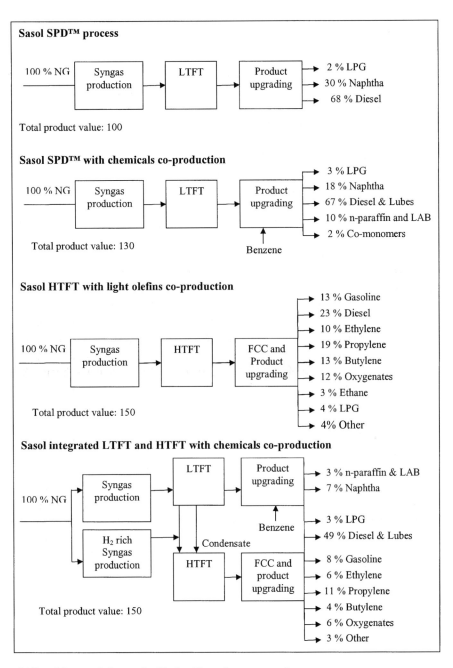

Figure 21 Possible mass balances for Fischer-Tropsch process options

6. COAL CONVERSION USING FT TECHNOLOGY

There are several different technology options for the conversion of coal to synthesis gas suitable for FT synthesis. These will all involve some form of gasification technology as discussed in Chapter 4. It is appropriate to first obtain some insight into the mass and energy balance considerations that apply to the coal conversion route. Coal is a carbon rich feedstock that varies in composition, e.g. the carbon to hydrogen ratio, but the idealized case of a pure carbon feed can be used as a departure point to gain some fundamental process insights. Thus the conversion of carbon to olefinic hydrocarbons can be examined using mass balance and thermodynamic considerations as follows:

Mass balance reaction for ideal steam gasification:
- Gasification: $1.5C + 2 H_2O \rightarrow CO + 2H_2 + 0.5 CO_2$
- FT: $CO + 2H_2 \rightarrow (-CH_2-) + H_2O$

(Note the carbon efficiency is 67%)

Thermodynamics for this ideal gasification are as follows:
- $\Delta H = 158$ kJ (endothermic reaction)
- $\Delta G = 104$ kJ (reaction is thermodynamically impossible)

Modify gasification to make it thermodynamically possible by adding oxygen:
- Gasification: $1.77C + 2 H_2O + 0.27 O_2 \rightarrow CO + 2H_2 + 0.77 CO_2$

(Carbon efficiency decreases to 56%)

Thermodynamics of 'possible' gasification:
- $\Delta G = -2$ kJ (reaction thermodynamically possible i.e. $\Delta G < 0$)
- $\Delta H = 51$ kJ (reaction still endothermic)

Now in reality the most effective method to provide heat for the endothermic reaction is by combusting additional carbon with oxygen in-situ. This leads to the following mass balances:

$C + 4/3 H_2O \rightarrow 2/3 CO + 4/3 H_2 + 1/3 CO_2 \rightarrow 2/3 (-CH_2-) + 2/3 H_2O + 1/3 CO_2$

$\uparrow \qquad\qquad\qquad\qquad\qquad\qquad \downarrow$

$1/3 C + 1/3 O_2 \rightarrow 1/3 CO_2 \qquad\qquad 120$ kJ

―――――――――――――――――――――――――――――――――

$4/3 C + 4/3 H_2O + 1/3 O_2 \rightarrow 2/3 CO + 4/3 H_2 + 2/3 CO_2$
$\rightarrow 2/3 (-CH_2-) + 2/3 H_2O + 2/3 CO_2 + 120$ kJ

Using two carbon atoms as the starting basis the above mass balance becomes:

$2C + 2 H_2O + 1/2 O_2 \rightarrow CO + 2 H_2 + CO_2 \rightarrow (-CH_2-) + H_2O + CO_2 + 180$ kJ

(Carbon efficiency decreases to 50% and product/feed thermal efficiency is about 75%)

The above equation can also be modifid by decreasing the steam fed to the gasifier to give:

$2C + H_2O + 1/2 O_2 \rightarrow 2 CO + H_2 \rightarrow (-CH_2-) + CO_2 + 240$ kJ

However, in this modified reaction, which produces a lower H_2/CO ratio syngas, the exothermic water gas shift reaction ($H_2O + CO \rightarrow H_2 + CO_2$) is eliminated from the gasification step and this requires the burning of additional carbon to form CO_2 to close the energy balance for this step. This decreases the carbon efficiency by about 5%, i.e. from about 50% to about 47%. Whether this is good or bad depends on the value that can be obtained from the extra heat produced by the water gas shift reaction that now takes place in the FT step (240 kJ versus 180 kJ).

The approximate heat content assumptions used are as follows:

C   = 400 kJ/mol
CH_4 = 800 kJ/mol
$-CH_2-$ = 600 kJ/mol
CO   = 300 kJ/mol
H_2 = 240 kJ/mol

The previous two mass balance equations essentially illustrate the difference between the Sasol-Lurgi gasifiers that produce a syngas with an H_2/CO ratio close to 2.0 compared to the high temperature entrained flow gasifiers that produce a syngas with this ratio close to 0.5. While there is a small difference in the theoretical process carbon efficiencies there are also other differences when the practical utility balances are taken into account. For example, energy is required to produce steam from water and to separate oxygen from air. Also from a capital cost point of view the Sasol-Lurgi approach requires more steam generation capacity for gasification and more carbon dioxide removal from the synthesis gas and more treatment facilities for FT reaction water.

This is obviously a highly idealized analysis, for example coal is perhaps better represented as CH rather than C. However, increasing the hydrogen content of the fuel does not significantly affect the process thermal efficiency (in terms of energy in the products as a proportion of energy in the feed) although it does improve the carbon efficiency. This can be appreciated by analyzing methane as the feed material:

$$CH_4 + 1/3 \ CO_2 + 2/3 \ H_2O \rightarrow 4/3 \ CO + 8/3 \ H_2 \rightarrow 4/3 \ (-CH_2-) + 4/3 \ H_2O$$

$$\uparrow \qquad\qquad\qquad\qquad\qquad \downarrow$$

$$1/3 \ CH_4 + 2/3 \ O_2 \rightarrow 1/3 \ CO_2 + 2/3 \ H_2O \qquad 240 \ kJ$$

$$\overline{4/3 \ CH_4 + 2/3 \ O_2 \rightarrow 4/3 \ CO + 8/3 \ H_2 \rightarrow 4/3 \ (-CH_2-) + 4/3 \ H_2O + 240 \ kJ}$$

Using one methane molecule as the starting basis gives:

$$CH_4 + 1/2 \ O_2 \rightarrow CO + 2 \ H_2 \rightarrow (-CH_2-) + H_2O + 180 \ kJ$$

The carbon efficiency is 100% but the thermal efficiency, ignoring the contribution from the FT reaction heat, is still 75%.

6.1 Combined production of hydrocarbon liquid and electrical power

The following is an extract of a paper by Steynberg and Nel published in FUEL [44]:

Any meaningful utilization of the vast coal resources in places like China, India, Australia, South Africa and the USA will involve conversion into some other form of energy. Coal, as a solid, has a high energy density and is therefore reasonably convenient as a heating fuel. The problem is that the associated pollutants negate any advantages compared to other cleaner burning fuels. Coal is rather converted into other cleaner forms of fuel such as liquid hydrocarbons, synthetic natural gas (SNG) and electric power. Initially these conversion plants simply concentrated the pollutants in one large scale conversion site while enabling the end user of energy to experience a cleaner fuel. With time, more and more success has been achieved in cleaning up the emissions from large scale coal conversion facilities.

One fact that cannot be avoided is that every ton of carbon in mined coal will sooner or later end up as 3.67 tons of carbon dioxide in the atmosphere (i.e. 44/12 being the molecular/atomic mass ratio of carbon dioxide and carbon). Although some attention has been paid to the possibility of carbon dioxide sequestration this is not currently, and may never be a viable option. Certainly the cost of sequestration seems likely to favor the use of other fossil fuel alternatives such as crude oil and natural gas while these are still readily available.

The increased production of carbon dioxide per unit of useful energy for coal relative to other fuels is inevitable [45]. Assuming this is acceptable, it becomes important to ensure that, when coal is used, it is used as efficiently as possible. Co-production of liquid hydrocarbons and electricity from coal using combined cycle power generation facilities has been proposed as a clean and efficient approach [15-22].

The coal fired power stations in South Africa and elsewhere produces some of the world's lowest cost electrical power but this large scale coal combustion has an environmental penalty. An alternative approach is to gasify the coal in order to produce a low heating value synthesis gas which may be cleaned prior to combustion. This gas is then combusted in a combined cycle power plant using both steam and gas turbines to produce electricity. There are alternative uses for this synthesis gas that may offer opportunities that are both economically more attractive and result in more efficient use of the coal. Where a region's economy is dependant on coal utilization, there is a strong case to be made to switch to more efficient and less polluting coal conversion technologies as these become available.

Synthesis gas can also be made from natural gas and the following comparison has been published previously by Shell [45]:

Syngas Manufacture + Fischer-Tropsch	Thermal Efficiency *	Relative Capital Cost
Coal: $2(-CH-) + O_2 \rightarrow 2CO + H_2 \rightarrow (-CH_2-) + CO_2$	60%	200
Natural Gas: $CH_4 + \frac{1}{2}O_2 \rightarrow CO + 2H_2 \rightarrow (-CH_2-) + H_2O$	80%	100

* Theoretical Maximum

This appears to paint a bleak picture for coal utilization but it ignores two factors. Firstly, the high energy density of coal allows the feed to be delivered to the conversion plant at a lower cost. Secondly, the lower energy efficiency ignores the opportunity to convert some of the byproduct heat into electrical power. (Also the analysis in Section 6 shows that by using an idealized approach that considers steam gasification reactions then the thermal efficiency, in terms of (product heating value)/(feed heating value), for both coal and natural gas gives the same theoretical maximum of about 75%.)

A potential use for some of the byproduct heat that is typically available in the hot outlet gases from high temperature gasifiers is to generate steam for the water gas shift reaction to be conducted downstream from the gasifier to modify the above coal conversion reaction as follows:

$$H_2O + 2CO + H_2 \rightarrow CO + 2H_2 + CO_2 \rightarrow (-CH_2-) + H_2O$$
$$\downarrow$$
$$\text{Removed}$$

This then becomes equivalent to the natural gas route for the synthesis gas conversion step but twice as much oxygen has been used to prepare the synthesis gas and the cost of generating steam and removing carbon dioxide provides a further cost penalty. Clearly, from the analysis in Section 6, where much less oxygen is required it is preferable for the steam reactions to occur inside the gasifier. In real life applications the ratio of oxygen consumption for high temperature gasifiers versus oxygen fired reformers is about 1.7. On the other hand the stem fed Sasol-Lurgi gasifiers use about one quarter of the amount of oxygen used by the high temperature gasifiers.

The cost penalty for generating steam is somewhat negated by the fact that steam generation is required anyway to remove the heat generated by the Fischer-Tropsch reaction and to cool the hot synthesis gas exiting the gasifier. Synthesis gas cleanup to remove acid gases is also required anyway for use in gas turbines for production of electricity.

The concept of producing Fischer-Tropsch liquid hydrocarbons and electrical power from coal derived synthesis gas has been around for quite a long time. A 1978 patent from Chevron pointed out the advantages of using this concept to cope effectively with power demand variations [17]. The US Department of Energy (DOE) supported by contracted studies by MITRE Corporation [15, 16] and companies such as Texaco [18], Air Products [19, 20] and Rentech [21, 22] showed that LTFT reactor technology can be used together with coal or petroleum coke gasifiers for the co-production of hydrocarbon liquids and electrical power. Fluor Daniel [46] has proposed the use of HTFT for this purpose.

The LTFT process [47] and the HTFT process [1] have been described in a 1999 publication on the topic of recent advances in Fischer-Tropsch synthesis. It is possible that the combined production of hydrocarbon liquids and electricity from coal can compete with natural gas conversion to only hydrocarbon liquids. Often there is no demand for electricity at the site where remote natural gas is located but there is demand near potential coal conversion sites. This coal conversion approach has been studied for two Fischer-Tropsch conversion options i.e. LTFT and HTFT, both using iron based catalysts.

6.2 Cases Studied

For both cases the use of a Texaco Gasifier was assumed to make a synthesis gas with the following composition (mole%):

H2	29.26
CO	37.36
CO2	13.30
CH4	0.16
H2O	19.43
Inert	0.49

This composition was obtained from a Texaco brochure for a coal with the following composition:

Dry Analysis (wt%)
Carbon   78.08
Hydrogen   5.26
Nitrogen   0.85
Sulfur   0.47
Oxygen   8.23
Ash   7.11

For the LTFT case this gas is fed to a two stage reactor system to form liquid hydrocarbon products. 83.3% of the reactants are converted using this approach. The tail gas is sent to a gas turbine to produce power. Other sources of power are the steam generated in the process of cooling the synthesis gas from the gasifier and in removing the heat generated in the Fischer-Tropsch reactors.

For the HTFT case, a similar quantity of hydrocarbon product can be produced with a single stage using only two reactors of the maximum size used by Sasol at its facility in Secunda, South Africa. 92.2% of the reactants are converted using this design approach. For this case the synthesis gas is subjected to a water gas shift reaction prior to acid gas removal in order to increase the hydrogen to carbon monoxide ratio in the synthesis gas. This was also found to significantly increase the utility requirements for the acid gas removal step.

The mass and energy balances for the two cases are shown in Tables 3 and 4. These are not necessarily fully optimized but are considered realistic to provide a fair basis for comparison.

The plant capacity and hence the feed rate is determined by the amount of synthesis gas required to fully utilize two full scale Sasol reactors for the HTFT case. For the LTFT case the two first stage reactors are followed by a further two second stage reactors with all four reactors at the maximum capacity offered by Sasol. This is likely to be the minimum feasible scale of operation. More likely, plants will be constructed with double this capacity.

The reported thermal efficiency is based on the lower heating value of the coal feed relative to the lower heating value of the primary liquid hydrocarbon products. These products are relatively pure so that only mild processing is required to convert them to final products. Further product loss in the upgrading process should not exceed 3% of the primary product heat value. A small amount of hydrogen will be required for the upgrading processes. This can be produced using a slipstream of the synthesis gas from the gasifiers. This is excluded from the above energy efficiency calculations. The product upgrading units have a negligible impact on the utility systems. The utility balances are shown in Tables 5 and 6.

6.3 Discussion of the results

Natural gas conversion processes can be expected to have thermal efficiencies in the range of 60 to 66%. For the current state of the technology, cost effective facilities will have thermal efficiencies closer to 60%. Coal conversion is clearly less efficient. It seems about 50% efficiency is achievable. The cost of feedstock per unit of energy must therefore be less in order to compensate for this efficiency deficit. However, this is not unrealistic. Existing coal conversion plants operating with thermal efficiencies closer to 40% are cash positive.

Compared to natural gas conversion the capital cost of coal conversion, based on similar quantities of energy product, are higher for a number of reasons. Firstly there is the need for acid gas removal from the synthesis gas. This step is not required for a sweet natural gas feed. The capital cost of synthesis gas generation is also significantly higher mainly due to the higher oxygen requirement (by a factor of nearly 1.7). The Fischer-Tropsch conversion step is marginally more expensive for the HTFT technology and much more costly for the LTFT route. Another factor that negatively impacts the coal conversion plant is that it is more utility intensive than a natural gas conversion plant. This is an inevitable consequence of the lower thermal efficiency.

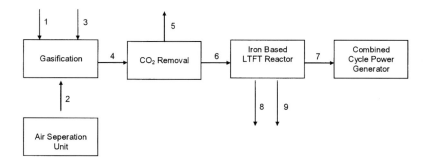

Table 3

Mass and energy balance for the LTFT case

Feed Stream Summery

Stream Number	Stream Name	Flowrate kg/h	LHV kJ/kg	LHV MW
1	Coal	928063	30506	7864
2	Oxygen	951597	0	0
3	Water	654404	0	0

Gas Stream Summery

Stream Number	Stream Name	Flowrate knm3/h	LHV kJ/nm3	LHV MW
4	Raw Gas	2243	9843	6133
5	CO2	364	0	0
6	Syngas	1879	11781	6148
7	Tailgas	695	7831	1513

Liquid Stream Summery

Stream Number	Stream Name	Bbl/day	LHV MJ/bbl	LHV MV	Ton/day (metric)	MJ/kg
8	Condensate	18248	4635	978	2090	40.5
9	Wax	35844	5809	2410	4740	43.8
	Total	54082		3388		

Thermal Efficiency

Stream	LHV MW	%
Coal	7884	100.0
Condensate	979	12.4
Wax	2410	30.6
Power	566	7.2
Overall	3955	50.3

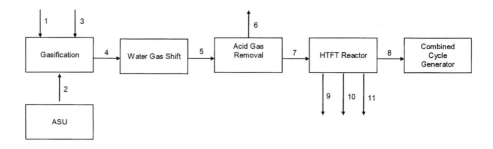

Table 4

Mass and energy balance for the HTFT case

Feed Stream Summery

Stream Number	Stream Name	Flowrate (kg/h)	LHV (kJ/kg)	LHV (MW)
1	Coal	928063	30506	7864
2	Oxygen	951597	0	0
3	Water	654404	0	0

Gas Stream Summery

Stream Number	Stream Name	Flowrate (knm3/h)	LHV (kJ/nm3)	LHV (MW)
4	Raw Gas	2243	9843	6133
5	Shift Gas	2880	7923	5898
6	CO2	799	0	0
7	Syngas	1881	11341	5925
8	Tailgas	360	14553	1453

Liquid Stream Summery

Stream Number	Stream Name	Bbl/day	LHV (MJ/bbl)	LHV MV	Ton/day (metric)	MJ/kg
9	Decant Oil	1799	6099	127	249	44.1
10	Stablised Light Oil	34070	4859	1918	3899	42.5
11	Condensate	25002	4112	1190	2334	44.1
	Total	60871		3233		

Thermal Efficiency

Stream	LHV MW	%
Coal to Gasifier	7884	100.0
Liquid Products	3233	41.1
Power Export	393	5.02
Overall	3626	46.1

Table 5
LTFT Utility Balances

Steam System

HP Saturated Steam Sources	Quantity (tonnes/h)
Gasifier outlet waste heat boiler	2449
Gas turbine heat recovery	814
HP superheated steam users	
Air separation unit turbine drives	2158
Other process compressor turbines	58
Electrical power generating turbine	1046
Saturated MP steam from FT reactor cooling	1607
Superheated MP steam electrical power generation	1051
Other MP Steam Users Including CO_2 removal	556

Electrical Power Sources

Source	Quantity (MW)
HP steam turbine	235
MP steam turbine	123
Gas turbine	309
Less: Internal use	100
Net power export	**566**

Tail Gas Fuel Users

Users	Quantity (MW)
HP steam superheater	438
MP steam superheater	16
Gas turbine	1075

Table 6
HTFT Utility Balances

Steam System

HP Saturated steam sources	Quantity (tonnes/h)
Gasifier outlet waste heat boiler	2449
FT reactor cooling	1119
Shift reactor waste heat boiler	373
Gas turbine heat recovery	787

HP Superheated Steam Users (63 bar 411°C)	Quantity (tonnes/h)
Air separation unit turbine drives	2157
Other process compressor turbines	53
Electrical power generating turbine	868

HP Saturated Steam Users (70 bar)	Quantity (tonnes/h)
Shift steam	343
Shift preheater	33
CO_2 removal	1061
Other process heating	241

Electrical Power Sources

Source	Quantity (MW)
HP steam turbine	243
Gas turbine	269
Less: Internal use	100
Net power export	**412**

HTFT Tail Gas Fuel Users

Users	Quantity (MW)
HP steam superheater	515
Gas turbine	938

Considering the comparison between the HTFT and LTFT options, it is clear that the LTFT route is more efficient when using an entrained flow gasifier. The main reason for the lower efficiency of the HTFT option is that the acid gas removal step becomes more utility intensive. As mentioned previously though, the capital cost for the LTFT Fischer-Tropsch section is higher. Only a detailed study beyond the scope of this paper will be able to determine whether the higher efficiency of the LTFT route can compensate for the higher capital cost. In addition the analysis in Section 6.4 shows that the combination of HTFT synthesis with Sasol-Lurgi gasifiers is an attractive option from a thermal efficiency perspective. The preferred solution may be different for different potential application sites and may be influenced by whether gasoline or diesel is the desired primary fuel product.

The economic success of the coal conversion plant will inevitably depend on the price received for the products. It seems unlikely that the price for liquid hydrocarbons will be sufficient to provide a suitable return on the capital invested while the energy resources of crude oil and natural gas are still readily available. However, if some assistance is provided for the initial capital costs, coal conversion to liquid hydrocarbons together with electrical power is an efficient option relative to other coal conversion options. The coal conversion efficiencies are higher when liquid hydrocarbons are produced than for facilities producing only electricity.

6.4 Coal conversion using Sasol-Lurgi gasifiers

A study was done using similar assumptions for a coal based facility using Sasol-Lurgi gasifiers instead of entrained flow gasifiers. The previous study was also updated with more optimistic assumptions for the efficiency of the gas turbines used in these schemes to generate power, in line with efficiencies published by SRI.

From the information provided below it can be seen that when LTFT synthesis is used entrained flow gasifiers result in the highest thermal efficiency, of about 52.5% compared to 49.6% when Sasol-Lurgi gasifiers are used. This assumes the use of radiant cooling in a waste heat boiler. If the Texaco quench design is used then the efficiency drops to 47.9%. The cost of syngas preparation using Sasol-Lurgi gasifiers is expected to be less than the Texaco design using radiant cooling and the Sasol-Lurgi gasifiers have a better record in terms of plant availability.

The overall most efficient approach is the use of Sasol-Lurgi gasifiers combined with HTFT synthesis providing an overall efficiency of about 53.5%. This is also the best proven approach since the syngas preparation and FT units are well proven at the Sasol plants in Secunda. The main reason for the higher efficiency is the ability of the Synthol reactors to convert both CO and CO_2 to hydrocarbon products at very high conversions in a single stage reactor. The lower temperature counter-current flow arrangement for the Sasol-Lurgi gasifier is efficient in spite of the fact that large volumes of steam are reacted to produce a relatively high H_2/CO ratio syngas. No downstream shift reactor is required to adjust the gas composition to be suitable for HTFT synthesis.

466

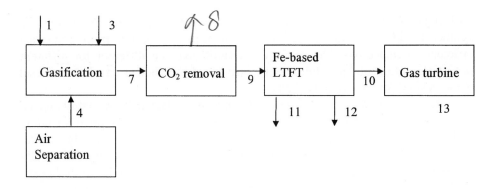

Table 7

Stream summary for Sasol-Lurgi gasification with LTFT synthesis

Stream no	Stream name	Flowrate
1	Coal	1348 t/h
2	HP steam	1725 t/h
3	Boiler feed water	263 t/h
4	Oxygen	374 kNm^3/h
5	Ash	175 t/h
6	Tar	46 t/h
7	Dry raw gas	2718 kNm^3/h
8	CO_2	750 kNm^3/h
9	Syngas	1968 kNm^3/h
10	Tailgas	699 kNm^3/h
11	Condensate	28896 bbl/day
12	Wax	24954 bbl/day
13	Steam from gas turbine heat recovery section	1649 t/h

FT product summary

FT products	53850 bbl/day
FT products + electricity	71913 bble/day

Thermal efficiency

Stream	LHV (MW)	Thermal %
Coal	9428	100.00
Tar	538	5.71
Condensate	1433	15.20
Wax	1652	17.52
Power	1052	11.2
Overall	**4137**	**49.6**

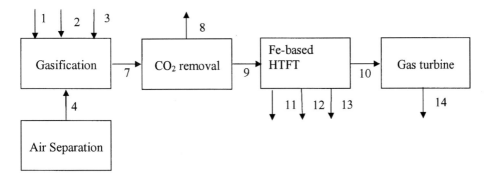

Table 8

Stream summary for Sasol-Lurgi gasification with HTFT synthesis

Stream no	Stream name	Flowrate
1	Coal	1348 t/h
2	HP steam	1725 t/h
3	Boiler feed water	263 t/h
4	Oxygen	374 kNm^3/h
5	Ash	175 t/h
6	Tar	46 t/h
7	Dry raw gas	2718 kNm^3/h
8	CO_2	750 kNm^3/h
9	Syngas	1968 kNm^3/h
10	Tailgas	226 kNm^3/h
11	Condensate	32359 bbl/day
12	Light oil	30468 bbl/day
13	Heavy oil	2419 bbl/day
14	Steam from gas turbine heat recovery section	1539 t/h

= 22,421 t/d

FT product summary

FT products	65246 bbl/day
FT products + electricity	78789 bble/day

= 9321 t/d

Thermal efficiency

Stream	LHV (MW)	Thermal %
Coal	9428	100.00
Tar	538	5.71
FT products	3744	39.71
Power	766	8.12
Overall	**5047**	**53.54**

Table 9

Texaco LTFT case with higher turbine efficiency, radiant cooler and scale-up for same coal flow as Lurgi:

Stream summary

Stream no	Stream name	Flowrate
1	Coal	1348 t/h
2	Oxygen	1406 kNm^3/h
3	Water	951 t/h
4	Slag	175 t/h
5	Dry raw gas	3258 kNm^3/h
6	CO$_2$	529 kNm^3/h
7	Syngas	2729 kNm^3/h
8	Tailgas	1010 kNm^3/h
9	Condensate	26505 bbl/day
10	Wax	52063 bbl/day
11	Steam from gas turbine heat recovery section	1182 t/h

FT product summary

FT products	78568 bbl/day
FT products + electricity	96940 bble/day

Thermal efficiency

Stream	LHV (MW)	Thermal %
Coal	11423	100
Condensate	1422	12
Wax	3500	31
Power	1070	9
Overall	**5992**	**52.46**

When the radiant boiler is excluded, that is, it is assumed that the raw gas will be **quenched** to cool it, the results change in the following way:

FT product summary

FT products	78568 bbl/day
FT products + electricity	87977 bble/day

Thermal efficiency

Stream	LHV (MW)	Thermal %
Coal	11423	100
Condensate	1422	12
Wax	3500	31
Power	548	5
Overall	**5470**	**47.89**

There are many unwanted byproducts from the Sasol-Lurgi gasification process such as bitumen and the aqueous gas liquor processed in the Phenosolvan unit. In addition, these gasifiers cannot handle fine coal. The flowsheet can be significantly simplified if the waste water from gasification together with the other unwanted gasification byproducts and fine coal are fed to Texaco type gasifiers operating in parallel with the Sasol-Lurgi gasifiers.

6.5 Comparison with the use of remote natural gas

The import of liquefied natural gas (LNG) is becoming increasingly popular as a fuel for the production of electricity. It is unlikely that LNG will be supplied at a price less than $3/GJ due to the capital intensive nature of this technology. Currently the price is higher than this. Taking into account a 60% conversion efficiency to electricity for natural gas and a 40% conversion efficiency for coal, then the coal price that leads to the same feedstock cost is $61/metric ton (using a heating value of 30506 kJ/kg). This is much higher than the prevailing price of coal. This demonstrates that coal is the preferred feedstock for the generation of electricity when viewed from a feedstock cost perspective. However, coal conversion plants cost more to build and have a larger environmental impact.

When it comes to the production of hydrocarbon liquids from coal or natural gas, a price range of between $0.5 and $1.0 can be expected for the lean natural gas at remote locations. Assuming a conversion efficiency of 60% for natural gas and 40% for coal, this translates to a coal price between $10 and $20/metric ton for the coal to have the same feedstock cost as natural gas. This is in line with coal prices obtainable in China but lower than the prevailing prices in the USA.

By combining the production of hydrocarbons with the production of electricity, the coal conversion efficiency may be enhanced from about 40% to about 50%. This increases the feedstock cost parity figures mentioned above by 25%. This is likely to make the production of hydrocarbons from coal viable with low cost coal from an operating cost perspective. The associated increase in efficiency for electricity generation provides a further competitive advantage for coal compared to imported LNG from a feedstock cost perspective.

According to Shell [1] the coal conversion capital cost may be double that for natural gas conversion to liquid hydrocarbons. By way of comparison, the following units proposed in Ref. 3 are eliminated if the Fischer-Tropsch tail gas is fed to a gas turbine to produce electricity:

- carbon dioxide removal,
- hydrocarbon recovery,
- hydrogen recovery and
- autothermal reforming.

With this approach it is still possible to exceed 80% conversion of reactants in the clean synthesis gas entering the Fischer-Tropsch unit. This is the basis for the plant designs presented above. In spite of the potential to decrease capital costs by process simplification, the operating cost advantage for the coal route can never be sufficient to offset the higher capital cost. However, governments may consider assistance with the initial high capital costs for coal conversion plants in order to secure a strategic longer term operating cost advantage as well as ensuring increased self sufficiency and price stability for their energy needs.

Production of liquid hydrocarbons from coal as described above has two further advantages. Firstly the market is huge and the facilities can take advantage of economics of scale for very large facilities producing hydrocarbon fuels together with electricity. However, the huge capital investment involved is another reason why state assistance will probably be necessary to fund such a coal conversion facility. The other benefit is that the final liquid hydrocarbon fuels will be ultra clean and will not contain sulphur or nitrogen. This will lead to local environmental benefits at the places where these fuels are used.

6.6 Conclusions for coal conversion using FT technology

Certainly coal conversion to hydrocarbon liquids and electricity is economically attractive from an income/cash cost viewpoint. However large capital investments are required to reap the benefits from economics of scale and the return on capital invested is not particularly attractive. It is therefore expected that these coal conversion plants will only become a reality if there is some form of state assistance with the initial capital investment, such as low interest loans. This would be motivated by the independence from imported energy and the more efficient and cleaner utilization of the local energy resource. The Chinese government has accepted this approach and China is likely to be the pioneer for this technology option. Places that have abundant coal are likely to use this resource. It is the duty of technology providers to ensure that technology is made available to allow the coal to be used as efficiently as possible.

7. ENVIRONMENTAL ASPECTS

In the production of syngas all sulphur and nitrogen compounds are removed from the syngas upstream of the FT reactors. The gasoline and diesel fuels produced in the FT process are therefore S and N free. Hence the exhaust gases from combustion engines are free of SO_2 and the NO_X levels are lower because the FT fuels themselves make no contribution to the formation of NO_X gases in the engine cylinders. The straight run HTFT gasoline contains about 5% aromatics and the benzene content is very low. This is an advantage over crude oil derived gasoline from an environmental point of view. The alkene content of HTFT gasoline, however, is about 70%, which is far too high. To make the gasoline environmentally acceptable the alkene level has to be decreased and since this lowers the octane rating this has to be raised by other means (see Chapter 6).

The efficiency of a diesel engine is about 44% as against only about 24% for a gasoline engine. From an environmental point of view the use of diesel engine vehicles should therefore be encouraged. The diesel fuel produced in the LTFT process is of excellent quality (see Chapter 6). The aromatic content may be less than 1% for the straight run diesel and diesel from a wax hydrocracker. Diesel obtained from the oligomerization of olefins in the naphtha cut using a zeolite catalyst will typically contain about 10% aromatics. Currently an aromatic content of about 30% is typical for diesel fuels produced from crude oil. Tests have shown that the exhaust emission levels of hydrocarbons, CO, NO_X and particulate matter were respectively 56%, 33%, 28% and 21% lower than crude oil derived diesel fuels [48]. In addition it has been shown that FT diesel fuel is readily biodegradable.

When GTL plants using FT technology are configured in the maximum diesel mode then there are significant overall environmental benefits compared to the crude oil route to transportation fuels. In addition, the environmental impact for the GTL approach is mostly confined to the plant location that will typically be at a remote sparsely populated site. At the places where the fuels are used there are tremendous environmental benefits that result from the ultra pure fuels. The main environmental concern at the plant site is the CO_2 emissions which is not detrimental to the local communities but remains a concern as a 'greenhouse gas'. Water treatment technology is available to avoid any water effluent issues. Ongoing development work to improve environmental performance is therefore aimed primarily at improving the efficiency of the gas conversion to liquid hydrocarbons in order to further reduce CO_2 discharge to the atmosphere. Studies to quantify the environmental benefits obtained using GTL technology tend to be complex and precise quantification is the topic of ongoing debate.

Fairly in depth life cycle assessment (LCA) studies have been done by PricewaterhouseCoopers for both the Shell (SMDS) and Sasol (SPD) GTL processes. An LCA compares the potential environmental impact of different technologies, taking into consideration all inputs (energy and resources) and outputs (emissions) over the complete lifecycle of each technology. It involves building a model to simulate both the technical details of the process, and the market dynamics with regard to the feedstock and products within the time frame of the study. Economic considerations are however excluded from an LCA.

In building the LCA model certain choices need to be made regarding the scope, methodology, input data and underlying assumptions – all of which could impact on the results. The Sasol-Chevron and Shell studies completed by PricewaterhouseCoopers were based on the system boundary expansion (SBE) method as recommended by the ISO 14040 standard. Critical external reviewers were used such as Dr Michael Wang (Argonne Laboratories, California).

The SBE methodology implies that **all** products related to a specific technology are taken into account. However, due to the complexity of a refinery system with its multiple products, most other transportation fuel LCA's executed to date used the allocation method, where only the impact of the diesel and/or gasoline streams was taken into account. The allocation methodology (comparing single products rather than technologies) has the advantage of being simpler and easier to understand, but it clearly does not give credit to technologies which has a more desirable overall product slate from both an environmental and market demand perspective.

Because the product slate of a GTL plant is different from that of a refinery (less products), a straight comparison is not possible. Since all of the refinery products do have a function and a market value, the GTL system needs to be expanded to include the environmental burden of supplying these additional functions, to enable a fair comparison. The GTL system has to meet the same **functions** as the refinery system, but not necessarily through the same **products**. The actual product selection is dictated by market forecast and industry trends.

In terms of greenhouse gas emissions some of the findings were:
- The higher efficiency of the refinery during the production phase is balanced by the better performance of the GTL system during the product use phase, and that as a result the two systems are equivalent from a greenhouse gas perspective.
- While the product use phase dominates the greenhouse gas results, the main difference between the two systems in this regard lies in the electricity generation function (comparing heavy fuel oil with natural gas), and not in the transportation fuel function.
- By narrowing the study boundaries to include only transportation fuels (diesel and gasoline) the balance is shifted in favour of the refinery products - this is the result that is typically reported by most well-to-wheels studies comparing GTL and refineries.

The study methodology and assumptions have an impact on the results. Although the SBE method, which gives credit to GTL for having a cleaner overall product slate, has only been applied by potential GTL producers to date (Sasol Chevron, Shell, ConocoPhilips), General Motors did acknowledge in their 2002 wells-to-wheels analysis for Europe that GTL would have a greenhouse gas benefit if the less desirable refinery products were included in the comparison.

Regulated Emissions for regional and local air quality:

In this category the GTL system provides substantial improvement over the refinery system as indicated below; however, the actual impact or relevancy of these emissions is strongly dependent on the location of the emission (unlike CO_2 which has an equivalent global warming impact, regardless of where the emission takes place). For this reason, the transport-related emissions typical to an urban environment were shown separately as well.

Table 10
Regulated emissions: GTL as % of refinery reference (100)

	SOx	NOx	VOC (potential for smog formation)	CO	Particulates
Lifecycle Results	39	84	20	18	58
Transport –related Results: Heavy Duty	-	72	42	67	78
Transport –related Results: Light Duty	-	100	60	74	67

Calculated on a g/km travelled basis

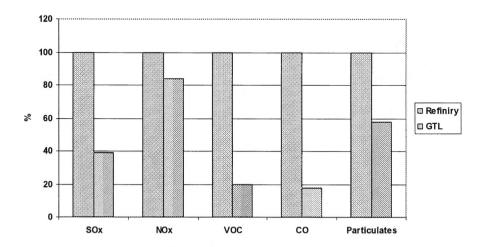

Figure 22 Regulated emissions: Lifecycle basis

474

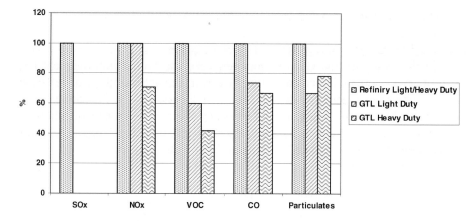

Figure 23 Regulated Emmissions: Tank-to-wheels (urban emissions)

Non-Regulated Emissions:

In addition to the above regulated emissions, GTL diesel also results in lower non-regulated emissions such as air toxics and polyaromatic hydrocarbons (PΛH's). This is due to the low aromatic content of GTL diesel, which is <0.5% compared to 26-30% in a typical refinery low sulphur diesel.

Toxicity and Biodegradability:

As a result of the high proportion of linear paraffins in GTL diesel, it has been shown to be readily biodegradable, and non-toxic/ not harmful to aquatic organisms. As a result of this, GTL diesel does not need to carry the dead fish and tree symbol like normal diesel, which has considerable advantages from a public perception point of view. Based on the favourable toxicity and biodegradability ratings, GTL diesel would be suitable for use in environmentally sensitive areas such as forests and estuaries.

Energy depletion

The GTL system is currently more energy intensive than the refinery system. For the current Sasol SPD process the total resource depletion is about 110% of that of the refinery. This comparison can be quite dependant on the quantity of natural gas liquids that are recovered upstream of the GTL conversion plant. If the future targeted improvements in energy efficiency are attained then the natural gas and crude oil routes will essentially be equivalent in terms of resource depletion.

In the case of GTL however, the main resource being depleted is natural gas which is the second most abundant fossil fuel (after coal). Natural gas also has favourable hydrogen to carbon ratio, compared to crude oil, from a greenhouse gas perspective.

The LCA analysis has lead to the following conclusions:

- In an urban environment GTL diesel offers substantial air quality benefits since it is already meeting all regulated emissions (nitrogen oxides, particulates, carbon monoxide and hydrocarbons) as well as unregulated emissions (aromatics, toxicity, biodegradability).
- Based on its superior product slate and product properties, GTL offers a wide range of attributes to leverage value for refiners who are under continuous pressure of reconfiguring to meet a changing demand pattern and quality imperative. This can be done by neat supply of GTL diesel or by upgrading lower quality middle distillates to high quality diesel by blending.
- GTL offers a substantial cost benefit over other alternative technologies being directly compatible with current infrastructure in terms of distribution networks and engine technologies. This issue is not reflected in the current LCA results.
- All these benefits are achieved without a greenhouse gas penalty.

A big advantage of using CH_4 instead of coal as the carbon source is that the production of syngas from CH_4 is much more carbon efficient and hence much less CO_2 is produced which has to be released into the atmosphere. As CO_2 is one of the greenhouse gases there is currently much emotional and political pressure to minimise CO_2 emmission. Some scientists, however, question whether the higher CO_2 levels do indeed significantly contribute to global warming [49]. They point out that the world has gone through previous temperature cycles, e.g. the 'Medieval Climate Optimum' a thousand years ago and the 'Little Ice Age' of three hundred years ago. Nevertheless, the responsible approach is to use the fossil fuel resources as efficiently as possible. Currently coal is used extensively to generate electrical power often in old facilities with low efficiencies. Replacement of these facilities with plants based on coal gasification and the co-production of electricity and liquid hydrocarbons could result in significant efficiency improvements and environmental benefits. See Section 6.1.

In a natural gas (methane) based FT plant the handling of sulphur contaminants and waste water streams is much simpler than for coal based plants. The methane usually contains only low levels of sulphur impurities and these have to be removed upstream of the CH_4 reformers. The process is standard, catalytic hydrogenation of any organic sulphur compounds to H_2S followed by absorption of the H_2S by ZnO. There is no highly contaminated gas liquor stream as in the case of coal gasification.

7.1 Fischer-Tropsch reaction water treatment

In addition to the production of hydrocarbon products, the FT processes also produce considerable volumes of FT reaction water. On a mass basis, approximately 1.3 tons of water is produced for every ton of hydrocarbon product. After separation from the hydrocarbon product, the FT reaction water typically contains numerous water soluble oxygenated compounds (e.g. alcohols, aldehydes, acids, ketones, esters), traces of other hydrocarbons and possibly traces of metals of catalyst origin. The type and degree of contamination is largely a function of the type of catalyst used in the FT synthesis, and the reaction conditions employed. Table 11 details typical water compositions for different FT synthesis operating modes.

Table 11
Typical composition of FT synthesis water from different synthesis operating modes

Component (mass %)	LTFT Co Catalyst	LTFT Fe Catalyst	HTFT Fe Catalyst
Water	98.89	95.70	94.11
non-acid oxygenated hydrocarbons	1.00	3.57	4.47
Acidic oxygenated hydrocarbons	0.09	0.71	1.41
Other Hydrocarbons	0.02	0.02	0.02
Inorganic components	< 0.005	< 0.005	< 0.005

The presence of the abovementioned contaminants generally limit direct application or disposal of this water, and purification of the FT reaction water is required in commercial processes. Primary treatment of FT synthesis water is typically by distillation. During distillation, the bulk of the non-acid oxygenated compounds can be removed, leaving a carboxylic acid rich bottom stream, commonly known as FT acid water. Depending on the scale of the FT synthesis process, the recovered oxygenated compounds can be worked-up to products, and/or used as an energy or fuel source.

The FT acid water stream, after extraction of the oxygenates, contains only low concentrations of organic acids which bio-degrade readily and can be purified in either aerobic or anaerobic digesters to produce good quality cooling water or even drinking water. An advantage of anaerobic digestion is that it produces methane which can be utilised whereas aerobic digestion produces a sludge which has to be incinerated.

For FT reaction water derived from LTFT applications, direct biological treatment (i.e. without a primary distillation step) using anaerobic processes can be employed as the primary treatment step [50]. During such processes, oxygenated hydrocarbons are converted to methane and carbon dioxide. For this reason, this process route is only applicable where work-up of oxygenated compounds to products is not desired.

Treatment of the FT acid water stream involves conventional biological treatment processes [51]. Aerobic, anaerobic process or combinations of these processes can be employed to produce a water product that is suitable for disposal to the environment or for reuse (e.g. as process cooling water) within the process.

Membrane-based and liquid-liquid extraction processes have also been developed to recover carboxylic acids from FT acid water [52, 53]. The resultant water stream from these processes is generally of a good quality and can be reused or disposed to the environment.

The production of high purity water (i.e. suitable as boiler feed water) is also achievable, but requires polishing of the water stream after primary and secondary treatment to remove suspended solids, dissolved salts (usually introduced as part of primary and/or secondary treatment) and traces of organic material [50-54].

7.2 Environmental issues for coal based plants

In a coal based FT plant using Sasol-Lurgi gasifiers there are two aqueous streams and one gaseous stream that have to be treated before re-using them or releasing them into the environment. The 'raw gas liquor' stream is the condensed water from the coal gasification step. For the major identified pollutants commercial control technology is available, including the following:

- Physical separation of oil and water streams.
- Steam stripping to remove volatiles.
- Phenol recovery by non-proprietary oil extraction or by Lurgi's proprietary Phenosolvan process.
- Sulphur recovery through a number of processes for removal, conversion to elemental sulphur, and cleanup of tail-gas streams.
- Biological oxidation to remove residual amounts of disolved salts, phenol, ammonia, etc.
- 'Polishing' operations, such as activated carbon adsorption, if required, to remove residual amounts of refractory organics which are not biodegradable.

Substantial quantities of sulphur and nitrogen compounds may remain in the liquefied product and, especially in the case of synthetic crude production, will require additional refining. This will increase the number of individual cleanup processes required and, hence, plant emissions. After recovery of the ammonia and phenols the water still contains low levels of oxygenated aliphatics/aromatics and these are destroyed in aerobic biodigestors.

The FT synthesis water stream, after extraction of the desired oxygenated compounds, still contains some organic acids and this stream together with rain water run-off from the FT complex is also treated in the bioworks. The biological sludge produced is incinerated. The product water from the digestors is recycled to the plant as it is suitable for use as cooling water. The blowdown stream from the cooling water systems contains inorganic salts, mostly originating from the gas liquor streams. The blowdown is treated with coal ash which chemically binds the bulk of the salts. The remaining water is evaporated. The raw synthesis gas is scrubbed to remove the CO_2 and the sulphur compounds (predominantly H_2S). The H_2S containing CO_2 stream is treated in a Sulfreen or similar process where the H_2S is oxidised to elemental sulphur.

The principal solid effluent from coal liquefaction will consist of the mineral matter present in the coal feedstock. Indirect liquefaction processes will discharge an ash with properties which depend on the process specifics. Landfill disposal is an alternative, but leaching characteristics of the solid wastes must be considered in selecting landfill procedures.

Other solid wastes will include spent catalysts, organic sludge from biological wastewater facilities, inorganic sludges from flue-gas desulfurization or evaporation ponds, and by-product sulphur if not recovered. These wastes may require specialized treatment to avoid transmission of soluble components, including trace elements (e.g. arsenic, chromium, mercury, molybdenum, and selenium) to local water supplies.

Coal storage and preparation will also require pollution control techniques for dust and rainwater runoff.

There are also significant environmental challenges related to the mining of the coal. Surface coal mining normally generates a lesser solid waste problem than underground mining because surface mining generally levels most of the overburden after the mining operation has been completed. Underground mining produces large quantities of solid wastes where preparation plants are associated with the underground mines in order to upgrade the coal. On the average, approximately 25 percent of the extracted coal is rejected as waste by a coal preparation plant. Some waste is produced from the sinking of shafts and the driving of entryways and tunnels.

Wind erosion, and hence, particulates, generate air pollution from surface mining. Because most of the coal surface-mining equipment is mostly electrically powered, diesel emissions are relatively low. Air pollution indirectly generated by surface mining occurs at the electric power station site rather than at the mine site.

Air pollution from underground mines arises from several sources. Blasting and other production operations generate some pollutants. Fires from coal refuse banks emit not only smoke and minute particulate materials, but also noxious and lethal gases. In the past, fires have occurred in abandoned deep mines and in un-mined underground areas.

REFERENCES

[1] A.P. Steynberg, R.L. Espinoza, B. Jager and A.C. Vosloo, High temperature Fischer-Tropsch synthesis in commercial practice, Applied Catalysis A: Gen. 186 (1999) 48.

[2] J.R. Rostrup-Nielsen, Catal. Today, 6 (3) 183 (1994).

[3] K.P. de Jong, Catal. Today, 29 (1996) 173.

[4] M.E. Dry, Catalysis Today, 6 (3) 183 (1990).

[5] K. Terblanche, Hydrocarbon Engineering, March/April 1997, page 2.

[6] G. Shaw, H. de Wet and F. Hohmann. AIChE Ammonia Symposium, Vancouver, BC, Canada. October 3-6 1994.

[7] H.M.W. van Wechem and M.M.G. Senden in Natural Gas Conversion II, H.E. Curry-Hyde, R.F. Rowe (eds.) Studies Surf. Sci. Catal., 81 (1994) 43.

[8] J. Eilers, S.A. Posthuma, S.T. Sie, Catalysis Letters, 7 (1990) 253.

[9] P.J.A. Tijm, J.M. Marriott, H. Hasenach, M.M.G. Senden and T. van Herwijnen. Alternative Energy symposium, Vancouver, Canada, May2-4 (1995) 228.

[10] M.E. Dry, The Sasol route to fuels, Chemtech (December 1982) 747.

[11] J.C. Hoogendorn and J.M. Salomon, Sasol: World's largest oil from coal plant, Brit. Chem. Eng., 2 (1957) 241.

[12] Ullmann's Encyclopedia of Industrial Chemistry, Fifth Completely Revised Edition, Vol. A12, 1989, p.203.

[13] W.S. Ernst, S.C. Venables, P.S. Christensen and A.C. Berthelsen, Push syngas production limits, Hydrocarbon Processing, March (2000).

[14] W. Stupin, R. Ruvikumar and B. Hook, Liquid Products from Petroleum Coke Via Gasification and Fischer-Tropsch Synthesis, NPRA 2000 Annual Meeting, San Antonio, Texas, AM-00-10 March (2000).

[15] D. Gray and G.C. Thomlinson, Contractor Report DE91-004668, Assessing the Economic Impact of Indirect Liquifaction Process Improvements: Volume 1: Development of the Integrated Indirect Liquifaction Model and Baseline Case, Oct.(1990).

[16] D. Gray and G.C. Thomlinson, A Technical and Economic Comparison of Natural Gas and Coal Feedstocks for Fischer-Tropsch Synthesis, M de Pontes, RL Espinoza, C.P. Nicolaides, J.H. Scholtz and M.S. Scurrell (eds.), Natural Gas Conversion IV, Studies in Surface Science and Catalysis, 107 (1997) 145.

[17] C.J. Egan, Chevron Research Company, US Patent No. 4,092,825 June (1978).

[18] C. Schrader, and L. Shah, Gasification Technologies 1: Applications, Texaco Gasification and Fischer-Tropsch (TGFT) Process September (2000).

[19] Sorensen et al, Air Products, US Patent 5, 865, 023, February (1999).

[20] Sorensen et al, US Patent 5, 666, 800, (September 1997) – Air Products.

[21] M.S. Botha and C.B. Benham, Coal-to-Liquids via Fischer-Tropsch Synthesis, 244 International Technical Conference on Coal Utilisation and Fuel Systems, March (1999).

[22] Bohn, et al, Rentech Inc., US Patent 6, 306, 917, October (2001).

[23] B.L. Bhatt, R. Frame, A. Hoek, K. Kinnari, V.U.S. Rao and F.L. Tungate, Catalyst and Process Scale-up for Fischer-Tropsch Syntehsis, Topics in Catalysis 2 (1995) 235-257.

[24] B.L. Bhatt, E.S. Schaub, E.C. Heydorn, D.M. Herron, D.W. Studer and D.M. Brown, Liquid Phase Fischer-Tropsch Synthesis in a Bubble Column, DOE Liquifaction Contractor's Review Conference, Pittsburg, 22-24 September (1992).

[25] B.L. Bhatt, E.S. Schaub and E.C. Heydorn, Recent developments in slurry reactor technology at the La Porte Alternative Fuels Development Unit, 18th International Technical Conference on Coal Utilisation and Fuel Systems, Clearwater, 26-29 April (1993).

[26] H. Kolbel and M. Ralek, The Fischer-Tropsch Synthesis in the Liquid Phase, Catal. Rev. – Sci. Eng. 21(2) (1980).

[27] A.P. Steynberg, W.U. Nel and M.A. Desmet, Large scale production of high value hydrocarbons using Fischer-Tropsch Technology, Natural Gas Conversion Symposium 7, Dalian, China (2004) in press.

[28] Anonymous, Fixed-bed Reactor Successful in Fuels from Coal Synthesis, Oil and Gas Journal (January 1992) 54 (based on article presented at Achema 91, June 9-15, 1991, Frankfurt.)

[29] G. Weber, J. Oil & Gas, March 24 (1949) 248.

[30] W. De Graaf and F. Schrauwen, World Scale GTL, 17th Annual International Pittsburg Coal Conference, Hydrocarbon Engineering, May (2002).

[31] B. Eisenberg, R.A. Fiato, C.H. Mauldin, G.R. Say, and S.L. Soled, Exxon's advanced gas-to-liquids technology, Stud. Surf. Sci. Catal. 119 (1998) 943-948.

[32] J.R. Rostrup-Nielsen, I. Dybkjaer and R.G. Coulthard, , Syngas for Large Scale GTL-plants, 2001 Spring National Meeting of AIChE, Houston, 22-26 April (2001).

[33] A.P. Steynberg, A.P. Vogel, J.G. Price and H.G. Nel, Technology Targets for Gas-to-Liquids Applications, 2001 Spring National Meeting of AIChE, Houston, 22-26 April (2001).

[34] R. Hansen, J. Sogge, M.H. Wesenberg and O. Olsvik, Selecting Optimum Syngas Technology and Process Design for Large Scale Conversion of Natural Gas into Fischer-Tropsch Products (GTL) and Methanol, J.J. Spivey, E. Iglesia and T.H. Fleisch (eds.), Studies in Surface Science and Catalysis (2001) 405-410.

[35] S.T. Sie, M.M.G. Senden and H.M.H. van Wechem, Conversion of Natural Gas to Transportation Fuels via the Shell Middle Distillate Synthesis Process (SMDS) Catalysis Today 8 (1991) 390.

[36] D. Schanke, R. Hansen, J. Sogge, K.H. Hofstad, M.H. Wesenberg and E. Rytter, Optimum Integration of fischer-Tropsch Synthesis and Syngas Production, WO 01/42175, 14 June (2001).

[37] D. Reinaido, A. Denking, P. Blankenstein, T. Meuris and J.G.M. Decker, EP 428223, Shell Internationale Research Maatschappi, BV, Process for the Preparation of Extradutes, and use of the Extrudates, November (1990).

[38] Studien- undVerwertungsgesellschaft mit Besschränkter Haftung UK Patent no. GB 454,948, Oct. 12 (1936).

[39] A.P. Steynberg, N.N. Clark, Thermally efficient process to produce liquid hydrocarbons, PCT Patent Application WO 02/38699, 10 May (2002)

[40] B. Ghaemmaghami and S.C. Clarke, Design and Engineering of a GTL plant, Forster Wheeler Review, Spring 2001, based on a paper presented at "The Sasol Slurry Phase Distillate Process Technology Seminar" held in South Africa in November 2000.

[41] R.L. Espinoza, A.P. Steynberg, B. Jager and A.C. Vosloo, Low temperature Fischer-Tropsch synthesis from a Sasol perspective, Applied Catalysis A: Gen. 186 (1999) 13.

[42] L.P. Dancuart, J.F. Mayer, M.T. Tallman and J. Adams, Performance of the Sasol SPD Naphtha as Steam Cracking Feedstock, General Papers symposium, Division of Petroleum Chemistry, ACS national meeting, Boston, MA, August 18-22 (2002).

[43] C.W. Quinlan, AGC-21 – ExxonMobil's Gas-To-Liquids Technology, Gastech 2002, Doha, Qatar (2002).

[44] A.P. Steynberg and H. Nel, Clean coal conversion options using Fischer-Tropsch Technology, Fuel 83 (2004) 765-770.

[45] M. van den Burgt, J. Van Klinken, and S.T. Sie, The Shell Middle Distillate Synthesis Process, paper presented at the 5th Synfuels Worldwide Symposium, Washington, D.C., USA, November 11-13 (1985).

[46] W. Stupin, R. Ruvikumar and B. Hook, Liquid products from Petroleum Coke via Gasification and Fischer-Tropsch Synthesis, NPRA 2000 Annual Meeting, San Antonio, Texas, AM-0010, March (2000).

[47] R.L. Espinoza, A.P. Steynberg, Jager, B. and A.C. Vosloo, Low temperature Fischer-Tropsch synthesis from a Sasol perspective, Applied Catalysis A: General 186 (1999) 13-26.

[48] P.W. Schwaberg, I.S. Myburgh, J.J. Botha, P.N. Roets, L.P. Dancuart, 11th World Clean Air Congress, Durban, South Africa, September (1998).

[49] A.B. Robinson, S.L. Baliunas, W. Soon, Z.W. Robinson, Medical Sentinel, 3 (5) (1988) 171.

[50] L.P. Dancuart, G.H. du Plessis, F.J. du Toit, E.L. Koper, T.D. Phillips and J. van der Walt, patent applications WO 03/106351, 18 Jun (2002).

[51] L.P. Dancuart, G.H. du Plessis, F.J. du Toit, E.L. Koper, T.D. Phillips and J. van der Walt, patent applications WO 03/106354, 18 Jun (2002).

[52] L.P. Dancuart, G.H. du Plessis, F.J. du Toit, E.L. Koper, T.D. Phillips and J. van der Walt, patent applications WO 03/106353, 18 Jun (2002).

[53] L.P. Dancuart, G.H. du Plessis, F.J. du Toit, E.L. Koper, T.D. Phillips and J. van der Walt, patent applications WO 03/106349, 18 Jun (2002).

[54] L.P. Dancuart, G.H. du Plessis, F.J. du Toit, E.L. Koper, T.D. Phillips and J. van der Walt, patent applications WO 03/106346, 18 Jun (2002).

Studies in Surface Science and Catalysis 152
A. Steynberg and M. Dry (Editors)

Chapter 6

Processing of Primary Fischer-Tropsch Products

L. P. Dancuart, R. de Haan and A. de Klerk

Sasol Technology R&D
P.O. Box 1, Sasolburg, 1947, South Africa

1. INTRODUCTION

The refining of Fischer-Tropsch (FT) products is very different from crude oil refining in terms of feed composition, refining focus and heat management [1]. Despite these differences, the refining of FT products is not widespread enough to have attracted FT specific refining technologies. The same basic technologies and commercial catalysts used in crude oil refining [2] have therefore been adapted for use in FT primary product refining.

The refinery configuration and choice of refining technologies depend largely on the split between chemicals and fuels production. Valuable products, like linear alpha-olefins, can be extracted from the FT product. The recovery (or non-recovery) of such chemicals would change the product slate and consequently the refining methodology. Historically the refining of FT products focused mainly on fuels. The evolution of FT refining is discussed with emphasis on the impact on the refinery design of changing fuel specifications; improvements in technology and the exploitation of chemical opportunities.

As for a crude oil refinery, the FT refinery design depends on the nature of the feed that must be processed. Some key properties of the primary FT product components are discussed with an explanation of their impact on the refining approach, most notably the hetero-atom constraints, compatibility with fuel requirements and carbon number distribution. This serves as an introduction to the detailed discussion on the preferred refining options for the various cuts. The focus is on fuels production from the primary product although some of the chemical extraction opportunities that exist are highlighted in this context.

Towards the end of this chapter some opportunities to target the large scale production of bulk commodity chemicals are presented.

Some notes on refinery integration opportunities are provided and there are pointers to the design of future FT refineries.

2. HISTORICAL PERSPECTIVE

The original Sasol 1 production facility in Sasolburg had both Arge Low Temperature FT (LTFT) and Synthol High Temperature FT (HTFT) synthesis technology operating in parallel. The Synthol refinery produced mainly petrol (also termed gasoline or mogas, in other words, spark ignition engine fuel), while the Arge product refinery produced mainly heavier products, i.e. diesel (compression ignition engine fuel) and wax [3, 4, 5]. The HTFT plant in Brownsville, Texas (Hydrocol process) [6, 7] that preceded the Sasolburg facility also targeted petrol as the main fuel product. Fuel specifications in the 1950's were not very demanding and product upgrading was mainly by distillation, hydrogenation, clay treatment (to decrease the oxygenates and increase the aromatic contents) and oligomerisation of light olefins. Lead containing additives were used to increase octane as was common practice in oil refineries at that time.

The Sasol production facilities in Secunda exclusively use HTFT technology. Fuel specifications in the 1980's were still not very demanding and the FT product refineries were consequently of a straightforward design (Fig. 1) [8]. Since coal is used as feed material, the overall facility is a bit more complicated [9]. In addition to the FT refinery, a tar refinery is required to process the pyrolysis products from coal gasification (not shown in Fig. 1). Tar refining is linked to coal processing and is similar to the refining required during coke production for the metallurgical industry.

The refinery designs not only reflected the prevailing fuel specifications, but also the fact that the HTFT refineries were seen as mainly fuels producers. The potential to extract chemicals was noted [10], but was not incorporated in the designs. The first change in this perspective started in the 1960's, with the announcement that ethylene from FT synthesis, and nitrogen from the air separation plant, would be recovered for the production of plastic and ammonia at the Sasol 1 facility [11].

The separation of ethylene and propylene was incorporated in the later Secunda designs. This was taken a step further in the 1990's, with the recovery and purification of linear alpha-olefins as co-monomers for the polymer industry. Similarly, heavier alpha-olefins are now recovered and purified for detergent alcohol production via hydroformylation.

484

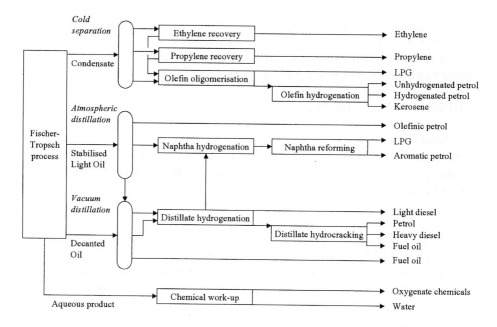

Figure 1 High Temperature Fischer-Tropsch refinery design originally used at Secunda

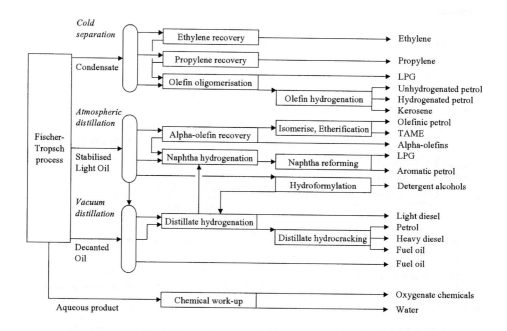

Figure 2 Current (2004) High Temperature Fischer-Tropsch refinery used at Secunda

The HTFT reactors at Sasol 1 were decommissioned in the 1990's and the facility was converted to a chemicals production site. The Secunda plants also started to lose their identity as fuels-only refineries as more and more chemicals were targeted for extraction [12]. Yet, most of the production volume still went to fuels and with the more stringent fuel specifications, some fuels refining units were added, such as etherification and skeletal isomerisation (Fig. 2).

With the construction of the first commercial HTFT gas-to-liquids (GTL) plant, an all-fuels approach was again followed. The PetroSA GTL plant started operation in 1993 as Mossgas in Mossel Bay, South Africa. The plant is an integrated facility for natural gas processing that makes use of Sasol Synthol technology. The feed is obtained from their own natural gas fields located offshore. The primary gas separation is done at the main production platform where Raw Natural Gas and Natural Gas Liquids (NGL) are produced and transported some 120 km to the GTL plant by pipelines. A secondary separation is done onshore where the natural gas composition is made compatible with the GTL process requirements and the NGL is separated into Liquefied Petroleum Gas (LPG), Naphtha and Diesel. The refinery design resembled that of the Secunda facilities, but some changes were made to the C_2-C_8 processing (Fig. 3) [13].

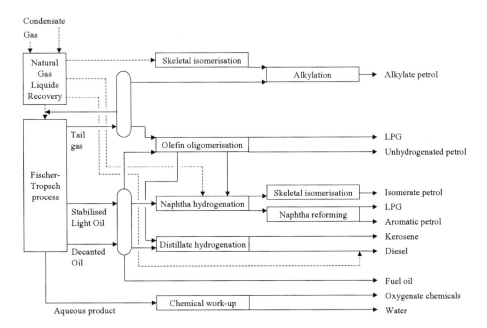

Figure 3 High Temperature Fischer-Tropsch refinery design of Mossgas (now PetroSA) at Mossel Bay

Chemical production at the Mossel Bay refinery is possible but lacks economy of scale relative to the Secunda plants. Although not implemented, some chemical production opportunities have been investigated [14]. Future HTFT plants are likely to target bulk commodity chemicals to decrease the reliance on high oil prices for satisfactory economic performance. Chemicals extraction is an attractive option to grow income [15] and this can be considered even after the FT plant is in operation provided the remaining plant life is sufficient. Since the Mossel Bay refinery uses natural gas as feed material and not coal, it does not have a tar refinery associated with it. However, the co-processing of natural gas offers its own advantages.

The historic evolution of the Sasol HTFT refineries was not only driven by technology advances and market requirements, but also by the strategic positioning of the company. Current thinking favours expansion in the field of polymers and future changes may well have to balance the demands of more stringent fuel specifications with growth in ethylene and propylene production. To meet such an eventuality a change in the refinery configuration to include a Superflex™ Catalytic Cracker (SCC) unit is being implemented [16].

3. UPGRADING OF HIGH TEMPERATURE FISCHER-TROPSCH (HTFT) PRODUCTS

3.1 Influence of feed properties on refining approach

To understand HTFT refining, it is necessary to take a closer look at the refinery feed material. Although the feed composition depends strongly on the nature of the FT catalyst and operating conditions [4, 5], a high temperature iron based FT primary product typically has the following attributes:

- The oil and aqueous products are essentially sulphur-free.
- The oil and aqueous products are low in nitrogen-containing compounds.
- The oil and aqueous products contain percentage levels of oxygenates (alcohols, acids, esters and carbonyls).
- The oil product is rich in olefinic material, especially linear alpha-olefins.
- The oil product is poor in aromatics and naphthenics (cyclo-paraffins and cyclo-olefins).
- The oil product is mostly linear or with a low degree of branching.
- The oil product follows a Schultz-Flory distribution, heavily weighted towards light hydrocarbons.
- The aqueous product contains most of the short chain oxygenates.

The discussion presupposes that the products are typical HTFT products and have not been modified in some way by the addition of a second catalyst type. It may seem desirable to use a second catalyst type to make the HTFT product more amenable to refining [17], but this train of thought will not be explored further.

3.1.1 Hetero-atom constraints

The most obvious advantage is the low sulphur content. There is no need to worry about deep hydrodesulphurisation or other sulphur reduction technologies to meet fuel sulphur specifications. There is also no need to worry about the sulphur sensitivity of catalysts and high activity unsulphided catalysts can in principle be used for hydrotreating.

The bugbear of a FT refinery is oxygenates and most refining processes are affected. Only the lighter than C_5 boiling range material is oxygenate free or have oxygenates present in very low concentration. Oxygenates can therefore be considered critical in HTFT refining, with carboxylic acids being especially troublesome. They restrict the metallurgy of processing equipment and cause metal leaching of some classes of unsulphided catalysts, most notably nickel promoted catalysts. Furthermore, the carboxylic acids tend to be difficult to remove by hydrotreating.

Unlike hydrodesulphurisation (HDS) and hydrodenitrogenation (HDN) that produces hydrogen sulphide and ammonia, hydrodeoxygenation (HDO) produces water. This not only complicates product work-up, but also creates problems in catalysis. Hydrothermal dealumination can lead to changed catalyst behaviour [18] and faster catalyst deactivation [19]. In chlorided catalyst systems severe corrosion may also result from the presence of water.

3.1.2 Petrol component properties

It is instructive to look at the hydrocarbon component classes in relation to their octane value and abundance in FT products. As an example, the C_7 hydrocarbon compound classes have been listed, with their relative abundance on an oxygenate-free basis in a high temperature FT product (Table 1).

It is clear that an unrefined HTFT product has a low octane and that the only saving grace is its high olefin content. Exploiting the high olefin content and redressing both the low degree of branching and low aromatics content, are therefore central to the fuels refining strategy. This can typically be done by skeletal isomerisation of the olefins [21, 22] (with the possibility for etherification), skeletal isomerisation of the paraffins [23, 24], selective aromatisation or reforming [25, 26, 27]. Similarly, the extraction of alpha-olefins as chemicals is desirable from a fuels point of view. In some cases it may also be useful to consider double bond isomerisation as a cheap way to improve the octane of alpha-olefin-rich material.

Table 1

Research octane value (RON) and motor octane value (MON) of some C_7 hydrocarbons [20] and their relative abundance in unrefined High Temperature Fischer-Tropsch product

Class	Compound	RON	MON	HTFT abundance
Linear paraffin	n-Heptane	0.0	0.0	8
Branched paraffin	3-Methylhexane	52.0	55.8	4
	2,3-Dimethylpentane	91.1	88.5	
	2,2,3-Trimethylbutane	112.2	101.3	
Linear olefin	1-Heptene	54.5	50.7	50
	trans-2-Heptene	73.4	68.6	
Branched olefin	3-Methyl-1-hexene	82.2	71.5	22
	trans-3-Methyl-2-hexene	91.5	79.6	
	2,3-Dimethyl-1-pentene	99.3	84.2	
	2,3-Dimethyl-2-pentene	97.5	80.0	
	2,3,3-Trimethyl-1-butene	105.3	90.5	
Cycloparaffins	Methylcyclohexane	74.8	71.1	<1
	Ethylcyclopentane	67.2	61.2	
Cyclo-olefins	1-Methylcyclohexene	89.2	72.0	6
	1-Ethylcyclopentene	90.8	71.4	
Aromatics	Toluene	120.0	103.5	10

Fuel specifications may vary from country to country, but migration to Euro-4 standards is likely. In comparison to historic fuel specifications, heavy metal additives are no longer acceptable to boost octane and requirements for RON ($\geq$ 95) and MON ($\geq$ 85) have been increased. A lowering of the aromatic content ($\leq$ 35% volume) and benzene content ($\leq$ 1% volume) will place pressure on crude oil refiners. The most likely future sulphur specifications ($\leq$ 10ppm) has already caused a flurry of activity and many new sulphur reduction technologies have come to market. The debate on which oxygenates (ethers and alcohols) are acceptable, and even mandatory, has not yet been resolved [28]. The limitation of total oxygen content ($\leq$ 2.7% mass) will probably remain unchanged for the time being. The Euro-4 specifications also limit the olefin content ($\leq$ 18% volume) and thereby put pressure on HTFT refiners.

3.1.3 HTFT Diesel properties

It can be said that 'what is bad for octane is generally good for Cetane' so the low aromatics content and low degree of branching are very beneficial for the Cetane number. Conversely the low degree of branching results in poor cold properties and the low aromatics content in a low density. However, hydrotreated, slightly hydroisomerised FT products in general make a good diesel and an excellent blending component to enhance the properties of crude oil derived diesel blending material. The primary product from HTFT processes is more branched and more olefinic than the LTFT material; moreovers the HTFT product contains some aromatics while aromatics are almost totally absent from the LTFT material. As implied above, some aromatic content is desirable to increase the diesel density.

Future diesel specifications focus mainly on sulphur reduction, with the sulphur content likely to be limited to 10 ppm. Moving from Euro-2 to Euro-4 specifications also require increased Cetane number (≥ 51), lower T95 boiling point (i.e. 95% volume recovered at $\leq 360°C$) and narrower density range (820-845 $kg.m^{-3}$ @ $15°C$).

3.1.4 Carbon number distribution

HTFT products follow a Schultz-Flory carbon number distribution that is heavily weighted towards lighter products. Dealing with the large quantity of material in the $<25°C$ boiling range (material lighter than C_5), is consequently very important. Some butane can be blended into the fuel and most of the propylene can be recovered for chemical use, but this still leaves some propane-rich C_3 material and a large amount of n-butene-rich C_4 material to be beneficiated. Due to the highly olefinic nature of the HTFT product, it is not difficult to shift the carbon number distribution to heavier hydrocarbons by oligomerisation [29, 30, 31]. The moderately branched olefins produced by oligomerisation have a high octane value and are prime blending materials. However, fuel stability issues limit the olefin concentration in the fuel and may require hydrogenation of some olefinic material to paraffins with a poor octane.

Unrefined HTFT product also has material in the $>360°C$ boiling range (material heavier than C_{22}). The direct inclusion of this heavy material in diesel fuel is limited by the T95 boiling point specification. The heavier material is fairly clean, with no sulphur and little polynuclear aromatics present. It is therefore well suited for hydrocracking, using standard commercial hydrocracking catalysts, as means of shifting the carbon number distribution to lighter products. The hydrocracked material has good diesel properties, but poor gasoline properties. Maximising the diesel to naphtha ratio during hydrocracking is consequently important.

3.2 HTFT fuels refining

From the preceding discussion it can be inferred that HTFT primary product upgrading requires specific types of refining:

• Oligomerisation for shifting light products to higher boiling material.
• Hydrocracking for shifting heavy products to lighter boiling material.
• Aromatisation and isomerisation for improving octane and density.
• Hydrogenation for removing unwanted oxygenates, olefins and dienes.

It will be noted that well-known refining technologies like catalytic cracking, thermal cracking (visbreaking), alkylation, etherification and coking have not been included in the list. This does not imply that these technologies cannot be used with FT products, but rather that these technologies neither have a specific refining advantage nor address a specific shortcoming. In some instances application of these technologies provides excellent refining opportunities. By looking at the HTFT product on a carbon number for carbon number basis, these opportunities will be highlighted.

The focus of this discussion remains on fuels refining, although opportunities for chemical extraction are briefly discussed where applicable. The carbon numbers discussed actually refer to the boiling range of the equivalent hydrocarbons in the FT primary product. Most oxygenates have higher boiling points than their hydrocarbon analogues. Within the same boiling range, oxygenates will have three to four carbon atoms less than the hydrocarbons. For example, hexanol will be relevant to the discussion of C_9/C_{10} hydrocarbons and will not be discussed with the C_6 hydrocarbons.

3.2.1 C_3 hydrocarbons

Propylene is the most abundant HTFT product. When purified, it can be used for chemicals and polymer production. However, the level of propylene recovery is limited by economic considerations and the refinery has to deal with propylene in a propane-rich mixture. Direct marketing as liquid petroleum gas (LPG) is not always possible and the preferred refining options are:

- Alkylation: the process involving benzene and propylene to produce cumene is a well-established one that is fairly insensitive to the propane content in the feed. The reaction takes place over an acidic catalyst, like solid phosphoric acid or a zeolite. Cumene has a high octane value (RON=113 and MON=99) and has value as a chemical commodity too. This is an excellent way to improve octane, increase fuel density and move propylene into the fuel boiling range. Aliphatic alkylation of iso-butane with propylene is possible, but will typically not be considered. FT material is not only low in iso-butane, but propylene also gives a poor quality alkylate when compared to 2-butene which is the industry standard (5-8 octane units difference) [32].

- Aromatisation: The aromatisation of C_3 hydrocarbons to give a high octane aromatic product has the advantage that it converts both propane and propylene. A number of commercial technologies exist, most being based on metal promoted ZSM-5 catalysts. The liquid product yield (C_5 and heavier) is typically in the order of 60-70%. It is therefore a good way to improve octane, increase density and move C_3 hydrocarbons into the fuel boiling range. It is especially beneficial for propylene lean streams, when olefin-based technologies like oligomerisation and alkylation would be able to upgrade only a small part of the C_3-fraction.

- Oligomerisation: Propylene can be converted by homogeneous or heterogeneously catalysed processes to heavier hydrocarbons. Zeolites, solid phosphoric acid and amorphous silica-alumina based catalysts are in use commercially. Although propane is not converted, it is a convenient heat sink for the reaction heat. The catalyst system and operating conditions determine the product quality. In general the olefinic petrol quality is good, but the octane value of the hydrogenated product is poor. It is preferable to move the propylene into the diesel boiling range to improve the cold flow properties of the FT diesel, but not all catalyst systems are able to achieve high diesel yields. Oligomerisation is the least preferred of the options presented from a fuel refining perspective.

The preferred processing route is to recover the propylene as a chemical feedstock for use on site (e.g. for the production of polypropylene). The remaining propane can then be processed for sale as LPG or converted to aromatics.

3.2.2 C$_4$ hydrocarbons

The refining of C$_4$ hydrocarbons is dominated by the chemistry of 1-butene. Despite the low iso-butene content, the purification of 1-butene as co-monomer for polymer production would still require an etherification step. Due to transportation difficulties this is generally not worthwhile unless the purified 1-butene is used on-site. Butane (RON=111 and MON=101) can be blended directly into the fuel and the amount is limited only by vapour pressure specification of the final fuel blend. Some butane can be marketed as liquefied petroleum gas (LPG), but the bulk of the C$_4$ material must be refined. The preferred refining options are:

- Alkylation: Alkylation of benzene with butene to produce mostly iso-butyl benzene can be done in an analogous fashion to alkylation with propylene. Iso-butyl benzene has a high octane value (RON=111 and MON=98) and most of the advantages listed for propylene alkylation also apply to butene alkylation. However, iso-butyl benzene is slightly inferior to cumene as fuel additive and is not a common chemical commodity. Butene also has better refining alternatives. Propylene alkylation of benzene is therefore preferred to butene alkylation. Aliphatic alkylation of iso-butane with 2-butene is a classic refining option and preferred to iso-butane alkylation with propylene. Yet, in a FT refinery the low iso-butane concentration and high olefin to paraffin ratio makes it unattractive, unless an external source of iso-butane is readily available (like natural gas condensate).

- Aromatisation: The aromatisation of C$_4$ hydrocarbons is analogous to the conversion of C$_3$ hydrocarbons. Unlike the C$_3$'s where propylene is recovered as a chemical, the C$_4$'s have a high olefin content and the co-conversion of paraffins would not be able to improve the liquid product yield compared to other process options. Since C$_3$'s and C$_4$'s can be co-converted, it would be an attractive refining option for combined processing, but a less preferred option for butenes per se.

- Oligomerisation: It has been shown that a 1-butene-rich feed yields a dimerisation product with better octane than a 2-butene-rich feed [33]. Although this is not expected based on a classical carbocation mechanism, dimerisation via some protonated cyclo-propane intermediate [34] could well explain this phenomenon. Oligomerisation of a butene-rich feed results in a fuel with high olefinic octane and a considerably better paraffinic octane (about 30 octane units) than that achievable with propylene. The less branched dimers, having a lower octane value, also have value as feed for hydroformylation to plasticizer alcohols. Specific homogeneous [35], heterogeneous [36] and bi-phasic ionic liquid [37] processes for the conversion of butenes for this purpose are commercially available. Oligomerisation can also be used to convert the butenes to liquid products suitable for reforming or selective aromatisation. The versatility of oligomerisation for fuel and chemical production makes it the preferred refining option for butenes.

In principle skeletal isomerisation of n-butenes to iso-butene can be considered in conjunction with either dimerisation or etherification. It would be difficult to justify the cost associated with butene skeletal isomerisation in relation to the benefit derived over 1-butene oligomerisation. Skeletal isomerisation followed by etherification to methyl tertiary butyl ether (MTBE) or ethyl tertiary butyl ether (ETBE) is more beneficial, since both ethers have high octane (RON=118 and MON=101). However, the production of MTBE has become a politically and environmentally sensitive issue [38, 39]. If the refinery is aimed at chemicals production, metathesis of 1-butene with ethylene to enhance the propylene production for use on-site as a chemical feedstock is worth consideration.

3.2.3 C$_5$ hydrocarbons

This is the heaviest carbon number that has almost no oxygenates associated with it. The product distribution is dominated by 1-pentene, which can be recovered for co-monomer use [40], but this has not yet found general acceptance in the marketplace. The refining strategy is influenced by the fact that the C$_5$'s are already in the liquid phase and that the C$_5$'s are very linear and mostly olefinic. Aromatisation is undesirable due to the loss in liquid volume (about 50% loss) through increased density although this increased density and the production of less volatile components can be beneficial for the fuel characteristics. If the C$_5$'s are blended directly into the fuel, 1-pentene (RON=91 and MON=77) has a high vapour pressure contribution. An even bigger drawback is the contribution to the olefin content of the fuel. The olefin content is limited by the fuel specifications and it is therefore important to preferentially include molecules with a large octane differential between the olefinic and hydrogenated (paraffinic) species. Although this does not preclude direct blending of the C_5's, the preferred refining options are:

- Skeletal isomerisation: Both pentane [23, 24] and pentene [21, 22] are readily converted to the methyl branched species and a number of commercial technologies exist. Metal promoted alumina, chlorided alumina, zeolites and metal oxide catalysts are used for pentane skeletal isomerisation, but pentene skeletal isomerisation is limited to alumina and non-zeolitic molecular sieve catalysts. The skeletal isomerisation reaction is equilibrium limited [41], but high conversions can be achieved by recycling the unconverted product. In a FT environment the high linearity of the feed implies that a large benefit can be derived from skeletal isomerisation. Iso-pentane has an acceptable octane (RON=92 and MON=90) for a paraffin. Unlike C_4's, pentene and pentane skeletal isomerisation generally requires less severe operating conditions, thereby favouring the equilibrium for isomerisation that can be closely approached in commercial operation. The acceptable paraffinic octane, large octane gain with respect to the linear feed, ease of conversion and further beneficiation possibilities of the olefin by etherification, makes it the most preferred refining option.

- Etherification: The etherification of 2-methyl-1-butene and 2-methyl-2-butene with methanol yields tertiary amyl methyl ether (TAME). The importance of TAME as octane booster was recognised in the late 1970's and has been thoroughly studied since then [42, 43]. A number of commercial technologies exist. Although TAME has a slightly lower octane (RON=115 and MON=100) compared to the iso-butene derived ethers, it has a lower vapour pressure. The low degree of branching in the FT material presupposes pentene skeletal isomerisation and etherification is therefore not a primary refining option. The decision to invest in etherification technology should be weighed up against purifying ethanol from the aqueous FT product stream as alternative oxygenate-containing fuel additive.

The high olefin content makes oligomerisation a possibility. This can be considered if it is necessary to increase the diesel to petrol ratio or reduce the vapour pressure of the petrol.

3.2.4 C_6 hydrocarbons

The most abundant C_6 hydrocarbon is 1-hexene, which is a high value co-monomer used in the polymer industry. It makes sense to extract and purify it, since 1-hexene also happens to be a poor fuel component (RON=76 and MON=63). Some oil soluble oxygenates start appearing in the FT product, but these are mainly non-acid chemicals (mostly carbonyl compounds) and do not pose a serious refining problem. In terms of fuel properties, vapour pressure is still on the high side, but the most important change is in the octane value. Only the di-branched C_6 isomers have an acceptable paraffinic octane and there is a noticeable octane difference between the olefins and the paraffins (about 15 octane units for the mono-branched species). Since this differential increases with increasing carbon number, it is better to reserve the olefin capacity of the fuel pool for heavier hydrocarbons. The direct inclusion of unrefined C_6 FT hydrocarbons into the fuel pool is consequently possible, but not optimal. The preferred refining options are:

- Skeletal isomerisation: Hexane and pentane use the same commercial skeletal isomerisation technologies and can consequently be processed in the same unit. However, there are advantages to separate processing. Hexane has to be isomerised to the di-branched species and therefore requires a longer residence time than pentane to achieve equilibrium. The column configuration for recycle operation is also less complicated if the hexanes are processed separately. There is no benefit in separate paraffin and olefin processing either. Hexene can be hydrogenated in the same unit and be processed as hexanes.

- Aromatisation: Reforming of C_6 material is very ineffective, but aromatisation over a non-acidic Pt/L-zeolite catalyst can be very efficient [25, 44]. The biggest drawback of technology based on this type of catalyst is its sulphur sensitivity [45]. In this respect FT material has a distinct advantage, since it is sulphur-free. The high yield of aromatics (>80%) and low gas yield, makes it a clean and attractive technology for aromatics production. Furthermore, since the catalyst is non-acidic, the aromatic product is mostly of the same carbon number as the feed. Hexane consequently produces benzene with high selectivity (>90%). Although benzene is not a desired fuel component, it can be sold as commodity chemical, or be used for alkylation to produce high octane fuel.

The etherification of hexene isomers has not been extensively studied [46]. In general the hexyl ethers have considerably poorer octane values than MTBE, ETBE and TAME [47] and should not be considered as replacements for the listed ethers. The oligomerisation of heavier olefins is possible [48], but it is a capital-intensive process and not a preferred refining option. Yet, the oligomerisation of HTFT hexene fractions is practised commercially and yields a high quality diesel product [49].

3.2.5 C_7 hydrocarbons

In theory the refining of C_7 hydrocarbons should not be difficult, yet, it remains one of the most troublesome FT cuts to upgrade. The product composition is again dominated by the α-olefin. Due to the high 1-heptene

(RON=55 and MON=51) content and the presence of oxygenates, it is not suitable for direct blending. Unlike the even numbered molecules, 1-heptene is not a high valued product either, although it is possible to convert 1-heptene to 1-octene [50]. Reforming does not convert C_7 molecules well and C_5/C_6 skeletal isomerisation technology can tolerate C_7's in low concentration only. Oligomerisation to produce diesel is practised commercially, but it is currently expensive. The alternatives are limited and the preferred refining option for petrol production is:

- Aromatisation: The advantages of non-acidic Pt/L-zeolite aromatisation cited for the conversion of C_6 hydrocarbons, apply equally well to C_7 hydrocarbon conversion. The only difference is that the main product is toluene (RON=120 and MON=103), not benzene. In commercial applications the C_6's and C_7's are processed together, with subsequent benzene and toluene separation. Despite the absence of sulphur in the feed, hydrogenation of oxygenates in the feed is required as pre-processing step. The large volume of hydrogen generated also makes hydrogen recovery desirable as a post-processing step.

Future processing of the C_7 hydrocarbons at the Sasol Secunda refineries will entail catalytic cracking [16]. Although this is a costly refining approach, it was justified based on the local conditions and existing on-site infrastructure for propylene processing.

3.2.6 C_8 hydrocarbons

The linear 1-octene present in high concentration in the C_8 hydrocarbons has the same commercial benefit as 1-hexene as a valuable co-monomer, but it is considerably more difficult to purify. The oxygenates are not only more concentrated, but they also contain some carboxylic acids. However, once a decision has been made to install a process to separate the oxygenates from the hydrocarbons (e.g. by extractive distillation) then the remaining olefins are also worth recovering. Since the FT product is very olefinic and has a low degree of branching, the olefins can alternatively be used as feed for hydroformylation to produce plasticizer alcohols. It is possible to upgrade more than 60% of the C_8 hydrocarbons to high value chemical products once the oxygenates have been removed. Irrespective of whether these opportunities are exploited, the oxygenates must be removed prior to further refining and the alternative is hydrotreating. The preferred fuel refining options are:

- Aromatisation: Like C_6's and C_7's it is possible to use non-acidic Pt/L-zeolite based technology to convert C_8 hydrocarbons to aromatics. The xylenes and ethyl benzene have high octane values and are good fuel blending components. Since the feed is low in naphthenics (cyclo-paraffins), better yields can be expected than with normal reforming, but this advantage is offset by the higher capital cost of such technology. There is consequently little overall benefit compared to reforming and unless such technology has already been selected for C_6-C_7, this would not be the refining option of choice.

- Reforming: Reforming is a well established technology for the upgrading of C_8 and heavier naphtha fractions. The product is rich in aromatics and has a high octane. FT feed has a disadvantage in reforming due to its high linearity and low cyclic content. Since many reforming technologies use chlorided metal promoted catalysts, the potential presence of water or oxygenates in the feed is a further disadvantage. Hydrodeoxygenation of the feed prior to reforming therefore required. Nevertheless, reforming remains a proven refining option and the fuels refining option of choice.

3.2.7 C_9-C_{10} hydrocarbons

The C_9-C_{10} FT product is rich in oxygenates, including carboxylic acids, although the main components remain the α-olefins. The extraction of 1-decene can be considered for the manufacture of poly-alpha-olefins (PAO) [51, 52], a synthetic lubricating oil. The odd numbered 1-nonene presently has no equivalent application, although it can in principle also be used for PAO manufacture. The complexity of the mixture makes purification difficult and it is easier to consider fuels refining. There is little choice in refining methodology and the preferred refining option for fuels production is:

- Reforming: The use of reforming presupposes hydrotreating of the feed. It has already been noted that FT material has some disadvantages compared to crude derived material for reforming. However, it is a proven technology and able to upgrade the C_9-C_{10} material without problem.

Depending of the fuel specifications, it could be possible to include some of the hydrotreated C_{10} material in kerosene, illuminating paraffin or diesel. It is sulphur free and due to the linearity of the material, its Cetane is high.

3.2.8 C_{11}-C_{22} hydrocarbons

The diesel boiling range is one of the strengths of the FT product. It is sulphur free and fairly linear, giving it a good Cetane value. The particulate emission requirements are easily met, because of the low aromatic content in general and the absence of polynuclear aromatics in particular. However, the material is rich in oxygenates, including carboxylic acids and due to its linearity it has poor cold flow properties. There is consequently some refining necessary and the preferred options are:

- Hydrotreating: The hydrogenation of oxygenates can be done with a sulphided Ni/Mo, Ni/W, Ni/Co/Mo or Co/Mo catalyst. This requires the addition of a small amount of sulphur to the sulphur free feed.

Hydrogenation with an unsulphided catalyst is more difficult, since the acids tend to leach the metal from the catalyst. Hydrotreating technology is therefore not the obstacle, but catalyst selection is difficult. The hydrogenated product responds well to the addition of commercial quality improving additives.

- Hydroisomerisation: The main difference between hydrotreating and hydroisomerisation is that the latter uses a catalyst that includes acid functionality. This enables the isomerisation of the feed to take place while doing the hydrotreating. This results in a product with better cold flow properties, albeit at the potential loss of some Cetane. The reactive nature of the feed needs to be taken into account when the catalyst and operating conditions are specified. The catalysts used for hydrocracking and hydroisomerisation are the same. At low processing severity hydroisomerisation is the dominant reaction pathway, but as processing severity increases more hydrocracking will take place, thereby increasing the production of low octane naphtha. Gum formation on the acidic sites of the hydroisomerisation catalyst is also possible if the feed in not hydrotreated. The use of a hydroisomerisation catalyst as an intermediate bed in a hydrotreater is therefore preferable to using hydroisomerisation as a refining option on its own.

It is possible to extract some of the linear α-olefins (typically C_{12}-C_{15} range) to produce detergent alcohols by hydroformylation [53]. This has been done in the Sasol Secunda refineries.

One shortcoming that is still left unaddressed is the low density of the diesel. This can only be redressed by aromatics addition. Blending with biodiesel improves the physical density but does not improve the energy density and high energy density is actually the desired product characteristic for the fuel user. Even Euro-4 does not impose a limit on aromatics for diesel, only polyaromatics ($\leq$ 11%). Some limitation on aromatics content is indirectly provided by the reduction in upper density limit to 845 kg.m^{-3}.

3.2.9 Heavier than C_{22} hydrocarbons

The fraction of the HTFT product with a boiling point higher than 360°C is small, but not negligible. Like the diesel fraction (C_{11}-C_{22}) it consists mainly of hydrocarbons, mostly olefins, with a low degree of branching and oxygenates. Size reduction of the hydrocarbon chains is the most important objective, since post-processing in the appropriate carbon number range is always possible. Another important issue is the density of the diesel. The preferred refining option is:

- Hydrocracking: Hydrogen addition, rather than carbon rejection, underpins the molecular size reduction. It is a clean, well-known and efficient technology that can be used to move heavier products into the less than 360°C boiling range, while improving the cold flow properties by isomerisation. The catalysts used are typically sulphided Co/Mo, Co/W, Ni/Mo and Ni/W on silica-alumina or acidic zeolitic carriers and unsulphided Pt, Pd and Pt/Pd on silica-alumina or acidic zeolitic carriers. Depending on the choice of catalyst and operating conditions it can also be used for lube oil production. Since the naphtha derived from hydrocracking is clean, but generally of a low octane, maximum diesel yield is preferred. In principle the unconverted heavy-end can be recycled to extinction or worked up as oils. Although it seems that hydrocracking only addresses the size reduction issue, it can also be used to produce aromatics for density improvement. By lowering the hydrogen pressure it is possible to produce aromatics to increase density with the additional benefit that the rate of hydrocracking also increases. Some hydrotreating before hydrocracking may be advisable due to the high oxygenate content, which can be done in the same reactor.

In crude oil refineries heavy-end upgrading is done by thermal cracking (visbreaking) or fluid catalytic cracking (FCC). These refining options are possible, but not preferred, because of the clean nature of the HTFT heavy-end product and small volume produced. Depending on the market and location of the refinery the heavy-end can also be sold as a fuel oil after hydrotreating. If this is possible, the hydrotreating of the C_{11}-C_{22} material and $>C_{22}$ material can be done in the same reactor before being fractionated.

4. REFINERY INTEGRATION

It has already been mentioned that coal based processes may require a tar refinery to deal with the coal pyrolysis products when low temperature gasification processes are used. Although this is a separate refinery with its own challenges, it offers interesting integration opportunities. Analogous opportunities exist when natural gas is used as feedstock and these have been used at the Mossel Bay plant. The use of low and high temperature FT processes at the same site, as for the original Sasol 1 refinery, also have synergies. Likewise the exchange of products with a crude oil refinery can be beneficial, as exploited using the Natref crude oil refinery in Sasolburg [11]. With the completion of the natural gas pipeline from Mozambique to the Sasol plants in South Africa in 2004, the integration of all the aforementioned products will be possible: natural gas, crude oil, coal tar as well as high and low temperature FT products. This is very exciting and considering that chemical extraction is practiced too, it makes for interesting refinery integration.

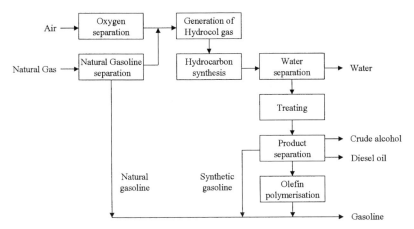

Figure 4. Integration of natural gas with the Hydrocol HTFT refinery.

4.1 Tar integration

The configuration of a tar refinery is dictated by the nature of the coal pyrolysis products. In general the coal pyrolysis products are aromatic and contain sulphur, nitrogen and oxygen as hetero-atoms [54]. Since it is especially rich in phenolic material, phenol, cresols and xylenols (also collectively known as tar acids) can be extracted economically. In a fuel refining context hydroprocessing of the tar must be done in such a way that the hetero-atoms are effectively removed, but that the aromaticity is retained. In the petrol boiling range aromatics are required for octane and in the diesel boiling range aromatics are required for density. The products from a tar refinery therefore compliment that from a FT refinery nicely in terms of redressing octane and density shortcomings. Integration can take place downstream during fuel blending [9], but also at refinery level. Hydroprocessing of the more refractory hetero-atom-rich tar fractions may result in aromatics saturation. This destroys much of the benefit, but since the naphthenes can improve reforming performance, such streams can be combined with the FT feed to a reformer.

4.2 Natural gas integration

The condensate from natural gas can be seen as a paraffinic feedstock to the refinery. When the Hydrocol process was developed, all the condensable natural gas was used as direct blending stock with the FT derived material (Fig. 4) [6]. Current fuel specifications would make such integration less likely. The condensate in the diesel boiling range can still be blended directly, but the condensate in the petrol boiling range requires refining.

The processing scheme used for natural gas integration at the Mossel Bay refinery was developed with significant influence from the Sasol plants in Secunda. It targets the production of motor gasoline (Fig. 3) [15]. Therefore, it includes conventional petroleum refinery units like catalytic reforming (C_7-C_{10} paraffins), alkylation (C_4 paraffins and olefins) and skeletal isomerisation (C_4-C_6 paraffins).

It is interesting to note the co-processing of the NGL naphtha fraction with its equivalent synthetic cut. These are hydrogenated together before being catalytically reformed to improve its octane. The hydrogen used in the hydrogenation stage is obtained from the syngas used in the Synthol units. In an analogous way the C_4 paraffins needed for alkylation are derived from the NGL. This is an efficient integration, but depending on the FT refinery configuration, other integration possibilities can also be considered.

The diesel obtained from the Mossel Bay plant is a blend of distillates from the HTFT Synthol Light Oil and NGL with product obtained by oligomerisation of the light hydrocarbons derived from the FT synthesis. This diesel has good fuel characteristics but there are some quality differences associated with its aromatics content (Table 2) [49]. The Mossel Bay plant also produces some oxygenates including ethanol.

Table 2
Composition of the PetroSA diesel fuel

			NG-derived Diesel Fuel	Zero-Sulphur Diesel (COD)
Density at 20°C	kg/l		0,8088	0,8007
Distillation		▪ IBP, °C	222	229
		▪ T90, °C	322	323
		▪ FBO, °C	360	361
Total sulphur	% mass		<0,001	<0,001
Cetane Number			53,0	51,4
Aromatics content	% volume		16,4	10,1

4.3 LTFT and HTFT integration

The integration of high and low temperature FT processes has been well documented [3, 4], since such a processing scheme was used for the original Sasol 1 refinery. Integration of LTFT products in the HTFT refinery causes the carbon number distribution to be shifted to heavier, more linear paraffinic products. This is good for diesel production.

4.4 Oil integration

Unlike tar and natural gas, crude oil is generally not associated with FT processing. However, the integration of an oil refinery with a FT refinery offers interesting synergies. Most crude oils are rich in aromatics, which is deficient in a FT product – moreover, the synthetic hydrocarbons are rich in olefins and low in sulphur, which can benefit oil refining. Integration needn't be at refinery level, but can be in the blending operation. Exchange of partially refined products can also be mutually beneficial.

4.5 Other integration schemes

The integration of products from chemical work-up (aqueous FT product), especially alcohols, can be considered as alternative to etherification products for octane enhancement. In future it may also become desirable to integrate bio-derived materials. A FT refinery would have an advantage over an oil refinery for the integration of bio-derived material, since bio-derived material is rich in oxygenates which are already dealt with in the FT product. In any event, it is possible to combine the integration schemes in various ways, including complete integration. Each combination offers opportunities to create unique product characteristics that would not have been possible from crude oil refining only, or FT refining only.

5. THE FUTURE OF HTFT REFINING

The evolution of the four HTFT refineries in South Africa over the past half-century has been instructive. During this period technology also evolved. Some of the technologies that are considered preferred refining options for FT products, were not yet fully developed when these refineries were built. The change in fuel specifications has been the main driver for change in the refining environment. In addition to this there was a realisation that the extraction of high value chemical commodities is beneficial.

One may ask what the best configuration for a HTFT refinery is? A similar question has been pondered for crude refining, considering the impact of environmental legislation and the increasingly stringent global fuel specifications [55, 56]. The answer is of course not straightforward and depends on factors like refinery location, size, capital constraints, feedstock cost and product pricing [57]. Some general trends can be noted though:

- Minimise wastage: Fuels refining is very sensitive to feed and product pricing. It is therefore better to select technologies that do not degrade products. Hydrogen addition is consequently preferable to carbon rejection.
- Upgrade paraffins. Paraffins are considered environmentally benign and are the only molecules not legislated by fuel specifications. Investing in technologies that upgrade paraffins or yield good paraffinic products is therefore prudent. Heavy paraffin hydroisomerisation and hydrocracking and light paraffin skeletal isomerisation fall into this category. Investing in such technologies would also make integration of natural gas condensates and LTFT products easier.
- Install clean technologies. The cleanest technology is the technology with the smallest environmental footprint to achieve a specific outcome. The

environmental footprint refers to energy use, quantity and nature of waste and by-product formation. Although technologies like HF alkylation and chlorided platinum catalyst skeletal isomerisation produce desirable paraffinic products, it may be better to consider solid acid alkylation and Pt/zeolite skeletal isomerisation even though it might be slightly more expensive. It is less likely that such technologies will be affected by future environmentally driven legislation. This can also be seen as a form of responsible engineering.

- Target high value products. When a choice can be made between producing a fuel or higher value chemical, it is generally better for the process economics to opt for chemical production [58]. However, this should not be done at the expense of refinery flexibility or in the absence of a proven market for the chemical. The aim of chemical production must always be to improve profitability of the refinery, which implies that the facility must remain viable even when a downturn of the economy depresses the chemicals business. The best chemical products to target are those that have a sustainable competitive advantage and those that can easily be reincorporated into the refinery if necessary (Table 3). In this case low cost refinery units are preferred to avoid large capital investment in idle capacity. An alternative may be to use an imported hydrocarbon feed (e.g. a natural gas condensate fraction) to use the spare refining capacity created by the extraction of chemical components.

Table 3

Chemicals extraction, their impact and their reincorporation pathways in a refinery

Chemical	Extraction impact	Refinery incorporation pathways
Ethylene	Marginal	Alkylate benzene to ethyl benzene for petrol. Recycle to the methane reformer. Metathesis with 1-butene to produce propylene.
Propylene	Refining less costly	Aromatisation to produce aromatics and LPG. Alkylate benzene to cumene (fuel or chemical). Oligomerisation to fuel.
1-Hexene	Octane gain	Aromatisation to produce benzene. Hydrotreating / skeletal isomerisation for fuel.
1-Octene	Octane gain	Aromatisation to produce aromatics for fuel. Reforming to produce aromatics-rich fuel.
Octenes mix	Octane gain	Aromatisation to produce aromatics for fuel. Reforming to produce aromatics-rich fuel.
1-Decene	Octane gain	Reforming to produce aromatics-rich fuel.
C_{12-15} α-olefins	Cetane loss	Hydrotreating to produce high Cetane diesel.

Using these pointers with the preferred refining options discussed in the preceding section, it should be possible to have some indication of how an "optimal" future HTFT refinery should look like.

The art of HTFT refining lays not so much in meeting the fuel specifications as in exploiting the unique feed advantages offered by the FT product. In conclusion it can be said that a HTFT product lends itself to an integrated fuels and chemicals refining approach. Refining with the aim to produce only fuels or only chemicals can be done, but this would be sub-optimal.

6. UPGRADING OF LOW TEMPERATURE FISCHER-TROPSCH PRODUCTS

6.1 Characterisation of the primary LTFT products

The typical LTFT plant would produce two primary products: a light fraction, usually liquid at room temperature, and a heavy fraction, usually solid at the same conditions. The former is often named hydrocarbon condensate or simply condensate and includes hydrocarbon species with a final boiling point around 370°C. The latter, also known as wax, includes the heavy paraffins. There are two other product streams: (i) light hydrocarbons gases, mostly generated during the FT synthesis, and (ii) reaction water, which include some dissolved oxygenates like alcohols and organic acids. The gas stream can have many applications as a fuel gas. The reaction water needs to be further processed and, at some locations, might even become a valuable product. There are several options for the purification of LTFT reaction water [59].

The primary LTFT products have been described in detail in Chapter 3 and in the literature [60]. Typical distillation ranges for the LTFT Condensate and Wax are presented in Table 4.

While the main species in the LTFT primary products are linear paraffins, smaller contents of olefins and oxygenates are present as well as some branched paraffins. The product slate for a particular system depends on the reactor configuration used, operating conditions and the catalyst that is employed [61]. FT Wax production is about double of that of the FT Condensate.

Table 4
Typical distillation range for LTFT syncrude fractions

Distillation Range	FT Condensate % vol	FT Wax % vol
C5-160°C	44	3
160-270°C	43	4
270-370°C	13	25
370-500°C	-	40
>500°C	-	28

The composition and yield immediately suggest that an effective approach to the refining of the LTFT primary products should include some form of hydroprocessing. The recovery of the oxygenated hydrocarbons is always an option depending on the scale of the operation. Further oxidation of the hydrocarbons can even be considered for niche applications [62].

Hydroprocessing is a term used generically for a number of processes that include hydrogen in heterogeneous reaction systems. This includes the technologies summarized in Table 5.

Table 5
Hydroprocessing based technologies

Technology	Purpose	Remarks
Hydrogenation	Saturation of C=C bonds	
Hydrodesulphurisation	Sulphur removal from CS bonds	
Hydrodeoxygenation	Oxygen removal from CO bonds	
Hydrodenitrogenation	Nitrogen removal from CN bonds	
Hydrodemetallation	Metal removal from hydrocarbons	
Hydroisomerisation	Modification of molecular structures	
Hydrocracking	Cracking of large molecules and saturation of the C=C bonds	Products are mostly isoparaffins
Catalytic dewaxing	Selective cracking of large molecules and saturation of C=C bonds	Some products are usable as base oils

Hydroprocessing often includes more than a single type of chemical reaction. In these cases the name of the most significant one is generally used to name the overall process [63]. One typical example is hydrocracking which is required to convert wax to middle distillates and also includes hydrogenation and hydroisomerisation. The FT primary products are sulphur free so hydrodesulphurization is obviously not applicable.

6.2 LTFT primary product refining

The LTFT primary products are ideally suited for upgrading to middle distillates with naphtha as the main co-product. The most suitable middle distillate product is diesel and the lighter and heavier fractions are usually undesirable due to limited markets and/or lower prices. The exception is the option to process wax for the production of lubricant base oils as discussed in Section 6.2.3. The production of kerosene/jet fuel as a co-product is optional. The kerosene cut is generally more valuable as a feedstock for the production of detergent alkylates and is then separated prior to processing the remaining hydrocarbons. Even if this cut is not separated upfront it is still possible to cut the diesel to meet a typical flash point specification and allow all the lighter material to report to a high quality naphtha product.

For middle distillate production, two types of processes could be used: hydrocracking of the heavy FT paraffinic wax and catalytic oligomerisation of light olefins, i.e. C_3-C_5 olefins. The latter is especially applicable to the products from the Sasol Synthol HTFT process, where the bulk of the product consists of these olefins. Only LTFT processes using fixed bed reactors would not consider oligomerisation due to the low olefin content of the primary product. If the olefins lighter than diesel are converted to the diesel boiling range then the remaining naphtha may not require hydrogenation to improve the storage stability. Oligomerisation has already been discussed in relation to the refining of HTFT products so the remainder of the discussion on the processing of LTFT products will focus on the hydroprocessing operations.

The hydrocracking of the heavy paraffins serves two purposes, to reduce the boiling range to middle distillates and to improve the cold properties as the hydrocracked products are mostly branched.

Currently the production of a high quality diesel fuel is a preferred option to the production of gasoline. This is because the very factors which count against FT gasoline, viz. product linearity and low aromatic content, are very positive factors in favour of high quality, i.e. high cetane number, diesel fuel. For maximum production of high quality diesel fuel the slurry bed reactor operating in the high wax selectivity mode with either iron or cobalt based catalysts, between 210 and 260 °C and about 3Mpa, is the recommended route. The straight run FT diesel makes up about 20% of the total FT product and because it is predominantly linear it has a cetane number of about 75. Note that at present the specified cetane number of diesel fuels varies from about 40 to 50, depending on the location. The FT slurry reactors are operated for maximum wax production because subsequent down stream hydrocracking of the wax under relatively mild conditions makes the largest contribution to the final diesel fuel pool. In the case of cobalt catalysts there are two reasons for operating the FT process at high pressures, the wax selectivity increases with pressure (see Chapter 3, Section 9.3.2) and the degree of branching decreases [64]. The hydrocracking of wax with standard bi-functional catalysts was investigated at Sasol during the 1970's [65, 66]. Mild catalytic hydrocracking of the wax yielded about 80% diesel, 15% naphtha and 5% C_1 to C_4 gases. The product cut heavier than diesel was recycled to extinction. Simple calculation shows that the above product yields are the result of random beta scission along the linear wax chains. There is therefore a big incentive to improve the selectivity of the wax hydrocracking operation in order to increase the diesel cut yield. Some chain branching does occur during the wax hydrocracking operation and so the cetane number of the diesel fuel produced is somewhat lower than that of the straight-run FT diesel. The final diesel pool nevertheless has a cetane number of above 70. The naphtha produced in the wax hydrocracking process consists only of alkanes. The naphtha produced in the FT process also consists predominantly of linear alkanes. To convert these two naphtha cuts to in- specification gasoline would require a considerable amount of further octane number upgrading. However, since these naphthas consist essentially of linear alkanes they would be an excellent feedstock for the production of ethylene by steam cracking, yielding a much higher selectivity of ethylene than would be obtained from normal crude oil naphtha.

6.2.1 Hydrocracking of heavy paraffins

In contrast with petroleum hydrocracking feedstocks, the LTFT Wax is predominantly paraffinic, sulphur free, metals free and practically aromatics free. These characteristics are the ideal for hydrocracking and, as a consequence, LTFT feeds can be processed under much milder conditions than typical crude oil derived feeds, e.g. vacuum gas oils. In the hydrocracking of crude oil derived feeds pressures of typically as high as 150 bar are required to prevent coking of the catalyst by the aromatic compounds. This is not necessary with paraffinic feeds and pressures between 35 and 70 bar are used to hydrocrack LTFT products using commercial hydrocracking catalysts. The relatively low levels of oxygenates present in FT waxes, mainly alcohols and lesser amounts of acids and carbonyls, are easily and completely hydrodeoxygenated.

The processing pressure is dependent on the hydrogenation capacity of the catalyst. When the hydrogen partial pressure is too low dehydrocyclisation of the paraffins starts to occur with the formation of polynuclear aromatics which eventually would lead to deactivation of the catalyst due to coking. Lower hydrogen/wax ratios lead to a decrease of conversion rate of the $C_{22}+$ fraction and, after adjustment of reaction conditions to achieve the same conversion levels, an increase of both the iso-paraffin content and the selectivity to products lighter than diesel. Iso-paraffin content also increases with operating pressure.

The hydrocracking process has to meet the following conditions:
- the chain length of the hydrocracked fragments should be predominantly in the desired products range,
- the components above said desired range should be hydrocracked in preference to those which are already in or below the desired range, and
- the production of less commercially attractive species, e.g. light hydrocarbon gases, should be minimized.

Due to the clean nature of the feed, non-sulphided hydrocracking catalysts containing a noble metal component, like platinum, can be considered. The use of noble metals catalysts leads to higher hydroisomerisation activity and, consequently, better low temperature characteristics for the diesel product.

Hydrocracking of FT Wax using a conventional catalyst has been studied over at least the last 30 years. Processing of Arge LTFT was reported in detail in a project sponsored by the US Department of Energy and completed by UOP and the Allied-Signal Engineering Materials Research Center in 1988 [67]. Sasol had earlier approached UOP to investigate the hydrocracking of FT waxes based on the promising results that Sasol had obtained from their in-house research (see Section 6.2). This included studying the effect of the reactor pressure on the hydrocracking performance from 35 to 70 bar. The feed contained some 14% of species boiling in the distillates range. It was found that as the pressure increased, the overall distillate yield first increased, passed through a maximum and subsequently decreased. The higher pressure inhibits secondary cracking and lighter product formation. The pour points of diesels produced from this program varied between -12°C and -37°C. UOP reported that the FT Diesel derived was of extremely high quality, with a very high Cetane number, which can be blended with low-value refinery products such as light cycle oil to increase the volume of the diesel pool. The hydrocracking catalyst used in this program was a commercial sulphided catalyst designed for petroleum refining.

Keeping some differences in mind, comparable trends were reported processing a comparatively light feed – it contained ca 61% wt of material already in the distillates range [68]. The catalyst used was platinum (0.3 mass%) on amorphous mesoporous silica-alumina. This program included testing over the same pressure range (35 to 70 bar) at temperatures between 330-355°C. The degree of isomerisation in the hydrocracked products increased with an increase in the conversion. As a consequence, the freezing point of the kerosene was a very low -50°C, and the pour point of the diesel -30°C, remarkable figures considering the high fraction of linear light species in the feed.

This approach of using a noble metal on amorphous silica alumina is likely to be the preferred approach for maximum production of highly desirable diesel blending material. It is likely that all the $C_{14}+$ material will be treated using such catalysts due to the enhancement in cold flow properties resulting from the simultaneous isomerisation. A high blending Cetane value is retained and very little of the diesel range material in the feed is degraded into naphtha and kerosene.

There are many technology licensors for hydroconversion processes including ChevronTexaco, UOP, IFP (Axens) and Haldor-Tøpsoe. Additionally catalysts can be obtained from these companies as well as from Akzo Nobel, Süd-Chemie and Axens among others.

6.2.2 Hydrotreating of FT paraffins

The quality of the primary FT products, both condensate and wax, can be improved by hydrotreating as Sasol has been doing commercially for many years. The primary objective of this process is to saturate olefins and oxygenates to the corresponding paraffins. This hydrogenation results in a hydrocarbon product that is stable when stored for long periods. Stability during storage is the primary objective for the hydrogenation of the hydrocarbon condensate or fractions thereof. In the case of LTFT wax, the resulting product has better colour and stability. It is also possible, depending on the reaction conditions, to transform some of the linear hydrocarbons into branched species. In this case the crystallization temperature of the wax can also be influenced. These results are similar to those attainable when hydrotreating petroleum wax in a similar way [69]. However, as it was mentioned for hydrocracking, the required process conditions are significantly less severe because of the chemical nature of the synthetic product.

6.2.3 Catalytic dewaxing / hydroisomerisation

Special hydroconversion operations may be used with fractions of the FT wax to produce lubricant base oils. The hydroisomerisation of petroleum-derived wax to base oil has been practiced for some time, and products from this process have been in the market for at least twenty years.

- In 1972 Shell commissioned their base oil manufacturing plant at the Petit-Couronne refinery in France. This plant uses severe hydrotreatment and solvent dewaxing to produce lubricating base oils. Feedstocks for this plant are waxy petroleum distillates, deasphalted oils and slack waxes, selected based on the desired base oil quality [70].

- Chevron, now ChevronTexaco, was the first to commercialize an all-hydroprocessing lube oil manufacturing process in 1984 at their Richmond refinery in California, USA. This design combined hydrocracking (for Viscosity Index upgrade) with catalytic dewaxing, which removed normal paraffins by cracking. Feedstocks to this process were petroleum vacuum gas oils [71]. Their ISODEWAXING process was brought into on-stream at the same refinery in 1993 [72]. This technology reaches the target pour point of the products in one step by isomerising part of the wax to base stocks and cracking part of it to fuels. In 2003 ChevronTexaco announced the commercial operation in Russia of an ISODEWAXING unit operating on petroleum wax [73].

- Mobil-developed technology, Mobil Selective Dewaxing (MSDW), started up commercially in their Jurong (Singapore) plant in 1997 [74]. More recently, ExxonMobil has announced the commercialization of a new wax isomerisation process at their Fawley refinery, UK [75]. The products from Fawley will have viscosity indexes of 140 and pour points of -18°C. This operation was scheduled to start by the end of 2003 using slack wax as feed.

All of these processes can be used for processing FT wax, and the derived base oils have extremely high VI's and low volatilities, making them ideal bases for future high-performance lubricants.

Fractionation into the various base oil products may be done upstream or downstream of the catalytic dewaxing step. This can be accomplished by using either high vacuum or short path distillation. The catalytic dewaxing operation may be followed by a hydrofinishing step to destroy any polynuclear aromatics formed in the first process step, stabilizing the base oil. The co-produced highly isomerised light fractions - diesel and naphtha, and optionally kerosene/jet fuel – have similar characteristics to the products from a hydrocracker but the yields are lower. A process scheme similar to that used with petroleum derived waxes and suitable for the catalytic dewaxing of FT Wax is shown in Fig. 5.

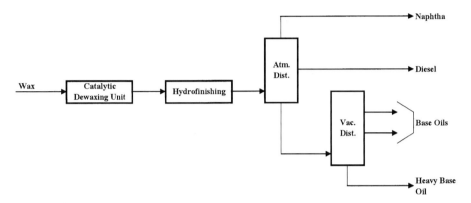

Figure 5 Process scheme for the production of synthetic base oils from wax

The catalytic dewaxing route to lubricant base oils is becoming the preferred processing option due to the lower capital and operating costs compared to conventional solvent dewaxing. Recently introduced catalytic dewaxing technology, e.g. ChevronLummus Global's ISODEWAXING and ExxonMobil's MWI (Mobil Wax Isomerisation), makes converting FT wax to base oils more economical than ever before.

6.3 Hydroprocesssing catalysts for LTFT primary products

6.3.1 Basic concepts

Although the type of hydrocracker feed from a FT process differs from the normal refinery hydrocracker feed, e.g. VGO, the basic hydrocracking mechanism is the same and standard hydrocracking catalysts can be used. As in a refinery hydrocracking process, the choice of catalyst will depend on the required product slate. When high middle distillate selectivity is desired amorphous carriers would be chosen while, on the other hand, when maximum kerosene and naphtha selectivity are wanted, a mixed zeolite/amorphous carrier will probably be used.

Commercial hydrocracking catalysts are dual function catalysts, containing a hydrogenation/ dehydrogenation function, provided by the active metal(s) and an acidic function derived from the carrier, typically amorphous silica-alumina or crystalline silica-alumina, i.e. zeolites.

According to generally accepted hydrocracking theory, the active metal sites promote dehydrogenation of the paraffins to olefins [76]. These are then protonated on the acidic carrier to a carbenium ion. Finally, the carbenium ion undergoes isomerisation and cracking on the acid site.

The ratio between the cracking function and hydrogenation function is of utmost importance and can be adjusted to achieve the optimum activity and selectivity. Generally speaking, other things being equal, it can be stated that the higher the acidity at a given hydrogenation activity, the higher the activity and degree of isomerisation of the cracked products and the lighter the product slate. On the other hand, increasing the hydrogenation activity at a given acidity will lead to a heavier and less isomerised product. A higher acidity to hydrogenation ratio will produce a more isomerised diesel with improved cold temperature characteristics and with a slightly lower Cetane number.

The amorphous carrier used in most commercial hydrocracking catalysts is silica-alumina, although other acidic carriers, like silica-magnesia, silica-titania and other mixed acidic oxides have been reported. Zeolite Y is the most frequently used zeolite, together with silica-alumina or alumina alone as a binder. The metals mostly used are either non-noble metal combinations like Co/Mo, Co/W, Ni/Mo or Ni/W as the sulphides or noble metals like Pt or Pd, either alone or in combination.

The hydrocracking of LTFT products can be compared to the second stage hydrocracker operation used in petroleum refineries because no feed pre-treatment is required. Moreover, noble metal catalysts can also be used due to the fact that the feed is sulphur free. LTFT feeds do contain oxygenates, mainly alcohols, carboxylic acids, aldehydes and esters. These oxygenates are easily decomposed and hydrogenated under hydrocracking conditions to give the corresponding paraffins and water. Carboxylic acids may also undergo decarboxylation to CO_2 plus a lower paraffin.

6.3.2 Catalyst carriers

The carrier functions both as a support for the metal and to provide the required acid functionality. As a carrier it has to have a high surface area for high dispersion of the metal as well as the right pore size to allow easy diffusion of the feed molecules into the catalyst and of the hydrocracked and isomerised products out of the catalyst. As the LTFT derived feed consists mainly of linear paraffins and does not contain polyaromatics, the requirement for sufficiently large pores is less stringent than in the case of petroleum feeds.

The acidity, in the case of amorphous silica-alumina, depends on the preparation method as well as mainly on the silica/alumina ratio. Brönsted acidity, which is accepted to be responsible for the formation of the carbenium ion intermediate, reaches a maximum at around 70% silica and is caused by tetrahedrally coordinated aluminium, which carries a negative charge, compensated by a proton which protonates the olefin, forming the reactive carbenium ion.

In the case of zeolites, the acidity depends also on the silica/alumina ratio as well as on the zeolite structure and the degree of exchange of sodium ions with protons. In mixed zeolite amorphous silica/alumina carriers the amorphous phase can either act simply as a binder at low level or at higher levels can also contribute to the activity.

6.3.3 Metal components

Most commonly the metals used in hydrocracking catalysts are a combination of group VIA (molybdenum, tungsten) and group VIIIA (cobalt, nickel). These metals are sulphided prior to use and kept in a sulphided state during the hydrocracking operation by sulphur compounds in the feed in the case of refinery feeds or by adding sulphur containing compounds in the case of sulphur free feeds, like FT derived feeds.

From hydrogenation studies it was found that the optimum ratio of the metals is:

$M_{VIII}/(M_{VIII} + M_{VI}) = \sim0,25$, where M_{VIII} is Co or Ni and M_{VI} is Mo or W.

The hydrogenation activity of the various couples decrease in the order:
Ni-W > Ni-Mo > Co-Mo > Co-W.

The hydrogenation activity is also function of the metal loading and dispersion. The required metal loading will depend on the desired balance between the cracking/isomerisation and hydrogenation functions.

The noble metal content of hydrocracking catalysts is generally around 0.5% or less. For non-noble metal the loadings are 3-8% nickel or cobalt oxides and 10-30% for molybdenum or tungsten oxides.

6.4 Hydroprocessing flow sheet options to produce diesel

The FT syncrude upgrading may be designed in a number of configurations. These can be selected by process synthesis optimisation depending on the desired product slate [77]. Four of the possible process configurations are described in Table 6 and shown in Fig. 6.

512

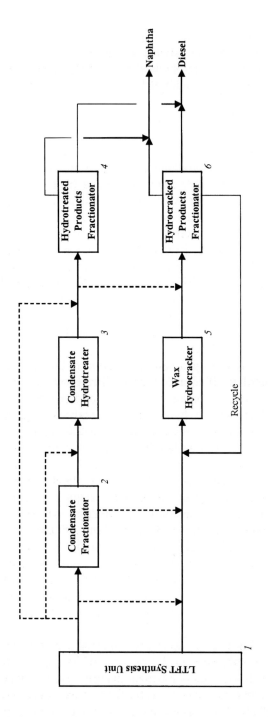

Figure 6   Possible process configurations for the upgrading of LTFT products

Table 6
Process configuration options for FT syncrude product upgrading

	Process Unit	Possible Process Configuration			
2[*]	Condensate Fractionator	X		X	
3[*]	Condensate Hydrotreater (HT)	X	X		
4[*]	HT Condensate Fractionator		X		
5[*]	Wax Hydroprocessing (HP)	X	X	X	X
6[*]	HP Products Fractionator	X	X	X	X

[*] Reference for the Fig. 6 numerals

The FT condensate may be hydrogenated to saturate the double bonds of olefins and eliminate oxygen by hydrodeoxygenation. After this process the product is a mixture of more chemically stable linear paraffins.

As indicated before, hydroprocessing is used to upgrade FT wax to distillates or lube base oils and distillates. This process also converts the olefins and oxygenates to paraffins. The FT wax undergoes hydrocracking reactions and the product is a mixture of isomerised paraffins with low levels of naphthenic and aromatic species. High quality distillates can be obtained by proper selection of the catalyst and processing conditions.

Hydrogenation is less costly than hydrocracking. For large scale plants a condensate hydrotreater should be used to process hydrocarbons that are already lighter than the desired middle distillate products (option 1) and also for the straight run middle distillates if cold flow properties are not an issue (option 2). Option 2 may also be chosen if it is desired to produce linear paraffin products. The condensate fractionator can also serve to remove light gases in order to stabilize the condensate for intermediate storage at atmospheric pressure. If a nearby naphtha cracker is available then the condensate fractionator will provide a naphtha product that need not be hydrogenated to ensure storage stability (option 3). Alternatively in this case, the straight run naphtha may be routed to an oligomerisation unit. As mentioned in section 6.2, the kerosene fraction should be separated using the condensate fractionator for use as a chemical feedstock. Finally option 4 may be selected due to its simplicity if the product flows can be handled by a single reactor.

If both hydrotreating and hydrocracking are used then they may be operated at the same pressure and use a common hydrogen loop. Alternatively, the hydrotreater can be operated in a once through mode at a lower pressure and the hydrogen tail gas is then subsequently pressurized to be fed into the hydrocracker loop.

The Sasol SPD™ process is configured in synthesis units whose capacity is set by the maximum practical capacity of some of its process units. The unit that defines the current 17 000 bpd-equivalent maximum capacity per train is the air separation unit. Recent advances in technology at Sasol give confidence for even larger Slurry Phase FT Reactor trains. The maximum capacity of the product upgrading unit is comparable to those in oil refineries. Table 7 contains a mass balance for a 34 000 bpd-equivalent Sasol SPD™ product upgrading unit corresponding to Option 4 [77].

Table 7
Mass balance for a 34 000 bpd Sasol SPD™ product upgrading unit.

	Feed MSCFD	Sasol SPD™ Products bpd
Natural Gas	170	-
▪ LPG	-	1000
▪ Naphtha	-	7000 – 9000
▪ Diesel	-	24000 – 26000
Total Products	-	34 000

6.5 Alternative process options

Processing of the LTFT primary products can be effected in many ways, following a process synthesis exercise. The final configuration will be selected considering:

- Site specific factors like plant capacity, available land and possible integration with other facilities.
- Market issues, including product(s) demand and logistics.
- Other factors like capital requirement and skilled labour availability.

A few possible process configurations based on two primary LTFT liquid products are shown in Fig. 6. While these schemes make reference to only three final products, i.e. LPG, naphtha and diesel, it is completely possible to configure a plant in different ways. Other products might include solvents, kerosene, jet fuel, illuminating paraffin and base oils, as well as some olefinic petrochemicals.

7. CHARACTERISTICS OF THE LTFT DISTILLATE PRODUCTS

The main fuel product obtainable from the LTFT process is the proven low emissions diesel fuel (i.e. fuel for a compression ignition engine). In addition, this process also delivers a significant fraction of naphtha, an excellent feed for the production of olefins via steam cracking.

7.1 LTFT diesel

The first LTFT diesels were commercially produced in Germany around 1935. This fuel was obtained by distillation of the light hydrocarbons from the FT reactors; there was no hydroprocessing and as a consequence it contained olefins and oxygenates. This scheme is presented in Fig. 7.

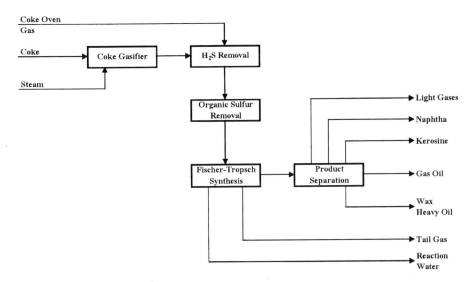

Figure 7   Typical process scheme of the 1935-1945 German FT plants

The first LTFT plants of Sasol included a more complex processing scheme. The original concept, shown in Fig. 8, showed a high degree of integration to maximize the benefits of the unique characteristics of the FT products [3]. This process scheme included eight separate units. While its objective was to produce fuels, gasoline and diesel, it also included the production of LPG and hydrogenated waxes. The quality of the gasoline was by improved by a Hot Refining step, a 400°C fixed catalyst bed treatment with two objectives: (a) conversion of oxygenates to other hydrocarbon species and (b) shifting the double bond in olefins to the centre of the molecule. Both changes improved fuel stability and octane number of the synthetic naphthas. The diesel was a blend of the straight run FT products and the cracked stock from the Paraformer, a thermal cracker unit – therefore still containing some olefins and oxygenates.

Table 8 presents the quality of a sample obtained and analysed by the Allied forces after the war [78]. It is interesting to note that this straight run synthetic diesel produced using a cobalt FT catalyst, contained aromatic species and a significant amount of olefins.

With time the processing of the LTFT primary products became more sophisticated. Most GTL technology licensors include different hydroprocessing arrangements that eventually will lead to comparable distillates. Therefore, in broad terms, the information related to the characteristics and performance of the Sasol Slurry Phase Distillate™ (SPD™) diesel that follows is also comparable to the other synthetic GTL LTFT-based distillates [79, 80].

Table 8
Quality of a World War II German FT diesel

		Reported value
Density	g/cm^3 (20°C)	0.7681
Distillation	• IBP, °C	193
	• T10, °C	218
	• T50, °C	248
	• T90, °C	291
	• FBP, °C	311
Flash Point	°C	78
Pour Point	°C	-1
Bromine Number		6.9

The LTFT synthetic diesel has several important environmental advantages over conventional fuels. It has superior combustion characteristics due to their high Cetane number. Moreover, it is practically sulphur free and has a very low aromatics content. The basic properties of the Sasol SPD™ diesel are compared in Table 9 with those of two fuels: a CARB specification fuel and a commercial US 2-D diesel. The lower aromatics and sulphur contents of the Sasol SPD™ diesel are immediately evident.

Table 9
Basic properties of diesel fuels [75]

		US 2-D	CARB Specification	Sasol SPD™ Diesel
ASTM D86	IBP, °C	184	203	189
Distillation	T50, °C	259	249	256
	T90, °C	312	290	331
Density	kg/L	0,855	0,831	0,777
Viscosity	cSt (40°C)	2,4	2,4	2,4
Cetane Number		40	49	>70
Aromatics	wt %	32,8	6,7	0,5
Sulphur	mass %	0,028	0,022	0,001

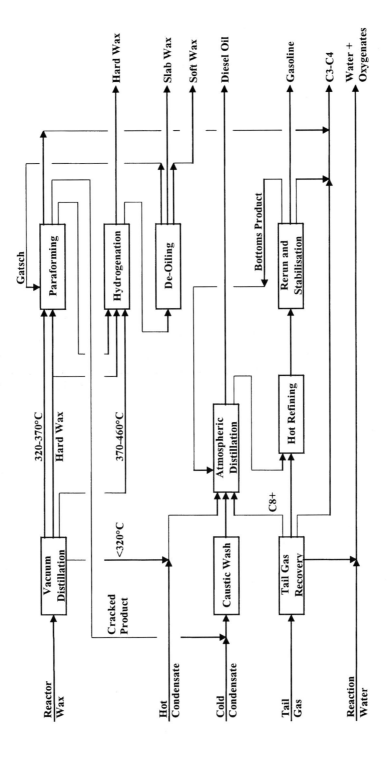

Figure 8   The original process scheme for the LTFT Sasol 1 plant (ca 1957)

7.1.1 LTFT diesel as a blending component

The Sasol SPD™ diesel was blended with the US 2-D specification fuel used for the emission tests to evaluate possible impact on quality. The typical quality for these blends is shown in Table 10 [79]. Blends of this fuel with the synthetic diesel resulted in improvements in Cetane number, and reductions in the sulphur content and the Cold Filter Plugging Point (CFPP), see Figs. 9 and 10. However, the synthetic fuel density is lower than that of petroleum fuels because of its very low levels of aromatics and naphthenics.

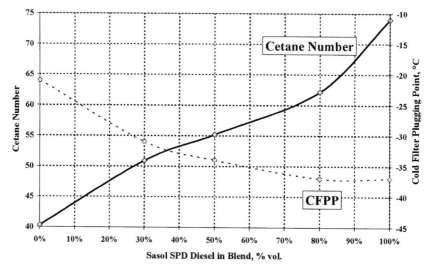

Figure 9   Sasol SPD™ Diesel US 2-D-Diesel Blends: Cetane Number and Cold Filter Plugging Point (CFPP)

7.1.2 LTFT diesel fuel performance

The emission tests were completed at the Southwest Research Institute (SwRI). The results show significant improvement when compared with the commercial US 2-D diesel and with the CARB specification fuel. The synthetic diesel produced lower emissions compared with the CARB fuel standard in all four groups: hydrocarbons, CO, NO_X and particulates. The improvement relative to the commercial US 2-D diesel performance was even more significant [79].

Blends of the Sasol SPD™ diesel with the US 2-D fuel showed reductions in emissions that were generally proportional to the amount of the synthetic fuel in the blend. Moreover, it was estimated that a blend of approximately 40% of the Sasol SPD fuel with the US 2-D fuel used in the tests would result in equivalent emissions to the CARB fuel.

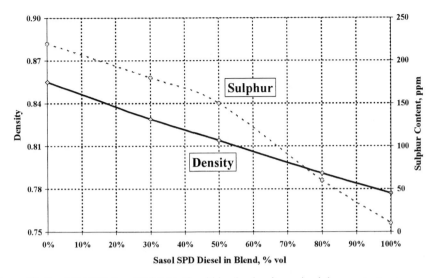

Figure 10 Sasol SPD™ diesel US 2-D-diesel blends: density and sulphur content

Table 10
Average results from the emission tests with CARB, Sasol SPD™ and US 2-D diesel fuels

| | | | Sasol SPD™ Diesel - US 2-D blend, % vol | | | | | CARB |
			0%	30%	50%	80%	100%	Diesel
Transient Emissions								
▪ Hydrocarbons	HC	g/kWh	0.22	0.13	0.11	0.09	0.09	0.13
▪ Carbon Monoxide	CO	g/kWh	3.83	3.17	2.9	2.62	2.57	3.34
▪ NO$_X$	NO$_X$	g/kWh	7.05	6.17	5.72	5.3	5.08	5.96
▪ Particulate	PM	g/kWh	0.277	0.278	0.265	0.241	0.218	0.276
▪ BSFC (a)		g/kWh	236	237	231	228	231	233
Test Engine			12,7 L DDC series 60 (1988), rebuilt to 1991 emission levels					
Rated Power			261 kW @ 1 800 rpm					
Peak Torque			1 830 Nm @ 1 200 rpm					

(a) Brake Specific Fuel Consumption (BSFC)

If a blend is aimed at meeting the CARB specification then the boiling point range of the LTFT diesel will need to be adjusted accordingly. LTFT diesel can be blended with any current diesel fuel on the market to produce a blend that conforms to the CARB specification. In the case of a blend with US 2-D diesel, 30% LTFT diesel in the blend is more than sufficient to meet the CARB specification in all respects except the aromatics content and the emissions performance is nearly equivalent. 80% LTFT diesel will be required to meet the CARB specification for aromatics but then a higher density third blend component e.g. biodiesel and/or some other heavy oxygenated hydrocarbons will be required to meet the density specification. If biodiesel is used then it is also free of aromatics so that the combined amounts of LTFT diesel and biodiesel in the blend will be about 80%. This blend will have an emission performance that is far superior to conventional CARB diesel.

7.1.3 LTFT diesel environmental characteristics

Sasol also evaluated the biodegradability and ecotoxicity characteristics of the Sasol SPD™ diesel, including some petroleum fuels for comparison.

The biodegradability was measured using the modified Sturm OECD 301B method, based on the carbon dioxide evolution. It was found that the synthetic diesel can degrade rapidly and completely in an aquatic environment under aerobic conditions. Petroleum fuels tested simultaneously behaved differently, with lower degradation rates under the same conditions.

The bacterial toxicity was determined using the OECD method 209 based on activated sludge respiration inhibition. From a toxicity perspective, the Sasol SPD™ diesel performed similarly to petroleum based fuels.

7.1.4 LTFT diesel product applications

The best use of the synthetic LTFT diesels will be to upgrade, by blending, the quality of conventional diesel fuels. Based on their superior quality, the synthetic distillates are the ideal, high value, blending component for upgrading of lower-quality stock derived from catalytic and thermal cracking operations, for example cycle oils [81, 82]. In 1948 JA Tilton from the Esso Standard Oil Company suggested that *"there is a possibility that the Fischer-Tropsch synthesis process may eventually be a source of premium quality Diesel fuels"* [83]. It is interesting to note that it was also anticipated that blends of the use of these high Cetane number diesels with lesser fuels were usable to meet a 50 Cetane specification, concluding that these high Cetane number fuels *"might be used either alone or as blending agents"*. The blend material from conventional processing includes product from FCC and Coker units. The blended final product will be low in sulphur and aromatics, a fuel compatible with demanding environmental legislation. Hence the products could enter a market where the LTFT diesel characteristics are valued as blend material to meet local requirements. It is also possible to use this fuel as a neat fuel in applications where its premium characteristics are desired.

7.2 LTFT naphtha

The characteristics and applications of the Sasol SPD™ naphthas have been the subject of presentations at recent international meetings [77, 84]. Four LTFT naphthas can be produced using the process configurations summarised in Table

11; all of these are products obtainable from a Sasol SPD™ plant. Their typical characteristics are presented in Table 12.

Table 11
Process schemes usable for the production of Sasol LTFT naphthas

FT Naphtha		Production Scheme
SR	Straight Run	Fractionation of FT condensate
HT SR	Hydrogenated Straight Run	Hydrotreating and fractionation of FT condensate
HX	Hydrocracked	Hydrocracking and fractionation of FT wax
SPD	Sasol SPD	Blend of HT SR and HX naphthas

Table 12
Typical characteristics of the Sasol SPD™ naphthas

	Naphtha				Notes
	SR	HT SR	HX	SPD	
ASTM D86 Distillation					
▪ IBP, °C	58	60	49	54	
▪ T10, °C	94	83	79	81	
▪ T50, °C	118	101	101	101	
▪ T90, °C	141	120	120	120	
▪ FBP, °C	159	133	131	131	
Density, kg/L (20°C)	0.710	0.683	0.688	0.685	
Cloud Point, °C - predicted	-51	-54	-35	-33	
Flash Point, °C	-9	-18	-21	-20	(a)
Cetane Number	n/a	42.7	30.0	39.6	
Total sulphur, mg/L	<1	<1	<1	<1	
Composition, % wt					
▪ n-paraffins	53.2	90.1	28.6	59.0	
▪ Iso-paraffins	1.2	8.3	66.7	38.2	
▪ Naphthenics	-	-	-	-	
▪ Aromatics	-	0.1	0.5	0.3	
▪ Olefins	35.0	1.5		2.5	
▪ Alcohols	10.7	-	-	-	

(a) Predicted from DSC analyses.

7.2.1 LTFT naphtha as petrochemical feedstock

The Sasol SPD™ naphtha is an excellent feed for the production of lower olefins, in particular ethylene, by steam cracking [84]. Cracking of highly paraffinic naphtha at the highest feasible severity maximises ethylene yields and reduces by-products. This is also an indication of environmental friendliness because for a fixed ethylene production paraffinic naphtha demands less feed and energy consumption than naphthas from other sources.

522

A comparison of the characteristics of the Sasol SPD™ naphtha with typical petrochemical naphthas is shown in Table 13. The synthetic product can be described as a highly paraffinic low sulphur naphtha with non detectable levels of aromatic species.

The Sasol SPD™ naphtha used in this research program was the fully hydrogenated product derived from the Isocracking® of the Sasol SPD FT syncrude. The program included testing the Sasol SPD™ naphtha performance at several cracking severities and a separate test designed to evaluate the coke build-up at high cracking severity. The results were compared with typical yields reported for two petroleum naphthas [85], and with that of a highly paraffinic product.

The results experimentally determined for the Sasol SPD™ naphtha are compared with those expected when cracking petroleum naphthas at the same severity level under the relatively long residence time attainable during the research runs. This comparison is presented in Table 14.

Table 13
Quality of the Sasol SPD naphtha and the petroleum reference naphthas

Naphtha	Sasol SPD	Typical Petroleum Light Naphtha	Highly Paraffinic Naphtha	Typical Petroleum Heavy Naphtha
Specific Gravity	0.687	0.672	0.660	0.717
Sulphur (ppm wt)	<1	nr	<2	nr
ASTM D86 Distillation				
▪ IBP, °C	51	33	37	33
▪ 10%, °C	54	49	43	52
▪ 30%, °C	57	56	47	71
▪ 50%, °C	75	62	52	96
▪ 70%, °C	93	71	65	124
▪ 90%, °C	107	83	94	170
▪ FBP, °C	125	101	148	207
Composition (%)				
n-paraffins	49.9	49.7	nr	30.7
i-paraffins	42.9	34.6	nr	40.1
▪ Total Paraffins	92.8	84.3	93.0	70.8
▪ Olefins	0.0	nd	nd	nd
▪ Naphthenes	7.2	13.3	5.5	17.8
▪ Aromatics	nd	2.5	1.5	11.4
Total	100.0	100.0	100.0	100.0

nr   not reported
nd   not detectable

Cracking severity was measured by calculating the propylene/ethylene (P/E) mass ratio. Note that P/E has an inverse relation to the cracking severity: high P/E values correspond to low severities and lower methane co-produced.

The optimum olefins yields obtainable from any naphtha depend on the maximum cracking severity attainable at commercial conditions. This in turn depends on the chemical composition of the naphtha and, up to certain extent, to the physical configuration of the cracker unit. The former is evident in Figs. 11 and 12 which show the ethylene and commercial olefin yields at different cracking severities for the synthetic and the two petroleum feedstocks.

It is of relevance to indicate that while this kind of performance was directionally predictable, it was not fully quantified. Indeed, hydroprocessing has been previously reported as a crucial pre-treatment stage to improve the properties of FT naphthas for conversion to lower olefins. [86].

Table 14
Research performance comparison of the Sasol SPD™ naphtha

	Typical Petroleum Light Naphtha			Typical Petroleum Heavy Naphtha			Sasol SPD™ Naphtha			
P/E mass ratio	0.77	0.68	0.59	0.82	0.74	0.65	0.71	0.59	0.50	0.40
Yield (wt %)										
Ethylene	24.1	27.0	29.3	19.4	21.9	24.0	26.5	32.3	35.7	37.5
Propylene	18.7	18.4	17.2	16.0	16.1	15.5	18.8	19.1	17.8	15.0
1,3-Butadiene	4.0	4.2	4.4	3.8	4.1	4.3	2.9	3.8	4.2	4.2
Light Olefins	46.7	49.6	50.8	39.2	42.1	43.8	48.2	55.2	57.7	56.7
Other Olefins	8.2	6.9	5.4	8.3	7.3	6.1	9.9	8.4	6.2	4.1
Hydrogen	0.6	0.7	0.8	0.5	0.6	0.7	0.6	0.7	0.8	0.9
Methane	12.8	14.6	16.3	10.9	12.6	14.2	10.2	13.0	14.9	16.2
Other Paraffins	11.8	9.4	7.5	11.8	9.4	7.3	5.1	5.3	4.8	4.4
Alkynes	0.4	0.7	0.9	0.4	0.7	1.0	0.4	0.7	1.1	1.4
C5+ Liquids	19.4	18.1	18.2	28.8	27.2	27.0	25.7	16.9	14.4	16.3
Total Products	100	100	100	100	100	100	100	100	100	100

The coking rate for LTFT Naphtha under commercial conditions is expected to be lower than that of conventional feeds, meaning that run lengths in commercial steam cracking operations using LTFT Naphtha can be expected to be longer than those expected using conventional naphthas at similar cracking severities.

524

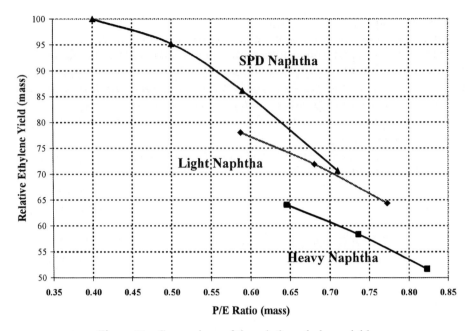

Figure 11   Comparison of the relative ethylene yields

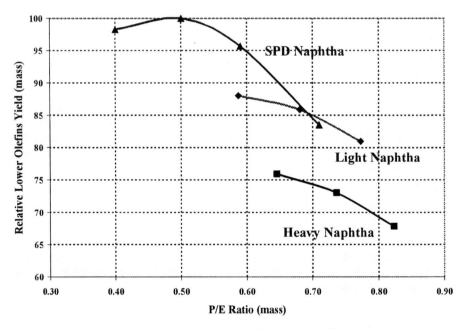

Figure 12   Comparison of the relative total olefin yields

7.2.2 LTFT naphtha fuel performance

Excluding the SR fuel, all these LTFT naphthas can be regarded as paraffinic. This suggested evaluating them as possible distillates. The Cetane numbers for the HT SR and the SSPD fuels are marginally below typical values found in commercial diesel. As expected, being more isomerised, that of the HX naphtha was even lower. The SR naphtha contains significant levels of both olefins and alcohols that reduced its Cetane number and it was decided not to test this fuel in the Cetane machine.

The emissions and specific fuel consumption tests were carried out at the Sasol Oil R&D facilities [77]. A commercial summer grade diesel fuel produced at a South African oil refinery was also tested for comparison purposes under identical conditions. The same standard diesel fuel engine settings were used for all fuels. The results from the engine test are shown in Table 15.

It is evident from their composition that these LTFT naphthas that they are far from being compatible with that high octane naphthas usable in gasoline blends.

Table 15
Diesel engine and emissions performance of the Sasol SPD naphthas

	Sasol FT Naphtha				South African Diesel
	SR	HT SR	HX	SSPD	
Engine	Mercedes Benz 407T				
Test condition	1 400 rpm				
Load	553 Nm				
BSFC (Brake Specific Fuel Consumption), g/kWh	216.4	206.2	201.5	206.8	216.8
CO emissions, g/kWh	0.87	0.79	1.03	0.83	1.11
CO_2 emissions, g/kWh	668.1	676.1	698.9	670.1	700.9
Hydrocarbon emissions, g/kWh	0.398	0.373	0.445	0.301	0.138
NO_X, emissions g/kWh	13.59	12.55	12.47	12.55	16.99
Bosch Smoke Number	0.32	0.37	0.31	0.37	0.67

7.2.3 LTFT naphtha as a blending component

Petroleum refiners producing diesel fuels for cold weather specifications usually change the boiling points of their fuels to improve their cold flow characteristics, making them compatible with low temperature operation. This results in lower production volumes, not only of diesel fuels but also of other light products. The characteristics of the Sasol synthetic naphthas suggested its possible use in blends with diesel fuels suitable for low temperature operation.

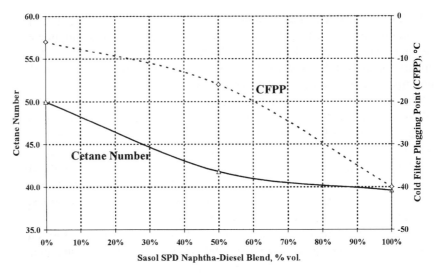

Figure 13   Sasol SPD™ Naphtha-Diesel blend – Cetane number & CFPP

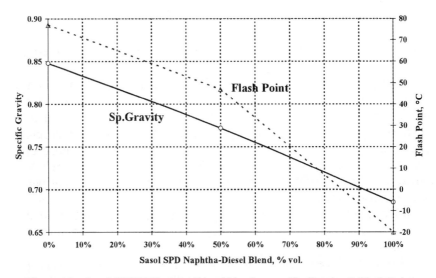

Figure 14   Sasol SPD™ Naphtha-Diesel blend – specific Gravity & Flash Point

Blends of Sasol SPD™ naphtha with conventional diesels would meet winter grade fuel specifications [77, 87]. The quality of the final product will depend on the fraction of naphtha included in the blend, as shown in Figs 13 and 14. As an additional benefit, this practice would help maintaining the production volumes of jet fuel. Maximising jet fuel recovery, while maintaining diesel volumes, should translate into better economic results for most oil refineries. All

FT naphthas have the added benefit of having sulphur contents below 1 ppm. Their inclusion in blends should assist refiners in meeting low sulphur content specifications.

7.3 Other applications for LTFT products

The applications for products derived from LTFT Syncrude are the same as those of most crude oils. However, due to its ultralow sulphur content, all FT-based products will need less processing to meet strict environmental specifications.

The low sulphur content and highly paraffinic nature of the FT derived products make them ideal candidates for fuel cell applications [88]. In this case, their low aromatic and naphthenic contents become an additional advantage that results in extremely low carbon deposition in the reforming catalysts.

High-octane naphthas could be obtained from FT syncrude after adequate processing. Although this is technically attainable, there are better and more cost-effective petroleum processes for this purpose.

The possible use of the LTFT wax as rocket propellant is another interesting option [89].

8. LARGE SCALE PRODUCTION OF HIGH VALUE LTFT PRODUCTS

Most of the recent interest for future LTFT GTL plants has been for the use of supported cobalt catalyst in slurry phase reactors. There are good reasons for this approach not least of which is the relative simplicity of the process that lends itself to successful application at remote locations. Work is in progress to construct plants at Ras Laffan, Qatar and Escravos, Nigeria. This concept uses the Sasol Slurry Phase Distillate™ (Sasol SPD™) process or the Shell Middle Distillate Synthesis (SMDS™) process to produce mainly diesel fuel with by-product naphtha.

Although these processes are ideally suited for the production of diesel, Sasol and Shell have recognized the opportunity to produce chemicals in a synergistic manner.

There was a time when it was thought that future large scale GTL plants will not be able to target higher value hydrocarbon products due to limitations imposed by the size of the markets. For example, Shell stated in 1995 that their future SMDS projects will be based on transportation fuels only [81, 82]. However, for at least three important products namely ethylene, propylene and lubricant base oils (also known as waxy raffinate), the markets are large enough to sustain large scale co-production of these products. In the recently announced Shell plant in Qatar, that is eventually expected to produce about 140 000 bbl/d of hydrocarbon products, the production of lubricant base oils and paraffins is included in the initial product slate.

Sasol has proposed two approaches for the large scale production of higher value hydrocarbons via FT technology. Firstly, there is the option to modify the product upgrading schemes for the primary products from supported cobalt catalyst to produce olefins and lubricant base oils in addition to diesel fuel [90]. Secondly, Sasol believes that its proprietary iron based FT processes can be used to further enhance or expand the production of higher value hydrocarbon products with fuels as significant by-products. Being highly olefinic, the nature of the products from iron catalysts is well suited to extraction and processing for the production of commodity chemicals. This is discussed in more detail in Chapter 5.

REFERENCES

[1]   A. de Klerk, Hydrotreating in a Fischer-Tropsch refinery, 2nd Sub-Saharan Africa Catalyst Symposium, Swakopmund, Namibia 5-7 Nov. (2001).

[2]   J. Stell, Worldwide catalyst report, Oil Gas J., 99:41 (2001) 56-76.

[3]   J.C. Hoogendoorn, J.M. Salomon, Sasol: World's largest oil-from-coal plant, British Chem. Eng., (1957) 238-244, 308-312, 368-373, 418-419.

[4]   M.E. Dry, Sasol's Fischer-Tropsch experience, Hydrocarbon Process., Aug. (1982), 121-124

[5]   M.E. Dry, Chemicals produced in a commercial Fischer-Tropsch process, Industrial Chemicals via C_1 Processes, ACS Symp. Ser., 328 (1987) 18-33.

[6]   P.C. Keith, Gasoline from natural gas, Oil Gas J., 15 Jun. (1946), 102-112.

[7]   M.L. Kastens, L.L. Hirst and R.G. Dressier, An American Fischer-Tropsch plant, Ind. Eng. Chem., 44:3 (1952) 450-466.

[8]   J.S. Swart, G.J. Czajkowski and R.E. Conser, Sasol upgrades synfuels with refining technology, Oil Gas J., 79:35 (1981) 63-66.

[9]   J.N. Marriott, Sasol process technology – The challenge of synfuels from coal, ChemSA, 12:8 (1986) 174-178.

[10] M.D. Schlesinger, H.E. Benson, E.M. Murphy, H.H. Storch, Chemicals from the Fischer-Tropsch synthesis, Ind. Eng. Chem., 46:6 (1954) 1322-1326.

[11] J. Meintjes, Sasol 1950-1975, Tafelberg, Cape Town, 1975.

[12] J. Collings, Mind over matter. The Sasol story: A half-century of technological innovation, Sasol, Johannesburg, 2002.

[13] C. Steyn, The role of Mossgas in Southern Africa, 2nd Sub-Saharan Africa Catalyst Symposium, Swakopmund, Namibia, 5-7 Nov (2001).

[14] O.R. Minnie, F.W. Petersen and F.R. Samadi, Effect of 1-hexene extraction on the COD process conversion of olefins to distillate, 2003 South African Chemical Engineering Congress, Sun City, South Africa 3-5 Sep. (2003).

[15] J. Terblanche, The Mossgas challenge, Hydrocarbon Eng., Mar. (1997) 2-4.

[16] KBR, The first Superflex Project – Fischer-Tropsch liquids to propylene, 5th EMEA Petrochemicals Technology Conference, Paris, France (25-26 Jun 2003).

[17] X. Song, A. Sayari, Sulfated zirconia as a cocatalyst in Fischer-Tropsch synthesis, Energy & Fuels, 10:3 (1996) 561-565.

[18] P.G. Smirniotis and W. Zhang, Effect of the Si/Al ratio and of the zeolite structure on the performance of dealuminated zeolites for the reforming of hydrocarbon mixtures, Ind. Eng. Chem. Res., 35 (1996) 3055-3066.

[19] A. de Klerk, Deactivation behaviour of Zn/ZSM-5 with a Fischer-Tropsch derived feedstock, in S.D Jackson, J.S.J. Hargreaves and D. Lennon (eds), Catalysis in application, Royal Society of Chemistry, Cambridge, 2003, 24-31.

[20] ASTM DS 4B, Physical constants of hydrocarbons and non-hydrocarbon compounds, ASTM, Philidelphia, 1988 (2ed).

[21] S.M. Ozmen, H. Abrevaya, P. Barger, M. Bentham and M. Kojima, Skeletal isomerisation of C_4 and C_5 olefins for increased ether production, Fuel Reformulation, 3:5 (1993) 54-59.

[22] J-L. Duplan, P. Amigues, J. Verstraete and Ch. Travers, Kinetic studies of the skeletal isomerization of n-pentenes over the ISO-5 process catalyst, Proc. Ethylene Prod. Conf., 5 (1996) 429-449.

[23] F.M. Floyd, M.F. Gilbert, M. Pérez and E. Köhler, Light naphtha isomerisation, Hydrocarbon Eng., Sep. (1998) 42-46.

[24] P.J. Kuchar, R.D. Gillespie, C.D. Gosling, W.C. Martin, M.J. Cleveland and P.J. Bullen, Developments in isomerisation, Hydrocarbon Eng., Mar. (1999) 50-57.

[25] P.W. Tamm, D.H. Mohr and C.R. Wilson, Octane enhancement by selective reforming of light paraffins, Stud. Surf. Sci. Catal., 38 (1988) 335-353.

[26] T.R. Hughes, R.L. Jacobson and P.W. Tamm, Catalytic processes for octane enhancement by increasing the aromatics content of gasoline, Stud. Surf. Sci. Catal., 38 (1988) 317-333.

[27] R.L. Peer, R.W. Bennett, D.E. Felch and E. Von Schmidt, UOP Platforming leading octane technology into the 1990's, Catal. Today, 18 (1993) 473-486.

[28] C. Hodge, Comment: More evidence mounts for banning, not expanding, use of ethanol in US gasoline, Oil Gas J., 101 6 Oct. (2003) 18-20.

[29] P.C. Weinert and G. Egloff, Catalytic polymerization and its commercial application, Petroleum Process., Jun. (1948) 585-593.

[30] S.A. Tabak, F.J. Krambeck, and W.E. Garwood, Conversion of propylene and butylene over ZSM-5 catalyst, AIChE J., 32:9 (1986) 1526-1531.

[31] F. Nierlich, Oligomerize for better gasoline, Hydrocarbon Process., Feb. (1992) 45-46.

[32] J-F. Joly, Aliphatic alkylation, in P. Leprince (ed), Petroleum Refining 3: Conversion Processes, Editions Technip, Paris, 2001, 257-289.

[33] M. Golombok and J. De Bruijn, Dimerization of n-butenes for high octane gasoline components, Ind. Eng. Chem. Res., 39 (2000) 267-271.

[34] S.T. Sie, Acid-catalyzed cracking of paraffinic hydrocarbons. 1. Discussion of existing mechanism and proposal of a new mechanism, Ind. Eng. Chem. Res., 31 (1992) 1881-1889.

[35] J.F. Boucher, G. Follain, D. Fulop and J. Gaillard, Dimersol X process makes octenes for plasticizer, Oil Gas J., 80, 29 Mar (1982) 84-86.

[36] R.H. Friedlander, D.J. Ward, F. Nierlich and J. Neumeister, Make plasticizer olefins via n-butene dimerisation, Hydrocarbon Process., Feb. (1986) 31-33.

[37] M. Freemantle, Designer solvents, Chem. Eng. News, 30 Mar (1998) 32-37.

[38] E.J. Swain, U.S. MTBE production at a record high in 1998, Oil Gas J., 97:24 (1999) 99-101.

[39] G. Parkinson, All sides pumped up for MTBE ban, Chem. Eng., (Jun 1999) 49.

[40] H. Becker, Process for the manufacture in pure form of 1-pentene or an alpha-olefin lower than 1-pentene, US patent No. 6 483 000 Nov. (2002).

530

[41] F.D. Rossini, Chemical thermodynamics in the petroleum industry, J. Inst. Petroleum, 58:564 (1972) 279.

[42] A.O.I. Krause and L.G. Hammarström, Etherification of isoamylenes with methanol, Appl. Catal., 30 (1987) 313-324.

[43] R.S. Karinen, J.A. Linnekoski and A.O.I. Krause, Etherification of C_5- and C_8-alkenes with C_1- to C_4-alcohols, Catal. Lett., 76 (2001) 81-87.

[44] J.D. Swift, M.D. Moser, M.B. Russ and R.S. Haizmann, The RZ platforming process: something new in aromatics technology, HTI Quarterly, Autumn (1995) 86-89.

[45] T. Fukunaga and V. Ponec, The nature of the high sensitivity of Pt/KL catalysts to sulphur poisoning, J. Catal., 157 (1995) 550-558.

[46] T. Zhang and R. Datta, Ethers from ethanol. 3. Equilibrium conversion and selectivity limitations in the liquid-phase synthesis of two tert-hexyl ethyl ethers, Ind. Eng. Chem. Res., 34 (1995) 2237-2246.

[47] Z-M. Marais, Sintese en evaluering van 'n reeks petrolmengkomponente (Engl., Synthesis and evaluation of a series of petrol blending components), M.Sc. dissertation, University of the Orange Free-State, Bloemfontein (1991).

[48] R.J. Quann, L.A. Green, S.A. Tabak and F.J. Krambeck, Chemistry of olefin oligomerisation over ZSM-5 catalyst, Ind. Eng. Chem. Res., 27 (1988) 565-570.

[49] C. Knottenbelt, Mossgas "gas-to-liquid" diesel fuels – an environmentally friendly option, Catal. Today, 71 (2002) 437-445.

[50] K. McGurk, 1-Heptene to 1-octene: A new production route, 2003 South African Chemical Engineering Congress, Sun City, South Africa, 3-5 Sep. (2003).

[51] A. Ranwell, Die oligomerisasie van Sasol alfa-olefienfraksies (Engl., The oligomerisation of Sasol alpha-olefin fractions), M.Sc. dissertation, Rand Afrikaans University, Johannesburg (1994).

[52] C.D. Chang, H. Scott, J.G. Santiesteban, M.M. Wu and Y. Xiong, Oligomerization process for producing synthetic lubricants, US patent No. 5 453 556 Sep. (1995).

[53] H. Naudé, C. McGregor and C. Grove, Production of oxygenated compounds from FT-derived olefins: Sasol ModCo™ technology, 2003 South African Chemical Engineering Congress, Sun City, South Africa, 3-5 Sep. (2003).

[54] G.T. Austin, Shreve's Chemical Process Industries, McGraw-Hill, New York, 5ed., 1988 p.p.70-88.

[55] N.Y. Chen, An environmentally friendly oil industry?, Chem. Innovation, 31:4 (2001) 10-21.

[56] S.N. Maiti, J. Eberhardt, S. Kundu, P.J. Cadenhouse-Beaty and D.J. Adams, How to efficiently plan a grassroots refinery, Hydrocarbon Process., 80:6 (2001) 43-49.

[57] J.R. Joiner and J.J. Kovach, Sasol Two and Sasol Three, Energy Progr., 2:2 (1982) 66-68.

[58] D.A. Linnig, P.F. Mako and W.A. Samuel, Coal to oil and gas, Sasol One, Two, and Three, Energy Process./Canada, 74 (1982) 49-54.

[59] L.P. Dancuart, G.H. du Plessis, F.J. du Toit, E.L. Koper, T.D. Phillips and J. van der Walt, patent applications WO 03/106354, 03/106346, 03/106353, 03/106351, and 03/106349,18 Jun (2002).

[60] M.E. Dry, Technology of the Fischer-Tropsch Process, Catal.Rev.Sci.Eng., 23(1&2), (1981) 265-278.

[61] M.E. Dry, The Fischer-Tropsch Synthesis, Cat.Sci. & Tech., ed. by J.R. Anderson, (1981) 159-255.

[62] A. de Klerk, Continuous-mode thermal oxidation of Fischer-Tropsch waxes, Ind. Eng. Chem. Res., 42:25 (2003) 6545-6548.

[63] J.H. Gary and G.E. Handwerk , Petroleum Refining Technology and Economics, Second Edition, Marcel Dekker, Inc., 1984.

[64] H. Pichler, H. Schultz, D. Kuhne, Brent. Chem., 49 (11) (1968) 344.

[65] M.E. Dry, Hydrocarbon Processing, 61 (8) (1982) 121.

[66] M.E. Dry, ChemSA, February (1984) 286.

[67] P.P. Shah, G.C. Sturtevant, J.H. Gregor, M.J.J. Umbach, F.G. Padrta and K.Z. Steigleder, FT Wax Characterisation and Upgrading: Final Report (Report DE88014638 by UOP Inc.), June (1988).

[68] V. Calemma, S. Peratello, S. Pavoni, G. Clerici and C. Perego, Hydroconversion of a mixture of long chain n-paraffins to middle distillate: Effect of the Operating Parameters and Product Properties, Proceedings of the 6th Natural Gas Conversion Symposium, E. Iglesia (ed.), Alaska, June (2001).

[69] K.M. Murad, M. Lal, R.K. Agarwal and K.K. Bhattacharyyal, Improve Quality of Wax by Hydrofinishing, Petroleum & Hydrocarbons, 7, 2 (1972) 144-147.

[70] S. Bull, Lube Oil Manufacture by Severe Hydrotreatment, paper presented at the 10th World Petroleum Congress, Bucharest – Romania, vol.4, 1979, pp.221-228

[71] T.R. Farrell and J.A. Zakarian, Lube facility makes high-quality lube oil from low-quality feed, Oil & Gas Journal, May 19, 1986

[72] S.J. Miller, J. Xiao, and J.M. Rosenbaum, Applications of ISODEWAXING, a New Wax Isomerization Process for Lubes and Fuels, Sci.Tech.Cat., 65 (1994) 379.

[73] C.C.J. Shih, L. de Bruyn and V.A. Ziazine, First ISODEWAXING Unit in Russia to Manufacture High Quality Base Oils with Chevron Technology, paper presented at the 3rd Russian Refining Technology Conference, Moscow (2003).

[74] J.E. Gallagher, K.E. Fyfe, N.M. Page and G. Sanchez, Slack Wax Isomerisation for High Performance Lube Basestocks, paper LW-02-127 presented at the NPRA International Lubricants & Waxes Meeting, Houston – TX, Nov. (2002).

[75] T. Sullivan, ExxonMobil Retrofitting Fawley for Group III, Lube Report, vol.2, n. 43, 23 Oct. (2002).

[76] J. Scherzer and A.J. Gruia, Hydrocracking Science and Technology, Marcel Dekker Inc., 1996.

[77] L.P. Dancuart, Processing of Fischer-Tropsch Syncrude and Benefits of Integrating its Products with Conventional Fuels, National Petrochemical & Refiners Association Annual General Meeting, Paper AM-00-51 (2000).

[78] C.C. Ward, F.G. Schwartz and N.G. Adams, Composition of Fischer-Tropsch Diesel Fuel (cobalt catalyst), Ind.Eng.Chem., 43, 5 May (1951) 1117-1119.

[79] P.W. Schaberg, I.S. Myburgh, J.J. Botha, P.N. Roets, C.L. Viljoen, L.P. Dancuart and M.E. Starr, Diesel Exhaust Emissions Using Sasol Slurry Phase Distillate Process Fuels, presented at the SAE International Fall Fuels & Lubricants Meeting & Exposition, Tulsa, Oklahoma, USA, Oct. (1997) Paper 972898.

[80] P.M. Morgan, C.L. Viljoen, P.N. Roets, P.W. Schaberg, I.S. Myburgh, J.J. Botha and L.P. Dancuart, Some Comparative Chemical, Physical and Compatibility Properties of Sasol Slurry Phase Distillate Diesel Fuel, presented at the SAE International Fall Fuels & Lubricants Meeting & Exposition, San Francisco, USA, Oct. (1998) Paper 982488.

[81] P.J.A. Tijm, H.M.H. van Wechen, and M.M.G. Senden, The Shell Middle Distillate Synthesis Project - New Opportunities for Marketing Natural Gas, Alternate Energy '93 Conference, Colorado Springs, April (1993).

[82] P.J.A. Tijm, J.M. Marriott, H. Hasenack, M.M.G. Senden and Th. van Herwijnen, The Markets for Shell Middle Distillate Synthesis Products, paper presented at Alternate Energy '95, Vancouver, Canada, May 2-4 (1995) 228.

[83] J.A. Tilton, W.M. Smith and W.G. Hockberger, Production of High Cetane Number Diesel Fuels by Hydrogenation, Ind.Eng.Chem, 40, 7, 1269-1273 (1948).

[84] L.P. Dancuart, J.F. Mayer, M.J. Tallman, and J. Adams, Performance of the Sasol SPD Naphtha as Steam Cracking Feedstock, American Chemical Society National Meeting, Boston, USA, August 2002, Pet. Chem. Div. Preprints, 48, 2 (2003) 132-138.

[85] R.L. Grantom and D.J. Royer, Ethylene, Ullmann's Encyclopedia of Industrial Chemistry, volume A10, (1987) 45-93.

[86] C.D. Frohning and B. Cornils, Hydrocarbon Proces., 143-146, 53, 11, Nov. (1974).

[87] L.P. Dancuart, US patent 6,656,343, 2 December (2003).

[88] J.H. Fourie and J.W. de Boer, US Statutory Invention Registration H1, 849, 2 May (2000).

[89] L.P. Dancuart and J. Beigley, Rocket Propellants, Document IPCOM000021747D (www.IP.com), Jan. (2004).

[90] A.P. Steynberg, W.U. Nel and M.A. Desmet, Large Scale Production of High Value Hydrocarbons using Fischer-Tropsch Technology, paper to be presented at the 7th Natural Gas Conversion Symposium, Dalian, China, June (2004).

Studies in Surface Science and Catalysis 152
A. Steynberg and M. Dry (Editors)

Chapter 7

FT catalysts

M. E. Dry

Catalysis Research Unit, Department of Chemical Engineering,
University of Cape Town, Rondebosch, 7701, South Africa

1. INTRODUCTION

Only the four group VIII metals, Fe, Co, Ni and Ru have sufficiently high activities for the hydrogenation of carbon monoxide to warrant possible application in the FT synthesis. Of the four metals ruthenium is the most active (see section 4) but its high cost (see Chapter 3, Table 13) and low availability rules it out for large scale application. Nickel is also very active but has two major drawbacks. Being a powerful hydrogenating catalyst it produces much more methane than Co or Fe catalysts (It is after all the industrial methanation catalyst of choice). Nickel forms volatile carbonyls resulting in continuous loss of the metal at the temperatures and pressures at which practical FT plants operate. From the above it is clear that only cobalt and iron based catalysts can be considered as practical FT catalysts. The three South African FT plants currently use iron based catalyst while Shell's Malaysian plant uses cobalt catalyst. Because Co is so much more active than Fe (see section 4) future plants aimed mainly at diesel fuel production (see Chapter 5) will probably use cobalt based catalysts. For the production of the high value linear alkenes, however, iron catalyst, operating at high temperatures in fluidized bed reactors, will remain the catalyst of choice (see Chapter 5). The LTFT iron catalyst may also find future applications for the conversion of coal-derived syngas (see Chapter 5).

2. CATALYST PREPARATION AND CHARACTERISATION

2.1 Iron based catalysts

2.1.1 Low temperature catalysts for wax production (LTFT)
2.1.1A Preparation and characterisation

The Sasol LTFT process is geared at the production of high molecular mass waxes. The catalyst was originally developed by Ruhrchemie [1] and was used in the Arge reactors installed at the first Sasol plant. In spite of a few changes the preparation method used at Sasol remains similar to that used by Ruhrchemie. These catalysts, prepared by precipitation techniques [2] are used in both the Sasol fixed and slurry bed reactors. Scrap iron, of suitable chemical composition, together with metallic copper is dissolved in nitric acid. The Ruhrchemie preparation procedure [1] was as follows. The near boiling solution of iron and copper nitrate (40 g Fe and 2 g Cu per litre) was rapidly poured into a hot solution of Na_2CO_3 with vigorous stirring until the pH reached just above 7. The hydrated ferric oxide was thoroughly washed with hot distilled water to remove the bulk of the sodium ions. The precipitate was then re-slurried with water and the appropriate amount of potassium waterglass was added to give the desired silica to iron ratio of 25 g SiO_2 per 100 g Fe. Sufficient nitric acid was then added to the slurry so that after filtration the desired amount of K_2O (5 g per 100Fe) was retained by the gel-like catalyst cake. The precipitation method described above is a batch process. An alternative process is the continuous precipitation process in which two steams, one of sodium carbonate and the other of the metal nitrates are fed into a mixing tank maintaining the pH at about 7.

The precipitated iron catalyst used by Rentech (see Chapter 5) was unsupported and was promoted with about 1 per 100 Fe of each of K and Cu [3]. Ammonium hydroxide (ambient temperature) was added to the hot Fe/Cu nitrate solution. The precipitate was washed and potassium carbonate was added to the slurry which was then spray dried to produce spherical particles in the range of 5 to 50 micron. These were then calcined in a fluidized bed at about 300°C to convert the hydrous oxide to hematite.

If the Sasol catalyst is to be used in the Sasol multitubular reactors (see Chapter 2) the catalyst cake described above is extruded and dried [2]. If it is to be used in Sasol slurry bed reactors the cake is re-slurried in water and spray dried [4]. In a fixed bed reactor there is little or no catalyst break-up. However, there is some break-up in the turbulent high throughput slurry reactor which may cause blockages in the wax/catalyst separation unit. To minimise this break-up the spray dried catalyst is calcined at about 400 to 500 °C to increase its mechanical strength. This results in about a 15% decrease in the BET area of the unreduced catalyst but does not result in any loss of the FT activity of the final operating catalyst.

A typical catalyst contains 25g SiO_2 5g Cu and 5g K_2O per 100g Fe [2]. As the Fe/Si ratio exceeds 4 the silica presumably does not act as a classical support but rather acts as a binder and as a spacer, the latter inhibiting sintering of the high area iron oxide. Investigation at Sasol indicated that when the potassium waterglass is added to the slurried hydrated ferric oxide precipitate all the silica is precipitated or adsorbed onto the high area iron precipitate as subsequent washing can remove all of the potassium but none of the silica. Thus it appears that the addition of nitric acid is to control the K_2O content of the catalyst and not to precipitate the silica [2].

Table 1
Influence of SiO_2 on precipitated Fe_2O_3 [2]

g SiO_2/ 100g Fe	Unreduced			H_2 reduced			
	Pore Vol $cm^3 g^{-1}$	Area $m^2 g^{-1}$	Area in pores > 4.5 nm $m^2 g^{-1}$	Pore vol $cm^3 g^{-1}$	Area / m^2 g^{-1}	Area in pores > 4.5 nm $m^2 g^{-1}$	Reduction %[a]
0	0.37	275	41	0.22	35	35	100
8	0.47	345	59	0.43	190	68	80
19	0.74	375	90	0.48	250	80	46
25	0.71	390	94	0.61	270	84	58
29	0.75	370	96	0.65	265	85	57
50	N.A.	405	N.A.	N.A.	280	N.A.	N.A.

a % of total Fe present in metallic state after a fixed time at a fixed temperature.
N.A. Not available.

Table 1 shows that both the total BET area and the total pore volume increase as the silica content is increased. This is so for both the dried unreduced catalyst and the reduced catalyst [2]. The area in the larger pore sizes also increases. The percent reduction (using H_2 for a fixed time and at a fixed temperature) decreases with increasing SiO_2 content. This is probably due to the slow reduction of the 'Fe_2O_3-SiO_2 complex' present. The porosity and pore size distribution is established in the very first stage of preparation, viz. the precipitation of the ferric oxide [2]. Thus if the precipitation conditions are such that the iron oxide has a low porosity or narrow pore size distribution then the final catalyst after silica and potassium promotion will also have these features. The variables that have been shown to affect the total pore volumes and the pore size distributions are the following: The concentration of the Fe and carbonate solutions; the temperature of precipitation; the order in which the two solutions are mixed (Fe into carbonate or the reverse); the precipitation time; and the final pH. Fig. 1 illustrates three extreme examples. Fig. 1(a) shows the N_2 adsorption isotherm for the catalyst prepared as described above, i.e. the acidic iron nitrate solution is added to the carbonate solution. In this case the precipitation therefore takes place in an alkaline environment. Fig. 1(b) is for the 'reverse' preparation, i.e. the carbonate solution was added to the nitrate solution and so much of the precipitation took place at a pH well below 7. The BET areas of the two catalysts were not all that different but as can be deduced from the shape of the isotherms the normal catalyst had a wide distribution of pore sizes while the

536

reverse case only appeared to have narrow pores present. Case 1(c) was prepared in the same way as case (a) except that excess nitrate solution was added to the carbonate solution resulting in a final pH of 2. As can be seen both the area and porosity had been markedly lowered.

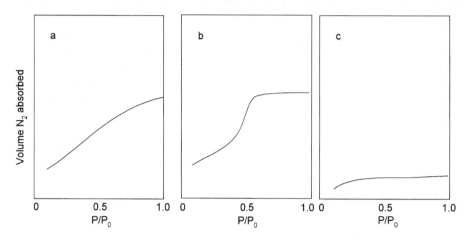

Figure 1 N_2 adsorption isotherm

The chemical nature of the alkali used for precipitation also has an effect on the pore size distribution of the catalyst. Table 2 shows that when using carbonates the catalysts are more porous than when hydroxides are used. (Note that for all four catalysts the silica contents were the same.) Since pore diffusion rates affect the overall kinetics of the FT reaction (see Section 4) it can be expected that the catalysts prepared by carbonate precipitation may be more active. In agreement with this it was reported earlier that the use of carbonates as precipitation agent resulted in better catalysts than when hydroxides were used [5].

Table 2
Influence of precipitating agent [2]

Precipitant	Average pore size (nm)	Pore volume in pores > 12.5 nm (%)
NaOH	2.9	1
Na_2CO_3	4.0	40
NH_4OH	2.6	1
$(NH_4)_2 CO_3$	3.1	18

The drying procedure is also important as the final total pore volume depends on the degree of shrinkage that occurs during drying. During subsequent hydrogen reduction further shrinkage occurs and this is why precipitated catalysts used in fixed bed reactors are pre-reduced in order to maximise catalyst loading in the FT reactors. For catalysts used in slurry reactors this pre-shrinkage is of course not needed. The presence of liquid water in the pores of the catalyst appears to enhance the degree of shrinkage during the drying process. If, before drying, the wet catalyst is treated with an excess of a low surface tension liquid, such as acetone, to displace the bulk of the water and the catalyst is then dried the pore volume can be doubled [2]. In general there must, however, always be a compromise between the opposing requirements of high porosity and adequate particle strength.

Another aspect to consider in the catalyst preparation process is the re-use of off-specification material. The spray-dried catalyst used for the slurry phase reactor must adhere to strict Particle Size Distribution (PSD) requirements. Inadvertently, due to these strict specifications, some off-specification product is produced.

The Arge catalyst production entails extrusion of wet catalytic material, which is subsequently dried to produce the final catalyst extrudates. During the extrusion and drying process steps, a significant amount of agglomerate material (over size) and fine material (undersize) is produced. The fines production is due to a significant amount of mechanical handling of the extruded product.

Both the oversize and undersize product cuts from the spray drying and extrusion processes are separated from the in-specification product. Summated this off- specification product constitutes a significant amount of about 8 - 9 % of the total production capacity.

The off-specification (size related) catalytic material can be wet milled and added after the impregnation step of the catalyst manufacturing process. The wet milling process produces a milled slurry product. This milled slurry product is restricted to a specified range of solid contents to accommodate solid content requirements in the normal catalyst production processes.

The milling step of the over size material is required to ensure complete integration with the freshly produced catalyst product, without detrimentally affecting the catalyst production processes (due to blockage of equipment, etc.) and to ensure minimal impact on the mechanical integrity of the final extrudate product. Corrective actions are taken to address potential differences in chemical make-up of the recycled and freshly produced catalytic materials.

Approximately 10 mass% (on an iron mass percentage basis) recycled milled catalyst product is added to the normal production capacity of the process in Sasolburg and the addition of up to 15 mass% recycled material has been proven to be commercially viable.

The spray dried catalyst material is physically significantly smaller than the extruded Arge catalyst and due to the critical mechanical integrity requirements of the spray-dried catalyst; the addition of milled material has to adhere to very strict PSD requirements. Ideally, the milled material PSD has to be comparable to the PSD of the fresh catalytic material normally sent to the spray drier.

Due to the increased milling requirements and the milling constraints associated with the wet milling process (limited milling capacity, time consuming and expensive) a new milling technology was sourced.

This new milling technology is similar to the well-known homogeniser mixer concept, with the inclusion of impact zones that facilitate breakage/milling in a slurry medium. This equipment is physically significantly smaller than the typical wet mill process, with a significantly lower capital requirement and operating expense, due to the mode of the milling operation.

Commercial evaluation on milled catalyst incorporation for the slurry phase reactor catalyst has been successfully completed, utilising up to 30 mass% recycled material.

2.1.1B Reduction and conditioning of precipitated iron LTFT catalysts

In the past (as well as at present) it was always considered necessary to pre-reduce the FT catalysts, whether they be Co or Fe based, under relatively mild conditions. The objective was to generate high metallic surface areas, i.e. active catalysts. Pichler reported that unreduced precipitated Fe catalysts were not very active but that reduction at the high temperature of 360°C did not improve matters much [5]. Scheuermann found that reduction at about 200°C yielded an active catalyst whereas reduction at 300°C resulted in a lower activity [5]. In these previous studies the influence of using different reduction gases: H_2, mixtures of H_2 and CO or CO only, were investigated [5]. The FT selectivity was also influenced by the reduction temperature. Thus it was found that pre-reduction with hydrogen at 300°C produced a more active catalyst than one reduced with syngas at 230°C but the former had a lower wax selectivity [5].

More recently Bukur [6, 7, 8] and Davis [9] have reinvestigated reduction procedures. Davis reported that, compared to activation with hydrogen rich syngas, activation with CO or CO rich syngas resulted in more active catalysts. Bukur reported that a catalyst reduced with H_2 at 250°C had a higher activity but a lower wax selectivity than the same catalyst reduced with CO at 280°C. When the H_2 reduction temperature was lowered to 220°C, however, there was a smaller difference in the wax selectivity. A catalyst reduced with H_2 at 250°C for four hours was found to be significantly more active than when it was reduced for eight hours at the same temperature. The activities of all the catalysts declined with time on steam. At steady state the activity of the catalyst reduced with CO at 280°C was the highest.

The reason for adding copper to the precipitated $Fe_2O_3/CuO/K_2O/SiO_2$ catalyst is to facilitate the reduction of the iron oxide at lower temperatures, typically 220°C [2]. Increasing the temperature of reduction of a Cu-free catalyst does of course increase the rate of reduction but it is then found that the FT activity of the catalyst is inferior. The activation of the catalyst used in the Sasol multitubular fixed bed reactors is carried out in several steps [2]. The extrudates are first pre-shrunk by H_2 reduction at atmospheric pressure. This initial reduction rate is high and about 20% of the theoretical amount of water expected from the reduction of Fe_2O_3/CuO is produced in this period. This initial rapid reduction phase is characterised by a marked exotherm, which is no doubt due to the exothermic reductions of the nitrates and copper oxide as well as the exothermic reduction of hematite to magnetite. In practice the reduction is not taken beyond this stage as further reduction only occurs very slowly. It is presumed that the catalyst preparation procedure results in the formation of an iron oxide-silicate complex, which reduces very slowly. In support of this concept note from Table 1 that as the silica content of the catalyst was increased the degree of reduction, attained at fixed conditions, decreased. Note that the precipitated catalyst containing no silica reduced to completion. In this case CO chemisorption measurements indicated that over 90% of the surface was metallic whereas for a typical SiO_2 containing catalyst the CO coverage was only about 5% after the rapid initial reduction referred to above [2]. It should be pointed out, however, that no XRD evidence of the presence of an iron silicate phase was observed, even after hyrotreating the catalyst in an autoclave at 280°C. The only diffuse patterns that could be identified were those of α- and γ-Fe_2O_3.

After the partial reduction procedure described above the catalyst is coated with wax to protect it from re-oxidation and then loaded into the fixed bed FT reactors. The temperature of the reactor is raised to about 200°C under a H_2 atmosphere in order to avoid the formation of volatile iron carbonyls which would otherwise contaminate the FT wax product. Synthesis gas is then slowly introduced and the system is taken to full FT operating conditions. During this 'conditioning' period the iron is converted to Hägg carbide (Fe_5C_2). This process takes place over several days during which the FT conversion increases by a few percentage points before it peaks and then slowly deactivates (see Section 3.2.1). From the data presented in Table 3 it can be seen that over the whole procedure, from unreduced to fully operational carbided catalysts, the BET area decreases while the pore volume and average pore size increases. Table 3 shows the results for two differently prepared catalysts having different initial pore volumes. Note that the differences in porosity between the two catalysts persist after the various treatment stages.

Table 3
Changes in area and pore structure of precipitated iron catalysts [2]

State	Catalyst	Total pore vol $cm^3 g^{-1}$	Total area $m^2 g^{-1}$	Area in pores > 4.5 nm (%)	Vol in pores > 16.0 nm (%)
Unreduced	A	0.39	355	5	0
	B	0.67	340	25	19
Partially reduced	A	0.30	195	20	1
	B	0.46	150	45	22
After some FT synthesis (All wax extracted)	A	0.22	51	100	10
	B	0.38	47	100	44

2.1.2 Iron based catalysts for high temperature synthesis (HTFT)

2.1.2A Catalyst preparation and characterisation for the HTFT process

The preparation of the catalyst currently used by Sasol is very similar to that of the ammonia synthesis catalyst, namely, fusion of iron oxide together with the chemical promoter K_2O and structural promoters such as MgO or Al_2O_3. In the presence of air molten iron oxide at about 1500°C should consist only of molten magnetite (Fe_3O_4). Theoretically wustite (FeO) should oxidise to Fe_3O_4 and hematite (Fe_2O_3) should lose oxygen. In practice, however, because of the use of carbon electrodes in the arc furnace the situation within the molten oxide pool becomes somewhat reducing and so some wustite can be formed (or survive if fed into the furnace). The molten mixture of oxides is poured into ingots and rapidly cooled. The ingots are then crushed in a ball mill to the particle size distribution required for effective fluidization in the HTFT reactors. For effective structural properties magnetite is the preferred iron oxide phase (see discussion below). If a hematite-rich ore is fed to the fusion furnace it has been observed that the cooled solid ingots contained small voids (bubbles) presumably as the result of the release of oxygen during the fusion process. This could increase the friability of the ingots.

In the 1950's Sasol imported a magnetite ore (Allenwood ore) from the United States for the production of the fused catalyst since this was the ore on which the design of the CFB reactors had been based. Due to the relatively short useful life of the catalyst in the HTFT operation importation of the ore all the way from the U.S.A. obviously increased the cost of catalyst manufacture. Research in the Sasol pilot plants was undertaken to evaluate the suitability of various locally available ores and oxides. It was found that mill scale from a nearby steelworks was in fact a suitable substitute for the imported Allenwood ore. The mill scale, however, was rich in wustite and also was contaminated with sulphur containing oils. It was therefore necessary to roast the mill scale at a high temperature in air to increase its ferric ion content and to burn off the contaminating oil and associated sulphur. Currently Sasol still uses this mill scale for the preparation of their HTFT catalysts. The regeneration of spent catalyst was investigated and it was found that a satisfactory catalyst could indeed be produced but this involved extensive air oxidation to burn off all the heavy oils and carbon deposits and to fully re-oxidise the iron. The remaining oxide particles were not in specification regarding particle size distribution and alkali content and so had to be re-fused with promoter top-up. In principle the direct feeding of spent catalysts to fusion furnaces or smelters that are used in metallurgical industries is possible but this has not yet been commercially demonstrated. Because of the low cost of the locally available mill scale, the regeneration of spent catalyst has not yet been considered to be a priority development. The spent catalyst is also suitable as a feed material to the iron and steel industry.

During the solidification of the molten mixture of iron and of promoter oxides the structural promoter cations enter into solid solution in the magnetite phase. For example, a Mg^{2+} ion being of similar size to the Fe^{2+} ion can replace it in the magnetite lattice, similarly Al^{3+} ions can replace Fe^{3+} ions [2]. These promoter cations are therefore atomically dispersed in the magnetite (i.e., in solid solution in the magnetite phase) and on subsequent hydrogen reduction small aggregates of Al_2O_3 or MgO are precipitated in between the reduced iron crystallites. These deposits of non-reducible, high melting point oxides, act as spacers which inhibit sintering resulting in iron catalysts with higher surface areas.

The relative effectiveness of various potential structural promoters such as the oxides of Ca, Al, Mn, Ti, and Mg were investigated [2]. In this study a highly pure iron oxide was used as a base stock for the preparation of a series of magnetite samples which contained different levels of various individual promoter oxides. If the cations of the added oxides did enter into solid solution in the magnetite then the unit cell size of the magneitite should decrease if the foreign ions were smaller than the iron ions that they were replacing. The reverse would be observed if the substituting ions were larger than the iron ions. The unit cell sizes of the prepared samples were measured by X-ray diffraction [10] and the results are shown in Fig. 2.

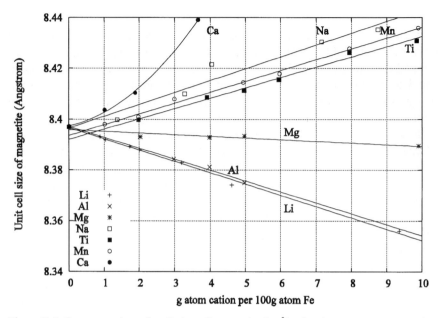

Figure 2 Influence on the unit cell size of magnetite (in Å) of various promoters entering
solid solution with magnetite. Promotor content is g atom cation per 100g atom Fe

As can be seen the unit cell size of the magnetite decreased with increasing
levels of the smaller Mg, Al or Li ions and increased with the larger Ca, Na , Mn
and Ti ions. These results support the contention that all of these ions can enter
into solid solution and thus potentially could be homogeneously distributed in
the magnetite phase. Potassium ion addition had no measurable effect on the
unit cell size presumably because they are too large to fit into the magnetite
lattice structure. Silica on its own did not go into solid solution but it had a
marked indirect effect on the ability of basic cations such as Li, Na and Ca to
enter into solution. The apparent reason was that the silica chemically combined
with these basic oxides to form separate silicate phases. This was supported by
microscopic investigations of polished sections of the various catalysts. It was
seen that when either potassium or silica was added small occlusions of some
other phase were present, whereas in the absence of these two components all
the other catalyst samples revealed the presence of only one phase (magnetite).
In the case of MgO promotion it appears that silica does not prevent to any
obvious extent the Mg ions from going into solid solution in the magnetite
phase.

As discussed above the presence of non-reducible, high melting point
oxides in solution in the magnetite is expected to result in higher surface areas of
the catalyst after reduction with hydrogen. Fig. 3 shows the influence of various

added oxides on the BET areas of the reduced catalysts [2]. It was postulated that the degree of area increase correlated with the ratio ionic charge to ionic radius of the of the added promoter cations [11]. Thus the cations of Al, Ti and Cr resulted in marked increases in surface areas. The effect of Mg cations was less marked while Mn, Ca and Li cations had little effect. Promotion with Na actually resulted in a decrease in BET area. The 'fluxing' effect of Na, however, does not appear to be related to the observation that Na ions are capable of going into solution with magnetite (see Fig. 2). This is because K ions (which, due to their large size, cannot fit into the magnetite lattice) also results in a decrease in area. The negative effect of the stronger alkalis is not due to pore blocking but rather to the enhanced crystal growth of the iron (i.e., sintering) as evidenced by X-ray line broadening studies. Silica, which on its own has little effect on the area, does have an indirect effect in that it chemically combines with the alkalis and thus depresses the 'fluxing' effect of the stronger alkalis.

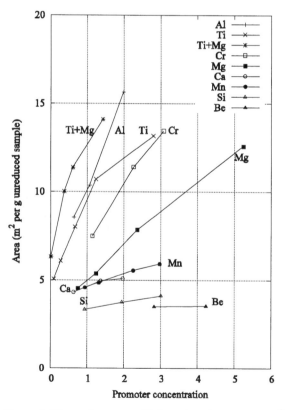

Figure 3 The surface area of fully reduced catalysts in m^2 (g unreduced sample)$^{-1}$ as a function of the promoter content (g atom promoter cation per 100 g atom Fe)

In wustite-rich fused catalysts the structural promoters MgO and Al_2O_3 are less effective for area promotion than for magnetite-rich fused catalysts. For the former the area at first increases as the promoter level is increased but then it levels off whereas for the later the area continues to increase [2]. At the same promoter levels magnetite-rich fused catalysts yield higher surface areas on reduction with hydrogen.

For metal catalysts the actual available metal surface area is obviously of more importance than the total BET area as part of the area is bound to be covered with the various non-metallic components present such as the added promoters or other extraneous compounds. It is common practice to evaluate the exposed metal surface area by measuring the amount of H_2 or of CO chemisorbed under appropriate conditions of temperatures and pressures. Since alkali plays such a vital role in both the activity and selectivity of iron catalysts in the FT process it is also of interest to be able to evaluate the coverage of the catalyst surface by the alkali. For this purpose the measurement of the amount of an acidic gas such as CO_2 has been used as it has been shown to chemisorb only weakly on pure iron surfaces, but the amount adsorbed increases as the alkali increases.

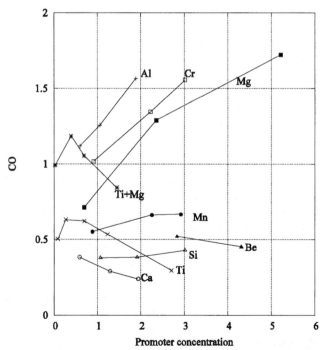

Figure 4 The volume of CO chemisorbed by fully reduced catalysts (cm^3 (STP) per gram unreduced sample) as a function of promoter content (g atom cation per 100 g atom Fe)

Fig. 4 shows the amounts of CO chemisorbed by variously promoted fully reduced fused iron catalysts. In general, as could be expected, those promoters which resulted in increased BET areas (Fig. 3) also yielded proportionately higher metal areas as reflected by the amounts of CO chemisorbed [2]. Titanium oxide promotion yielded rather peculiar results. Whereas the BET areas increased with increasing TiO_2 the CO chemisorption at first increased and then decreased. As a percentage of the total (BET) area MgO gave the highest metal area while CaO and TiO_2 yielded the lowest values.

The basicity of the iron catalyst's surface determines the activity and, in particular the selectivity of the catalyst in the FT reaction (see Chapter 3, Section 9.2.2). As mentioned above CO_2 chemisorption does appear to reflect the basicity of the catalyst. This is evidenced by the observation that not only does the amount of chemisorbed CO_2 increase with increasing amounts of alkali added but also the chemisorbed coverage of catalysts promoted with the stronger alkali K_2O is higher than when the weaker alkali Li_2O is used. Also it has been found that the presence of silica lowers the chemisorption of CO_2 on alkali promoted catalysts [2]. These findings are in keeping with the observations that lithium is an inferior chemical promoter to potassium and also that silica lowers the heavy hydrocarbon selectivity of alkali promoted iron catalysts. It has been found that chemisorbed nitrogen lowers the amount of CO_2 that can subsequently be chemisorbed which appears to indicate that surface nitrides lower the basicity of the catalyst. This is in keeping with the observation that nitriding iron catalysts results in a lower heavy hydrocarbon selectivity [5]. Carbon dioxide chemisorption data nevertheless needs to be interpreted with care. For instance, promotion with CaO, being alkaline, increases the CO_2 chemisorption but it has no observable effect on the selectivity of the catalyst.

As mentioned previously all the promoters, including the key alkali chemical promoters, are fed together with the iron oxides to the fusion furnace. Most iron oxides used are contaminated with some silica. It was shown above that neither the K^+ nor the Si^{4+} ions enter into solid solution with the magnetite and so on cooling small occlusions of alkali/silicates are present as separate phases. The cooled ingots are milled to a fine powder. To ensure that the powder has good fluidization properties it is important to control the particle size distribution. Because the alkali/silicate occlusions are separate phases they will tend to break away from the magnetite phase during the milling process. Microscopic investigations of the powdered catalyst showed that whereas the larger particles still contain alkali occlusions the finer particles consist of a mixture of separate alkali-silicate and magnetite particles. Hence, the distribution of alkali in the milled fused catalyst is apparently very heterogeneous. During reduction and FT synthesis, however, the alkali does to some extent spread over the catalyst surfaces [12, 13] and so the distribution of this key promoter is improved. In confirmation it has been shown that when finely ground potassium silicate is added separately to milled fused magnetite its performance in the FT synthesis is fairly similar to that of a catalyst prepared in the normal way, i.e. the addition of the potassium promoter during the fusion of magnetite. See also the discussion in Chapter 3, Section 9.2.2.

2.1.2B Reduction and conditioning of fused iron catalysts

Whereas unreduced precipitated iron catalysts can have high BET areas, fused iron oxides are non-porous and hence their areas are virtually zero. Pre-reduction is therefore essential to develop the area required for acceptable FT activity. Reduction is normally carried out with H_2 in the region of 350 to 450°C. The temperature used to ensure a reasonable rate of reduction depends on the type and quantity of structural promoter added during the fusion process. The presence of structural promoters retards the reduction rate. Typical surface areas of fully reduced fused catalysts are shown in Fig. 3. Precipitated catalysts are readily activated for FT synthesis by mixtures of H_2 and CO. However, the reduction of fused iron oxide is markedly retarded when CO is present in the H_2 [2]. This is illustrated by the results shown in Table 4. The presence of only 2% CO in the H_2 lowered the degree of reduction attained under set conditions by a factor of four. From thermodynamics it is expected that reduction with CO should be more favourable than reduction with hydrogen. However, as can be seen in Table 4 this is far from the case at 400°C. The negative effect of CO may possibly be linked to the fact that instead of metallic iron, Hägg carbide is formed during the reduction process, as shown in Table 4. The diffusion of the oxygen ions to the surface of the crystals may be retarded by the presence of an outer surface layer of carbide. In any event at 400°C the Boudouard reaction producing elemental carbon proceeds rapidly (see Section 2.1.2C) and so the use of CO containing gas for reducing fused catalysts is not recommended.

Table 4
Degree of reduction at 400°C and 8 hours at fixed space velocity. Promoted fused Fe_3O_4 [2]

Gas	Reduction (%)[a]	Phases present
100% H_2	80	α-Fe, Fe_3O_4
98% H_2 + 2% CO	20	Fe_5C_2, Fe_3O_4
100% CO	4	Fe_5C_2, Fe_3O_4

a Amount of iron in the oxide form converted to either metallic or carbidic iron.

Water vapour, which is the product of H_2 reduction, markedly depresses the rate of reduction. The effect of adding different small amounts of water vapour to the hydrogen on the initial rate of reduction is illustrated in Fig. 5. For practical purposes the measured reduction rate is insignificant at a H_2O/H_2 molar ratio of only about 20% of the thermodynamic equilibrium value for the reduction of magnetite to metallic iron under the conditions of the experiments.

Thus to ensure a high reduction rate it is necessary to use a high hydrogen gas linear velocity to maintain a low water vapour pressure in the reduction reactor.

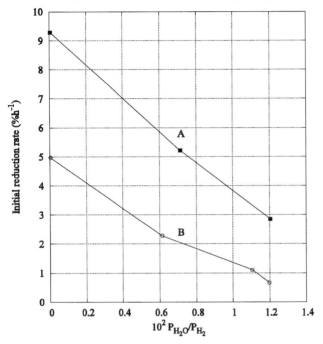

Figure 5 The retarding effect of water vapour on the initial reduction rate.
Samples A and B contain 0.012 and 0.030 g atom Al per g atom Fe respectively

The effect of the hydrogen space velocity on the reduction rate is illustrated in Fig. 6. Because the presence of water vapour also enhances the rate of sintering of the metallic iron crystals, a high H_2 flow rate results in higher final surface areas. It is assumed that continuous surface oxidation-re-reduction cycles in the presence of both water and hydrogen is the cause of the sintering. In general there appears to be an inverse relation between the initial rate of reduction and the area of the fully reduced catalyst. This is illustrated in Fig. 7 where the ratio of water to hydrogen in the exhaust gas is taken as a measure of the reduction rate. The promoters such as the oxides of aluminium, chromium, titanium or magnesium yield high surface area reduced catalysts (see Fig. 3) but they also retard the reduction rate. In the fused magnetite these promoters are in solid solution in the iron oxide phase (see Section 2.1.2A). If the reduced catalysts are deliberately fully re-oxidised to magnetite with water vapour at 400°C and then re-reduced with hydrogen then the rates of reduction are much higher and the areas much lower than for the original fused catalyst. Thus these promoters were not re-incorporated into the magnetite lattice when re-oxidised under the relatively mild conditions.

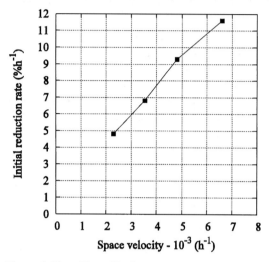

Figure 6 The effect of hydrogen space velocity on reduction rate

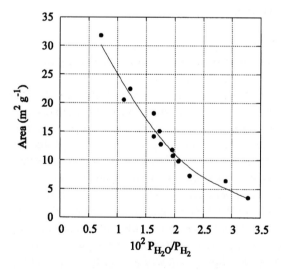

Figure 7 The relation between the final surface area developed and the initial reduction rate as depicted by the value of P_{H2O}/P_{H2} of the exhaust gas. The reduction temperature was the same in all cases. The samples contained different contents of various promoters.

In Section 2.1.1B it was mentioned that the addition of copper oxide promoted the reduction of the precipitated iron catalyst. For the fused magnetite catalysts, however, the addition of copper did not appear to have any significant effect on the reduction rate.

When reduced iron is exposed to FT synthesis gas the iron metal is rapidly converted to iron carbide (Hagg carbide). This is a highly exothermic reaction and localised high catalyst surface temperatures could result in damage to the catalyst such as sintering, fouling and excessive carbon deposition. The normal procedure is then to start up the reactors pressurised with hydrogen, operating in the recycle mode, and then over a predetermined time replace the hydrogen with synthesis gas. The carbon atoms originating from the CO (see section 2.1.2C) migrate into the iron metal lattice and the interstitial carbides are formed. After 'saturation' of the metal lattice, i.e. after completion of the carbiding process, no further elemental carbon is deposited when the Fe catalyst is operated in the LTFT wax producing mode. At the higher temperature HTFT process, however, the deposition of elemental carbon continues.

It should be noted that when Co is exposed to only CO, at the normal FT operating temperature used for Co, carbides are formed. When, however, hydrogen is also present, i.e. during normal FT operation, any Co carbide present is hydrogenated back to the metal and hence there is no build-up of either carbidic or 'free' elemental carbon.

2.1.2C Carbon deposition on HTFT iron catalysts

At FT operating temperatures below about 240°C, irrespective of whether Ni, Co, Ru or Fe based catalysts are used little or no elemental carbon is deposited on the catalysts. During high temperature (about 280 to 350°C) operation with iron based catalysts, however, elemental carbon, as distinct from carbonaceous coke, is deposited throughout the run at a fairly constant rate. The two source reactions could be:

$$2CO \rightarrow C + CO_2 \quad \text{(the Boudouard reaction)}$$
$$\text{or} \quad CO + H_2 \rightarrow C + H_2O.$$

The Gibbs free energy change for the Boudouard reaction is more negative than the alternative reaction (about -66 as against -49 kJ/mole at 600K). Furthermore, as will be seen below, the rate of carbon deposition is markedly depressed by higher hydrogen partial pressures. The Boudouard reaction is thus considered to be the key reaction resulting in carbon deposition. The activation energy of the Boudouard reaction, of about 113 kJ/mole [14], is higher than that of the FT reaction and the former is therefore more sensitive to increases in temperature. (Under typical HTFT conditions the rate of carbon deposition

550

increased by about 50% when the temperature was increased by 10°C.) This
agrees with the findings, mentioned above, that catalysts such as Co, Ru or Fe
operating below about 240°C do not deposit elemental carbon during normal FT
operations.

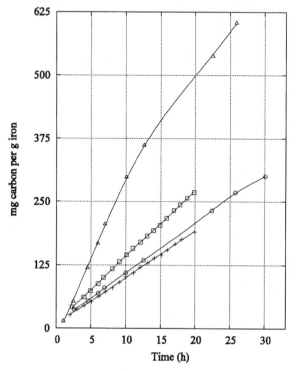

Figure 8 The effect of water vapour on the Boudouard reaction
o fluidized bed with no water added
Δ with 2.9 % water vapour added
+ Fixed bed with no water added
□ with 2.9 % water added

When elemental carbon forms in iron catalysts the density of the particles
is lowered and catalyst fines are produced. The indirect effect of this on the FT
performance of the catalysts is discussed in section 3.2.2. During studies of the
Boudouard reaction the following results were observed. When a small amount
of hydrogen, or of water, or ethanol or acetic acid was added to the CO feed gas
the rate of carbon deposition increased [14]. Fig. 8 shows the effect of adding
about 3 mole % water vapour to the CO. The presence of small amounts of
ammonia lowered the rate of carbon deposition. On ceasing the addition of
ammonia the rate returned to what it had been prior to ammonia addition. The

addition of nitrogen or paraffins or olefins had no obvious effect on the carbon deposition rate.

In another series of experiments the influence of various promoters on the rate of carbon deposition over fused, and subsequently reduced magnetite was investigated [15]. It was found for instance that promotion with alumina markedly increased the rate whereas calcium oxide promotion lowered the rate of carbon deposition. However, when the rates were calculated on a unit Fe metal surface area basis, the rates of the various samples were not significantly different. These results are illustrated in Fig. 9. Promoting fused magnetite with Cu, or Mo, or Zn or with Mn appeared to have little, if any, effect on carbon deposition rates during HTFT synthesis [2]. It has been claimed that promotion with Cr oxide decreased the rate of CO decomposition over iron catalysts [16].

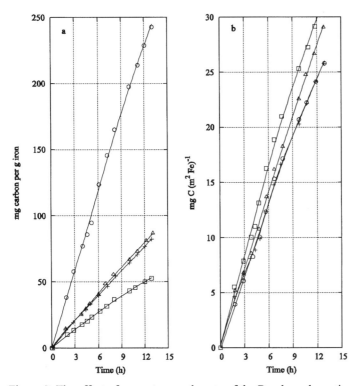

Figure 9 The effect of promoters on the rate of the Boudouard reaction.
a – represents the absolute rates and
b – the intrinsic carbon deposition rates.
+ unpromoted sample
o promoted with about 2.2 g Al_2O_3 ;
Δ promoted with about 2.3 g SiO_2 and
□ promoted with 1.35 g CaO per 100 g Fe

For the Boudouard reaction alkali promotion increases the intrinsic rate of carbon deposition, with K_2O having a greater effect than the weaker base Na_2O [5, 15]. When silica is also present it lowers the basicity of the alkali and in keeping with this the rate of carbon deposition is also lowered. In general this behaviour is in keeping with the FT performances of such catalysts (see Chapter 3, Section 9.2.2).

Evaluation of the rates of carbon deposition during the FT reaction over reduced fused iron catalysts (which were operated at normal HTFT temperatures but at different pressures, different fresh feed gas compositions and different recycle ratios) lead to the conclusion that the rate of carbon deposition was proportional to the value of P_{CO}/P_{H2}^2 inside the reactor [2, 17]. The inverse dependence on the partial pressure of hydrogen was also reported by previous investigators [18, 19]. The results obtained in the HTFT Sasol pilot plant studies are shown in Tables 5, 6 and 7. As can be seen the carbon deposition rate does not correlate with the simple H_2/CO ratio but does correlate with the factor P_{CO}/P_{H2}^2. A speculative derivation of the carbon deposition factor was presented elsewhere [17]. The partial pressure of CO_2 appears to have no effect on the rate as it was observed that a four-fold increase appeared not to influence the carbon deposition rate [17]. The data in Table 5 are for two catalysts of different basicities and are illustrated in Fig. 10. The higher basicity catalyst clearly has a markedly higher rate of carbon deposition.

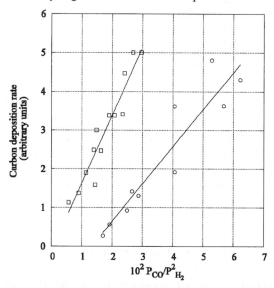

Figure 10 The carbon deposition rate (arbitrary units) as a function of the ratio P_{CO}/P^2_{H2}
The two sets of results are for two differently promoted catalysts operating at two different temperatures. The two plots are for cases B and A in Table 5. Catalyst B has a higher basicity than catalyst A.

Table 5
The influence of reactor entry gas composition on carbon deposition rate [2]

Set	Entrance partial pressure bar abs.			P_{CO}/P_{H2} ratio	P_{CO}/P^2_{H2} x 100	Carbon deposition rate[a]
	H_2	CO	CO_2			
	11.2	2.2	2.2	0.20	1.7	0.24
A	9.6	1.8	1.9	0.19	1.9	0.53
	8.3	1.7	1.8	0.20	2.5	0.86
	7.8	1.8	0.7	0.23	2.9	1.3
	7.2	1.4	1.5	0.19	2.7	1.4
	7.5	2.3	3.9	0.31	4.1	1.9
	6.6	1.8	1.0	0.27	4.1	3.6
	6.9	2.7	6.8	0.39	5.7	3.6
	6.3	2.5	2.7	0.40	6.3	4.3
	5.5	1.6	1.2	0.29	5.3	4.8
	32	6.4	6.4	0.20	0.6	1.2
B	24	5.1	4.6	0.21	0.9	1.4
	16.4	4.0	4.1	0.24	1.5	1.6
	17.0	3.6	1.3	0.21	1.2	1.9
	15.6	3.4	3.7	0.22	1.4	2.5
	12.5	2.6	2.5	0.21	1.7	2.5
	13.4	2.7	1.1	0.20	1.5	3.0
	9.2	1.6	0.6	0.17	1.9	3.4
	11.5	2.8	3.4	0.24	2.1	3.4
	8.6	1.8	0.7	0.21	2.4	3.4
	8.5	1.7	1.5	0.20	2.4	4.5
	8.1	1.8	1.7	0.22	2.7	5.0
	8.8	2.3	1.6	0.26	3.0	5.0

a g C per 100g Fe per unit time

A very significant feature of the carbon deposition factor P_{CO}/P^2_{H2} is that if the production capacity of a HTFT reactor is increased by raising the operating pressure (and simultaneously increasing the feed flow to keep the same linear gas velocity) the rate of carbon deposition actually decreases despite the higher hydrocarbon production rate. This is shown in Table 6. The FT production rate, at a fixed reactor pressure, can also be increased by increasing the fresh gas feed flow and appropriately decreasing the recycle flow. Again it will be found that the rate of carbon deposition decreases as the FT production increases (see Table 7). If elemental carbon were to be considered as part of the FT product selectivity spectrum then as the overall production increased it would be expected that the carbon formed would also increase, but as found this is not the case. Another advantage of operating at higher pressures is that lower ratio H_2/CO fresh feed synthesis gas can be fed without the formation of excessive carbon deposition. Thus it becomes easier to maintain higher average particle densities which compensates for the negative effect of higher gas density on the maximum allowable gas velocity in the fluidized bed reactor.

Table 6
The influence of pressure on carbon deposition [2]

Set	Total pressure bar abs.	Total feed		CO + CO$_2$ converted mol hr^{-1}	Carbon[a] deposition rate
		P_{CO}/P^2_{H2} x 100	H$_2$/CO ratio		
A	9.4	7.7	4.3	49	5.9
	12.9	3.7	5.2	68	3.3
	14.6	3.3	5.3	80	2.1
	18.1	2.3	5.8	94	1.6
	21.5	1.7	5.9	120	1.1
B	20.8	2.3	5.0	106	4.5
	30.8	1.7	4.8	157	2.5
	60.8	0.9	4.7	320	1.4
	75.8	0.6	5.0	377	1.2

a g C per 100g Fe per unit time

Table 7
The influence of recycle/fresh feed ratio on carbon deposition [2]

Recycle/fresh feed ratio	Entrance partial pressures bar abs..		P_{CO}/P^2_{H2} x 100	Total feed H$_2$/CO ratio	CO + CO$_2$ converted/ mol hr^{-1}	Carbon[a] deposition rate
	H$_2$	CO				
1.2	10.7	1.7	1.5	6.3	117	0.4
2.0	7.9	1.4	2.2	5.6	103	1.1
3.0	5.9	1.0	2.9	5.9	78	1.4
4.0	4.8	0.8	3.5	6.0	63	2.8

a gC per 100 g Fe per unit time

2.1.3 Alternative iron based catalysts

Instead of using a fusion process iron oxides can be sintered together with the desired promoters at temperatures typically ranging from about 400 to 800°C. The actual temperature used will largely depend on the required strength of the catalyst particles. In general these catalysts' FT performances are similar to their fused counterparts [5]. As fusion can be considered as the extreme case of sintering the above finding should not be surprising.

Instead of fusing or sintering, the iron oxide powders can be bound together by suitable compounds such as aluminium nitrate, alkali borates or silicates (waterglass). It has been reported that these preparations are less effective than those prepared by fusion or sintering [5]. One possible reason for this is that the binder compounds chemically react with the key K_2O promoter to form less basic, and hence less effective, compounds.

A common method of preparing metal catalysts is to impregnate the catalyst precursor onto a high area support such as alumina or silica. This, for instance, is the way currently used for the preparation of commercial cobalt catalysts. In the case of iron the amount and basicity of the alkali promoter is a vital factor in determining the FT performance of the catalyst. The alkali is also added to the support by impregnation and so the bulk of the alkali could be distributed on the high area support and in many cases would also preferably chemically interact with the support. The overall result is that the effectiveness of the alkali promoter with respect to the iron component could be markedly lowered. The activities as well as the wax selectivity, of impregnated catalysts are inferior to those prepared by precipitation methods. The use of chemically inert supports like wide pore charcoals could possibly be better choices. The type of iron salt used for impregnation is also important. It has been found that using the sulphates or chlorides results in inferior catalysts compared to using the nitrates. Catalysts prepared by thermal decomposition of iron nitrate are less active than those prepared by precipitation [5].

2.1.4 Phase changes of iron catalysts during FT synthesis

The ease of reducibility of the oxides increases in the order Fe, Co, Ni and Ru. The ease of re-oxidation of the metals in a H_2O/H_2 atmosphere can therefore be expected to decrease in the same order. By analogy bulk Fe metal oxidises slowly in air at ambient temperatures while the other three do not, Ru after all is a noble metal. In keeping with this it is found that under typical FT conditions Ru is not oxidised and neither is Co, as long as the Co crystals are larger than about 5 nm [92]. For iron based catalysts, however, some iron oxide, as magnetite, is always found to be present in catalysts which have been operating for some time. The situation regarding the formation of carbides during FT synthesis is similar. If Co metal is treated with only CO, cobalt carbide is formed but when hydrogen is also present, as is the case for the FT synthesis, no carbides are formed. For iron catalysts carbides are always formed and they remain present during the entire length of the FT run. In pure hydrogen, i.e. in the absence of CO, the carbides of both Co and Fe are rapidly reduced to the metals. Overall thus the stability of carbides under normal FT conditions depends on the relative rates of the carbiding and reducing reactions.

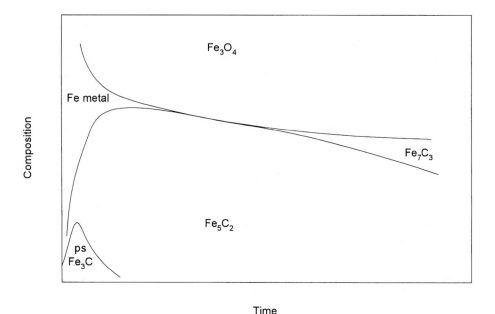

Figure 11 The change in composition of iron catalysts during FT reaction. The phases are those which are present according to X-ray diffraction analysis. At time zero the catalyst is 100 % metallic Fe. The units are undefined since the rates of phase changes depend on alkali content. The figure is only intended to illustrate the trends.

Fig. 11 illustrates the phase changes that occur with time on stream for a fully reduced iron catalyst loaded into a commercial scale HTFT fluidized bed reactor operating at about 330°C and at about 22 bar [2]. During the 'conditioning' stage when the hydrogen is being replaced by synthesis gas (see section 2.1.2B) the Fe metal is rapidly converted to a mixture of carbides. In the early stages of full FT operating conditions a significant portion of the Fe is rapidly oxidised to magnetite after which the rate of further oxidation decreases. The initial carbide formed was designated as 'speudo cementite' (ps Fe_3C). Its XRD pattern is somewhat similar to that of cementite (Fe_3C). This carbide phase is unstable and is converted to Hagg carbide (Fe_5C_2) within a day. At a later stage some Eckstrom-Adcock carbide (Fe_7C_3) is slowly formed. The overall effect is that with time on stream the carbides become richer in carbon content. (It should be noted that carbides are intersticial compounds.) It was found that in the 5 cm ID pilot plant reactors, operating under similar conditions to that of the commercial units, no Fe_7C_3 carbide was formed. At operating pressures above 60 bar, however, this carbide was also formed in the pilot plant runs.

Iron carbides are much more resistant to oxidation than metallic iron. This has been demonstrated in the laboratory by passing a mixture of hydrogen and water vapour at a fixed ratio over a fully carbided catalyst and also, separately, over a fully reduced iron catalyst. The latter immediately started to oxidise to magnetite while in the case of the carbide sample the oxidation commenced only after a significant delay. In another experiment two pilot plant reactors were operated in tandem, both charged with the same catalyst which had been previously used in FT runs [2]. The first reactor was operated at the normal high conversion level and the total effluent was fed to the second reactor. The phase composition of the catalyst in the second reactor did not appear to change. Feeding this same gas to a reduced iron catalyst would result in rapid oxidation.

The catalyst used in the FT fluidized bed units have a particle size distribution ranging from below 10 to above 100 micron. Analysis of the various particle size fractions showed that the smaller particles only consisted of carbide and that the oxide content increased with increasing particle size. An explanation of this observation could be as follows. As the synthesis gas diffuses into the porous catalyst particles the FT reaction occurs, consuming hydrogen and carbon monoxide (both reducing gases) and producing water vapour (an oxidant). Thus the deeper the penetration into the particle the more oxidising the gas mixture becomes. It thus follows that the cores of the larger particles would be oxidised while those of the smaller particles would not be oxidised. Microscopic investigation of polished sections of used FT catalysts confirmed the above [21]. The question then arises whether there is any point in reducing the HTFT catalyst fully when the cores of the larger particles will in any event re-oxidise during the FT reaction. Pilot plant tests at Sasol indeed showed that reduction beyond a certain level did in fact not result in any further increase in FT activity. This was in line with previous studies which showed that reduction of 6 to 8 mesh fused magnetite particles beyond about 25% did not improve the catalyst's performance in fixed bed reactors [22]. It should be noted that reduction of fused catalyst particles proceeds from the outside towards the centre, i.e. the shrinking core model applies. Hence it follows that at a certain overall degree of reduction the smaller particles will be fully reduced while the cores of the larger particles will still be unreduced.

Early studies showed that K_2O promotion of iron catalysts increased the rate of carbide formation [23]. This is in keeping with the effect of alkali promotion on the rate of carbon deposition (see Section 2.1.2C). Studies carried out in the Sasol pilot plant fluidized bed units found that the higher the amount of K_2O promoter the higher was the carbide level in the catalyst at the end of the FT run [2]. The ability of the alkali to keep the iron in the active carbide state is obviously an important aspect of alkali promotion. Anderson [5] reported that pre-nitriding iron catalysts made them more resistant to oxidation than carbided catalysts.

2.2 Cobalt based catalysts for LTFT operation

2.2.1 Preparation of cobalt catalysts

The catalysts used in the original German FT plants were prepared by co-precipitating the nitrates of cobalt and thorium (or zirconium or magnesium) with a basic solution in the presence of kieselguhr to yield an intimate mixture of the oxides supported on the kieselguhr. A typical mass ratio of the components was 100 Co/18 ThO_2/100 kieselguhr [5]. It was found that the addition of copper (up to about 2%) enhanced the rate of reduction which meant that the temperature required to reduce the cobalt could be lowered [24]. Unfortunately, however, it was found that the presence of copper increased the rate of activity decline [25] and so copper promotion was not applied.

Because of the high cost of cobalt it is important to minimise the cobalt content and at the same time have a high cobalt metal surface area. This is achieved by supporting the cobalt on stable oxides having the required surface areas and pore size distributions. A common technique is incipient wetness impregnation of the support with a cobalt salt solution of the appropriate concentration, drying, calcining to decompose the nitrate to the oxide and finally reduction with hydrogen. In order to obtain a high cobalt metal area, i.e. a high dispersion, each of these preparation steps needs to be optimised.

Using titania as a support Baerns [26] has studied various ways of adding the cobalt and also using different cobalt salts. The catalysts were tested at 2.0 MPa and 200°C and it was found that the catalysts prepared from cobalt oxalate by "spreading" (heating the mechanically mixed components to 250°C) yielded the most active catalyst. Incipient wetness impregnation with cobalt (III) acetyl acetonate produced a more active catalyst than the commonly used nitrate.

The effect of using different Co precursors for catalyst preparation by incipient wetness impregnation of alumina was investigated by van de Loosdrecht et al [27]. When loading only about 2.5% Co and using cobalt EDTA or ammonium cobalt citrate as precursors, very small Co oxide particles were formed. On thermal treatment under reducing conditions these small particles reacted with the alumina support to form the aluminates which were inactive for the FT reaction. The catalysts prepared from Co nitrate formed larger oxide particles. This catalyst did reduce and the resulting catalyst was active for the FT reaction. When the Co loading was increased to 5% using the citrate salt the oxide particles were larger and they were reducible. FT synthesis was carried out at 250°C and at atmospheric pressure, both factors which would account for the high methane selectivities obtained, namely 36 to 64%.

Niemela et al compared the use of Co nitrate and two Co carbonyls as precursors for the preparation of Co/silica catalysts. The dispersion of Co metal on the reduced catalysts decreased in the order $Co_2(CO)_8 > Co_4(CO)_{12} >> Co$

nitrate and consequently the carbonyl derived catalysts had a higher initial FT activity [28]. They also report that the FT activity of the Co/silica catalyst reduced at 450 °C was markedly lower than when reduced at 400°C [29]. In contradiction to this Moon [30] reported that the FT activity increased with increasing reduction temperature in the range 325 to 525°C.

At the University of Cape Town the use of silica and of alumina as supports for Co catalysts was investigated [31]. The preparation procedures were all similar, hydrogen reductions were carried out at 360° C for 14 hours and the FT reaction runs at 220°C, 15 bar with 2/1 H_2/CO syngas. TPR studies showed that after the above standard reduction the Co/silica catalyst contained little or no unreduced cobalt whereas the Co/alumina catalyst was only about 35% reduced (the balance of the Co being present as some form of "Co-aluminate"). Despite this the FT activity of the Co/alumina catalyst was significantly higher than the Co/silica catalyst. TEM and chemisorption studies of the reduced catalyst showed that the Co/silica catalyst contained smaller Co metal crystals. Under FT conditions the small Co crystals are apparently oxidised (see Section 3.3) and this could possibly account for the lower activity of the Co/silica catalyst in the above study. As could be expected, as the Co loading was increased the metallic area per gram of catalyst increased and consequently the FT activity increased but the metallic area per gram of Co decreased. Multiple-step impregnation of the carrier to a certain Co content gave similar results to a one-step impregnation. Silinisation of the surface of the alumina support had little effect on the FT activity but pre-treating the alumina with a KOH solution resulted in a very inactive catalyst. The use of different Co salts was also investigated using alumina as support. The acetate yielded smaller crystals than the nitrate and the initial FT activity was higher but it declined faster than the nitrate case. The acetyl acetonate case yielded a poor catalyst apparently due to pore blockage by undecomposed acetyl acetonate deposits.

Shell's commercial multi-tubular fixed bed FT plant at Bintuli uses a cobalt based catalyst and both Exxon and Sasol have successfully operated demonstration scale slurry phase FT units with cobalt catalysts. Several other companies are now actively pursuing this approach (e.g. Syntroleum, IFP/ENI/Agip, Conoco and Statoil). The method of preparation of these catalysts has not been revealed but the patents taken out by various companies should give some indication of the catalysts being developed. Recently Goodwin [32] has comprehensively reviewed many patents including those of Exxon, Shell, Statoil and Gulf and the reader is referred to this publication and the references therein. Besides being supported on a high area stable oxide, the cobalt catalysts usually also contained another metal or oxide promoter.

Exxon has apparently concentrated on using TiO_2 (rutile) as the support. Due to the relatively low area of rutile, about 15 m^2/g, the cobalt loading was limited to about 12 mass%. The addition of low levels of Ru was found to enhance both the initial reduction and the in-situ regeneration of the cobalt catalyst [33]. The patents assigned to Gulf [34, 35] apparently stimulated research on the promotion of cobalt catalysts by several metals, particularly Ru. The Gulf patents were eventually assigned to Shell. Researchers at Exxon thoroughly investigated and discussed the influence of low level Ru promotion of Co on supports such as TiO_2 and SiO_2 [33]. In agreement with the above patents it was found that at Ru/Co atomic ratios of about 0.008 the overall activity as well as the C_5^+ selectivity increased relative to un-promoted Co. Thus for Co/TiO_2 operating at 200°C and 20 bar promotion by Ru increased the site-time yield from about 18×10^3/s to about 56×10^3/s and the C_5^+ selectivity increased from 85 to 91% with the CH_4 selectivity decreasing from 7 to 5%. The key role of Ru appears to be its enhancement of the reduction of bulk and surface oxides and of hydrocarbon contaminants leading to a higher active site density. This results in higher activity and also in higher long-chain hydrocarbon selectivity because FT chain growth involves 'polymerization' of neighbouring 'CH_2' surface intermediates. To maximise the positive effect of Ru intimate contact with Co is required and this is achieved by calcining in air at about 300 °C prior to H_2 reduction. A spinel ($Co_2 Ru O_4$) is apparently formed which is isostructural with Co_3O_4. A further advantage of Ru promotion was that deactivated catalysts could be regenerated by treatment with hydrogen at normal FT synthesis temperatures whereas un-promoted Co catalysts could not be regenerated in this manner. It was previously reported that, for both Co and Ru on SiO_2, Al_2O_3 or TiO_2 supports, the overall FT activities were proportional to the metal dispersion but the rates per metal site were independent of metal dispersion or of support material [36].

Shell appears to have concentrated on SiO_2 as a support with a noble metal promoter and Zr, or Ti or Cr oxide included (see section 3.4). Both impregnation and kneading methods were used in the preparations.

In the Statoil patents alumina was used as the support. Promoters tested included rhenium and rare earth oxides. Hilmen et al [37] reported that Re promotes Co reduction whether co-impregnated with Co on Al_2O_3 or whether Re/Al_2O_3 is intimately mixed with Co/Al_2O_3. It is postulated that hydrogen spillover from Re is responsible. However, the same group found that when water vapour was added to the syngas feed the 21% Co 1% Re/Al_2O_3 catalysts deactivated more rapidly than the un-promoted catalyst. The tests were carried out at 210 to 220°C and 8 to 13 bar [38]. Since H_2O is the main by-product in FT these results suggest that Re would not be a satisfactory promoter. These authors also showed that in the presence of H_2O and H_2 mixtures no bulk oxidation of Co occurs but there was significant oxidation of the surface [39]. This study supports the findings of Exxon that a promoter such as Ru which helps keep the surface 'clean' is required. The Trondheim group reported that Pt promotion of Co on Al_2O_3 or SiO_2 enhanced reducibility and increased the FT activity about four fold (at 210°C, 1 bar, $H_2/CO = 7$) but the selectivity was <u>not</u> influenced [40]. By the use of SSITKA they deduced that the true turnover numbers were the same for the promoted and for the un-promoted catalysts and that the apparent higher turnover number for the Pt promoted catalyst was due to a higher surface coverage of reactive intermediates. Noronka et al [41] report that Pt, Pd and Rh promotion of Co/Nb_2O_5 catalysts promoted surface reduction of the cobalt as well as the FT activity at 260°C and 1 bar. Only Rh increased the C_5^+ selectivity. At a Rh:Co ratio of 9:91 the C_5^+ selectivity was 62% as against 49% for Co only. This high level of Rh promotion, however, would obviously be very costly.

Bianchi et al [42] studied the effect of Ru on Co ion exchanged titanium silicate. The Ru/Co ratios were about 0.015 and 0.03. H_2 reduction at 350°C was enhanced by Ru but the FT tests at 275°C and 5 bar resulted in an increase in CH_4 selectivity from about 50 for Co to above 75% for Ru/Co. Goodwin et al [43] also reported that for $Ru/Co/Al_2O_3$ (Ru/Co atom ratio about 0.01) the Ru facilitated the H_2 reduction of the cobalt oxide at 350°C. FT tests at 220°C and 1 bar showed that Ru increased activity threefold but not the selectivity (CH_4 selectivities were all at about 30%).

Ruthenium is not unique as other metal promoters have apparently been shown to have similar effects on reducibility and activity. Eri et al [44] found that 1% Re addition to a 30% Co on Al_2O_3 catalyst increased the activity about 2.7 fold without significantly altering the selectivity. The CH_4 selectivities were about 11%. The promotional effect of Re on Co supported on TiO_2 or on SiO_2 was much less marked. Tests were carried out at 1 bar and 185 to 205°C.

The promotion of a Co/silica catalyst with La markedly decreased the FT methane selectivity [45] and also enhanced the reducibility of the Co [46].

Promotion with Zr increased the FT activity of the Co/silica catalyst but did not affect the hydrogen reduction rate of the Co [47]. Khodakov el al [48] studied the reducibility of Co/silica using in situ XRD, EXAFS and FTIR spectroscopy. They concluded that after reduction at 450°C that the catalyst contained metallic Co, Co^{2+} ions (as CoO) and a Co ionic species in an amorphous phase. They confirmed that the ease of reduction decreased with decreasing Co oxide crystal size. In apparent contraction to the above, but in agreement with the work of Chirinos [31], Lapidus et al [49, 50, 51] reported that on silica there was little chemical surface interaction of the Co with the support while for alumina there was more extensive interaction. This resulted in a higher degree of reduction of the former catalyst. Calcination in air resulted in a decrease in reducibility.

Goodwin and co-workers [32] prepared and tested a series of cobalt catalysts supported on TiO$_2$, SiO$_2$ and Al$_2$O$_3$ and variously promoted with Ru, Re metals and or La, Zr oxides. From the FT tests they concluded that overall the Al$_2$O$_3$ supported Ru promoted catalyst had the best performance while the TiO$_2$ catalysts, due to their lower surface areas had the poorest performance. Ru or Re increased the activity of the Al$_2$O$_3$ or TiO$_2$ supported cobalt and ZrO$_2$ did the same for the Co/SiO$_2$ catalysts. TPR studies also showed that Ru promotion of a Co/alumina catalyst lowered the reduction temperature by about 100°C and complete reduction was attained at 350°C as against only about 60% for the un-promoted catalyst [52, 53]. It was found that for the Ru/Co/alumina catalyst the overall activity decreased markedly when, prior to the hydrogen reduction, the calcination temperature was increased from 300 to 400°C. However, irrespective of the calcination or of the reduction temperature in the above temperature range, the intrinsic activity was the same [53]. Neither the calcination nor the reduction temperature had any clear influence on the FT selectivity spectrum.

Bartholomew et al [54, 55] compared the FT activity and selectivity of unsupported Co and Co supported on alumina, silica, titania, carbon and magnesium. The tests were carried out at one atmosphere and at low conversion levels, i.e. far removed from practical FT synthesis conditions. The order of decreasing FT activity for catalysts containing 3% Co at 225°C was Co/titania, Co/silica, Co/alumina, Co/carbon and Co/magnesia. This order is quite different from that found by others under more realistic FT conditions [31, 32] and with catalysts containing higher Co loadings. A 15% Co/alumina catalyst is about 20 times more active than a 3% case and the gasoline selectivity is about 7 times higher which again emphasises the need to compare catalysts at realistic process conditions and conversion levels. Bartholomew [56] also studied the FT activities and selectivities of Co/alumina catalysts as a function of catalyst preparation method, Co loading and reduction temperature and significant effects were observed.

Sasol uses alumina as the support for their slurry bed FT catalyst [57, 58, 59]. The aqueous slurry phase impregnation using cobalt nitrate is their preferred technique for cobalt loading. The commonly used technique is incipient wetness, e.g. a volume of an aqueous solution of cobalt nitrate equal to the total pore volume of the catalyst support is added to the support and the impregnated sample is then dried. In the aqueous slurry phase impregnation an amount of the metal nitrate solution in excess of the pore volume of the support is added and the slurry is then dried under sub-atmospheric pressure using a set procedure. It was found that this procedure produced a catalyst which was about twice as active as a catalyst made using the incipient wetness procedure [60]. The incipient wetness procedure was found to yield a Co-rich encapsulating outer layer whereas the slurry impregnation procedure resulted in a more uniform distribution of the Co throughout the catalyst particles. This presumably explains the difference in activities.

The relative activities of the Sasol catalyst is compared with some other patented catalysts below [60, 61]:

Company and patent number	R.I.A.F[*]
Sasol	2.1
Gulf US 4413064 (1983)	1.1
Exxon WO 99/39825 (1999)	1.8
Shell EP 167215 (1986)	0.7
IFP GB 2258414 (1993)	1.6
Statoil US 4801573 (1989)	4.8

[*]Relative intrinsic activity factor is a dimensionless parameter which is a measure of the activity per unit mass of catalyst taking into consideration all the synthesis conditions such as temperature, pressure etc [62,63].

The R.I.A.F. of the Sasol catalyst is three fold higher than the catalyst used by Krishna et al in their reactor modelling study [64, 65].

Sasol found that during the slurry impregnation procedure some of the alumina is dissolved and re-precipitated. This results in an inferior bonding of the cobalt with the support which leads to some physical loss of cobalt-containing fines during the FT process which not only results in contamination of the wax product but also increases the rate of activity decline. This problem was solved by pre-coating the support with silica, by impregnating tetra ethoxy silicate dissolved in ethanol (TEOS), drying under vacuum and calcining in air at 500°C [59, 66]. (When the incipient wetness procedure was used in the preparation of the catalyst the rate loss of Co from the reactor was much higher.)

In research carried out at Sasol [57] many differently promoted and prepared cobalt catalysts were tested at 18 bar and about 220°C in both fixed and slurry bed units. In general it was found that the wax selectivity was dependent on both the activity and on the average pore size of the catalyst. The higher the activity the lower was the wax selectivity. The wax selectivity increased as the average pore size increased and for one series of catalysts it levelled out at pore sizes above 14 nm. It was reported that the Ru promoted catalysts (Ru/Co atom ratios 0.008 and 0.005) were not superior in performance to cobalt catalysts promoted with very low levels of Pt (Pt/Co atom ratio 0.0005). The Re promoted catalysts (Re/Co atom ratio 0.03) was inferior to the Pt or Ru promoted catalysts. Unfortunately this particular series of tests did not include a precious metal-free cobalt catalyst for comparison. The main claim of the patent was that drying impregnated catalysts under vacuum resulted in higher conversion activity.

It has been reported that several different oxides also affect reducibility, activity and selectivity. Lanthanum promotion of Co/SiO_2 (La/Co ratio of 0.75) enhanced cobalt oxide reducibility and increased FT activity (at 220°C and 1 bar pressure) about two fold and lowered the CH_4 selectivity from about 32 to 17%. This was attributed to an increase in the concentration of active Co metal sites [67]. In apparent contradiction Ernst et al [68] report that promotion with La or Ce showed no clear effect on activity but resulted in a marked increase in CH_4 selectivity. The catalysts all contained 25% cobalt with the Si content decreasing as the La (or Ce) content was increased from 5 to 38%. FT tests were carried out at 20 bar and 220°C. Zirconium appeared not to promote reduction but it did increase FT activity at 220°C and 1.3 bar. On the whole the selectivity was adversely affected with the CH_4 selectivity increasing from 22 to 28% [69].

With the objective of producing light alkenes Das et al [70] studied the effect of Mn addition to Co loaded onto silicalite. Mn formed a spinel with Co oxide and it decreased the reduction temperature. Mn increased the FT activity at 21 bar and 250°C but, strangely, not at 275°C. For practical purposes the CH_4 selectivities at about 20 % are too high and the olefinity of the C_2 to C_4 products are inferior to those achieved by standard high temperature iron catalysts [2]. The effect of Mn promotion on 11% Co supported on ZrO_2 coated SiO_2 was studied by Zhang et al [71]. Tests were carried out at 20 bar and 200°C and it was found that 1.6% Mn lowered the CH_4 selectivity to 9% (from 14% with zero Mn) but higher levels of Mn increased the CH_4 selectivity (to 28% at 4% Mn).

Adesina [72] found that molybdenum improved the olefinity and decreased the CH_4 selectivity of a Co/K/SiO_2 catalyst at 280°C and 1 bar. At this high temperature, however, the CH_4 selectivities were unacceptably high at about 70 mole percent. The effect of Mo needs to be investigated at lower temperatures. An IFP patent [73] describes the preparation of a series of cobalt catalysts supported on SiO_2 and promoted with two other metals, mostly Mo and K. The catalysts were tested at 20 bar, 200 to 220°C with a syngas having a H_2/CO ratio of 2. The most active catalyst contained about 30% Co, 2% Mo and 0.6% K. The C_5^+ selectivities were mostly between 85 and 90%. From the data it is not clear whether the third promoter (K, Na, Cu or V) was in fact required at all, and even the need for Mo appears to be questionable.

Note that several of the FT studies carried out in various laboratories were carried out at low pressures. Since the heavy hydrocarbon selectivity increases markedly from 1 to about 20 bar (see Chapter 3, Section 9.3.2) it remains an open question whether the selectivity improvements ascribed to the promoters would still be significant at the higher practical pressures.

Cobalt catalysts are normally pre-reduced separately with hydrogen before being loaded into the FT reactors. In the case of slurry bed reactors the FT wax which is used as the liquid medium could be hydrocracked if the reduction were carried out at high temperatures in the FT reactors. Hydrogen reduction is carried out in the range 250 to 400°C, preferably at low pressures and high linear gas velocities to minimize the vapour pressure of the product water which enhances sintering of the reduced metal. TPR studies show that reduction of Co_3O_4 starts at about 230°C. It has been claimed that the activity of a cobalt catalyst could be increased by reduction in hydrogen, then re-oxidising the catalyst followed by re-reduction in hydrogen [74]. In another patent it was claimed that hydrogen reduction in the presence of hydrocarbon liquids enhanced the initial FT performance of the catalyst [75]. Alternatively it was claimed that the selectivity of the heavier FT products were increased if the catalyst was first oxidised and then reduced with carbon monoxide [76]. In contradiction to this, work at Air Products showed that reduction in hydrogen produced a more active catalyst with a higher liquid fuel selectivity than a catalyst reduced with syngas having a H_2/CO ratio of 1.0 [77, 78, 79].

Bessell [80] compared the FT performance of Co supported on kieselguhr, silica, alumina, bentonite, Y-zeolite, mordenite and on ZSM-5. The alumina, Y-zeolite and mordinite supported catalysts did not undergo complete reduction. At normal FT conditions, 240°C and 20 bar, all the catalysts produced the normal carbon number FT product distribution. Whereas the low acidity supports such as silica and alumina produced linear products, the acidic supports produced more branched products, and at higher temperatures produced aromatics as well. It is presumed that the isomerisation and aromatisation are secondary, acid promoted, reactions of the FT olefins. In other words, this can be seen as a combination of the FT and the Mobil olefins to gasoline (MOG) process. (Note that with iron based catalysts this approach to gasoline production is unlikely to be successful as with iron catalyst alkali promotion is essential and the alkali would neutralise the required acid sites on the zeolite support.) Bessell also investigated the performance of Co on a series of ZSM zeolites, -5, -11, -12 and –34 and found that, as could be expected from diffusion rate considerations, the FT activity increased with increasing size of the zeolite pores [81].

Calleja et al [82] studied the FT performance of Co/HZSM-5 prepared by incipient wetness of the zeolite. They found that promotion with thorium, which is basic in character, increased the heavy hydrocarbon selectivity but it decreased the acidity of the zeolite and so less aromatics were formed. They also compared the performance of this catalyst with a physical mixture of a 'commercial' Co FT catalyst and HZSM-5. As expected from the effect of the basic thorium they found that for the mixture the amount of aromatics formed was somewhat higher but the amount of heavier hydrocarbons formed was significantly lower. Stencel et al prepared Co/ZSM-5 by incipient wetness with Co nitrate and reported the presence of highly dispersed, ion exchanged, non reducible Co ions inside the zeolite pores and large reducible Co oxide particles on the outside of the zeolite crystals [83].

2.3 Metal catalysts other than cobalt or iron

Ruthenium is a very active FT catalyst as evidenced by the fact that at low temperatures it is much more active than Ni, Co or Fe. It also is very versatile in that at high temperatures it is an active methanation catalyst (i.e. the only product is methane), whereas at low temperatures it produces large amounts of waxes in the low polyethylene range [84. 85]. The activity as well as the wax selectivity increases as the pressure is increased [85]. The waxes are free of oxygenated products and are essentially paraffinic. At low conversion levels, however, the lighter products do contain both olefins and oxygenates [2]. As mentioned previously the high price and the low availability of Ru make its commercial application in large FT plants very unlikely.

Of the other Group VIII noble metals only Rh and Os have some moderate FT activity [84]. The FT products of Rh contain relatively large fractions of oxygenated molecules. The US Bureau of Mines investigated Mo as an FT catalyst, the concept being that synthesis gas containing sulphur compounds could then be used, i.e. purification of the gas would not be required [86]. Even though the Mo catalyst was active in the presence of H_2S it was nevertheless more active in its absence, but still much less active than iron catalysts. Studies at Sasol found that to achieve reasonable conversions the reaction temperature required was about 400°C and then the methane selectivity was about 90%. Promotion with alkali lowered it to only about 50% [2]. A Cr based catalyst was also investigated but its activity was even lower than that of the Mo catalyst. Due to present day very strict environmental regulations any sulphur compounds in the fuels, as would inevitably be the case if sulphur-containing synthesis gas were to be converted over Mo catalysts, would in any event still have to be removed in downstream units. This does diminish the incentive to develop S-resistant FT catalysts.

3. CATALYST DEACTIVATION DURING FT SYNTHESIS

The factors involved in lowered activity, and in the decline of FT activity with time on-stream are the following:
- The presence of high molecular mass waxes and or aromatic coke precursors in the catalyst pores, resulting in diffusion restrictions.
- Fouling of the catalyst surface by coke deposits.
- Poisons in the feed gas such as H_2S and organic sulphur compounds.
- Hydrothermal sintering.
- Oxidation of the active metal/carbide to the inactive oxide.
- Boudouard carbon deposition.

High molecular mass hydrocarbon products are present as liquids in the pores of all normal FT catalysts under typical operating conditions, even for the high temperature operation with Fe catalysts (see Section 2.1.2). These compounds lower the FT conversion by lowering the rates of diffusion in and out of the catalyst particles but should only cause activity decline with time on stream if there is a continuous build-up of these products in the catalyst pores. When large amounts of waxes are produced as in the case of Fe or Co catalysts operating at the lower temperatures little or no build-up in the pores is expected as there should be a continuous flow of liquid out of the catalyst pores. At higher temperatures, where the bulk of the products exit the catalyst pores in the gas phase, build-up of the heavy products, e.g. coke precursors, could occur. In the low temperature wax producing operations no aromatics, and hence no coke precursors and coke deposits are formed.

All catalysts, at all FT operating conditions, will be poisoned by sulphur compounds present in the synthesis gas fed to the reactors. Similarly hydrothermal sintering and oxidation of the small crystallites of the metals/carbides will, even if only very slowly, occur with time on stream. Elemental carbon deposition will only occur at high temperatures (see Section 2.1.2C).

3.1 Poisoning by sulphur and other compounds

It was established long ago that sulphur compounds in the synthesis gas resulted in rapid activity decline of Ni, Co and Fe catalysts. Fischer [87] recommended that the sulphur content be kept below 2 mg m^{-3} but for present day commercial operation this level is unacceptable. Table 8 illustrates the effect of various sulphur levels on the performance of a fluidized iron catalyst. In a fixed bed Fe catalyst the conversion was noticeably steadier when the S content was lowered from 0.2 to 0.02 mg m^{-3} [2]. The Lurgi Rectisol purification process used at Sasol lowers the S content to about 0.03 mg m^{-3} but even at this level sulphur poisoning occurs (see Section 3.2.1).

The extent of sulphur poisoning also depends on the type of commercial FT reactor used. For fluidized bed reactors all of the catalyst would be poisoned whereas for a top fed fixed bed reactor the bulk of the S would be absorbed by the upper sections of the bed leaving the lower sections relatively unscathed (see section 3.2.1). The relative effectiveness of C_2H_5SH and H_2S as poisons of iron catalysts in fixed bed reactors was investigated [2]. It was found that the organic sulphur resulted in a higher rate of activity decline than the inorganic sulphur. This appears to indicate that at about 230°C hydrogen sulphide reacts with iron more readily than ethyl hydrosulphide. The former would thus be more completely absorbed in the upper layers of the catalyst bed whereas the latter would penetrate deeper into the bed before reacting and in so doing would poison a larger portion of the iron catalyst in the reactor. In the case of short catalyst beds as used in laboratory scale tests the opposite effect would be observed, namely that H_2S is a more powerful poison that C_2H_5SH.

Table 8
Influence of the sulfur content of synthesis gas on the rate of activity decline for fluidised iron catalyst at about 320°C [2]

Sulfur content of synthesis gas (mgS m^{-3}_n)	Drop in the % conversion per day
0.1	Very low
0.4	0.25
2.8	2.0
28	33

It has been found that when an iron catalyst operating in the HTFT fluidized mode has been poisoned by sulphur it is difficult to regenerate it. Reduction with hydrogen at 450°C did not reactivate the catalyst. Oxidation, followed re-fusion and re-reduction were also unsatisfactory unless the oxidation process was very thorough so as to burn off all the sulphur. This has also been observed in previous investigations [5]. In fixed bed reactors it was found that hydrogen re-reduction at 375°C adversely affected the catalyst and this was ascribed to redistribution of the sulphur over the entire bed depth [5]. Chemisorption studies showed that the amount of CO chemisorbed on an alkali promoted reduced fused iron catalyst decreased linearly with the amount of H_2S pre-adsorbed on the catalyst [2]. This indicated that alkali did not act as a sulphur getter as had been previously proposed, i.e. the S did not combine with the alkali in preference to adsorbing on the iron surface.

The strongly basic alkalis promote both the activity and selectivity of iron catalysts by increasing the strength of CO chemisorption. The bases act as electron donors to the Fe surface (see Chapter 3, Section 9.2.2). Based on this concept it could be expected that highly electronegative elements, which will act as electron acceptors, would deactivate the catalyst. This has indeed been found to be the case for O, S, Cl and Br (but not for F).

3.2 Deactivation of Iron FT catalysts

The contribution of the different deactivating factors depend markedly on the temperature at which the FT synthesis is carried out and so the low (220 to 240°C) and high (320 to 350°C) operations will be dealt with separately.

3.2.1 Low temperature FT (LTFT) operation

The low temperature fixed bed process is geared at the production of wax and under normal running conditions the wax is in the liquid phase and so the reactor operates as a trickle bed. Periodic in situ washing of the catalyst with a suitable solvent immediately results in a big increase in conversion but the effect is short lived as the catalyst pores are again filled with the wax produced in the FT reaction [2]. This suggests that diffusion restrictions in the wax filled pores lower the rate of the reactions. This is confirmed by the observation that the conversion is inversely proportional to the catalyst particle size. In the low temperature slurry phase reactor the catalyst particles are much smaller and so diffusion rates within the catalyst pores are less of a restriction. Fouling by ultra high molecular mass waxes is unlikely in systems where liquid wax is being produced within the catalyst particles as they would be continuously flushed out. The low temperature FT products contain no aromatics. Furthermore, when the used catalyst is exhaustively extracted with powerful solvents no aromatic materials, i.e. coke precursors, have been found in the extracted material [21].

In the fixed bed LTFT reactors the conversion as well as the wax selectivity declines with time on-stream. The total surface area of a silica supported precipitated catalyst which has been operating for only a short while is about 200 m^2/g. The XRD pattern of the catalyst is broad and indistinct confirming the presence of very small carbide crystallites. The area of the catalyst after it has lost about 20% of its original activity is about 50 m^2/g and the XRD pattern is sharper [2]. (Note that the BET determination gives the total area and not the area of the active Fe sites.) Both the changes in BET areas and in XRD patterns indicate that crystal growth had occurred over the time of the run. In confirmation it was found that if the amount SiO_2 , which acts as a support/spacer, added to the catalyst is lowered, the rate of activity decline is faster. When water vapour was deliberately added to the reactor feed gas, thereby resulting in a higher water partial pressure throughout the reactor, the rate of activity decline was increased. This could have been due to both enhanced hydrothermal sintering and to oxidation. All the above observations indicate that sintering contributes to loss in activity [2]. Using the kinetic calculations presented in section 4.1 it is estimated that if half of the active Fe surface sites are lost due to sintering then the overall activity will drop by a factor of 0.3.

In order to obtain more detailed information about the possible causes of catalyst deactivation in tubular reactors, tests were carried out in pilot plant reactors which were single commercial tube reactors. The runs were carried out for different lengths of time and then carefully unloaded to produce about twenty samples down the length of the tubes for each test run. Each sample was then analysed and the FT activity and wax selectivity determined under fixed conditions in laboratory units [21]. The results of the FT tests are illustrated in Figs. 12, 13 and 14. As can be seen, for young catalysts there was relatively little difference in the intrinsic activity and wax selectivity of the catalyst in the different sections of the bed. With time, however, large changes in performance took place.

570

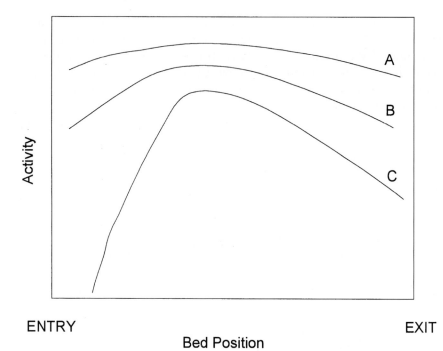

ENTRY   EXIT

Bed Position

Figure 12 Catalyst activity profiles as run time increases from A to C

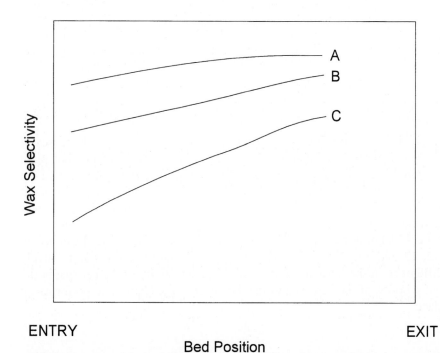

ENTRY   EXIT

Bed Position

Figure 13 Wax selectivity profiles as run time increases from A to C

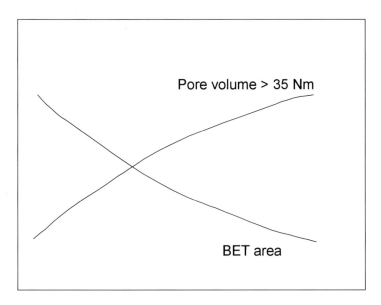

ENTRY   EXIT

Bed Position

Figure 14 Catalyst pore volume and BET area as a function of bed position

The activity of the catalyst near the entry of the bed markedly declined with time ending up completely inactive (see Fig. 12). The catalyst samples near the reactor exit also declined significantly but the catalysts from the middle-sections declined the least. The BET areas (after thorough wax extraction) of the samples declined progressively from the entry to the exit of the reactor tube. Fig. 14 illustrates the shifts in area and pore sizes for one of the test runs. X-ray line broadening studies confirmed that the average crystallite sizes increased from the entry to the exit of the reactor. The percent oxidation of the iron to Fe_3O_4 also increased in the same direction. Analysis of individual catalyst particles showed that the cores of the particles were more oxidised than the outer layers [2]. This is in line with expectation. As the synthesis gas diffuses into the porous catalyst and FT synthesis occurs on route, H_2 is consumed and H_2O is produced resulting in a more oxidising atmosphere as the gas moves towards the centre of the particle.

Chemical analyses showed a high S content in the very top sections of the bed, the S content rapidly decreasing with bed depth. Sulphur poisoning thus explains the rapid decline in activity with time on stream of the catalyst at the reactor entrance. Previous studies had shown [5] that H_2S interacted with iron catalysts more strongly than organic sulphur compounds. As discussed in section 3.1 the consequence of this is that the organic sulphur compounds can penetrate deeper into the catalyst bed before reacting with the iron. From these observations it thus appears that sulphur poisoning is the main cause of loss of activity of the front sections of the catalyst bed. Note that the BET areas of the samples near the reactor entrance were the highest of all the samples in the reactor and so clearly sintering was not the main factor causing deactivation of the catalyst at the reactor entrance.

As the FT reaction progresses down the reactor the partial pressure of water increases and thus the likelihood of hydrothermal sintering increases. This matches the observation of decreasing areas (Fig. 14) and increasing average crystal size. In addition to the increasing H_2O concentrations those of H_2 and CO decrease which means that the system becomes more oxidising down the tube as was in fact observed from the analysis of the catalyst samples taken. The increased sintering and degree of oxidation explain the progressive loss in FT activity in the lower sections of the catalyst bed.

As the temperature and feed gas compositions were fixed for the short catalyst bed laboratory FT tests carried out on the above catalyst samples it was assumed that the observed increases in wax selectivity (Fig. 13) were due to catalyst 'basicity' changes down the bed. Analysis, however, showed that the K_2O contents per unit mass of iron were unchanged. It was postulated that the basicity changes might be due to two factors. Firstly, the presence of electronegative sulphur on the catalyst surface in the upper sections of the bed would lower the basicity by, for example, chemically reacting with the alkali. Secondly, if the alkali promoter is mobile and the surface areas of the catalysts in the lower sections are decreasing this could translate to a higher K_2O surface coverage, i.e. a more basic surface which would result in a higher wax selectivity. It should, however, be noted that over the entire bed the wax selectivity declined with time on stream. This suggests that some other factors may also be involved.

In a slurry bed reactor the catalyst is fluidized and hence back-mixing of the solids occurs freely in the bed. This means that on average all the catalyst particles are exposed to the various gas compositions along the bed height and so all the catalyst particles should deactivate at the same rate. The factors causing deactivation would be the same as those discussed above, namely sulphur poisoning, hydrothermal sintering etc.

3.2.1.1 Handling of spent LTFT iron catalyst

The spent catalyst product from both the fixed and slurry bed reactors contains about 50 – 60 m% solid catalyst material, the rest being LTFT reactor wax. Historically all the spent catalyst products were dumped on the local ash heaps in Sasolburg, being comparatively minute quantities compared to the ash production volumes.

Due to the introduction of natural gas and the associated discontinuation of coal gasification, as well as increased environmental awareness, this catalyst dumping practise is no longer acceptable. In addition, the pyrophoric nature of the spent catalyst product inherently complicates the dumping practise.

The spent catalyst product is supplied to steam stations to be blended into the coal feed used in these coal-fired boilers, with a beneficial contribution to heating value. The effective oxidation of the catalyst product during this process inherently removes all the wax from the product and the pyrophoric nature of the catalyst is effectively negated. The coal-ash diluted product from steam stations make dumping significantly safer, less prominent and more environmentally acceptable.

Recovery of the metals from the spent catalyst by treatment with hot aqueous acid solutions is, in principle, also possible but has not yet been proven to be cost effective.

3.2.2 High temperature FT (HTFT) operation

As in the case of LTFT, the activity declines and the selectivity shifts towards lower molecular mass products with time on-stream. Despite the high operating temperatures, of about 340°C, the individual catalyst particles contain significant amounts of 'waxes'. In contrast to the situation in the LTFT process these waxes are highly aromatic. Stepwise extraction of the catalyst with heptane, xylene and pyridine showed that the atomic H/C ratio of the extracted hydrocarbons decreased from 1.7 to 0.9 to 0.8. On subsequent hydrogen reduction above 350°C of the pyridine treated catalyst, hydrocarbon gases and oils were evolved. This showed that there were still pyridine insoluble hydrocarbons present in the catalyst pores. Hydrogen treatment of carbidic carbon or of Boudouard carbon at 350°C does not produce any hydrocarbon oils and hence these products must have been the result of hydrocracking of the pyridine insoluble carbonaceous deposits. It therefore seems likely that for the HTFT process fouling by 'coke' contributes to activity decline. The presence of liquid hydrocarbons inside the catalyst pores is also likely to lower the overall FT reaction rate by lowering diffusion rates. This of course would not in itself result in activity decline with time on stream unless there is continuous build up of heavier more viscous oils in the pores.

As described in Section 2.1.4 the reduced Fe loaded into the fluidized bed reactors undergoes several phase changes with time on stream. At the high operating temperatures the Boudouard reaction

$$2\,CO \;\rightarrow\; CO_2 + C$$

results in the continuous increase in the elemental ('free') carbon content in the catalyst particles. After several weeks of operation there is, on an atomic basis, in fact more carbon present than iron. Since the decline in activity is relatively small it appears unlikely that carbon deposition as such is responsible for activity decline. As will be discussed below there is, however, a strong <u>indirect</u> influence of carbon deposition on activity decline.

A freshly carbided catalyst has the same BET area as the reduced catalyst but with time on stream there is a marked and continuous increase in the measured area. This is due to the deposition of the high area 'free' carbon. Because of this complication the BET area measurements cannot provide any evidence of loss of active area due to sintering and oxidation. Using the kinetic calculations discussed in Section 4.1 it will be found that, for the HTFT operation (assuming all sites are equally active), that loosing half the number of sites (due to oxidation or sintering) will lower the activity by only a factor of 0.13. (For the LTFT process the factor was 0.3, see Section 3.2.1.)

Analysis of aged catalyst shows that the smaller catalyst particles have higher levels of carbide and of Boudouard carbon and lower levels of oxide than the larger sized particles. SEM/EDX examination of polished sections revealed that the small particles consisted of small crystals of iron carbide embedded in a matrix of Boudouard carbon whereas the larger particles had cores of magnetite surrounded by the carbide/carbon mixture [21]. The formation of the oxidised cores has been explained in Section 3.1.4.

Because of the vigorous movement of the catalyst particles in the high gas velocity fluidized beds, scouring of the carbon rich carbide outer layers of the catalyst particles as well as possible break up of the particles takes place. Thus more of the fine carbide-rich particles are generated. Since these small particles have low densities, due to the high free carbon content, they pass through the cyclones and so are lost from the reactors. Hence, indirectly, carbon deposition does contribute to activity decline. The loss of carbide rich fines is probably the main reason that the oxide content of the catalyst remaining in the reactor increases with time on stream. As mentioned in Section 2.1.2A and in Chapter 3, Section 9.2.2, the finer fractions of the catalyst contain separate particles of alkali silicates. The loss of the fines hence also means that alkali is lost from the reactors which would contribute to the observed decline in activity and the shift in the selectivity to lower molecular mass hydrocarbons.

Overall the observed selectivity shift and activity decline with time on stream are possibly due to three factors. Firstly, the loss of the small loose alkali silicate promoter particles mentioned above. Secondly, the poisoning of the alkali/Fe sites by sulphur compounds in the feed gas. Thirdly, and probably more importantly, the preferential fouling by 'coke' of the more active alkali/Fe centres. The latter concept is supported by the observation that H_2 re-reduction at 350°C of used catalyst largely restores both the activity and selectivity [2]. Hydrogen treatment at 350°C does not reduce iron sulphide, i.e. any S-poisoning cannot be reversed. The observation that the regeneration of the activity by hydrogen re-reduction is as marked for a catalyst containing 5% oxide as for one containing 60% oxide also indicates that catalyst oxidation during synthesis is probably not the main cause of activity loss with time on stream [2].

Catalyst activity decline as measured by testing a fixed mass of catalyst in a laboratory reactor is not the only reason for reactor performance decline. Another reason is the decline of catalyst mass in the reactor. As carbon forms at the grain boundaries the catalyst swells and the density declines. This has the result that for a fixed reactor volume in the fluidized bed a lower mass of catalyst can be accommodated. In practice equilibrium catalyst properties are maintained within a fairly narrow range using periodic on-line partial removal and addition of catalyst. More catalyst mass must be removed than is added. Also the stresses created by carbon build-up cause the particles to break up generating 'fines' i.e. particles smaller than 22 microns. This increases the gas fraction or voidage for the aerated powder in the fluidized bed which is another reason why the catalyst mass in the reactor decreases with average catalyst age.

3.3 Deactivation of cobalt catalysts

It is commonly observed that during the first few hours the decline in the FT rate is relatively rapid and then later declines much more slowly [88, 89]. As the operating temperatures are low, around 220°C, the deposition of foulants such as 'coke' or 'free' carbon seems unlikely. The absence of CO_2 production indicates the absence of the Boudouart reaction which produces free carbon. One of the plausible reasons for the activity decline is the build up on the surface and in the catalyst pores of very long chain waxes which inhibit adsorption and slow down diffusion rates. It has been shown that Ru promoted Co catalyst can be regenerated by H_2 at reaction temperatures whereas the unpromoted catalyst cannot [33]. This Ru related feature is very probably the same one that accounts for the higher FT activity, C_5^+ selectivity and higher paraffin content of the products. Ru promotion results in a more hydrogenating surface thereby lowering the surface coverage by heavy waxes by hydrogenation and/or hydrocracking [90]. Hydrogen treatment at 220°C has also been shown to regenerate La-promoted catalysts [67]. Note that La had similar effects as Ru on Co oxide reduction and on FT performance. Metals like Re, Hf, Ce, Th and U have also been reported to assist catalyst re-reduction and increase activity [4].

Because iron based catalysts are relatively cheap the decline in activity in an operating slurry bed FT reactor can be compensated for by on-line partial catalyst removal and addition of fresh catalyst. This option is unattractive for cobalt based catalysts due to the much higher price of cobalt. Minimising the rate of activity decline in order to ensure a long on-stream lifetime is therefore very desirable. As in the case of iron any sulphur containing gas will result in permanent deactivation and so the purification of syngas needs to be very effective. In commercial practice cobalt catalysts are only likely to be used in the wax producing mode under which conditions the temperatures will be mild (about 220°C) and hence no aromatics should be co-produced. In addition, the use of a well operated slurry bed unit should eliminate the likelihood of any localised hot-spots in the catalyst particles. These factors make it unlikely that aromatic carbonaceous coke can be formed which will foul the catalyst surface. The wax produced is in the liquid phase under FT conditions and so in practice there must be a liquid flow from within the porous catalyst particles to the outside surfaces. The build up of very long chain waxes, which could clog up pores and thereby decrease the reactant/product diffusion rates, therefore seems unlikely to occur. Nevertheless Iglesia [33] found that when Co was promoted with Ru, treatment with H_2 at normal FT temperatures was accompanied to a greater extent by the evolution of light hydrocarbons than in the absence of Ru. The Co/Ru catalyst was apparently fully reactivated while the Co-only catalyst was not. The Ru appeared to enhance the rate of hydrocracking and it could be concluded that the build up of high molecular mass wax/'carbonaceous' deposit was at least in part responsible for the slow deactivation with time on stream. Alternatively, or in addition, the Ru could enhance the re-reduction of oxidised cobalt (see discussion that follows).

There is good evidence which indicates that an important factor in the deactivation of cobalt is the water vapour produced by the FT reaction. Three related aspects may be involved, sintering of the cobalt metal particles, surface or total oxidation of the cobalt particles and the formation of FT inactive cobalt/support compounds (e.g. cobalt silicates, titanates, aluminates).

Evidence was found for the formation of Co silicates on Co/silica catalyst during the FT process. With time on stream there was a loss of activity and an increase in the amount of silicates formed [91]. The deactivation of Co/silica catalysts was also investigated by Neimela and Krause [89]. They report that the higher the dispersion the higher the deactivation rates. In addition the active sites were blocked by wax and 'coke' deposits and some particle agglomeration occurred. Thermodynamically the bulk phase oxidation of cobalt metal by water under typical FT conditions (temperature and H_2O/H_2 ratios) is not possible. The Co atoms on the surface, however, are highly unsaturated/reactive and partial/temporary surface oxidation can occur. A popular concept of the FT mechanism is the dissociation of chemisorbed CO to adsorbed C and O atoms. The latter is equivalent to a surface oxide and can readily be hydrogenated to water. Alternatively water vapour oxidation of surface cobalt metal may be an 'equilibrium' reaction. The highly electronegative adsorbed oxygen atoms are bound to be associated with positively charged, i.e. oxidised, cobalt atoms. Whatever the detail, the continuous cyclic oxidation and reduction of the active surface cobalt atoms amounts to continuous re-arrangement/movement, i.e. sintering, which results in the loss of surface area and so loss in activity. The smaller the cobalt crystallites, the higher the degree of unsaturation of the exposed cobalt atoms and so probably the higher the coverage of the surface by inert 'oxide'. This amounts to loss of active surface. Crystallites below a critical size, estimated at 5 micron [92], could be fully oxidised and hence rendered completely inactive. Since the reaction of cobalt with the support material, e.g. the formation of cobalt silicate or aluminate, apparently requires cobalt to be in the oxidised state, the presence of water vapour should enhance this chemical process. These compounds are not reducible under normal FT conditions and so this translates to a loss of active surface. If regeneration with air at high temperatures is practised these chemical reactions will be accelerated.

It has been demonstrated on laboratory scale that for a Co/Al_2O_3 catalyst the addition of water to the syngas feed results in surface oxidation of the cobalt and in permanent loss in FT synthesis activity [93, 94]. Rapid deactivation occurred on the Re promoted catalyst whereas the unpromoted Co/alumina catalyst deactivated at a slower rate. XPS and gravimetry analyses indicated that only a small fraction of the bulk Co metal was oxidised but significant oxidation of the surface Co atoms occurred and /or highly dispersed Co reacted with the support. These two reasons are the likely cause of the decline in activity. In situ Mössbauer studies showed that oxidation of cobalt metal supported on alumina does occur under FT conditions [95]. The formation of both reducible and less reducible oxides was observed.

The effect of water on the FT activity of supported Co catalysts appears to depend on the type of support and also on the level of water vapour pressure inside the FT reactor. As pointed out above the addition of water to Re promoted or to unpromoted Co/alumina resulted activity loss [93, 94, 96]. Davis et al observed the same effect for Pt promoted Co/alumina [97]. From XANES studies Davis et al [98] concluded that on noble metal promoted Co/alumina the small Co metal clusters on the support oxidised in the presence of water and that during the initial deactivation stage cobalt metal cluster growth took place resulting in lower Co metal surface area.

On Co/silica catalysts the FT water, or water deliberately added to the reactor feed gas, has been found to increase [99] or to decrease [100] the FT activity. At high partial pressures of water, whether as the result of high FT conversion levels or due to high levels of water added to the feed, the FT rate deactivated rapidly whereas at low partial pressures of water the deactivation rate was low [100, 101]. At high water partial pressure conditions the activity loss was permanent due to the formation of irreducible Co silicates [101]. It has been reported that when the water partial pressure was increased over Co/titania catalysts the FT activity increased [92, 102, 103, 105] or decreased [105]. For an unsupported Co catalyst [106] and for a magnesia /thorium oxide/aerosil supported Co catalyst [107] water addition increased the FT activity but no effect was found for a zirconium oxide/aerosil supported Co catalyst [108].

The effect of water on the activity of the different Co catalysts is obviously confusing and as yet a satisfactory explanation for the diverging results is lacking. As far as the effect of added water on the selectivity of the FT reaction, however, the consistent finding to date is that the methane selectivity is depressed with of course an increase in the oil+ selectivity.

It appears that by removing poisons such as sulphur from the syngas feed to very low levels and by operating the FT reactor so as not to exceed a threshold water partial pressure that very long catalyst lifetimes can be attained at a high and stable catalyst activity. Highly attrition resistant catalysts have been developed that do not break during normal synthesis conditions in slurry phase reactors. It is known that the use of nickel catalyst in an adiabatic preformer upstream of an autothermal reformer removes any remaining sulphur compounds that pass through the typical upstream zinc oxide guard beds. Also with recent advances in catalyst activity the initial catalyst load to a slurry phase reactor need not be at the upper limit of acceptable catalyst concentrations. Thus any small activity decline can be compensated for by simply adding a small amount of catalyst on-line. It can be envisaged that a time will come when slurry phase reactors will operate for more than twenty years without a need to unload catalyst.

The cobalt metal can be recovered from spent catalyst for re-use in catalyst preparation by treating the spent catalyst with aqueous nitric acid. Methods from the noble metal refining industry can be used to recover the other metals associated with the typical industrial cobalt catalysts. The cost of the cobalt and other metals are the main constituents of the catalyst cost. If the initial catalyst cost is included in the plant capital cost then the contribution of catalyst cost to the plant operating costs is extremely low.

3.4 Recent Co catalyst patents

Table 9 provides a list of relevant patents that is not necessarily complete. Both Shell and Sasol have appear to have made significant recent advances to their initial, commercially applied, catalysts that are not yet reflected in published patents. The major players are similar to those found in Chapter 2, Table 5 that lists patents relevant to FT reactor technology. Shell has concentrated more on catalyst rather than reactor innovations keeping to the traditional fixed reactor design. The egg shell catalyst approach only provides benefits when larger catalyst particles are used i.e. for fixed bed or ebullating bed reactors. ExxonMobil appears to favour continuous catalyst reactivation to maintain a high catalyst activity and their patents favour the use of a titania support. There are many more patents on this topic that have not been listed in Table 9. BP appears to have concentrated on zinc oxide as a support. Other companies appear to favour the use of silica, alumina or silica-alumina supports. It is necessary to modify the support to be suitable for use in commercial slurry phase reactors if alumina is used. This was not recognised in the earlier patents. The initial Gulf patents (assigned to Shell after Gulf was taken over by Chevron) appear to have stimulated the interest in alumina as a support. Sasol now has a particularly extensive portfolio dealing with various aspects relating to supports comprising alumina or silica-alumina. Alumina also features prominently in the IFP/Agip and Statoil patent portfolios.

Many of the patents deal with the use of promoters. They cover a wide range of promoters and there is a high degree of inter-linkage between them.

Table 9

Patent No.	Assignee	Brief Description	Feature
US 5,733,839	Sasol	Composition (Co/Pt on alumina) and preparation method	Composition
US 5,939,350	Sasol	Ruthenium promoter	Composition
US 6,100,304	Sasol	Palladium promoter	Composition
US 4,585,798	Shell (Gulf)	Alumina supported cobalt-ruthenium catalyst	Composition
US 4,613,624	Shell (Gulf)	Co on alumina with one or more Group VIII or Group IVB metal oxide promotors	Composition
US 4,857,497	Shell	Co with zirconium on silica with noble metal promoter	Composition
EP 455308	Shell	Alumina-based extrudates containing Co, Fe or Ni	Composition
EP 466268	Shell	Group VIB, VIIB and VIII metal alloys	Composition
EP 510771	Shell	Extruded SiO_2 catalyst with promoter from Group IVB	Composition
EP 188304	Shell	Co and one or more of Zr, Ti, Cr and Ru deposited on silica, alumina or silica-alumina support	Composition
EP 142887	Shell	Co (15-50) on silica (100) and a second metal (chosen from Zr, Ti, Ru or Cr)	Composition
EP 142888	Shell	$Co:Zr:SiO_2 = 25:18:100$	Composition
EP 109702	Shell	$Co:Zr:SiO_2 = 25:0.9:100$	Composition
WO 97/00231	Shell	Supported (e.g. titania) Co+Mn and/or V catalyst	Composition
WO 98/25870	Shell	Supported (e.g. titania) Co/Mn catalyst	Composition
EP 5162284	ExxonMobil	Cu promoted Co-Mn spinel	Composition
EP 319625	ExxonMobil	Co with Ru on titania	Composition
US 5,169,821	ExxonMobil	Co on $Ti_xM_{1-x}O_2$ where x is Si, Zr or Ta	Composition
US 4,801,573	Statoil	Alumina supported cobalt-rhenium catalyst	Composition
US 5,102,851	Statoil	Co with a second metal (Pt,	Composition

Patent No.	Assignee	Brief Description	Feature
		Ir and/or Rh) on alumina	
US 4,880,763	Statoil	Co, Re and alkali supported on alumina	Composition
US 4,801,573	Statoil	Co, Re on alumina which may include a metal oxide promoter	Composition
US 4,857,559	Gas-to-Oil Inc.	Co, Re on alumina which may include a metal oxide promoter	Composition
EP 756895	IFP/ Agip	Co with Ru and scandium or yttrium – 3 step calcination, reduction and passivation	Composition
EP 800864	IFP/ Agip	Co with Ti and a third metal	Composition
EP 857513	IFP/ Agip	Co with Ru and tantalum	Composition
EP 934115	IFP/ Agip	Co-scandium catalyst preparation procedure	Composition
EP 935497	IFP/ Agip	Co and tantalum	Composition
EP 1094049	IFP/ Agip	Raney metallic alloy dispersed in a liquid phase	Composition
US 6,255,358	Sasol	La/Ba doped catalyst	Composition/characteristics
WO 01/96014	Sasol	Preparation to modify catalyst physical characteristics	Composition/characteristics
WO 01/96015	Sasol	Preparation to modify catalyst physical characteristics	Characteristics
EP 979673	IFP/ Agip	Metal particles on the support of controlled size with narrow distribution	Characteristics
EP 979807	IFP/ Agip	Metal particles of controlled size and distribution on the support	Characteristics
EP 987236	IFP/ Agip	Co particles spread as aggregates on the support	Characteristics
EP 1129776	IFP/ Agip	Support particle size, surface area and pore size	Characteristics
EP 1239019	IFP/ Agip	Co on alumina with cobalt oxide crystals of a certain size	Characteristics
US App.2003/ 0105170	ConocoPhilips	Increased cobalt surface area on silica support using cobalt amine carbonate precursors	Characteristics
WO 01/96017	Sasol	Partial reduction	Preparation
US 6,455,462	Sasol	Drying procedure in Co	Preparation

Patent No.	Assignee	Brief Description	Feature
		impregnation process	
WO 01/39882	Sasol	Calcination procedure	Preparation
WO 03/35257	Sasol	Reduction procedure	Preparation
US 6,191,066	Sasol	Non-promoted catalyst	Preparation
ZA 835606	Shell	Kneading technique	Preparation
EP 583837	Shell	Solvent and peptizing agent	Preparation
EP 421502	Shell	Drying and calcination	Preparation
US 4,605,679	Shell	Impregnation of Co carbonyl on alumina or silica	Preparation
EP 127220	Shell	3 step impregnation of silica with zirconium tetrapropoxide in propanol-benzene, then aqueous nitrate, kneading and then calcination	Preparation
EP 152652	Shell	Order for impregnation for cobalt and the promoter	Preparation
EP 168894	Shell	Reduction procedure	Preparation
US 4,623,669	ConocoPhilips	Impregnation using metal-solvent slurry	Preparation
WO 2000/10704	ConocoPhilips	Forming a catalyst gel by destabilizing an aqueous colloid and drying the gel	Preparation
WO 2000/10705	ConocoPhilips	Forming a catalyst gel by mixing at least one dissolved compound of a catalytic metal in water and/or ethanol, at least one dissolved alkoxide and optionally an aluminium compound and to hydrolyse, water supplement and dry	Preparation
US 5,728,918	BP	Reduction of zinc oxide supported Co catalyst with mixtures of carbon monoxide and hydrogen	Preparation
WO 02/097011	BP	Reduction (activation)	Preparation
WO 02/083817	BP	Reduction (activation)	Preparation
EP 581619	IFP/Agip	Co-Ru-Cu catalyst prepared by gel formation route	Preparation
EP 764465	IFP/Agip	Partial reduction of Co impregnated support then deposition of the remainder of the metals	Preparation
EP 1101531	IFP/Agip	Activation procedure using hydrogen and carbon	Preparation

Patent No.	Assignee	Brief Description	Feature
		monoxide	
EP 1126008	IFP/Agip	Preparation procedure for a silica-alumina support	Preparation/support
EP 1233011	IFP/Agip	Preparation procedure for a silica-alumina support	Preparation/support
EP 1058580	Sasol	Support modification	Support
WO 02/07883	Sasol	Support modification	Support
WO 03/12008	Sasol	Support modification	Support
US 6,262,132	Sasol	Support modification for increased attrition resistance	Support
WO 03/044126	IFP/Agip	Support modification	Support
US 6,515,035	IFP/Agip	Support modification	Support
US 4,794,099	ExxonMobil	Support modification	Support
US 4,960,801	ExxonMobil	Support modification	Support
EP 370757	ExxonMobil	Inorganic metal oxide binder	Support
US 4,595,703	ExxonMobil	Titania support	Support
EP 196124	Shell	Small catalyst particles 'encapsulated' in permeable meshes for use in fixed bed reactors	Support
EP 398420	Shell		Support
EP 455308	Shell	Alumina based extrudates	Support
WO 9811037	Shell	Titania support	Support
UK Res. Dscl. (1991) 323,180	Anon.	Honeycomb support	Support
US 5,648,312	Intevep S.A., Venezuela	Homogeneous mixture of two refractory inorganic carbides nitrides or mixtures	Support
US 4,826,800	BP	Zinc oxide supported Co catalyst	Support
WO 86/ 01499	BP	Co or Fe on a carbon support	Support
US 5,531,976	Sasol	Alumina with increased porosity	Support preparation
US 5,593,654	Sasol	Stabilized alumina	Support preparation
US 4,634,581	Sasol	High purity alumina	Support preparation
US 5,075,090	Sasol	Particulate mixed oxide	Support preparation
US 5,837, 634	Sasol	Stabilized alumina	Support preparation
EP 157503	Sasol	Reducing impurity content in alumina	Support preparation
US 5,045,519	Sasol	Aluminosilicate support	Support preparation
US 5,055,019	Sasol	Boehmmite alumina	Support preparation
EP 180269	Shell	Silica support obtained from	Support preparation

Patent No.	Assignee	Brief Description	Feature
		a Si halogen, alkoxy and/or acyloxy	
US 6,462,098	Sasol	Particle size optimized for both synthesis performance and catalyst/wax separation	Particle size for slurry phase synthesis
ZA 8505317	Shell (Gulf)	Average particle diameter of 10 to 110 micron	Particle size for slurry phase synthesis
EP 510770	Shell	Helical lobed shape by extrusion	Shape for fixed bed synthesis
EP 266898	ExxonMobil	Thin film of Co with Re on titania	Egg shell catalyst
EP 434284	ExxonMobil	Support contacted with a molten salt	Egg shell catalyst
US 4,977,126	ExxonMobil	Thin film of Co on especially titania	Egg shell catalyst
US 4,542,122	ExxonMobil	Impregnation of Co in the external portion of the support particle	Egg shell catalyst
EP 370757	ExxonMobil	Hydrogen treatment assisted by promoter metals	Reactivation
US 6,512,017	Sasol	Catalyst handling during start-up and shut-down	Operating conditions
WO 02/18663	Sasol	Metal recovery from spent catalyst	Spent catalyst handling

4. FT KINETICS

Early studies reported that the activities of metal catalysts for the FT reaction declined in the order Ru, Ni, Co and Fe [109]. In the 1970s Vannice reviewed the specific activities of the Group VIII metals [110]. With alumina as carrier the activities declined in the order Ru, Fe, Ni and Co [111]. With silica as carrier the order was Co, Fe, Ru and Ni [112]. The discrepancies may have been due to the different preparation methods and also to the probability that the number of active sites functioning under actual FT conditions were different from those measured on the freshly reduced catalysts. Overall, as far as the commercially viable metals are concerned, Co appears to be more active than Fe. Iglesia [113] has reported that the FT turnover rates on Co catalysts are about three times higher than on Fe catalysts. In general the observed rates will not only depend on the intrinsic rate of the FT reaction on the catalyst surface but also on the diffusion rates of reactants in and products out of the porous catalyst particles. These diffusion rates will in turn depend on the porosity and pore sizes, the particle size and whether or not liquid wax is present in the catalyst particles.

4.1 Iron catalysts

The performance of iron catalysts is very dependent on the strong alkali promoters added. Not only is the selectivity markedly dependent on the effective amount of alkali present but it also has a direct and/or indirect effect on the overall rate of the reaction (see Chapter 3, Section 9.2.2). For fused iron catalysts operating at high temperatures increasing alkali level result in higher conversions (see Chapter 3 Table 17). At lower temperatures the wax selectivity is higher and then it is found that increasing the alkali results in a lower activity (see Chapter 3 Table 18). For precipitated iron catalysts the activity also decreases when the alkali content is raised above a certain level (see Chapter 3, Section 9.2.2). Besides the speculative discussion in Chapter 3, Section 9.2.2 regarding the relative coverage of the catalyst surface by monomers and growing hydrocarbon chains the observed decline in activity at higher alkali levels is also probably due to diffusion restrictions in the porous catalysts. At higher alkali levels the wax formed within the catalyst pores becomes progressively longer chained and thus more viscous. The diffusion rates of the reactants through the wax to the active surface sites are therefor lowered and consequently the overall reaction rate is lowered. When wax together with the total gas feed is pumped into a fixed catalyst bed reactor operating in a high wax selectivity mode there is no resultant effect on activity. This is not surprising as the catalyst particles are already saturated with liquid wax. When, however, the catalyst bed is periodically treated with a light solvent which extracts the wax from the catalyst there is an immediate large increase in activity which then rapidly declines to its previous level as the catalyst particles are again filled with wax being produced [2]. These findings clearly support the concept that diffusion restrictions, due to the presence of wax in the catalyst pores, cause a lower overall reaction rate. When reduced iron catalysts were nitrided prior to FT synthesis the activity was higher and the wax selectivity was lower than for the same catalyst pre-treated in the normal way [5]. Studies at Sasol confirmed these findings, the activity of the nitrided catalyst increased by a factor of 1.3 while the wax selectivity decreased by a factor of 7 [2]. These findings again support the concept that the activity is influenced by the amount, and also by the chain length of the wax on the catalyst.

Further supporting evidence for the presence of diffusion restrictions is the observation that for the fixed bed LTFT process the activity increases as the catalyst extrudate size is decreased or the shape is altered to increase the external area of the extrudates [2]. When either fused or precipitated catalysts were operated in fixed bed reactors at near differential conditions (high feed gas flows resulting in low percentage conversions) at about 220°C the activity increased linearly with the amount of external surface area of the catalyst particles. For the HTFT process, which uses catalyst particles of less than 100 micron, the effect of particle size is also in evidence. The smaller particles have much higher deposited carbon content than the larger particles indicating that per unit mass of iron the smaller particles have done much more FT work than the larger particles [2]. As mentioned in Chapter 2, Section 1.2 when using precipitated catalysts of identical composition a slurry bed reactor is as active as a fixed bed reactor despite the much lower catalyst loading of the former. This is due to the much smaller catalyst particle size employed in the slurry bed operation and so this again indicates the influence of diffusion restrictions on the conversion activity. It is well known that the activation energies (E) of diffusion processes are normally lower than those of chemical reactions. Precipitated iron catalysts, which produce high yields of wax, have E values of 55 to 62 kJ /mol in the temperature range of 200 to 250°C [114]. Fused iron catalysts which produce much less wax had an E value of 71kJ/mol [115]. At very low wax selectivities, Vannice reported an E value of 113 kJ/mol for an iron catalyst [110, 112]. All these results again indicate the presence of diffusion restrictions in FT catalysis.

It should be borne in mind that there is a link between the kinetics of diffusion and those of the FT chemical reactions occurring on the catalyst surface. The rate of diffusion of reactants into the catalyst pores, and through the liquid wax in the pores, is proportional to the difference in concentration of the reactants in the gas phase and at the active catalyst surface. The concentration at the latter is of course dependent on the rate of the FT reaction. The higher the rate of the FT reaction, the greater is the concentration gradient and so, the rate of diffusion is higher. In practice the only partial pressures of the reactants and products that can be measured are those in the gas phase outside the catalyst particles and so these pressures are the ones that have to be used to derive the overall kinetics.

Quite often kinetic studies are carried out in high flow, i.e. low conversion, single pass laboratory reactors. This approach does indeed simplify the interpretations but it does ignore the possibility that one or more of the products, which would be present at relatively high partial pressures at higher conversion levels, could have an influence on the kinetics. To avoid these possible pitfalls it is recommended that laboratory studies should be carried out in high gas velocity, high recycle 'stirred tank' type units such as Berty reactors. By varying the fresh feed gas flow to these units the partial pressures and the kinetics at different conversion levels, i.e. at different bed depths, can be determined.

Various kinetic equations have been proposed to describe the rate of the FT reactions over iron based catalysts. A few of these equations are given below.

$$r = kP_{H2}^{0.6} P_{CO}^{0.4} - fr^{0.5} P_{H2O}^{0.5} \qquad [116] \qquad (1)$$
$$r = mP_{H2}P_{CO} / (P_{CO} + nP_{H2O}) \qquad [5, 114] \qquad (2)$$
$$r = aP_{CO}P_{H2}^{2} / (P_{CO}P_{H2} + bP_{H2O}) \qquad [117] \qquad (3)$$
$$r = a P_{CO}P_{H2} / (P_{CO} + bP_{H2O} + cP_{CO2}) \quad [118] \qquad (4)$$

At Sasol a large number of measurements and experiments were carried out in commercial and pilot plant fixed and fluidized bed reactors using the relevant commercial iron based catalysts [2]. Table 10 shows examples of some of the experimental pilot plant results.

In the pilot plant experiments the effect of various parameters on the amount of $(CO + CO_2)$ converted, which equates to the $(H_2O + CO_2)$ produced, i.e. the FT activity, were studied. The variables included different catalyst bed heights at fixed feed gas flow rates (i.e. different residence times and conversion levels), total pressures (from 0.8 to 7.6 MPa) at fixed gas linear velocities (i.e. fixed residence time in the catalyst bed), different recycle ratios and different feed gas compositions. The effect of individual partial pressures (of CO, CO_2, H_2 and H_2O) were studied by varying only one at a time and, as far as possible, keeping the others fixed at the reactor entrance. The temperatures were not varied but were at the values normally used in the commercial HTFT and LTFT processes.

Table 10

Influence of entry partial pressures on activity of Fe catalysts (integral reactors) [2]

Variable	Fixed bed precipitated catalysts 223°C					
	Partial pressure (bar abs.)				Total pressure (bar abs.)	Activity[a]
	H_2	CO	CO_2	H_2O		
P_{H2}	12.2	5.7	0.2	0	20	97
	15.2	5.6	0.2	0	23	109
P_{CO}	15.2	5.6	0.2	0	23	109
	15.7	8.0	0.2	0	26	134
P_{CO2}	14.3	6.4	1.2	0	27	77
	12.8	6.1	4.1	0	27	72
P_{H2O}	8.4	4.3	0.1	0	15	69
	8.4	4.1	0.1	0.35	15	51

	H₂	CO	CO₂	H₂O	Total pressure	Activity
Total	5.9	2.4	0.7	0	11.4	44
pressure	14.1	5.9	1.4	0	27.6	114

Fluidized bed fused catalyst 330°C

	Partial pressure (bar abs.)				Total pressure (bar abs.)	Activity[a]
	H_2	CO	CO_2	H_2O		
P_{H2}	7.2	1.7	1.0	0	18	89
	10.1	1.6	0.8	0		100
P_{CO}	11.6	1.5	0.1	0	13.8	41
	11.7	3.0	0.1	0	15.8	66
	11.9	4.6	0.1	0	18.8	99
P_{CO2}	7.2	1.7	1.0	0	18.0	98
	7.5	1.3	2.2	0		96
	7.2	1.3	3.1	0		98
	8.1	1.0	5.1	0		97
P_{H20}	9.8	1.8	0.7	0	20.8	107
	9.6	1.8	0.9	0.7		105
	10.0	1.8	1.2	1.4		98
	10.3	1.8	1.6	2.8		83
Total	4.3	1.0	0.9	0	10.8	54
pressure	8.5	2.0	1.5	0	20.8	106
	14.2	3.6	2.8	0	40.8	207
	30.7	6.7	4.6	0	75.8	375

a Moles ($H_2O + CO_2$) produced per hour.

The key observations were as follows:

- The rate was strongly dependent on the hydrogen partial pressure. In previously reported work it was found that at low percentage conversion levels the rate was in fact only dependent on the hydrogen partial pressure [115].
- The rate increased with the partial pressure of CO.
- The partial pressure of CO_2 did not appear to have a direct effect on the FT rate. It can of course affect the partial pressures of the other components via the water gas shift (WGS) reaction and so can have an indirect effect. If the WGS is taken into account the effect of CO_2 falls away.
- The rate was markedly depressed by the partial pressure of water.
- The level of hydrocarbon products in the reactor had no apparent effect on the FT rate.

Based on the above observations a satisfactory rate equation for iron based catalysts is:

$$r = mP_{H2}P_{CO} / (P_{CO} + aP_{H2O})$$

since it is in keeping with all the experimental findings.

Note that when the conversion level is low, i.e. the partial pressure of water is low, the equation simplifies to:

$$r = mP_{H2}.$$

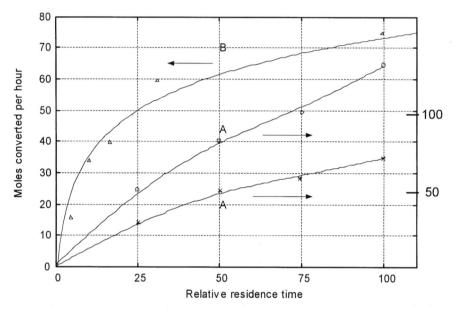

Figure 15 Typical reaction profiles in fixed bed, (A) and fluidized bed, (B), iron catalysts. The symbols show the actual measured amounts of moles converted. The solid lines are for the profiles calculated using $r = mP_{H2}P_{CO} / (P_{CO} + aP_{H2O})$

The simplified form of the rate equation is in agreement with previous findings [115] mentioned above. Note that both rate Eqs. (2) and (3) comply with this finding. Note also that the Sasol deductions, in effect, 're-invented' the equation proposed by Anderson about twenty years prior to the Sasol re-investigation! When applying this equation in multi-step calculations, and at each step also taking account of the simultaneously occurring WGS reaction, the conversion profile along the length of the reactors can be calculated. It was found to satisfactorily match the actually measured profiles in both the LTFT fixed bed and the HTFT circulating fluidized bed commercial reactor [2].

The value of m in the Eq. (2) incorporates the temperature, the activation energy and the catalyst loading per volume of the reactor section. The measured and calculated profiles are reproduced in Fig. 15 for a few pilot plant runs [2].

With regards to the WGS reaction, at the high temperature in the fluidized bed (about 330 to 340°C) this reaction is very rapid and it appears to be at equilibrium all along the reactor length. For the fixed bed reactors operating around 225°C a WGS rate equation is required to calculate the changes in gas composition along the reactor as the result of this simultaneously occurring reaction. A simple satisfactory equation under the conditions prevailing in the LTFT units was found to be:

$$r_{WGS} = cP_{CO} .$$

The negative effect of water on the FT rate equation is common to the all four proposed equations mentioned earlier. This effect is probably due to the fact that iron is sensitive to oxidation by water vapour and so the higher the H_2O partial pressure the higher the occupancy of surface iron sites by O atoms/ions. This lowers the amount of active surface sites available and thus lowers the FT rate.

An attempt to derive the kinetic equation from a 'theoretical' approach was made [114]. It was assumed that the slow step in the FT reaction sequence was the reaction of an undissociated hydrogen molecule from the gas phase with a chemisorbed CO molecule to form the chemisorbed HCOH 'monomer'. The rate of reaction then would be:

$$r = mP_{H2}\theta_{CO}$$

where θ_{CO} is the fractional coverage of the catalytically active iron surface sites by chemisorbed CO. If CO has to compete for adsorption sites with H_2O, CO_2 and H_2 then according to Langmuir's adsorption theory:

$$\theta_{CO} = k_{CO}P_{CO} / (1 + k_{CO}P_{CO} + k_{H2O}P_{H2O} + k_{CO2}P_{CO2} + k_{H2}P_{H2}).$$

Based on prior knowledge of the relative strength of adsorption of the gases on iron catalysts [119] as well as the observation that CO_2 has little or no influence on the FT reaction rate, θ_{CO} simplifies to:

$P_{CO} / (P_{CO} + aP_{H2O})$ where $a = k_{H2O}/k_{CO}$.

The rate equation then becomes:

$$r = mP_{H2}P_{CO} / (P_{CO} + aP_{H2O}) .$$

There is no doubt that the above derivation is speculative and can be questioned since the objective was to match the empirically deduced equation with a "theoretical" approach. Nevertheless it has been shown that the equation can be used not only to estimate the overall conversions for both the HTFT and LTFT reactors, but also to calculate the conversion profile along the catalyst bed length. Thus the equation is of practical use for design and for performance evaluation.

As pointed out previously this equation is identical to the one previously derived by Anderson [5] even though the 'mechanistic' assumptions were different. This, as has often been pointed out by others, again shows that kinetic information cannot be used as convincing 'proof' of a particular mechanism.

The kinetic equation also predicts that if the total pressure is increased and the gas feed rate is increased by the same factor, i.e. the residence time in the catalyst bed remains the same, then the degree of conversion remains unchanged. This means that the production rate is increased in proportion to the increase in pressure. Based on this prediction new pilot plants were constructed at Sasol and tests up to 6.0 Mpa and 7.5 Mpa for the fixed bed LTFT and for the fluidized bed HTFT operations respectively, were carried out. These tests confirmed the kinetic predictions. As stated in chapter 2 section 1.1 a 4.5 MPa fixed bed commercial reactor was subsequently built and it performed as per prediction. (Note that the older Sasol commercial fixed bed units operate at 2.7 MPa.) The HTFT CFB reactors subsequently constructed at Secunda and later at Mossel Bay also operated at pressures higher than the original units built at Sasolburg. It has been estimated that HTFT fluidized bed units can be viable up to 4.0 Mpa [120]. The use of even higher pressure is probably constrained by economic considerations (e.g. lower maximum gas velocities, increased compression and piping costs etc.) and heat exchange capacity requirements inside the units. (Note that pilot plant runs were carried out up to 7.5 MPa without any adverse activity or selectivity effects.)

Fig. 16 and Table 11 show the calculated conversion profiles along the length of the lean phase fluidized section of a commercial HTFT CFB reactor. The calculated conversion profile matches the measured profile. Table 11 also shows that the selectivity factor does not change much over the length of the reactor (see discussion in Chapter 3, Section 9.3.1). The carbon deposition rate factor, however, changes significantly with bed length (see Section 2.1.2C).

Table 11
CFB Commercial reactor data profiles

Relative bed length	Relative Conversion	Selectivity factor	Carbon deposition factor
0.1	0.61	0.46	0.010
0.2	0.72	0.45	0.008
0.4	0.84	0.43	0.006
0.6	0.91	0.42	0.006
0.8	0.96	0.41	0.005
1.0	1.00	0.40	0.005

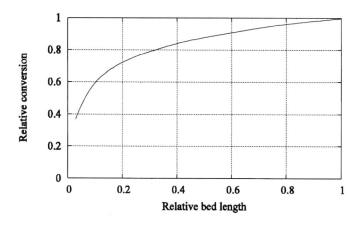

Figure 16 CFB reactor conversion profile

Tables 12 and 13 show examples of calculations for the HTFT and the LTFT processes using the FT kinetic Eq. (2) and taking into account the simultaneously occurring WGS reaction as the gas moves through the reactors.

Table 12

Calculated data for HTFT process with different H_2/CO ratio feed gases (320°C, 24 bar, fixed volume gas feed)

Feed gas H_2/CO ratio	Overall % conversion[1]	FT production[2]	Carbon deposition factor	
			Inlet	Outlet
0.5	35	5.5	0.254	0.477
1.0	50	6.1	0.080	0.017
2.0	57	4.6	0.028	0.004
4.0	90	4.3	0.011	0.001

   1. % $(CO + CO_2)$ converted
   2. Amount of FT products formed.
Note: Feed gas only consists of H_2 and CO for Tables 11 and 12.

Table 13

Calculated data for LTFT process with different H_2/CO ratio feed gases (230 °C, 30 bar, fixed volume gas feed).

Feed gas H_2/CO ratio	Overall % conversion[1]	FT production[2]
0.5	16	3.1
1.0	25	3.7
1.5	32	3.8
2.0	37	3.7

3.0	46	3.4

1. % $(CO + CO_2)$ converted
2. Amount of FT products formed

.For both sets of calculations the feed gas flows were kept constant, i.e. the residence times were the same. The gas consisted of only H_2 and CO, but with different ratios of these two gases. No recycling of gas was applied. For the HTFT process (Table 12) it can be seen that the calculations indicate that as the feed gas H_2/CO ratio increases the % conversion increases but the actual production rate of FT products formed peaks in the vicinity of the 1.0 ratio feed gas. With feed gas ratios below 2.0, however, the rate of carbon deposition will probably be unacceptably high (see Table 11 for typical commercial HTFT operation and also section 2.1.2C). For the LTFT operation (Table 13) it can again be seen that as the feed gas ratio increases the % conversion increases but there is a broad production peak for the feed gases with ratios from about 1.0 to about 2.0. Because of the low operating temperature carbon deposition is in any event low or insignificant and so this is not a constraint. The results confirm the expectation that for the LTFT operation the low ratio gas produced by high temperature coal gasifiers can be fed to either fixed bed or slurry bed LTFT reactors utilising iron based catalysts.

4.2 Cobalt kinetics

Currently there is only one industrial plant using cobalt based catalyst in multi-tubular fixed bed reactors, the Shell plant in Malaysia. Both Sasol and Exxon have operated large demonstration slurry phase FT units with supported cobalt catalysts. However, it appears that no kinetic information for these operations has been published. Various kinetic equations based on work carried out in laboratory reactors have, however, been generated. These equations have been reviewed elsewhere [121, 122, 123. 124]. A few examples of the diverse proposed equations are given below:

$$r = kP_{H2}^2 / P_{CO} \qquad [125] \qquad (5)$$
$$r = aP_{CO}P_{H2} / (1+bP_{CO})^2 \qquad [121] \qquad (6)$$
$$r = aP_{CO}P_{H2}^{0.5} / (1+ bP_{CO} + cP_{H2}^{0.5})^2 \qquad [126] \qquad (7)$$
$$r = aP_{CO}P_{H2}^2 / (1 + bP_{CO}P_{H2}^2) \qquad [5] \qquad (8)$$
$$r = c(P_{H2}^{.5}/P_{CO}^{.2})/(1+.93P_{H2}/P_{H2O}) \qquad [127] \qquad (9)$$
$$r = d(P_{H2}P_{CO})/(P_{CO}+.27P_{H2O}) \qquad (10)$$

Eq. 6 is often cited as a suitable rate equation. This equation responds to an increase in pressure in a similar way as does Eq. 2 for iron catalysts (although not to the same extent), namely, as the pressure increases the FT rate increases.

Despite the known fact that Co is more resistant to oxidation than Fe it has been reported by several investigators that under FT conditions Co crystals smaller than a certain size, e.g. < 5 microns [92], are oxidised by water vapour. It is feasible that the smaller the Co crystals the lower the H_2O/H_2 ratio at which they will oxidise and so become inactive. If this process is reversible then it seems feasible that the effect of the water vapour pressure can also be incorporated into the kinetic equation. Thus Eqs. 9 and 10 could make sense.

4.3 Comparison of the FT activity of iron and cobalt catalysts

It is important to note that, contrary to iron based catalysts, many of the proposed kinetic equations for FT over cobalt based catalysts do not contain a water partial pressure term. It is well known that cobalt is much more resistant to oxidation, whether by oxygen or by water, than is iron. It can therefore be presumed that under FT conditions the occupancy of the cobalt surface by oxygen atoms/ions will be much lower than in the case of iron catalysts and so this could account for the absence of a P_{H2O} term in kinetic equations for cobalt catalysts. This aspect could in principle give cobalt a big activity advantage over iron catalysts.

To evaluate the situation the conversion profiles for four cases were calculated using Eq. 2 for iron and Eqs. 6 and 9 for cobalt based catalysts:

2) $\quad r = mP_{H2}P_{CO} / (P_{CO} + nP_{H2O})$

6) $\quad r = aP_{CO}P_{H2} / (1+bP_{CO})^2$

9) $\quad r = c(P_{H2}{}^{.5}/P_{CO}{}^{.2})/(1+.93P_{H2}/P_{H2O})$

In order that the Co and Fe catalysts start off on the same footing the value of the constants m, a and c in these three equations were chosen so that the 4% conversion levels were achieved at the same position near the entrance of the catalyst bed. (Note that at this low conversion level the partial pressure of water is low and hence its influence on the kinetics is minimal at the bed entrance.) This was taken to mean that the 'initial' activities were the same, i.e. the product of the number of active surface sites per unit volume of catalyst bed and the intrinsic activity per site were the same. Note that since Co is about three times more active per surface site than Fe [113], the above means that there are effectively about three times more available active sites per unit volume of catalyst bed in the case of Fe. For both cases the reactors operated at 3.0 Mpa and the same volumetric rate of syngas was fed, i.e. the residence times were the same. The feed gas consisted only of H_2 and CO, the ratio being 2/1. Tail gas recycling was not applied, i.e. the calculations were on a once-through basis. The reactors operated in the LTFT mode. For the iron catalyst case allowance was made for the WGS reaction, but not for the cobalt case as cobalt has a very low WGS activity at LTFT conditions. The results of the calculations are shown in Fig. 17. (Note that the 'Bed length' is in arbitrary units.) Included in Fig. 17 is a case for a Fe catalyst with an initial activity five times higher than that of the

594

other Fe case. (In this case the Fe catalyst now has fifteen times more active surface sites available per unit volume of catalyst bed than the Co catalyst.)

As can be seen from Fig. 17, even though the Fe and Co catalysts have the same initial activity the Co catalyst increasingly outperforms the Fe catalyst as the syngas moves through the reactor. For example, at a bed 'length' of 30 units the conversion attained is over 80% for the Co catalyst using Eq. 6 and over 60% using equation 9 and just under 40% for the Fe catalyst using equation 2. If the Fe catalyst is made five times more active then it does outperform the Co case (Eq. 6) up to about the 60% conversion level, after which the Co catalyst again is superior. Should Eq. 9 for Co apply, however, the five times more active Fe catalyst outperforms the Co catalyst. From this there appears to be a big incentive to improve the activity of Fe catalysts, i.e. to increase the number of active sites per unit volume of catalyst. However, if the number of active sites for Co catalysts can be increased then of course the Co catalysts will again outperform the Fe catalysts.

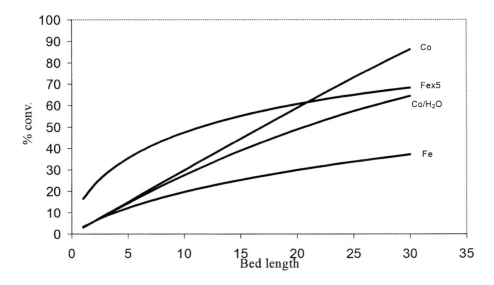

Figure17   Co is for equation 6, Co/H2O is for equation 9, Fe is for equation 2 and Fex5 is for Eq. 2 with a Fe catalyst 5 times more active.

It is of interest to point out that when using Eq. 10 the reaction profile obtained is identical to that predicted by Eq. 9. A further general point is that when applying the various kinetic equations for both Fe and Co catalysts it is found that a change in the partial pressure of hydrogen has a greater effect on the reaction rate than a similar change in the partial pressure of carbon monoxide. This possibly explains why so many 'apparently different' yet apparently satisfactory kinetic equations have been reported in the literature.

When using the kinetic Eqs. 2 (for Fe) and 6 (for Co) the response of cobalt catalysts to changes in total operating pressure is different to that observed with iron catalysts. The calculated results shown below are for once though operation (i.e., no recycle). When the total pressure was changed the total feed gas flow rate was changed by the same factor, thus keeping the entrance gas linear velocity the same (i.e., the residence times in the reactors were similar).

Catalyst	Pressure(bar)	Conversion %	Relative production
Co	60	47	109
Co	30	86	100
Co	3	99	12
Catalyst	Pressure(bar)	Conversion %	Relative production
Fe	30	37	43
Fe	60	37	86
Fe (5x more active)	30	68	79
Fe (5x more active)	60	68	158

The calculations show that, for iron catalysts, when the operating pressure is doubled the degree of conversion stays the same, which means that the FT production rate is doubled. This is in keeping with actual pilot plant results (see Chapter 2, Section 1.1 and Chapter 3, Table 20). For cobalt catalysts the calculations indicate that the degree of conversion decreases as the pressure is increased and so the extent of the increase in production with increasing pressure is not as large as for iron catalysts, especially at the higher pressures. Note again that even if the iron catalyst's activity is increased five fold the FT production is much lower than for the cobalt catalyst at the same operating pressure of 30 bar. At 60 bar, however, the production (and the percent conversion) of the higher activity iron catalyst exceeds that of the cobalt catalyst. If of course the activity of the cobalt catalyst were to be increased the situation would again shift in favour of the cobalt catalyst.

Although cobalt catalysts clearly appear to have a big advantage it must be pointed out that this does not mean that high conversions cannot be achieved with existing Fe catalysts. In the Sasol pilot plants, using the commercial Fe LTFT catalyst, it has been demonstrated that over 90% conversion can be attained at LTFT operating conditions. This entails using two reactors (fixed or slurry bed) in series with water knock-out between the two reactors and also each reactor operating in the recycle mode (which involves feeding some tail gas, after water knock-out, together with fresh syngas feed to the reactor). This mode of operating lowers the average water vapour pressure in the reactors and thus increases the FT reaction rate. This scheme obviously, however, increases construction and operating costs. For the HTFT process (fluidized bed at about

340°C) conversions over 90% are readily obtained in single stage reactors operating with recycle.

It should be noted that Fig. 17 depicts the performance of fresh catalysts. With time on stream (TOS) Fe catalysts (and to a lesser extent Co catalysts) deactivate. If the FT reactors are slurry bed units then in the case of iron catalysts, because of their relatively low cost, high conversion levels can be maintained by regular, on line, addition of fresh catalyst. This should narrow the difference in performance between the two types of catalysts. The superior performance of cobalt relative to iron catalysts nevertheless does mean that high conversions can be achieved in a single stage reactor without the need for large tail gas recycle flowrates. Thus fewer or smaller reactors would be required to produce a given amount of hydrocarbon product in the case of cobalt catalysts whereas with iron catalyst more reactors operating in tandem would be required. In the case of cobalt catalysts high conversion levels do of course result in high H_2O/H_2 ratios in the reactor. At 90% conversion the ratio at the reactor exit will be about 4.2. It is known that these high ratios accelerate the rate of deactivation due to enhanced sintering rate, oxidation of the smaller cobalt crystals present and also to the formation of inert compounds with the catalyst support material (see Section 3.3). To minimise these effects a two stage operation, with water knock-out between the stages, may have to be applied. Alternatively a larger reactor with recycle of a portion of the tailgas (after water knock-out) could be applied. The number or cost of the reactors will, however, still be lower than that required when using iron based catalysts.

For both iron and cobalt catalysts, operating at higher pressures means higher production rates per reactor can be achieved. The maximum practical pressure will, however, be determined by the heat exchange capacity of the reactor and by the cost of feed gas compression. Effective removal of the heat of reaction, particularly in the case of cobalt catalysts, is vital to avoid the production of high levels of methane. At high pressures and low temperatures volatile cobalt carbonyls are formed and this would lead to loss of cobalt metal from the reactor. For iron based catalyst this does not occur. In the case of Co catalysts in multi-tubular fixed bed reactors, operating once through at high conversion levels would mean a large drop in linear gas velocity in the lower sections of the tubes. This could result in poorer heat exchange rates in those sections and so some recycle of tail gas may in any event be required.

REFERENCES

[1] C.D. Frohning in Fischer-Tropsch-Synthese aus Kohle, J. Falbe (ed.), Stuttgart, Thieme (1977).
[2] M.E. Dry in Catalysis Science and Technology, 1, J.R. Anderson and M. Boudart (eds.), Springer-Verlag, (1981) 159.

[3] U.S. Patent No. 5,324,335, June 28 (1991).

[4] B. Jager, R.L. Espinoza, Catal. Today, 23 (1995) 17.

[5] R.B. Anderson in Catalysis, Vol IV, P.H. Emmett (ed.), Reinhold (1956).

[6] D.B. Bukur, L. Nowicki, R.K. Manne and X. Lang, J. Catal., 155 (1995) 366.

[7] D.B. Bukur, L. Nowicki and S.A. Patel, Canadian J Chem. Eng., 74 (1996) 399.

[8] D.B. Bukur, X. Lang and Y. Ding, Applied Catal. A: Gen., 186 (1999) 255.

[9] B.H. Davis, R.J. Obrien, L.G. Xu and R.L. Spicer, Eng & Fuels, 10 (1996) 921.

[10] M.E. Dry and L.C. Ferreira, J. Catal., 7 (1967) 352.

[11] M.E. Dry, J.A. du Plessis and G.M. Leuteritz, J. Catal., 6 (1966) 194.

[12] G. Ertl, D. Prigge, R. Schlogl and M. Weiss, J. Catal., 79 (1983) 359.

[13] P.M. Loggenberg, M.E. Dry and R.G. Copperthwaite, Surface and Interface Analysis, 16 (1990) 347.

[14] M.E. Dry, T. Shingles, L.J. Boshoff and C.S. Botha, J. Catal., 17 (1970) 347.

[15] M.E. Dry, T. Shingles and C.S. Botha, J. Catal., 17 (1970) 341.

[16] W. Bauklok, J. Hellbrugge, Brennst. Chem., 23 (1942) 87.

[17] M.E. Dry, Hydrocarbon Process, 59(2) (1980) 92.

[18] C.C. Hasll, D. Gall, S.L. Smith, Inst. Petrol, 38 (1952) 845.

[19] J.H. Arnold and P.C. Keith, Amer. Chem. Soc. Adv. Chem. Ser., 5 (1951) 120.

[20] E. Iglesia, S.C. Reyes, R.J. Madon and S.L. Soled, Adv. Catal., 39 (2) (1993) 345.

[21] M.E. Dry, Catal. Letters, 7 (1990) 241.

[22] R.B. Anderson, F.S. Karn and J.F. Schulz, U.S. Bur. Mines Bull., (1964) 614

[23] H. Pichler and H. Merkel, U.S. Bur. Mines Tech. Paper (1949) 718.

[24] F. Fischer, Brentstoff-Chemie, 16 (1935) 1.

[25] J.F. Schultz, L.J. Hofer, E.M. Cohn, K.C. Stein and R.B. Anderson, Bureau of Mines Bull. (1959) 478.

[26] M. Kraum and M. Baerns, Appl. Catal. A: Gen., 186 (1999) 189.

[27] J. van de Loosdrecht, M. Haar, A.M. Kraan, A.J. Dillen and J.W. Geus, Appl. Catal., 150 (1997)
    365.

[28] M.K. Niemela, A.O. Krause, T. Vaara, J.J. Kiviaho, M.K. Rienikainen. Appl. Catal., 147 (1996) 325.

[29] M.K. Niemela, Backman, A.O. Krause, T. Vaara, Appl. Catal., 156 (1997) 319.

[30] S.H. Moon and K.E. Yoon, Appl. Catal., 16 (1985) 289.

[31] A.Chirinos, PhD thesis, University of Cape Town (2003).

[32] R. Oukaci, A.H. Singleton and J.G. Goodwin, Appl. Catal. A: Gen., 186 (1999) 129.

[33] E.Iglesia, S.L. Soled, R.A. Fiato and G.H. Via, J. Catal., 143 (1993) 345.

[34] H. Beuther, T.P. Kobylinski, C.L. Kibby and R.B. Pannell, US Patent No. 4,585,798 (1986).

[35] T.P. Kobylinski, C.L. Kibby, R.B. Pannell and E.L. Eddy, US Patent No. 4,605,676 (1986).

[36] E. Iglesia, S.L. Soled and R.A. Fiato, J. Catal., 137 (1992) 212.

[37] A.M.; Hillmen, D. Schanke and A. Holmen, Catal. Letters, 38 (1996) 143.

[38] D. Schanke, A.M. Hillmen, E. Bergene, K. Kinnari, E. Rytter, E. Adanases and A. Holmen, Catal. Letters, 34 (1995) 269.

[39] D. Schanke, A.M. Hillmen, E. Bergene, K. Kinnari, E. Rytter, E. Adanases and A. Holmen, Energy and Fuels, 10 (1996) 867.

[40] A. Holmen, D. Schanke, S. Vada, E.A. Blekkan, A.M. Hillmen and A. Hoff, J. Catal., 156 (1995) 85.

[41] F.B. Noronka, A. Frydman, D.A.G. Aranda, C. Perez, R.R. Soares, B. Morawek, D.

Caster, C.T. Cambell, Frety and M. Schmal, Catal. Today, 28 (1996) 147.

[42] C.L..Bianchi, R. Carli, S. Merlotti and C. Ragaini, Catal. Letters, 41 (1996) 79.

[43] J.G. Goodwin, A. Kogelbauer and R. Oukari, J. Catal., 160 (1996) 125.

[44] S. Eri, J.G. Goodwin, G. Marcelin and T. Riis, US Patent No. 4,801,573 (1989).

[45] G.P. Huffman, N. Shah, J. Zhao, F.E. Huggins, T.E. Hoost, S. Halvorsen and J.G. Goodwin, J. Catal., 151 (1995) 17.

[46] G. J. Haddad, B. Chenand J.G. Goodwin, J. Catal., 160 (1996) 43.

[47] S. Ali, B. Chen, and J.G. Goodwin, J. Catal., 157 (1995) 35.

[48] A.Y. Khodakov, J. Lunch, D. Bazin, B. Rebours, N. Zanier, B. Moison and P. Chaumette, J. Catal., 168 (1997) 16.

[49] A. Lapidus, A. Krylora, V. Kazanskii, V. Borovkov, A. Zaitsev, J. Rathousky, A. Zukal and M. Jancalkova, Appl. Catal., 73 (1991) 65.

[50] A. Lapidus, A. Krylova, J. Rathousky, A. Zulkal and M. Jancalkova., Appl. Catal., 80 (1992) 1.

[51] J. Rathousky, A. Zukal, A. Lapidus and A. Krylova, Appl. Catal., 79 (1991) 167.

[52] A. Kogelbauer, J.G. Goodman and R. Oukaci, J. Catal., 160 (1996) 125.

[53] A.R. Belambe, R. Oukaci and J.G. Goodwin, J. Catal., 166 (1997) 8.

[54] R.C. Reuel and C.H. Bartholomew, J. Catal., 85 (1984) 78.

[55] C.H. Bartholomew, R.B. Pannell and J.L. Butler, J. Catal., 65 (1980) 335.

[56] L. Fu and C.H. Bartholomew, J. Catal., 9+2 (1985) 376.

[57] R.L. Espinoza, J.L. Visagie, P.J. van Berge and F.H. Bolder (Sasol Technology), European Pat. Appl. EP 0736326A1 (1996).

[58] R.L. Espinoza, J.L. Visagie, P.J. van Berge and F.H. Bolder, U.S. Patent No. 5733839, March 31 (1988).

[59] P.J. van Berge, J. van de Loosdrecht, E.A. Caricato and S. Barradas, Patent PCT/GB 99/00527, February 19 (1999).

[60] P.J. van Berge, S. Barradas, E.A. Caricato, B.H. Sigwebela and J. van de Loosdrecht Worldwide Catalyst Industry Conference, Houston, U.S.A., June (2000).

[61] P.J. van Berge, S. Barradas, J. van de Loosdrecht and J.L. Visagie, Erdol Erdgas Kohle, 117 (3) (2001) 138.

[62] P.J. van Berge, S. Barradas, E.A. Caricato and J. van de Loosdrecht, Patent No. WO 99/42214 (1999).

[63] P.J. van Berge, S. Barradas, E.A. Caricato, B.H. Sigwebela and J. van de Loosdrecht, Patent No. WO 00/20116 (2000).

[64] R. Krishna, C. Maretto. Studies in Surface Science and Catalysis, 119 (1998) 197.

[65] S.T. Sie and R. Krishna, Appl. Catal. A: Gen., 186 (1999) 55.

[66] J. van de Loosdrecht, S. Barradas, E.A. Caricato,] P.J. van Berge and J.L. Visagie, Div. Fuel Chemistry, 220th ACS meeting, USA. August (2000) 587.

[67] G.J. Haddad, B. Chen and J.G. Goodwin, J. Catal., 161 (1996) 274.

[68] B. Ernst, A. Kiennemann and P. Chaumette, Clean Fuels Symposium, ACS meeting San Francisco, April (1997).

[69] J.G. Goodwin, S. All and B. Chen, J. Catal., 157 (1) (1995) 35.

[70] D. Das, G. Ravuchandran and D.K. Chakrabarty, Appl. Catal. A: Gen., 131 (2) (1995) 335.

[71] Y. Zang, B. Zhong, and Q. Wang, Clean Fuels Symposium, ACS Meeting, San Francisco, April (1997).

[72] A.A Adesina and H. Chen, Appl. Catal. A: Gen., 112 (2) (1994) 87.

[73] P. Chaumette and C. Verdon, UK Patent Appl. No. GB 2258414 A (IFP) (1993).

[74] T.P. Kobilinski, C.L. Kibby, R.B. Pannell, E.L. Eddy, US Patent No. 4,605,679 (1986).

[75] W.N. Mitchel. US Patent No. 5,292,705, March (1994).

[76] B. Nay, M.R. Smith and C.D. Telford, US Patent No. 5,585,316, December (1996).

[77] P.N. Dyer, R. Pierantozzi and H.P. Withers, US Patent No. 4,681,867, July (1987).

[78] P.N. Dyer, R. Pierantozzi, US Patent No. 4,619,910, October (1986).

[79] P.N. Dyer, R. Pierantozzi and H.P. Withers, US Patent No. 4,670,472, June (1987).

[80] S. Bessell, Appl. Catal., 96 (1993) 253.

[81] S. Bessell, Appl. Catal., 235 (1995) 126.

[82] G. Calleja, A.D. Lucas and R van Grieken, Appl. Catal., 68 (1991) 11.

[83] J.M. Stencel, V.U.S. Rao, R.J. Diehl, K.H. Rhee, A.G.Dhere and R.J.de Angelis. J. Catal. 84 (1983) 109.

[84] H. Pichler, Advan. Catal., Vol. 5, N.J. Rideal (ed.) Academic Press (1952).

[85] H. Schulz. Chemierohstoffe aus Kohle, Falbe (ed) Thieme Verlag, Stuttgart, 1977.

[86] J. F. Schulz, F. S. Karn and R.B. Anderson, U.S. Bur. Mines rep. 6974.

[87] F. Fischer, Brennt. Chem., 16 (1935) 1.

[88] P.J. van Berge, R.C. Everson, Natural Gas Conversion IV, Studies Surface Sci. Catal., 107 (1997) 207.

[89] M.K. Niemela and A.O. Krause, Catal. Letters, 42 (1996) 161.

[90] K. Takeuchi, T. Matsuzaki, H. Arakawa and T. Hanaoko, Appl. Catal. 48 (1989) 149.

[91] A. Kogelbauer, J. C. Weber and J.G. Goodwin, Catal. Letters, 34 (1995) 259.

[92] E. Iglesia, Applied Catal. A: Gen., 161 (1997) 59.

[93] D. Schanke, A.M. Hilmen, E. Bergene, K. Kinnari, E. Rytter, E. Adnanes, A. Holmen, Catal. Letters 34 (1995) 269.

[94] A.M. Hilmen, D. Schanke, K.F. Hanssen and A. Holmen, Appl. Catal. A: Gen, 186 (1999) 169.

[95] P.J. van Berge, J. van de Loosdrecht, S. Barradas and A.M. van der Kraan, Syngas to Fuels and Chemicals Symposium, Div. Petr. Chem. 44 (1) (1999) 84.

[96] A.M. Hilmen, O.A. Lindvag, E. Bergene, D. Schanke, S. Eri and A. Holmen, Stud. Surf. Sci. Catal., 136 (2001) 295.

[97] J.Li, X. Zhan, Y. Chang, G. Jacobs, T. Das and B. H. Davis, Appl. Catal. A: Gen., 228 (2002) 203.

[98] G. Jacobs, P.M. Patterson, Y. Chang T. Das, J. Li, B. H. Davis, Appl. Catal. A: Gen., 233 (2002) 215

[99] S. Krishnamoorthy, M. Tu, M.P. Ojeda, D. Pinna and E. Iglesia, J. Catal., 211 (2002) 422.

[100] J.Li, Y. Chang, G. Jacobs, T. Das and B. H. Davis, Appl. Catal. A: Gen., 236 (2002) 67.

[101] G.W. Huber, C.G. Guymon, T.L. Conrad, B.C. Stephenson and C.H. Bartholomew, Stud. Surf. Sci. Catal., 139 (2001) 423.

[102] C.J. Bertole, C.A. Mims and G. Kiss, J. Catal., 210 (2002) 84.

[103] C.J. Kim, European Patent No. 339,923 (1989).

[104] C. J. Kim, US Patent No. 5,227,407 (1993).

[105] J.Li, G. Jacobs, T. Das and B. H. Davis, Appl. Catal. A: Gen., 233 (2002) 255.

[106] C.J. Kim European Patent Appl. (1990).

[107] H. Schultz, E van Steen and M. Claeys, Stud. Sur. Sci. Catal., 81 (1994) 455.

[108] H. Schultz, M. Claeys, S. Harms, Stud. Sur. Sci. Catal., 107 (1997) 193.

[109] H.H. Storch, N. Golumbic and R.B. Anderson, The Fischer-Tropsch and related synthesis, John Wiley, N.Y., 1951.

[110] M.A. Vannice, Catal. Rev., 14 (2) (1976) 153.

[111] M.A. Vannice, J. Catal., 37 (1975) 449.

[112] M.A. Vannice, J. Catal., 50 (1977) 228.

[113] S. Li, S. Krishnamoorthy, A. Li, G. Meitzner and E. Iglesia, J. Catal., 206 (2002) 202.

[114] M.E. Dry, Ind. Eng. Chem. Prod. Res. Develop., 15 (4) (1976) 282.

[115] M.E. Dry, T. Shingles and L. J. Boshoff, J. Catal., 25 (1972) 99.

[116] R.B. Anderson, F. S. Karn and J. F. Schulz, U.S. Bur. Mines Bull., 614 (1964).

[117] G.A. Huff and C.N. Satterfield, Ind. Eng. Process Des. Dev., 23 (1984) 696.

[118] S. Ledakowicz, H. Nettelhoff, R. Kokuun and W.-D. Deckwer, Ind. Eng. Chem. Process Des. Dev., 24 (1985) 1043.

[119] M.E. Dry, T. Shingles, L.J. Boshoff and G.J. Oosthuizen, J. Catal., 15 (1969) 190.

[120] A.P. Steynberg, R.L. Espinoza, B. Jager and A.C. Vosloo, Applied Catal. A: Gen., 186 (1999) 41.

[121] I. Yates and C.N. Satterfield, Energy and Fuels, 5 (1991) 168.

[122] B. Jager and R. Espinoza, Catal. Today, 23 (1995) 17.

[123] E. van Steen and H. Schulz, Appl. Cat. A: Gen., 186 (1999) 309.

[124] G.P. van der Laan and A.A. Beenackers, Catal. Rev.-Sci.Eng., 41 (3, 4) (1999) 255.

[125] W. Brotz, Z. Elektrochem, 5 (1949) 301.

[126] B. Sarup and B.W. Wojciechowski, Can. J. Chem. Eng., 67 (1989) 62.

[127] T.K. Das, W. Conner, G. Jacobs, X. Zhan, J. Li, M.E. Dry and B.H. Davis, Appl. Catal. A: Gen., in press (2004).

Studies in Surface Science and Catalysis 152
A. Steynberg and M. Dry (Editors)

Chapter 8

Basic studies

M. Claeys and E. van Steen

Catalysis Research Unit, Department of Chemical Engineering,
University of Cape Town, Rondebosch, South Africa

1. SURFACE SPECIES, REACTION INTERMEDIATES AND REACTION PATHWAYS IN THE FISCHER-TROPSCH SYNTHESIS

The understanding of the fundamental processes taking place on metal surfaces during the Fischer-Tropsch synthesis will lead to improved catalyst design and improved, macroscopic description of the Fischer-Tropsch process. Reaction parameters can then be selected to optimise product formation. Knowing the possible surface species on the catalyst surface during the Fischer-Tropsch (FT) synthesis, and their reactivity, enables the formulation of reaction pathways. This insight leads to mechanistic descriptions for the rate of product formation in the FT synthesis.

1.1 Surface species and reaction intermediates

A large number of surface species may exist on the catalyst surface at steady-state in the Fischer-Tropsch synthesis. A summary of some of the experimentally observed and postulated surface species is given in Fig. 1. All these surface species can be generated from the reactants hydrogen and carbon monoxide on metal surfaces, which are catalytically active in the FT synthesis.

The chemisorption of hydrogen is an activated process [1] yielding mono-atomic hydrogen ($\underline{1}$). Chemisorbed hydrogen is preferentially adsorbed on three-fold hollow sites on all metal surfaces active in the FT synthesis Ru(001) [2], Fe(110) [3], Co(0001) and Ni(111) [4]). Subsurface hydrogen has been postulated as a possible source of hydrogen in the FT synthesis [5], but is slightly less stable than hydrogen adsorbed at the surface [4]. Hence, subsurface hydrogen is not likely to play an important role in the FT synthesis. Chemisorbed hydrogen has a high surface mobility [6]. It is therefore assumed that chemisorbed, mono-atomic hydrogen reacts with other surface species during the FT synthesis [7]. The high mobility of hydrogen implies that adsorption of hydrogen and its consumption for the formation of organic surface species does not necessarily take place at the same site and may even spillover to another metal crystallite (see e.g. Ref. [8]).

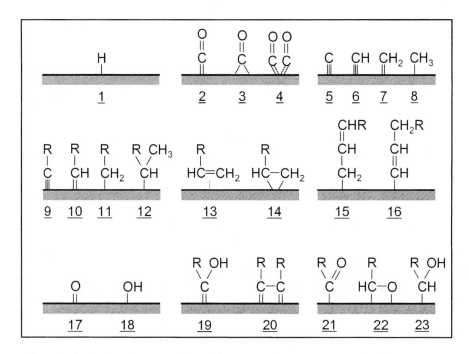

Figure 1 Some of the experimentally observed and postulated surface species on the catalyst surface during the Fischer-Tropsch synthesis (R= C_nH_{2n+1} with n≥0)

The chemisorption of carbon monoxide on metal surfaces has been well studied in the past. The adsorption of carbon monoxide on metal surfaces is mainly due to the interaction of the filled 5σ orbital and the double degenerated 2π* orbital of CO [9-11] and the centre of the metal d-band [12]. Three different modes of chemisorbed carbon monoxide are known, i.e. atop (2), bridged (3) and vicinal (4). Carbon monoxide adsorbed on atop and bridged sites has been shown using IR-measurements [13]. These modes of adsorption have also been confirmed using low energy electron diffraction (LEED) and photon electron spectroscopy (PES). The preferential adsorption site for carbon monoxide on metal surfaces capable of catalysing the FT synthesis is the on-top site (Co(0001) [14]; Ru(0001) [15]), except on Ni(111) where the bridge site seems to be preferred [16].

Chemisorbed carbon monoxide can dissociate yielding surface carbon (5) and surface oxygen (17). The dissociative adsorption of carbon monoxide seems to be suppressed going from left to right and from 3d to 5d in the periodic table of transition metal atoms [16-18]. Ciobîcă and van Santen [19] investigated various pathways for CO dissociation on planar and stepped Ru surfaces, viz. direct CO dissociation, hydrogen assisted CO dissociation and surface Boudouard reaction. Ruthenium was investigated, since this metal represents an intermediate case between associative CO-chemisorption and dissociative chemisorption. On stepped surfaces direct CO dissociation seemed to be favoured, whereas on planar Ru(0001) surface CO dissociation via hydrogen insertion seems to be the preferred reaction pathway. Blyholder and Lawless [20] suggested hydrogen assisted CO dissociation on a Fe_{12} cluster yielding surface CH and surface oxygen species. This dissociation pathway had a lower calculated activation energy than the unassisted dissociation of carbon monoxide.

Surface carbon (5), which is preferentially located in hollow sites [17], may diffuse into the bulk of the metal yielding carbidic carbon or agglomerate yielding graphitic carbon. The formation of stable carbidic phases has been observed with iron based Fischer-Tropsch catalysts [21-23]. The formation of graphitic carbon under Fischer-Tropsch conditions has been observed by Bonzel and Krebs [24] and Dwyer and Hardenbergh [25].

The sequential hydrogenation of surface carbon (5) yields surface methylidyne (6), surface methylene (7) and surface methyl (8) species. The existence of CH_x-species (x=1-3) on a nickel surface after CO hydrogenation was demonstrated using secondary ion mass spectroscopy (SIMS) [26] and on ruthenium based catalysts using IR-studies [27]. The existence of methylene and methylene species was detected after CO hydrogenation over Fe(110) using electron energy loss spectroscopy (EELS) [28]. Furthermore, the addition of pyridine to synthesis gas as a scavenger [29], revealed the presence of alkyl groups on the surface, since the product stream contained 2-methyl-pyridine, 2-ethyl-pyridine and 2-propylpyridine. Most quantum-chemical calculations indicate that methyl is preferentially located on atop sites, methylene on bridge sites and methylidyne on hollow sites (e.g. [30]), although some report methylene to be most strongly adsorbed on hollow sites [31]. The importance of surface methylene and methylidyne species in the FT synthesis was highlighted by van Barneveld and Ponec [32, 33]. They observed in the conversion of CH_xCl_{4-x} in the presence of hydrogen long chain hydrocarbons for x=1 or 2, whereas only methane was obtained with methylchloride (x=3). This shows that the presence of surface methylene or methylidyne species may result in chain growth. Brady and Pettit [34, 35] came to a similar conclusion after their studies on the conversion of diazomethane over transition metal surfaces.

Surface species 9 to 11 can be generated from reaction between surface species 5, 6 or 7 with a surface alkyl group (species 8 or 11). The alkylidyne species (9) has been identified as a species present on the metal surface during olefin hydrogenation (especially ethene hydrogenation, see e.g. [36]). The alkylidyne species is preferentially adsorbed on hollow sites with its C-C axis perpendicular to the surface. This adsorption mode is analogous to the adsorption of the surface methylene species (6). The surface alkylidene species (10) has been postulated as an intermediate in the conversion of adsorbed ethene into ethylidyne, although some studies regard this reaction pathway as unfavourable [37]. The surface alkyl species (11) has been observed under Fischer-Tropsch conditions [29].

The surface alkyl species (12) might be a precursor for the formation of branched products in the FT synthesis. It might be generated by the reaction of a surface alkylidene species (10) with a surface methyl species or by re-adsorption of a primarily formed α-olefin. The first reaction pathway is indirectly supported by the results obtained by Calvalcanti et al. [38], who observed increased formation of branched hydrocarbons in the synthesis gas conversion over Ru/KY upon addition of methyl iodide. Schulz et al. [39] observed upon addition of ^{14}C-propene to synthesis gas in the FT synthesis with cobalt the formation of radioactive labelled methyl branched products. They concluded that re-adsorbed propene acts as an intermediate for the formation of branched compounds.

Surface species 13 and 14 represent π-coordinated adsorbed olefin and the di-σ-coordinated olefin complex. Analogues of this type of bonding are well known in coordination chemistry [40]. These surface species might be important in the re-adsorption of reactive olefins under Fischer-Tropsch conditions.

The surface allyl species (15) and vinyl species (16) are key intermediates in the alkenyl mechanism [41].

Surface oxygen (17) will be generated by the unassisted dissociation of chemisorbed CO. Surface oxygen can react with adsorbed hydrogen yielding a surface hydroxyl group (18), react with adsorbed CO yielding CO_2, diffuse into the bulk of the metal causing the generation of an oxide phase (e.g. with iron based catalysts the formation of magnetite under Fischer-Tropsch conditions from α-Fe is well documented [22]). The removal of surface oxygen is generally assumed to be fast, since under Fischer-Tropsch conditions the surface is mainly covered with carbon [42]. Surface oxygen (17) and surface hydroxyl (18) species are preferentially adsorbed on hollow sites (high coordination sites) and form mainly ionic bonds [43]. The strength of chemisorption increases with decreasing d-band filling. Johnston and Joyner [44] proposed the involvement of surface hydroxyl species (18) in the formation of oxygenated product compounds.

The surface species (<u>19</u>) and (<u>20</u>) are key intermediates in the enol-mechanism [45], but these surface species have not yet experimentally been observed.

The oxygen containing surface species (<u>21</u>) to (<u>23</u>) have been proposed as intermediates in the CO-insertion mechanism [46]. Analogue homogeneous complexes to surface species (<u>21</u>) have been reported [47].

1.2   Reaction pathways in the Fischer-Tropsch synthesis

The Fischer-Tropsch synthesis is a polymerisation reaction, in which the monomers are being produced in-situ from the gaseous reactants hydrogen and carbon monoxide. Thus, all reaction pathways proposed in literature will have three different reaction sections, i.e.

1. generation of the chain initiator
2. chain growth or propagation
3. chain growth termination or desorption

It is generally assumed that not a single reaction pathway exists on the catalyst surface during the FT synthesis, but that a number of parallel operating pathways will exist. Numerous reaction pathways have been proposed to explain the observed product distribution in the FT synthesis. The four most popular mechanisms will be discussed in more detail.

1.2.1 'Alkyl' mechanism

The alkyl mechanism is presently the most widely accepted mechanism for chain growth in the FT synthesis. Fig. 2 shows the proposed reaction pathways for this mechanism. Chain initiation takes place via dissociative CO-chemisorption, by which surface carbon and surface oxygen is generated. Surface oxygen is removed from the surface by reaction with adsorbed hydrogen yielding water or with adsorbed carbon monoxide yielding CO_2. Surface carbon is subsequently hydrogenated yielding in a consecutive reaction CH, CH_2 and CH_3 surface species. The CH_3 surface species is regarded as the chain initiator, and the CH_2 surface species as the monomer in this reaction scheme. Chain growth is thought to take place by successive incorporation of the monomer, CH_2 surface species. Product formation takes place by either β-hydrogen abstraction or hydrogen addition yielding α-olefins and n-paraffins as primary products. The surface species involved in the 'alkyl'-mechanism have been found on the catalyst surface during the FT synthesis (see Section 1) [26-29].

This mechanism developed out of the so-called 'carbide'-mechanism originally proposed by the early workers in this area [48, 49]. In this early mechanism the reaction was thought to proceed via the formation of metal carbides. Carbon in the carbide phase was thought to hydrogenate to CH_2 species, which polymerise. Iron-based catalysts do form stable carbides under

Fischer-Tropsch synthesis, but other metals active in the FT synthesis, such as cobalt and ruthenium are not known to form carbides under typical Fischer-Tropsch conditions. Furthermore, a study by Kummer at al. [50] showed using an iron catalyst pre-carbided with radioactively labelled ^{14}CO that only a small fraction of the products in the gasoil range obtained in the FT synthesis using unlabelled CO contained the radioactive label (less than 4.2%). This indicates that the carbide phase might be dynamically involved in the FT synthesis, but that carbon in the product compounds does not solely originate from carbide. Hence, this mechanism was subsequently rejected.

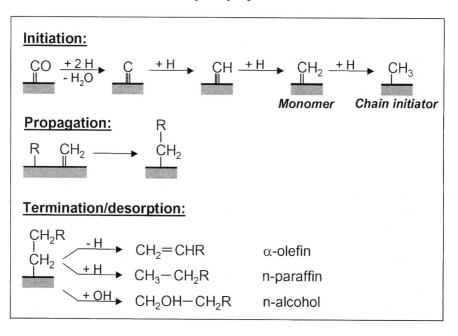

Figure 2 Reaction steps in the Fischer-Tropsch synthesis according to the 'alkyl' mechanism (R= C_nH_{2n+1} with n≥0)

The alkyl mechanism developed on the basis of the 'carbide'-mechanism regained prominence again since early 1980. Brady and Pettit [34, 35] converted diazomethane in the presence and absence of hydrogen over transition metal catalysts. Ethene was the sole product in the absence of hydrogen implying a CH_2-coupling and subsequent desorption. In the presence of hydrogen a product spectrum was obtained similar to that in the Fischer-Tropsch synthesis. This shows that CH_2 surface species can act as the monomer or are involved in the formation of monomer species for the formation of long chain hydrocarbons. Furthermore, the result of these experiments stresses the importance of hydrogen, which may enhance the formation of a chain starter, e.g. a CH_3-

species. The absence of formation of long chain hydrocarbons in the absence of hydrogen might also be taken as an indication that the transformation of CH surface species into CH_2 surface species is irreversible. The decomposition of CH_2 surface species into CH surface species would yield surface hydrogen, which may facilitate the formation of long chain hydrocarbons as seen upon the addition of hydrogen to the conversion of diazomethane. This was not observed and hence the transformation of CH surface species into CH_2 surface species might be regarded to be irreversible.

The experiments by Brady and Pettit [34, 35] were substantiated by experiments performed by van Barneveld and Ponec [32, 33], who converted CH_xCl_{4-x} (x=1-3) in the presence of hydrogen over typical Fischer-Tropsch catalysts. They observed the formation of methane in the conversion of CH_3Cl, whereas a typical Fischer-Tropsch product distribution was obtained in the conversion of $CHCl_3$ and CH_2Cl_2. The formation of long chain hydrocarbons from the hydrogenation of CH_3Cl was expected, if the transformation of a CH_2 surface species into a CH_3 surface species was reversible.

The formation of C-C bonds in the Fischer-Tropsch synthesis is still under debate. In the 'alkyl'-mechanism it is proposed that C-C bond formation takes place through a CH_2-alkyl coupling reaction. This route has been investigated by Zheng et al. [30], who observed a substantial energy barrier for both proximation of the surface species involved and for the actual coupling reactions on a Co(0001) surface. De Koster and van Santen [51] investigated theoretically the co-adsorption of CH_3 surface species and CH_2 surface species on different sites on a Rh(111) surface. The occurrence of C-C bond formation could not be found. Hence, Ciobîcă et al. [52] formulated two in parallel operating reaction pathways with some similarity to the 'alkyl'-mechanism. The low energy reaction pathway for C-C coupling on Ru(0001) involves the reaction between an alkyl surface species and a methylidyne surface species yielding an alkylidene surface species. The alkylidene surface species is subsequently hydrogenated to a surface alkyl species. In this reaction pathway the CH_2-insertion in the 'alkyl'-mechanism is replaced by a CH-insertion with subsequent hydrogenation. As an alternative reaction pathway Ciobîcă et al. [52] proposed C-C coupling through the reaction of an alkylidene surface species with a methylidyne species. The formed complex is subsequently hydrogenated yielding an alkylidene surface species. The desorption pathway involves the hydrogenation of the alkylidene surface species to the corresponding alkyl surface species, which can then undergo hydrogen addition yielding n-paraffin or β-hydrogen elimination yielding an α-olefin. An alternative reaction mechanism was proposed by Dry [53], in which CH_2-CH_2 coupling leads to the formation of a di-σ bound surface species as the growing chain (see also Chapter 3).

The chain propagation in the 'alkyl' mechanism is generally considered to be irreversible as well, since the reverse reaction, hydrogenolysis, is strongly inhibited by CO [46, 54]. It should however be noted that Schulz et al. [39] observed some cracking of added radioactively labelled n-hexadecene-(1) under Fischer-Tropsch conditions with a cobalt catalyst indicating some hydrogenolysis activity.

Chain termination occurs in the 'alkyl'-mechanism through hydrogen addition yielding n-paraffins or through β-H-elimination yielding α-olefins. The β-hydrogen elimination is a reversible reaction allowing for the re-incorporation of primarily formed olefins into longer chain hydrocarbons. α-Olefins and n-paraffins are the main products of the FT synthesis. Both have been identified as primary products in the FT synthesis by a large number of studies. The classical method to identify primary products is to follow the selectivity of the reaction (or the selectivity in a certain carbon number fraction) as a function of space velocity, i.e. conversion. At high space velocity (and thus low conversion) the selectivity of secondary products will tend to zero, whereas the selectivity of primary products will assume the value of the primary selectivity. In this manner n-paraffins and α-olefins were identified as primary products of FT synthesis [55-57]. Tau et al. [58] investigated the secondary reaction of 1-pentene during the FT synthesis at 7 bar, 260°C, $H_2/CO=2$, in a slurry reactor over a fused iron catalysts. The C_5-fraction in the FT synthesis contained 20% n-pentane. Hydrogenation of co-fed 1-pentene was not observed implying a primary selectivity for the formation of n-paraffins of 20%. A primary selectivity of 20% for paraffins was also suggested by Schulz and Gökcebay [59] from their studies over iron-manganese catalysts.

Branched hydrocarbons are also formed in small amounts in the FT synthesis. They are partially formed from re-adsorbed olefins, such as propene [39]. Lee and Anderson [60] concluded that the amount of branched hydrocarbons formed is larger than would be expected based on re-incorporation of re-adsorbed olefins. Hence, additional reaction pathways for the formation of branched hydrocarbons need to be considered. The main reaction pathway of the 'alkyl'-mechanism does not include the formation of branched hydrocarbons. Schulz et al. [61, 62] proposed a reaction pathway for the formation of branched hydrocarbons involving the reaction of an alkylidene surface species and a methyl surface analogue to the 'alkyl'-mechanism (see Fig. 3). The alkylidene surface species may originate from a reaction between an alkyl species and methylidyne. The branched alkyl species can undergo similar desorption reactions to those proposed for n-alkyl species.

The 'alkyl'-mechanism cannot explain the formation of oxygenates in the FT synthesis. However, Johnston and Joyner [44] proposed the involvement of surface hydroxyl groups in the formation of oxygenates. The coupling of a surface hydroxyl group with an alkyl group may lead to the formation of alcohols. Experimental evidence for the participation of surface hydroxyl groups in the formation of oxygenates is still lacking.

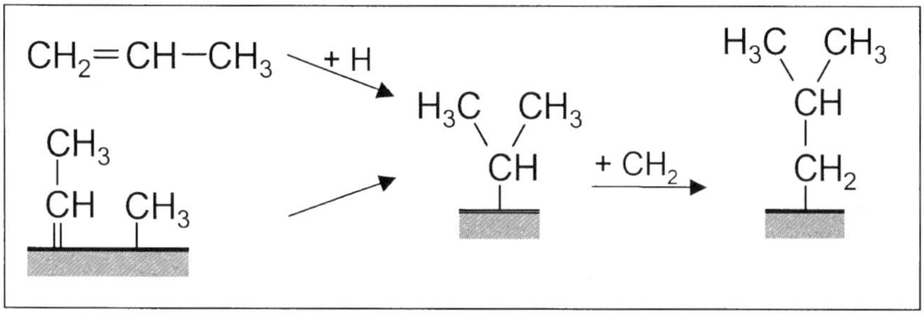

Figure 3 Proposed reaction pathways for the formation of branched hydrocarbons in the Fischer-Tropsch synthesis according to Schulz et al. [61, 62]

1.2.2 'Alkenyl' mechanism

Maitlis and co-workers [41, 63] proposed an alternative reaction pathway to predict the formation of olefins in the Fischer-Tropsch synthesis negating the sp^3-sp^3 coupling of methylene species in bridged position with methyl species adsorbed at atop position as proposed in the 'alkyl'-mechanism (see Fig. 4). The initial activation of carbon monoxide and its transformation into CH_x surface species are identical to the proposed 'alkyl'-mechanism. The formation of the first C-C bond occurs through the coupling of methylidyne (CH) and methylene (CH_2) to form a vinyl surface species ($CH=CH_2$), which is considered as the chain initiator. This coupling reaction is predicted to have an activation energy of 55.9 kJ/mol over Co(0001) and 116.5 kJ/mol over Ru(0001) [31].

Chain propagation involves the addition of a methylene species to a surface alkenyl-species (vinyl species) yielding a surface allyl species. This is followed by an allyl-vinyl isomerization forming an alkenyl species.

Product desorption involves the hydrogen addition to an alkenyl species yielding α-olefins. This mechanism fails to explain the primary formation of n-paraffins, and the co-existence of an alternative chain growth pathway is required [64], e.g. the one proposed by Ciobîcă et al. [52]. According to this reaction scheme the formation of branched product compounds might be via the isomerization of an allylic intermediate η^1-$CH_2CH=CHR$ to η^1-$CHRCH=CH_2$. The subsequent allyl-vinyl isomerization and reaction with surface methylene will lead to branched products.

610

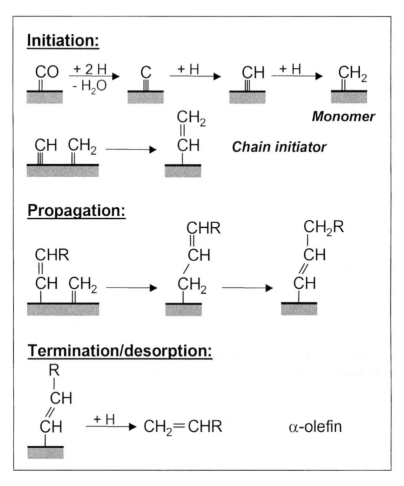

Figure 4 'Alkenyl' mechanism according to Maitlis et al. [41,63] (R= C_nH_{2n+1} with $n \geq 0$)

1.2.3 'Enol' mechanism

Workers at the Bureau of Mines [45] proposed an alternative reaction scheme involving oxygen containing (enol) surface species (see Fig. 5). According to this mechanism chemisorbed CO is hydrogenated to enol surface species. The formation of an adsorbed complex with H_2:CO=1:1 finds some support in the observed ratio of adsorbed hydrogen and carbon monoxide in adsorption experiments involving mixtures of H_2 and CO on cobalt [65] and results of desorption experiments on an iron based catalyst [66].

Chain growth occurs through a condensation reaction between enol species under elimination of water. The occurrence of branched hydrocarbons is thought to originate from the involvement of a CHROH surface species.

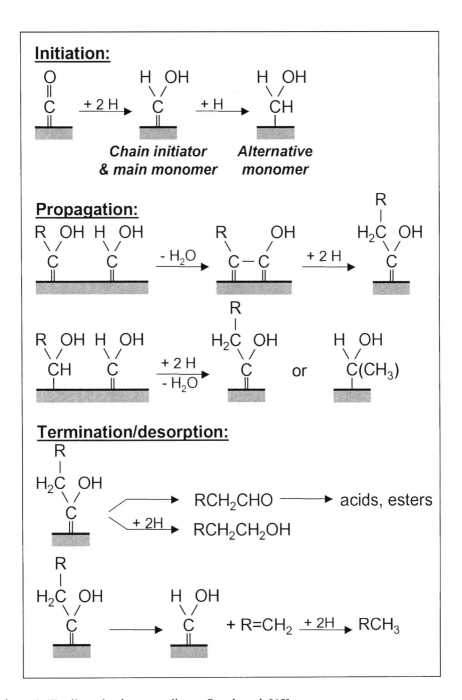

Figure 5 'Enol' mechanism according to Storch et al. [45]

Termination of the chain growth process or desorption yield oxygenates (aldehydes and alcohols) and α-olefins. According to this reaction mechanism n-paraffins are formed secondarily by hydrogenation of primarily formed olefins. The primary formation of n-paraffins would require an alternative reaction pathway. The formation of acids in the Fischer-Tropsch synthesis was postulated as a secondary reaction of primarily formed aldehydes as a result of Cannizzaro reactions. Acids reacting with an alcohol yields esters.

1.2.4 'CO-insertion' mechanism

The 'CO-insertion'-mechanism, which was originally proposed by Sternberg and Wender [67] and Roginski [68], was fully developed by Pichler and Schulz [46]. The mechanism is based on the known CO-insertion from coordination chemistry and homogeneous catalysis (see Fig. 6).

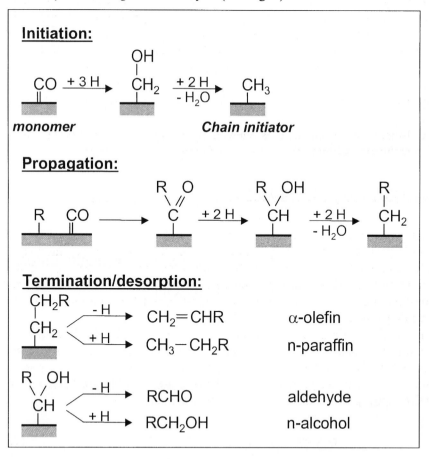

Figure 6 'CO-insertion' mechanism for chain growth in the Fischer-Tropsch synthesis

In the 'CO-insertion' mechanism chemisorbed CO is the monomer. The chain initiator is thought to be a surface methyl species. The reaction pathway leading to the formation of the surface methyl species differs from the 'alkyl'-mechanism at the time of the oxygen elimination from the surface species.

Chain growth takes place by CO-insertion in a metal-alkyl bond leading to a surface acyl species. This reaction step is well known in homogeneous catalysis [47]. The elimination of oxygen from the surface species leads to the formation of the enlarged alkyl species.

Various reaction pathways are proposed for product desorption. The reaction steps leading to the formation of n-paraffins and α-olefins are identical to those proposed in the 'alkyl'-mechanism, i.e. hydrogen addition and β-hydrogen elimination. Chain termination reactions may also involve oxygen containing surface species leading to the formation of aldehydes and alcohols. Hydrogen addition to an acyl species (not shown in Fig 6) will lead to aldehyde formation. These product compounds can also be formed by β-H-elimination of the RCHOH species yielding an enol (R=CHOH), which is known to isomerise quickly in the absence of a catalyst to an aldehyde. Alcohol formation can take place by hydrogen addition to a RCHOH surface species. Formation of ketones is proposed to occur through the addition of a surface alky group to the acyl species.

The 'CO-insertion' mechanism is viewed by many researchers as the main reaction pathway leading to the formation of oxygenates [69, 70].

2. RATE OF CO-CONSUMPTION IN THE FT SYNTHESIS

Adequate rate expressions are needed for up-scaling and process optimisation of Fischer-Tropsch reactors and, if related to the formation of the individual product compounds or classes, for directing selectivity. The kinetics of the Fischer-Tropsch synthesis has been studied extensively. The rate of synthesis gas consumption or CO consumption is usually correlated with the gas phase concentrations of carbon monoxide, hydrogen and/or water.

In the FT synthesis carbon monoxide is consumed for the formation of organic compounds and for the formation of carbon dioxide. From an elemental balance it thus follows that the rate of consumption of CO equals the rate of formation of organic compounds in the FT synthesis on carbon basis (units: mmol C/(g·min)) plus the rate carbon dioxide formation:

$$- r_{CO} = r_{C,org} + r_{CO_2} \tag{1}$$

The rate of formation of organic product compounds on carbon basis equals the rate of CO consumption for the formation of organic product compounds. The rate of CO consumption equals the rate of formation of organic product compounds on carbon basis, if the rate of CO_2 formation is negligible (e.g. for cobalt or ruthenium catalysed FT synthesis).

2.1 Rate of CO consumption for the formation of organic products

Many correlations for the rate of CO consumption or synthesis gas consumptions have been published. The rate expressions developed before 1974 have been summarised by Vannice [71]. An overview of rate equations for iron catalysts is given by Huff and Satterfield [72] and van der Laan and Beenackers [73] and for cobalt catalysts by Yates and Satterfield [74].

Most kinetic expressions have been developed empirically fitting the data to a power-law relationship. This is a powerful technique to gain some insight in the actual processes taking place on the catalyst surface, but hardly adequate for scale-up. Some rate expressions have been developed based on certain assumptions regarding the elementary reaction steps taking place on the catalyst surface. For the development of the rate expression, a rate-determining step is postulated.

Anderson [21] postulated the rate of desorption of growing chains on the catalyst surface to be the rate determining step. The rate of consumption of carbon monoxide for cobalt catalysts was assumed to be proportional to the rate, at which growing hydrocarbon chains desorb from the surface. The concentration of the growing chains was empirically correlated to $p_{H_2}^2 \cdot p_{CO}$. The rate of desorption in the Fischer-Tropsch synthesis must be slow, since otherwise low molecular weight hydrocarbons (especially methane) would only be formed. Inhibition of desorption of the growing chains makes chain growth feasible. It is however possible to decrease the rate of desorption of the growing chains forming longer chain hydrocarbons. The rate of CO consumption will then have increased. Hence, the rate of desorption cannot correlate with the rate of consumption of carbon monoxide. The rate of desorption in conjunction with the rate of CO-consumption for the formation of organic product compounds controls the average chain length and thus the product distribution.

Rautavuoma and van der Baan [75] and Huff and Satterfield [72] concluded from their measurements, using a cobalt, and a reduced fused magnetite, catalyst that the rate of consumption of synthesis gas can be modelled assuming the rate of formation of the monomer to be the rate determining step.

Cobalt [75]:
$$-r_{CO+H_2} = \frac{c \cdot p_{H_2} \cdot p_{CO}^{\frac{1}{2}}}{\left[1 + d \cdot p_{CO}^{\frac{1}{2}}\right]^3}$$

Fused iron [72]:
$$-r_{CO+H_2} = \frac{a \cdot p_{H_2}^2 \cdot p_{CO}}{p_{H_2O} + p_{H_2} \cdot p_{CO}}$$

The rate equations developed could describe the obtained experimental results reasonably well in the range of experimental conditions. This assumption used in the derivation for these rate expressions implies that all previous and consecutive steps in the Fischer-Tropsch synthesis are assumed to be much faster. All elementary reactions, except the rate-determining step can thus be regarded to be close to equilibrium. However, assuming all steps following the rate-determining step being close to equilibrium implies that the product distribution will be almost thermodynamically controlled. This is clearly not the case, since the product spectrum usually contains much less methane and much higher molecular weight organic compounds than predicted on the basis of thermodynamics, as desired for the practical application of the Fischer-Tropsch synthesis.

The Fischer-Tropsch synthesis is a polymerisation reaction with the successive addition of C_1-monomer units [76]. It was recently pointed out that therefore polymerisation principles rather than the rate determining step should be used for the development of rate expressions describing CO-consumption in the Fischer-Tropsch synthesis [77].

Three classes of reactions are distinguished in polymerisation reactions *viz.* initiation, propagation and termination. The rate of reaction is the sum of the rate of initiation, propagation and termination. Initiation, propagation and termination steps can also be found in the Fischer-Tropsch synthesis (see Section 1.2). The rate of formation of organic product compounds on carbon basis is therefore equal to the sum of the rate of formation of chain initiator, rate of propagation and rate of desorption with all rates expressed in rate on carbon basis in the specific reaction step:

$$r_{C,org} = r_{C,initiation} + r_{C,propagation} + r_{C,termination} \qquad (2)$$

In normal polymerisation reactions, in which high molecular weight products are formed, the rate of termination can be neglected, since the rate of propagation must be much larger than the rate of termination. Hence, the rate of termination can be neglected. The average molecular weight of the organic products formed in the Fischer-Tropsch synthesis is too low to neglect the rate of termination in comparison to the rate of propagation. However, the proposed termination reactions in the Fischer-Tropsch synthesis according to most mechanisms do not involve carbonaceous species (see Section 1.2). The formation of the main products in the Fischer-Tropsch synthesis, hydrocarbons, is thought to involve either H-addition or β-H-abstraction. Hence the rate of termination based on carbon basis equals zero.

$$r_{C,org} = r_{C,initiation} + r_{C,propagation} \tag{3}$$

In order to develop from here rate expressions correlating the rate of formation of organic CO-consumption with gas phase concentrations of reactants and product compounds, a certain mechanism has to be assumed. In the further development described here, it will be assumed that the 'alkyl'-mechanism is correct (see Fig. 2), although similar rate expressions can be derived based on the other proposed reaction mechanisms. The formation of oxygenates, such as n-alcohols and methyl alkyl ketones, is thought to be formed through CO-insertion. The amount of oxygenates formed is relatively small compared to that of the hydrocarbons. In the development of the rate expression it is assumed that the carbon incorporation in growing chains through CO insertion is negligibly small.

The methyl surface species is regarded as the chain initiator in the 'alkyl'-mechanism. Thus, the rate of formation of the chain initiator equals the rate of formation of the methyl surface species.

$$r_{C,org} = r_{CH_3} + r_{C,propagation} \tag{4}$$

In the further derivation it is assumed that the major reaction pathway for the formation of surface methyl species is the hydrogenation of surface methylene species, i.e. the contribution of hydrogenolysis to the formation of methyl surface species can be neglected [46, 54].

From the evidence in literature it seems clear that chain growth involves incorporation of CH_2 surface species into growing chains (chain growth through a sequential CH insertion followed by hydrogenation as proposed by Ciobîcă et al. [52] will lead to identical rate expression). For the derivation of the model it is assumed that CH_2 insertion is the only reaction for chain growth.

According to the 'alkyl' -mechanism surface methylene (CH_2) species are formed by hydrogenation of surface methylidyne (CH) species. The surface methylene species is being consumed for the formation of the chain starter (methyl species) or for the incorporation into growing chains. The rate of incorporation of monomer units has been shown to be dependent on the carbon number of the growing chain [62, 78, 79]. The rate of incorporation into growing chains represents here the sum of the rate of incorporation into chains of all sizes.

At steady state the net rate of formation of CH_2 surface species equals zero. Therefore, the rate of formation of CH_2 surface species by hydrogenation of CH species ($r_{CH \to CH_2}$) equals the rate of formation of the chain starter (r_{CH_3}) plus the rate of incorporation ($r_{C,propagation}$). Substituting this back into the rate of formation of organic product compounds on carbon basis leads to the conclusion that the rate of formation of CH_2 surface species equals the rate of CO consumption for the formation of organic compounds:

$$r_{C,org} = r_{CH \to CH_2} \qquad (5)$$

Treating the FT synthesis as a polymerisation reaction leads to a similar conclusion, which was drawn by other authors [72, 75] from their kinetic experiments without invoking the existence of a rate-determining step and its necessary consequences.

Since CH surface species are not accumulated on the surface at steady state conditions, the net rate of formation of CH_2 surface species equals the net rate of formation of CH species. It is assumed here that CH-species can only be hydrogenated to CH_2-species. The rate of consumption of carbon monoxide for the formation of organic compounds can thus be expressed as the rate of formation of CH surface species:

$$r_{C,org} = r_{C \to CH} = r_{CH \to CH_2} \qquad (6)$$

It has been pointed out, that CH surface species might also be incorporated into growing chains [61, 62]. The incorporation of these units into growing chains would lead to the primary formation of branched compounds. Although the primary formation of branched organic products was not taken into account in this derivation, its inclusion will lead to the same result.

If the formation of CH surface species can be regarded to be an *irreversible* hydrogenation of surface carbon, the net rate of formation of these surface species is proportional to the fractional coverage with surface carbon and the fractional coverage with surface hydrogen:

$$r_{C,org} = r_{C \to CH} = k_{C \to CH} \cdot \theta_C \cdot \theta_H \qquad (7)$$

The rate of formation of organic compounds on a carbon basis can now principally be derived using the steady-state approximation for C, H, CO, O and OH surface species. For simplicity, it might be assumed that these species are in equilibrium with the gas phase compounds

$$H_2 \; + \; 2* \; \rightleftharpoons \; 2\,H* \qquad K_H = \frac{\theta_H}{p_{H_2}^{0.5} \cdot \theta_*} \tag{8}$$

$$CO \; + \; * \; \rightleftharpoons \; CO* \qquad K_{CO} = \frac{\theta_{CO}}{p_{CO} \cdot \theta_*} \tag{9}$$

$$CO* + \; * \; \rightleftharpoons \; C* \; + \; O* \quad K_C = \frac{\theta_C \cdot \theta_O}{\theta_{CO} \cdot \theta_*} \tag{10}$$

$$O* \; + \; 2\,H* \; \rightleftharpoons \; H_2O \; + \; 3* \quad K_{H_2O} = \frac{p_{H_2O} \cdot \theta_*^3}{\theta_O \cdot \theta_{H*}^2} \tag{11}$$

(* denotes an active site on the catalyst surface)

It is well known that the various surface species bind differently to the catalyst surfaces. Nevertheless, it is assumed in the equilibrium between surface species and gas phase compounds postulated above, that each surface species occupies a single active site. The catalyst surface consists of coordinatively unsaturated metal atoms. A metal atom can principally bond to more than one surface species. The surface species do occupy a certain geometric space and limit in this way the access to surface metal atom. The number of active sites is therefore not directly proportional to the number of surface metal atoms, nor to the number of coordinatively unsaturated bonds. In the above postulated equilibrium, it is assumed that all surface species occupy approximately the same geometric space. It is further assumed that the dissociative adsorption of hydrogen requires two adjacent metal atoms.

With the assumed equilibrium, the coverage with surface carbon and surface hydrogen can be estimated and the rate of consumption of carbon monoxide for the formation of organic compounds can be expressed as:

$$r_{C,org} = k_{C \to CH} \cdot K_H^3 \cdot K_{CO} \cdot K_C \cdot K_{H_2O} \cdot \frac{p_{H_2}^{1.5} \cdot p_{CO}}{p_{H_2O}} \cdot \theta_*^2 \tag{12}$$

Under realistic Fischer-Tropsch conditions the catalyst surface is mainly covered with surface carbon. The rate of formation of organic compounds can then be expressed as:

$$r_{C,org} = \frac{k_{C \to CH} \cdot K_H^3 \cdot K_{CO} \cdot K_C \cdot K_{H_2O} \cdot \dfrac{p_{H_2}^{1.5} \cdot p_{CO}}{p_{H_2O}}}{\left(1 + K_H^2 \cdot K_{CO} \cdot K_C \cdot K_{H_2O} \cdot \dfrac{p_{H_2} \cdot p_{CO}}{p_{H_2O}}\right)^2} \qquad (13)$$

The high reaction order with respect to hydrogen reflects the double function of this compound, i.e. the hydrogenation of surface carbon yielding organic product compounds and the removal of surface oxygen from the surface. The influence of water as described here is indirect. At high water partial pressures the coverage with surface carbon will decrease due to an increase in the surface coverage of oxygen, which will suppress the formation of surface carbon via the dissociation of surface CO.

This rate expression was successfully correlated with experimental conditions [77]. Fig. 7 shows the principal dependency of the partial pressure of hydrogen, carbon monoxide and water on the rate of formation of organic product compounds. All partial pressures cause an initial increase in the rate of formation followed by a decrease at high partial pressures. The decrease at high partial pressures is caused by the depletion of one of the two or both of the necessary surface species for the formation of hydrocarbons, i.e. surface carbon and surface hydrogen. For instance, an increase in the partial pressure of hydrogen will initially increase the surface coverage with hydrogen and thus to an increase of the rate of formation. Increasing the hydrogen partial pressure further will lead to a depletion of surface carbon leading to a decrease in the rate of formation. A more indirect effect is caused by water.

The effect of water on the rate of formation of organic product compounds is indirect [80]. According to this model water affects the surface concentration of oxygen and over the CO-dissociation equilibrium the surface coverage of carbon. Increasing the water partial pressure will therefore reduce the amount of surface carbon. At low water partial pressure, the surface coverage of carbon species is high and inhibits the adsorption of hydrogen. The catalyst surface is cleaned with increasing the water partial pressure by indirectly reducing the amount of surface carbon. This will thus lead to an increase in the rate of formation of organic compounds on a carbon basis. Increasing to high water partial pressure leads to a depletion of surface carbon and hence a reduction in the rate of formation of organic compounds on a carbon basis.

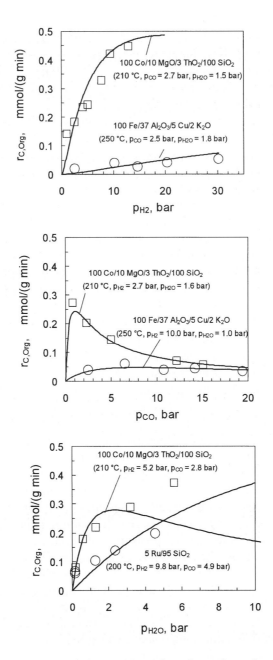

Figure 7 Principal dependency of rate of formation of organic product compounds (in mmol C/(g_{cat} min) as a function of partial pressure of hydrogen (top), carbon monoxide (middle) and water (bottom) for iron, cobalt and ruthenium based catalysts [77,80] (solid lines: model prediction)

The rate expression predicts zero rate of formation of organic compounds in the absence of water. However, Fischer-Tropsch activity is observed in the absence of water (e.g. at the entrance of fixed bed reactors). Thus, the rate expression developed by van Steen and Schulz [77] cannot predict the rate of reaction under all Fischer-Tropsch conditions.

The influence of water partial pressure on the rate of formation of organic product compounds on a carbon basis was introduced through the assumption of equilibrium between surface oxygen and water in the gas phase, which controls the removal of oxygen from the surface. Water is not necessarily in equilibrium with surface oxygen under Fischer-Tropsch conditions.

The rate of removal of surface species equals the rate of formation of organic compounds on a carbons basis. The rational behind this is the balance of carbon and oxygen in CO, which is being consumed for the formation of organic product compounds. It is assumed here that the incorporation of 'CO'-units in organic product compounds and thus the amount of oxygenated products formed is minimal. The coverage of the surface with oxygen can then be obtained by appropriate modelling of the rate removal of surface oxygen. For instance, if the reaction of surface oxygen and surface hydrogen can be considered to be irreversible:

$$r_{C,org} = r_{O-removal} = k_O \cdot \theta_O \cdot \theta_H \tag{14}$$

Substituting this result in Eq. (10) yields an expression for the surface coverage with carbon in terms of the rate of formation of organic product compounds on a carbon basis and the surface coverage with carbon monoxide and hydrogen:

$$\theta_C = \frac{K_C \cdot \theta_{CO} \cdot \theta_*}{\theta_O} = \frac{k_O \cdot K_C \cdot \theta_{CO} \cdot \theta_H \cdot \theta_*}{r_{C,org}} \tag{15}$$

Substituting this result back into the expression describing the rate of formation of organic compounds on carbon basis (Eq. 7) yields:

$$r_{C,org}^2 = k_{C \to CH} \cdot k_O \cdot K_C \cdot \theta_{CO} \cdot \theta_H^2 \cdot \theta_* \tag{16}$$

Assuming adsorption equilibrium between dissociatively adsorbed hydrogen and gas phase hydrogen, associatively adsorbed carbon monoxide and gas phase carbon monoxide, and assuming that the catalyst surface is mainly covered with carbon, leads to:

$$r_{C,org} = k_{C \to CH}^{0.5} \cdot k_O^{0.5} \cdot K_C^{0.5} \cdot K_{CO}^{0.5} \cdot K_H^{0.5} \cdot p_{CO}^{0.5} \cdot p_{H_2}^{0.5} \cdot \theta_*^2 \tag{17}$$

with $\theta_*^2 = \dfrac{1}{4 \cdot Q^2} \cdot \left(2 + 4 \cdot Q - 2 \cdot \sqrt{1 + 4 \cdot Q}\right)$ \quad and \quad $Q = b' \cdot p_{H_2}^{0.5} \cdot p_{CO}^{0.5}$

2.2 CO_2 formation

The Fischer-Tropsch synthesis over iron-based catalysts produces as a co-product significant amount of carbon dioxide. In the iron-catalysed high temperature Fischer-Tropsch synthesis (HTFT) the water-gas shift reaction is close to equilibrium. The amount of carbon dioxide present in the iron-catalysed low temperature Fischer-Tropsch synthesis (LTFT) is less than predicted by the water-gas shift equilibrium. Hence, secondary reactions involving hydrogen, carbon monoxide, carbon dioxide and water will produce carbon dioxide. With cobalt-based catalysts the formation of carbon dioxide can often be neglected.

Relatively few studies have been published to model the formation of CO_2 in the Fischer-Tropsch synthesis. Dry [81] showed that a reactor model incorporating carbon dioxide formation being proportional to the partial pressure of CO, describes the concentration profiles in the reactor adequately.

Iron-based catalysts consist of different phases, i.e. magnetite and carbide. Some studies assume the formation of CO_2 to be a secondary reaction, which takes place on the magnetite phase, whereas the hydrocarbon formation takes place on the carbide phase [82, 83]. The formation of carbon dioxide on the magnetite phase might be modelled assuming the reaction between chemisorbed carbon monoxide and a surface hydroxyl to be the rate-determining step [83]:

$$r_{CO_2} = \dfrac{k_W \cdot \left(p_{CO} \cdot p_{H_2O} - \dfrac{p_{CO_2} \cdot p_{H_2}}{K_{WGS}}\right)}{\left(1 + K_{CO,ads} \cdot p_{CO} + K_{H_2O,ads} \cdot p_{H_2O}\right)^2} \tag{18}$$

Although magnetite is known to be an active catalyst for the water-gas shift reaction, it is not clear from literature whether CO_2 is formed solely by the secondary water-gas shift reaction from the product water. Carbon dioxide might be formed by oxygen removal from the carbide surface with chemisorbed carbon monoxide. Hence, future thinking on the carbon dioxide formation should take into consideration both formation of carbon dioxide on the carbide phase and on the magnetite phase.

3. PRODUCT DISTRIBUTIONS

3.1 Products of Fischer-Tropsch synthesis

A huge variety of products of different chain length and different functionality is formed in Fischer-Tropsch synthesis. The actual composition/product distribution of a Fischer-Tropsch process depends on many reaction variables such as reaction conditions (temperature and partial pressures of the reactants and product water), the reactor system used, as well as the catalyst formulation and physical properties of a catalyst. The main products of Fischer-Tropsch synthesis are:

- n-Olefins (mainly α-olefins, also olefins with internal double bond)
- n-Paraffins

Typical side-products are:

- Oxygenates (1-alcohols, aldehydes, ketones, carboxylic acids)
- Branched compounds (mainly mono-methyl branched)

Product compositions of industrial processes are given in Chapter 3. Only in high temperature processes are fairly large amounts of branched compounds and aromatic compounds are formed.

Fig. 8 depicts a typical gas chromatogram of a low temperature Fischer-Tropsch product obtained with a cobalt catalyst [84]. Although highly complex the product spectrum shows a remarkable degree of order with regard to class and size of the molecules, which changes from a mainly olefinic product at low carbon numbers into a purely paraffinic product at higher carbon numbers.

The high degree of order with repeating selectivity patterns in different carbon number fractions suggests a strict kinetic basis of this surface polymerisation with stepwise addition of a C_1 monomer species (see Section 1), which is well suited for selectivity modelling. Many mathematical models have been developed to describe Fischer-Tropsch product distributions. These are described in this chapter starting with a very basic model, which can be modified stepwise to account for formation of individual products and their possible consecutive reactions. As complete predictive models for Fischer-Tropsch product distributions are not available it will at least be attempted to give guidelines as to how some reaction parameters can be expected to impact on the kinetic steps involved and therefore on the final composition and distribution of the products. A more detailed discussion on the effect of reaction conditions is given in Section 4. Understanding these concepts will be of crucial importance for any engineer, who takes part in designing a Fischer-Tropsch process aiming at maximum selectivity of desired products.

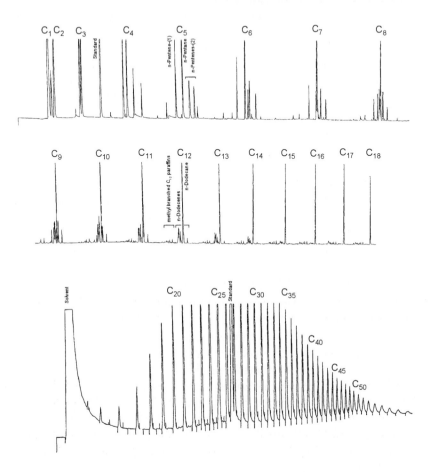

Figure 8 Typical gas chromatogram of a low temperature Fischer-Tropsch product obtained with a Cobalt catalyst in a laboratory scale stirred slurry reactor [84]. Top: carbon number range C_1-C_{18}; bottom: from carbon number C_{18} ("wax").
(Reaction conditions: catalyst: 100 Co / 15 ZrO_2 / 100 SiO_2 / 0.66 Ru (mass ratios), T=190°C, p_{H2}=5.1 bar, p_{CO}=2.2 bar, p_{H2O}=1.1 bar)

3.2   Basic model: Ideal chain growth with one sort of products

As a first approximation the molar amount of the sums of products in individual carbon number fractions declines exponentially with carbon number. This behaviour, which is indicative of a polymerisation reaction that proceeds via stepwise addition of a C_1 monomer, was originally noticed by Herrington

[76] and Friedel and Anderson [85]. A simplified growth scheme of this polymerisation, which occurs on the surface of the catalysts, is given in Fig. 9 assuming the formation of only one sort of products.

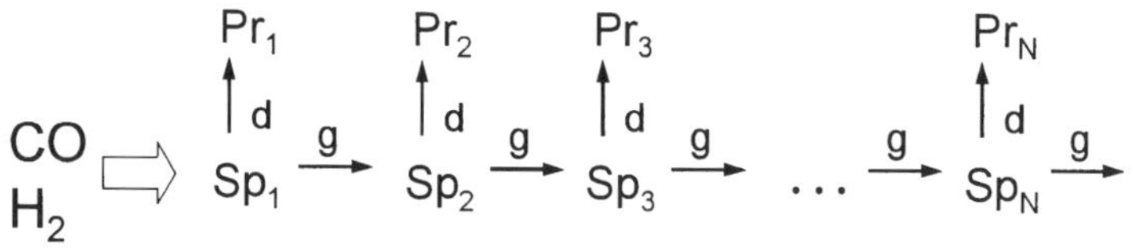

Figure 9 Ideal chain growth assuming formation of one sort of products (Pr$_n$) only. (Sp: surface species, N: carbon number, d: desorption, g: growth)

Carbon monoxide and hydrogen are adsorbed on the catalyst surface. After dissociation they form monomers and chain starters (C$_1$ surface species, Sp$_1$), which can either desorb to form a C$_1$ product molecule (Pr$_1$) or successively grow via insertion of C$_1$ monomers. Note that no assumptions are made in the above scheme regarding the nature of the monomer which – in contrast to conventional polymerisations – is formed in situ. The FT synthesis is therefore also termed a 'non-trivial surface polymerisation' [61, 86]. In the ideal case of carbon number independent chain growth probability (p$_g$) of surface species, i.e. ideal chain growth, the molar product distribution may be presented as:

$$x_N = (1-p_g) \cdot p_g^{N-1} \tag{19}$$

A complete derivation of this equation, which was first developed by Schulz [87] and Flory [88] for homogeneous polymerisations, is given in Fig. 10. The only parameter in Eq. (19) is the chain growth probability p$_g$, which is also often referred to as α. When plotted logarithmically straight lines are observed, and the chain growth probabilities can be determined from their slopes:

$$\lg x_n = \lg \frac{(1-p_g)}{p_g} + n \cdot \lg p_g \tag{20}$$

Fig. 11 shows logarithmic molar product distributions with varied chain growth probabilities. These plots are generally called Anderson-Schulz-Flory (ASF) distributions and are commonly used to characterise FT synthesis products. Assuming ideal ASF kinetics places constraints on theoretically achievable selectivity of product weight fractions, as shown in Fig. 12 (after Dry [70]). Only methane can be produced with a selectivity of 100 wt.%, i.e. at a chain growth probability of 0 ('methanation').

A mass balance around a surface species Sp_N at steady state results in:

$$r_{g,N-1} = r_{g,N} + r_{d,N} \tag{10.1}$$

Assuming first order kinetics with respect to the concentration of a surface species $(c_{s,Sp,N})$:

$$k_{g,N-1} \cdot c_{s,Sp,N-1} = k_{g,N} \cdot c_{s,Sp,N} + k_{d,N} \cdot c_{s,Sp,N} \tag{10.2}$$

and rearranged:

$$c_{s,Sp,N} = \frac{k_{g,N-1}}{k_{g,N} + k_{d,N}} \cdot c_{s,Sp,N-1} \tag{10.3}$$

The formation rate of a product compound with N carbon atoms r_N (i. e. its desorption rate) is given by:

$$r_N = r_{d,N} = k_{d,N} \cdot c_{s,Sp,N} = \frac{k_{d,N}}{k_{g,N} + k_{d,N}} \cdot k_{g,N-1} \cdot c_{s,Sp,N-1} \tag{10.4}$$

and with the definition of the desorption probability $p_{d,N}$

$$p_{d,N} = \frac{k_{d,N}}{k_{g,N} + k_{d,N}} \quad \text{(analogousely: chain growth probability}$$

$$p_{g,N} = \frac{k_{g,N}}{k_{g,N} + k_{d,N}}) \tag{10.5}$$

$$\Rightarrow r_N = p_{d,N} \cdot k_{g,N-1} \cdot c_{s,Sp,N-1} \tag{10.6}$$

An analogous balance can be made for the surface species Sp_{N-1}, Sp_{N-2}, etc. to Sp_1, which leads to:

$$r_N = p_{d,N} \cdot p_{g,N-1} \cdot p_{g,N-2} \cdot p_{g,N-3} \cdot \dots \cdot p_{g,2} \cdot k_{g,1} \cdot c_{s,Sp,1} \tag{10.7}$$

Since all product compounds are formed starting from a C_1 species, the sum of the formation rates of all product must equal the consumption rate of the species Sp_1, the chain starter:

$$\sum_N r_N = (k_{d,1} + k_{g,1}) \cdot c_{s,Sp,1} \tag{10.8}$$

It should be noted, that the sum of formation rates of all organic compounds on a carbon basis equals the rate of CO-consumption, which links the model to the actual conversion of synthesis gas.

The molar content of a product compound x_N with N carbon atoms in the total organic product spectrum is:

$$x_N = \frac{r_N}{\sum_N r_N} = p_{d,N} \cdot p_{g,N-1} \cdot p_{g,N-2} \cdot p_{g,N-3} \cdot \dots \cdot p_{g,2} \cdot p_{g,1} = p_{d,N} \cdot \prod_{N=1}^{N-1} p_{g,N} \tag{10.9}$$

In case of carbon number independent probabilities the model turns into Eq. 19.

Figure 10 Derivation of Eq. 10 ('ideal polymerisation kinetics')

The maximum straight run middle distillate yield (carbon number range C_{10}-C_{20}) according to these kinetics is only around 40 wt.% and can be obtained at $p_g{\approx}0.85$. If higher overall yields of middle distillate are required the FT process should be designed in a way that very high chain growth probabilities (high wax yields) are achieved and the wax can then be transferred into middle distillate via mild hydrocracking. In doing so a middle distillate selectivity of ca. 80 wt.% can for example be theoretically achieved when operating a Fischer-Tropsch process at a chain growth probability of 0.95 [89, 90].

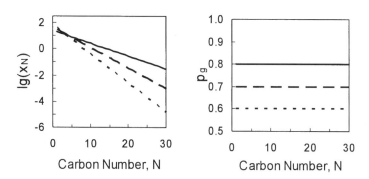

Figure 11 'Ideal product distributions': Left: logarithmic molar product content versus carbon number (ASF); Right: corresponding chain growth probabilities

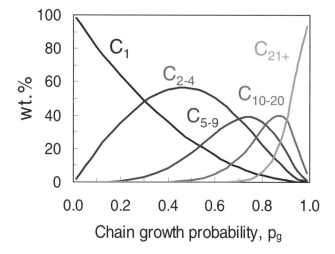

Figure 12 Product composition (wt.%) as function of chain growth probability assuming ideal ASF kinetics

The desired chain growth probability can – within limits – be selected by varying process parameters such as reaction temperature, H_2/CO ratio and catalyst composition. Low temperatures result in a FT product with a higher average carbon number (high chain growth probability), as is thermodynamically expected in an exothermic reaction. Longer chain products are generally found at low temperatures in polymerisation reactions, which can mechanistically be interpreted as an inhibition of the desorption step relative to the chain growth step. In LTFT processes (220-240°C) chain growth probabilities as high as 0.92 to 0.95 have been reported with cobalt or iron based catalysts respectively [91]. Whereas in HTFT synthesis chain growth probability is typically around 0.65 to 0.70, which is sufficiently low in order to minimise formation of excessive amounts of liquid products at reaction conditions a crucial requirement for operation of fluidized bed reactors [92].

The H_2/CO ratio is often not a process parameter as it is fixed and cannot be varied over a wide range. Generally lowering H_2/CO reactor inlet ratios leads to a FT product with a higher average carbon number.

The composition of a catalyst used in a process also has an effect on chain growth. Among all promoters the effect of potassium promotion of iron-based catalysts has to be highlighted. Potassium is considered a chemical promoter (see Chapter 3), it is believed to have an electronic effect on the catalyst by enhancing the electron-donor effect of iron catalysts, therefore facilitating CO adsorption and the dissociation of the C-O bond, while lowering the strength of the metal-hydrogen and the metal-oxygen bond [93]. As a consequence all carbon consuming reactions including chain growth are enhanced. All commercial iron catalysts contain small quantities of potassium (up to ca. 3 wt.%).

In practice ideal ASF distributions – in particular in LTFT processes - are not obtained as deviations from ideal behaviour are found; also different groups of products are present in the total product. The ideal chain growth model with formation of only one sort of product therefore needs to be extended.

3.3   Ideal chain growth: Desorption as olefin and paraffin

The main products of Fischer-Tropsch synthesis are linear alpha-olefins and linear paraffins. They are believed to be formed by desorption of a terminally bonded surface alkyl species either via dissociative β-hydrogen abstraction to form the olefin or via associative hydrogen addition to form the paraffin [78, 93] (see Fig. 13 left). Fixation of the alkyl chain at the chain end appears to be a feature of the FT regime caused by spatial constraints at FT sites [94]. The reaction of desorption as an olefin is faster than desorption as a paraffin. Primary molar olefin contents in corresponding hydrocarbon fractions (as not affected by secondary olefin reactions, see below) are believed to be around 70-90 mol% (see Fig. 13, right). Such high and carbon number independent olefin contents have for example been observed by Schulz and Gökcebay [59] when testing highly manganese-promoted iron catalysts. Only a small fraction (<5 mol%) of these olefins had internal double bonds. Normally olefin contents decrease with increasing carbon number (see Section on secondary olefin reactions, 3.3.1). Exclusive primary formation of olefins only as predicted by e.g. the vinyl mechanism appears unlikely (see Section 1 on mechanisms).

Desorption of alkyl surface species:

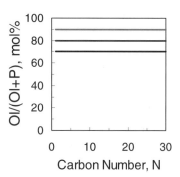

Figure 13 Primary formation of olefins

To account for the formation of olefins and paraffins the kinetic scheme can therefore be extended as follows:

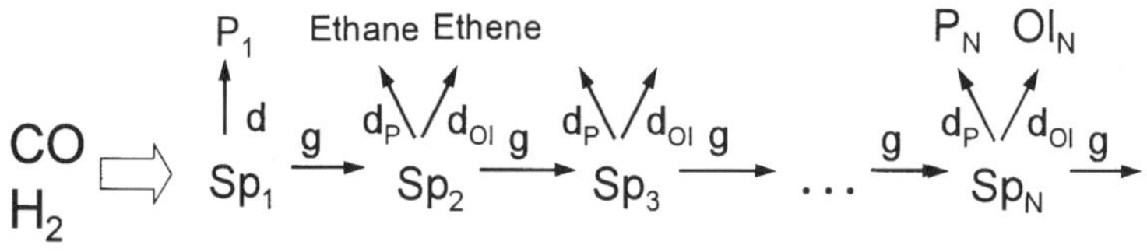

$$\text{CO} \atop \text{H}_2 \Longrightarrow \begin{array}{c} P_1 \quad \text{Ethane Ethene} \qquad\qquad P_N \ OI_N \\ \uparrow d \quad\quad d_P \diagdown d_{OI} \quad d_P \diagdown d_{OI} \qquad\qquad d_P \diagdown d_{OI} \\ Sp_1 \xrightarrow{g} Sp_2 \xrightarrow{g} Sp_3 \xrightarrow{g} \cdots \xrightarrow{g} Sp_N \xrightarrow{g} \end{array}$$

Figure 14 Ideal chain growth: Desorption as olefin and paraffin

In analogy to Eq. 10.9 in Fig. 10 the molar content of e.g. an olefin with carbon number n can be calculated as being:

$$x_{OI,N} = p_{d,OI,N} \cdot \prod_{N=1}^{N-1} p_{g,N} \tag{21}$$

or $x_{OI,N} = p_{d,OI} \cdot p_g^{N-1}$ for carbon number independent chain growth probability (ideal chain growth). The desorption probability $p_{d,ol,N}$ is defined as:

$$p_{d,Ol,N} = \frac{k_{d,Ol,N}}{k_{d,Ol,N} + k_{d,P,N} + k_{g,N}} \tag{22}$$

with

$$p_{d,Ol,N} + p_{d,P,N} + p_{g,N} = 1 \tag{23}$$

3.3.1 Deviations from ideal distributions: Secondary reactions of olefins

Real product distributions – in particular those obtained in low temperature FTS – often show deviations from ideal distributions with a curvature in the ASF diagram, which is equivalent to carbon number dependent growth probabilities and strongly carbon number dependent olefin contents. A typical product distribution, calculated from data from the gas chromatograms shown in Fig. 8 is given in Fig. 15.

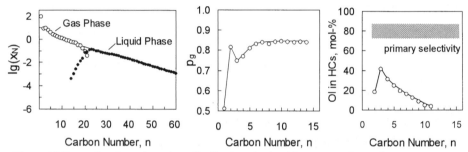

Figure 15 Example of a real product distribution obtained in FT-experiment in slurry reactor [84]; Left: ASF distribution, middle: Chain growth probability, right: olefin content in linear hydrocarbon fractions (Reaction conditions: catalyst: 100 Co / 15 ZrO$_2$ / 100 SiO$_2$ / 0.66 Ru (mass ratios), T=190°C, p$_{H2}$=5.1 bar, p$_{CO}$=2.2 bar, p$_{H2O}$=1.1 bar)

A method to determine chain length dependent growth probability using the model described above is shown in Fig. 16.

Commonly observed deviations from ideal distributions are:

- a relatively high molar methane content

- an anomaly in C$_2$ with relatively low molar contents in the ASF diagram and low olefin contents in the C$_2$ fraction

- a curvature of the ASF distribution at low carbon numbers, chain length dependent p$_g$ reaching asymptotic values

- chain length dependent olefin contents, decreasing with increasing carbon number

Possible reasons for these observations are discussed in the following text.

The calculations are started at (high) carbon numbers, where no changes of the slope in the ASF-plot (asymptotic end value of the chain growth probability (see eq.1)) and is proceeded towards lower carbon numbers

$$\lim_{N\to\infty} \frac{x_{N+1}}{x_N} = \lim_{N\to\infty} \frac{\dot{n}_{N+1}}{\dot{n}_N} = \lim_{N\to\infty} \frac{p_{g,N} \cdot p_{d,N+1}}{p_{d,N}} = p_{g,N} \qquad (1)$$

($\dot{n}_N$ = experimentally obeserved molar formation rate of compounds with carbon number N)

Reaction probabilities can then be calculated using:

$$p_{g,N-1} = \frac{1}{1 + p_{d,N} \cdot \left(\dfrac{\dot{n}_{N-1}}{\dot{n}_N}\right)} \qquad \text{and} \qquad p_{g,N} + p_{d,N} = 1 \qquad (2)$$

Figure 16   Algorithm to calculate carbon number dependent chain growth probabilities from experimentally observed FT product

It has been noticed in the very early days of Fischer-Tropsch research [e.g. 39, 96-99] that the desorption of olefins is reversible as they can undergo secondary reactions such as:

- hydrogenation to the corresponding paraffin
- incorporation into growing chains (chain start)
- formation of olefins with internal double bonds via double bond shift

and to a minor extent:

- hydroformylation to form alcohols and aldehydes
- hydrogenolysis

There is a wealth of literature dealing with secondary olefin reactions, which has been reviewed by e.g. Iglesia et al. [100] and Schulz and Claeys [101]. Being of consecutive nature these reactions have been evidenced by studies with variation of reactor residence time [59, 78, 101, 102] or olefin co-feeding [39, 96-113]. At short residence times secondary reactions are forced back and almost primary product distributions with carbon number independent molar olefin contents as shown in Fig. 13 (right) have been obtained [59], whereas at longer residence times generally lower and carbon number dependent olefin contents have been found. Co-feeding studies revealed that olefins can also act as chain starters (for example see Fig. 17) and double bond shift of the primarily formed α-olefin can occur.

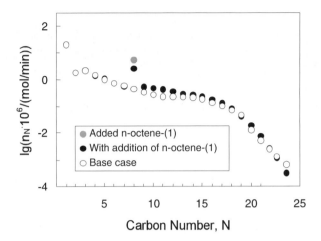

Figure 17 Effect of addition of n-octene-(1) on molar product formation rates during FT synthesis in a slurry reactor.
(Reaction conditions: catalyst: 100 Co / 15 ZrO$_2$ / 100 SiO$_2$ / 0.66 Ru (mass ratios), T=190°C, p$_{H2}$=5.1 bar, p$_{CO}$=2.2 bar, p$_{H2O}$=1.1 bar)

Hydroformylation and hydrogenolysis of olefins normally only play a negligible role, although it may be mentioned that the discovery of this side reaction of FT synthesis led to the development of the so-called 'oxo synthesis' by Roelen [114]. Hydrogenolysis is completely inhibited by carbon monoxide under typical FT reaction conditions [94, 115]. The three major secondary reactions of α-olefins are shown is Fig. 18.

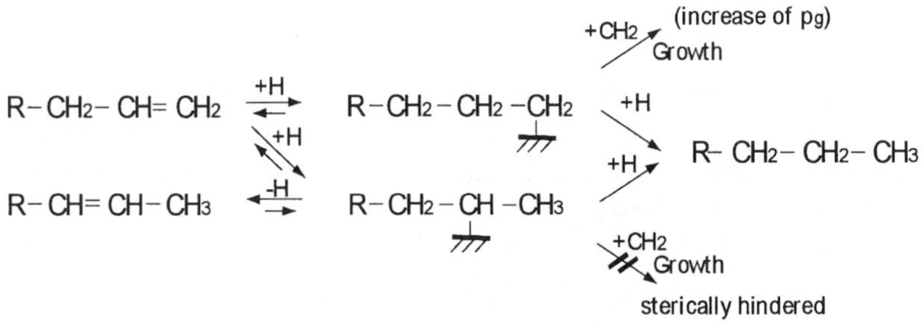

Figure 18 Kinetic scheme of main secondary reactions in FT synthesis [7, 94, 101]

The step of olefin re-incorporation effectively increases the probability of chain growth at a given carbon number and therefore affects the total product distribution. As α-olefins are the main primary organic products of FT synthesis, it is obvious that the extent and the selectivity of their secondary conversion will determine the total product distribution and possibly explain the deviations often observed in real products as in Fig. 15. It is generally accepted that secondary olefin reactions depend on chain length, the possible underlying processes causing this are:

- carbon number dependent diffusivity in liquid products [100, 102, 117, 118]

- carbon number dependent solubility in liquid products [58, 73, 86, 119-123]

- carbon number dependent physisorption strength of products on catalyst surface [73, 116, 119, 124]

These effects have been incorporated into selectivity models describing FT product distributions including the formation of olefins and paraffins.

3.3.1.1 Diffusion enhanced olefin readsorption model

Iglesia et al. [100, 102, 117, 118] realised the possible impact secondary olefin reactions might have on chain growth from studies on bed residence time effects. An example of such effects as obtained in studies using a ruthenium catalyst in a fixed bed reactor is shown in Fig. 19. Clearly deviations from ideal ASF distributions were found in the experiments. These were more pronounced at higher residence times, where readsorption events of primarily formed olefins are more likely. Correspondingly the olefin contents (expressed as olefin to paraffin ratios) show a strong chain length dependence and are much lower at a contact time of 12 s indicating secondary olefin consumption.

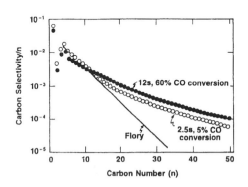 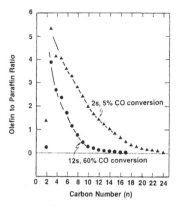

Figure 19 Effect of contact time / conversion on total molar product distribution and α-olefin to paraffin ratio [100]. (Reaction conditions: 1.2% Ru/TiO$_2$, T=203°C, (H$_2$/CO)$_{in}$=2.1, p$_{tot}$=6 bar, 5-60% conversion, Fixed bed reactor)

Iglesia et al. [100, 117] developed a model, in which they assumed diffusional limitations within the liquid-filled pores of a catalyst to slow down removal of long chain α-olefins, therefore increasing their residence time and enhancing their secondary conversion to form heavier and more paraffinic products. This scenario, which is depicted in Fig. 20, would ultimately account for the observed chain length dependencies in FT product.

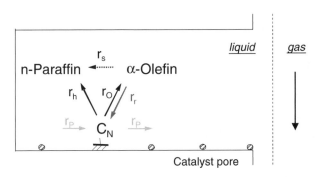

Figure 20 Diffusion enhanced olefin readsorption in liquid-filled catalyst pore, after [100, 117]

The kinetic scheme includes a step of direct hydrogenation of the olefin to the corresponding paraffin (r_s) as per Fig. 18. According to the authors this step is however negligible at typical FT reaction conditions, namely at conversion levels larger than 5% and reaction pressures of above 5 bar. This would ensure sufficiently high concentrations of CO and or H_2O, which would inhibit the direct hydrogenation pathway completely as demonstrated experimentally by them [100, 117]. Furthermore the authors imply in their model that the primary relative desorption rates of paraffins and olefins ('primary olefin selectivity') do not depend on local concentrations of H_2, CO, and H_2O). The model couples the chain growth kinetic scheme with diffusion and convection equations that describe transport processes in liquid filled spherical catalyst pellets and gas-phase interstices in the catalyst bed of a fixed bed reactor. The main mass conservation and transport equations are given in Fig. 21 using the nomenclature of the authors. The authors use virtual pressures (fugacities) of reactive species in the liquid phase as the kinetic driving force for the catalytic reaction. For details of the model the reader is referred to the original publications [100, 117].

All rate constants used in the model are assumed to be independent of chain length, except that of methane desorption and that of readsorption of ethene, which was assumed to be ten times more reactive than all other olefins

$(k_{r,2}=10\cdot k_{r,N})$. The carbon number dependent diffusivity of olefins was expressed as:

$$D_N = D_{N,0} \cdot e^{-0.3N} \qquad \text{(with } D_{N,0} \text{ reference liquid diffusivity)} \qquad (24)$$

Steady-state mass conservation of chain growth sites on catalytic surface:

$$-k_P C_1^* C_N^* + k_P C_1^* C_{N-1}^* - k_O C_N^* - k_h C_N^* + k_r P_N^v / RT = 0 \qquad (1)$$

(with k rate constant, C_N^* surface concentration of growing chain, P_N^v virtual pressure (fugacity) of olefin in liquid phase)

Steady state mass conservation equations for α-olefins and paraffins within particle voids:

$$D_N \nabla P_N^v - \rho_P[(k_r + k_s)P_N^v - k_O C_N^*] = 0 \qquad \text{(olefins)} \qquad (2)$$

$$D_N \nabla Q_N^v + \rho_P[k_h C_N^* + k_s P_N^v] = 0 \qquad \text{(paraffins)} \qquad (3)$$

(with D_N Hydrocarbon diffusion coefficient (assumed to be identical for olefins and paraffin of a given size, ρ_P catalyst pellet density, Q_N^* virtual pressure of paraffin in liquid phase)

Balance of mass fluxes at the pellet surface:

$$U\frac{dP_N^v}{dz} = a_v \frac{D_{ng}}{\delta}(P_N^v - P_{ng}) \qquad (4)$$

$$U\frac{dQ_N^v}{dz} = a_v \frac{D_{ng}}{\delta}(Q_N^v - Q_{ng}) \qquad (5)$$

(with U superficial gas velocity, z axial reactor position, a_v bed interfacial area, D_{ng} hydrocarbon diffusivity in reactor gas phase, P_{ng} olefin partial pressure in reactor gas phase, Q_{ng} paraffin partial pressure in reactor gas phase)

Figure 21 Mass conservation and transport equations used in "Diffusion enhanced olefin readsoprtion model [100, 117]

Dimensionless analysis of the coupled kinetic-transport equations show that a carbon number dependent Thiele modulus (Φ) can describe the diffusion effects on reactive processes:

636

$$\Phi_N^2 = \left(\frac{k_{r,N}}{D_N}\right) \cdot \left(\frac{L^2 \cdot \epsilon \cdot \theta_{Me}}{R_p}\right) = \Phi_0^2 \cdot e^{0.3N} \tag{25}$$

(with L: pellet diameter, ϵ: pellet porosity, θ_{Me}: metal site density and R_P: pore radius). Further input parameters are probabilities β_O, β_H, β_R and β_s, which characterise the ratio of the rate of desorption as an olefin and a paraffin, olefin readsorption and direct hydrogenation, respectively, to the rate of the corresponding chain growth step (e.g. $\beta_O = r_O/r_P$)[1]. Carbon number distributions of FT synthesis products can then be accurately described. An example of a model fit with experimental data obtained during FT synthesis using a cobalt catalyst in a fixed bed reactor is given in Fig. 22.

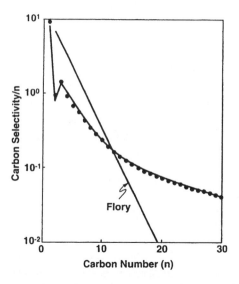

Figure 22 Comparison of model and experimental data. Carbon number distribution [100]. (Reaction conditions: 11.7% Co/TiO₂, T=200°C, p_{total}=20 bar, X_{CO}=9.5%, Fixed bed reactor); Model (solid line): ϕ_O^2=8, β_O=0.4, β_h=0.058, β_r=1.2, β_s=0

All deviations from ideal distributions including high methane contents, the anomaly in C_2, the curvature in the ASF plot and the carbon number dependent olefin contents are well described and a good fit with experimental data has been obtained. The effect of convection is indicated in Fig. 23, where

---

[1] Note that β is not strictly a probability and that its definition is not equal to the definition of reaction probabilities given above. They can be converted using e.g.: $\beta_O = p_{d,Ol}/p_g$

simulating an increase in bed residence time or respectively CO conversion from 5 to 60% results in a drastic decrease of olefin to paraffin ratios due to enhanced secondary conversion of primarily formed olefins. Note, that in this model the step of direct hydrogenation is normally neglected (β_s=0) and the deviation from ideal distributions with purely paraffinic heavy products is exclusively due to olefin re-insertion.

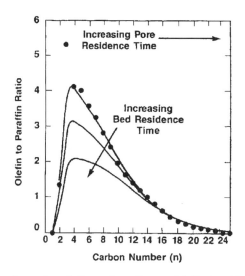

Figure 23 Transport-enhanced α-olefin readsorption model. Bed residence time effects on α-olefin to paraffin ratios [117].
(Reaction conditions: 1.2% Ru/TiO$_2$, T=203°C, (H$_2$/CO)$_{in}$=2.1, p$_{tot}$=6 bar, 5-60% conversion, Fixed bed reactor); Model simulations (solid lines)

The model also accounts for effects of the catalyst structure on selectivity of liquid products (C$_{5+}$) and CH$_4$ selectivity, as the second term of equation (25), which was named the structural parameter χ, exclusively contains catalysts properties:

$$\chi = \frac{L^2 \cdot \varepsilon \cdot \theta_{Me}}{R_p} \qquad (26)$$

The effect of this structural parameter on C$_{5+}$ selectivity and methane selectivity is shown in Fig. 24 for a series of supported cobalt catalysts tested at similar CO conversion levels (50-60%) in a fixed bed reactor. An increase of χ can for example be realised by an increase of a catalyst pellet diameter (pore length L). Olefin residence times and therefore their probability of secondary reinsertion

638

are much higher the longer the pores and therefore the longer their diffusion paths are. This results in an increase of C_{5+} selectivity to asymptotic values of around 90 wt.% and a corresponding decline of methane selectivity with increasing χ.

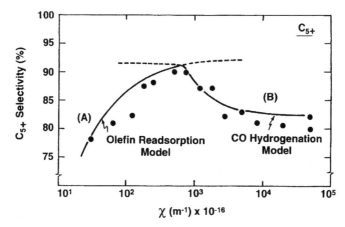

Figure 24 Effect of structural parameter on C_{5+} selectivity [100]
(Co catalysts: SiO_2, TiO_2 and Al_2O_3 supports, T=200°C,(H$_2$/CO)$_{in}$=2.1, p$_{tot}$=20 bar, 50-60% conversion, Fixed bed reactor); Model simulations (solid lines)

At very large values of χ experimental values of C_{5+} selectivity however start to decline again. The authors assume that here transport restrictions begin to also hinder the rate of arrival of the reactants CO and H_2 to the active sites. Due to the relatively lower diffusivity of CO, its concentration in a liquid filled catalyst pore declines faster than that of hydrogen, therefore leading to increasing H_2/CO ratios in the centre of a catalyst pellet [100, 115, 117, 125, 126]. By assuming Langmuir-Hinshelwood type rate equations for methane formation with exponents different to that of total CO disappearance (larger reaction order with respect to hydrogen) this would result in enhanced formation of methane and a corresponding decrease of C_{5+} selectivity. This 'primary CO hydrogenation reaction' is described in a similar kinetic-transport model by Iglesia et al. (for details see e.g. [100]). Results of the 'CO hydrogenation model' are also shown in Fig. 24. The increase in methane selectivity as a function of the structural parameter coincides with a decline of the catalyst effectiveness factor [100, 115]. Decreased FT reaction rates due to limitations of reactant transport have, for example, been experimentally obtained when testing iron [84, 127-129] and cobalt [126, 127] based catalysts with particle diameters larger than 300 to 500 μm at low temperature Fischer-Tropsch conditions (below 250°C). In the extreme case of complete depletion of carbon monoxide in the centre of e.g. a very large catalyst pellets, hydrogenolytic decomposition of hydrocarbons may occur and lead to increased methane selectivity with correspondingly low C_{5+} selectivity [115].

The model(s) of Iglesia et al. provide the researcher in the field of FT synthesis with a catalyst design parameter, which pre-determines a catalysts performance with respect to product selectivity and can therefore be of high importance when for example aiming at high selectivity of long chain products as required in GTL processes. Although only applied on ruthenium and cobalt based catalysts the authors note that their model is in qualitative agreement with product distributions obtained in commercial processes with iron catalysts. Here no curvature of the ASF plot (carbon number range C_1 to C_{14}) was obtained on the small catalyst particles in the commercial high temperature Synthol process [130] and a strong deviation is found with the large catalyst pellets used in the commercial low temperature fixed bed process [131], suggesting the absence of diffusion enhanced secondary olefin re-insertion in the pores of the small particles of the high temperature process – at least in the carbon number range reported. With respect to the extent of carbon number effects on olefin content and the curvature of the ASF plots the authors give the following order reflecting the relative rates of olefin readsorption:

$$Co > Ru > Fe$$

It has however been noted by Kuipers et al. [116] that the chain length dependency of the olefin diffusivity, the only chain length dependent parameter used in the model, is not in agreement with data available in literature, which predicts a much weaker carbon number dependency proportional to $N^{-0.6}$. Furthermore experimental data including co-feeding of olefins of different chain length [101] indicate that diffusional effects cannot exclusively explain chain length dependencies obtained in Fischer-Tropsch products (see below).

3.3.1.2 Effect of solubility on olefin readsorption
Background

It was earlier noted by several authors [58, 79, 86, 132, 133] that the chain length dependency of olefin contact time in the pore system of a catalyst pellet and the reactor system may be due to chain length dependent solubility of olefins in the liquid reaction product (wax). This solubility increases exponentially with carbon number, with a much stronger carbon number dependency compared to that of diffusivity [116], effects of solubility would therefore normally dominate effects diffusivity on secondary olefin reactions. Indeed, when testing iron catalysts in a laboratory scale slurry reactor Claeys and Schulz [129] found virtually the same carbon number dependent chain growth probabilities and olefin contents (see Fig. 25) on two hugely different catalyst pellet sizes, namely a powder fraction (particle size <0.1 mm) and a fraction of large cylindrical particles (1.7 x 1.7 mm).

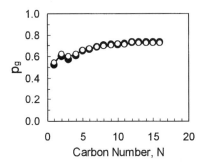 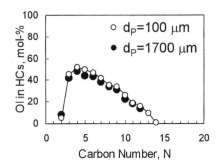

Figure 25   Chain growth probability and olefin content in linear hydrocarbon fractions obtained in FT experiments in slurry reactor using different catalyst particle sizes d_P [84,129]. (Reaction conditions: Cat: 100 Fe / 13 Al_2O_3 / 10 Cu, T=250°C, p_{total}=10 bar, $(H_2/CO)_{in}$=1.8, SV=60 ml/g/min, $X_{CO}(d_P$=100 μm)=31%, $X_{CO}(d_P$=1700 μm)=29%)

It was shown that these studies were not affected by external mass transfer limitations and no internal mass transfer limitations of the reactants were present as the same conversions were obtained at a fixed syngas flow rate with both catalysts fractions. The catalyst fractions used correspond to an increase of the structural parameter χ in the above described diffusion enhanced olefin readsorption model of more than 2 orders of magnitude, which – according to the model - should result in an increase in the formation of heavy product and a decrease of olefin contents. As this was however not observed, the chain length dependencies in Fig. 25 are unlikely due to diffusional effects.

In order to clarify these findings and to obtain a better insight into the kinetics of secondary olefin reactions and their carbon number dependency Schulz and Claeys [84, 101] conducted extensive studies, in which α-olefins of different chain lengths (C_2-C_{11}) were added during FT synthesis in a gradientless stirred slurry reactor at elevated pressures. Effects of reaction conditions such as CO partial pressure as well as reaction temperature on olefin reactions were investigated on powder fractions (<0.1 mm) of several cobalt and iron based catalysts. Hydrogenation to the corresponding paraffin, double bond shift and incorporation were the main reactions obtained in these experiments. The steady state conversion of the co-fed olefins showed increasing conversion with increasing carbon number (except C_2, see Fig. 26 left) and increasing yields of e.g. hydrogenation to the corresponding paraffin (see Fig. 26 right) and re-incorporation directly reflecting carbon number dependencies obtained in the distributions of the corresponding base case experiments without olefin co-feeding. As these olefins were externally added during the synthesis a carbon number dependent increase of their conversion is not to be expected if any transport limitations including within the pores of the catalysts were present. These findings are again in clear contradiction to the diffusion enhanced readsorption model.

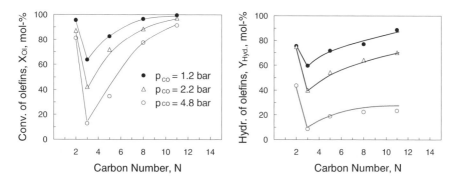

Figure 26: Conversion and hydrogenation yield of a-olefins co-fed during FT synthesis in slurry reactor at varied CO partial pressure [101]. (Reaction conditions: 100 Co / 15 ZrO_2 / 100 SiO_2 / 0.66 Ru, T=190°C, p_{H2}=5.1 bar, p_{H2O}=1.0)

Other major observations from the above olefin co-feeding studies are listed below:

- Reactions of added olefins (hydrogenation, isomerisation and incorporation) reflect selectivity of base case experiments and therefore reactions of primarily formed α-olefins
- Reaction conditions have a strong effect on the extent and selectivity of secondary olefin reactions (e.g. inhibition by CO (reaction order -2) and alkali in iron based FTS)
- Hydrogenation and isomerisation must not be neglected as secondary reactions
- First order dependency of olefin consumption with respect to olefin concentration
- Reactivity (rate constant) of olefins is not chain length dependent (exception ethene: 10-25 times more reactive)
- Selectivity of olefin incorporation is not chain length dependent
- Olefins with internal double bonds and paraffins do not participate in secondary chain growth

- Diffusional restrictions cannot account for observed chain length dependencies

Based on these important findings Schulz and Claeys [84, 121] extended the kinetic model described in Fig. 14 accounting for secondary olefin reactions, their chain length dependency being introduced by the chain length dependent solubility causing chain length dependent product residence time τ_n in the reactor, which – in a gradientless slurry reactor - can be expressed as:

$$\tau_N = \frac{n_N}{\dot{n}_N} = \frac{n_{g,N} + n_{l,N}}{\dot{n}_{g,N} + \dot{n}_{l,N}} \tag{27}$$

with n_N total moles of component with carbon number N in reactor and $\dot{n}_N$ molar flow of the same component leaving the reactor via the gas (g) and the liquid (l) phase. Assuming these phases are in thermodynamic equilibrium equation (x) can be rearranged introducing the partition ratio K_N, $(=c_{l,N}/c_{g,N})$:

$$\tau_N = \frac{c_{g,N} \cdot V_g + c_{l,N} \cdot V_l}{c_{g,N} \cdot \dot{V}_g + c_{l,N} \cdot \dot{V}_l} = \frac{V_g + K_N \cdot V_l}{\dot{V}_g + K_N \cdot \dot{V}_l} \tag{28}$$

with c: concentrations in either gas or liquid phase, V: volumes of total reactor gas (g) or liquid (l) phase respectively and $\dot{V}$: total volumetric flow rates of gas or liquid phase respectively (see Fig. 27 right).

Using Henry's law the partition ratio, which is a measure for product solubility, can be expressed as [101]:

$$K_N = \frac{\rho_{wax}(T) \cdot RT}{\gamma_N(T) \cdot p_N^{sat}(T)} \tag{29}$$

with ρ_{wax}: density of wax at temperature T, γ: activity coefficient and p_N^{sat}: saturation pressure of a component. Using literature data this equation predicts an exponential increase of the solubility of hydrocarbons in FT wax ($K_N = K_2 \cdot b^{(N-2)}$, e.g. at 190°C: $K_2 = 1.91$ and $b = 1.53$, at 250°C: $K_2 = 1.09$ and $b = 1.46$).

Fig. 27 left shows the mean reactor residence time of a hydrocarbon as per Eq. (28) in the experimental set-up used by Schulz and Claeys for a typical set of reaction conditions. As the convective removal of products depends on removal of gas and liquid phase mean reactor residence times of product compounds increase drastically with increasing carbon number, possibly explaining the preferred consumption of long chain olefins and the therefore the deviations from ideal ASF kinetics in FT synthesis.

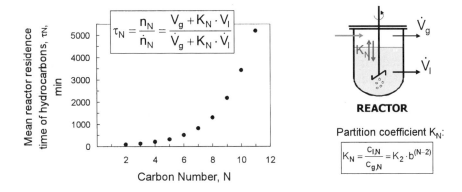

Figure 27 Calculated average residence times of product compounds (Eq. (28)) at FT
synthesis in a gradientless slurry reactor. (Parameters: V_g=210 ml, V_l=270 ml,
$\dot{V}_g$=5.38 ml/min, $\dot{V}_l$=6.28·10^{-4} ml/min, T=190°C)

Extended Model

The extended model by Schulz and Claeys [121], taking the above
observations into account, is shown in Fig. 28.

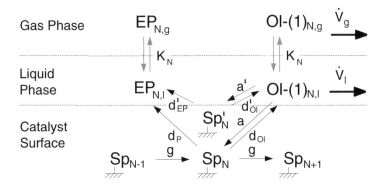

Figure 28 Kinetic model of secondary olefin reactions taking into account solubilities of
olefins in liquid Fischer-Tropsch product [121]

It assumes thermodynamic equilibrium of compounds in gas in liquid
phase as expressed by the partition ratio K_N, both phases continuously leaving
the reactor. The following assumptions were made in the model:

- Ideal chain growth, i.e. carbon number independent

- Desorption of surface species as α-olefin Ol-(1) or non-reactive end product EP (i.e. paraffin or olefin with internal double bond); primarily only α-olefins and paraffins are formed
- Readsorption of α-olefins as species Sp or Sp'; Sp' does not participate in chain growth
- All rate constants k are assumed to be carbon number independent (except $k_{a,2}=10 \cdot k_a$ to $25 \cdot k_a$, accounting for high reactivity of ethane as experimentally obtained [101])
- Thermodynamic equilibrium of gas and liquid phase (carbon number dependent partition ratio K_N)
- No effects of diffusional limitations, or temperature or concentration gradients (CSTR)
- Steady state conversion, convective product removal via gas and liquid phase

---

- **Balance around Sp_N:** (1)

$$r_{g,Sp,N-1} \cdot m_{Me} = (r_{g,Sp,N} + r_{d,P,N} + r_{d,Ol-(1),N} - r_{a,Ol-(1),N}) \cdot m_{Me}$$

$$k_g \cdot c_{s,Sp,N-1} \cdot m_{Me} = (k_g \cdot c_{s,Sp,N} + k_{d,P} \cdot c_{s,Sp,N} + k_{d,Ol-(1)} \cdot c_{s,Sp,N-1} - k_a \cdot c_{l,Ol-(1),N}) \cdot m_{Me}$$

- **Balance around $Ol-(1)_N$:** (2)

$$(r_{d,Ol-(1),N} + r'_{d,Ol-(1),N}) \cdot m_{Me} = (r_{a,Ol-(1),N} + r'_{a,Ol-(1),N}) \cdot m_{Me} + \dot{n}_{l,Ol-(1),N,out} + \dot{n}_{g,Ol-(1),N,out}$$

$$(k_{d,Ol-(1)} \cdot c_{s,Sp,N} + k'_{d,Ol-(1)} \cdot c'_{s,Sp,N}) \cdot m_{Me} =$$

$$(k_a + k'_a) \cdot c_{l,Ol-(1),N} \cdot m_{Me} + \dot{V}_l \cdot c_{l,Ol-(1),N} + \frac{\dot{V}_g}{K_N} \cdot c_{l,Ol-(1),N}$$

- **Balance around EP_N:** (3)

$$(r_{d,P,N} + r'_{d,EP,N}) \cdot m_{Me} = \dot{n}_{l,EP,N,out} + \dot{n}_{g,EP,N,out}$$

$$(k_{d,P} \cdot c_{s,Sp,N} + k'_{d,EP} \cdot c'_{s,Sp,N}) \cdot m_{Me} = \dot{V}_l \cdot c_{l,EP,N} + \frac{\dot{V}_g}{K_N} \cdot c_{l,EP,N}$$

(remark: the solubilities of EP_N are assumed to be equal to the solubilities of Ol-(1)$_N$)

- **Balance around Sp'_N:** (4)

$$r'_{a,Ol-(1),N} \cdot m_{Me} = (r'_{d,EP,N} + r'_{d,Ol-(1),N}) \cdot m_{Me}$$

$$k'_a \cdot c'_{l,Ol-(1),N} \cdot m_{Me} = (k'_{d,EP} \cdot c'_{s,Sp,N} + k'_{d,Ol-(1)} \cdot c'_{s,Sp,N}) \cdot m_{Me}$$

Figure 29 Steady state mass balances around species in Fig. 28.

- Effective chain growth probability:

$$p_{g,eff,N} = \cfrac{1}{1 + \cfrac{k_d}{k_g}\left(\cfrac{1}{\left(1 + \cfrac{k_{d,Ol-(1)}}{k_{d,P}}\right)} + \cfrac{1}{\left(1 + \left(\cfrac{k_{d,Ol-(1)}}{k_{d,P}}\right)^{-1}\right)} \cdot A\right)} \tag{1}$$

with:

$$A = 1 - \cfrac{k_a\,\dfrac{m_{Me}}{\dot{V}_g}}{k_a\,\dfrac{m_{Me}}{\dot{V}_g} + k'_a\,\dfrac{m_{Me}}{\dot{V}_g}\left(1 - \left(\dfrac{k'_{d,Ol-(1)}}{k_{d,Ol-(1)}} + \dfrac{k'_{d,EP}}{k_{d,P}} \cdot \left(\dfrac{k_{d,Ol-(1)}}{k_{d,P}}\right)^{-1}\right)^{-1}\right) + \dfrac{\dot{V}_1}{\dot{V}_g} + \dfrac{1}{K_N}}$$

and: $\qquad p_{d,eff,N} = 1 - p_{g,eff,N}$ (2)

The total carbon number distribution can again be derived by multiplication of the single (effective) reaction probabilities, that lead to the formation of a regarded product (see figure 10 eq.(10.9)).

- Molar α-olefin content in linear hydrocarbons:

$$\cfrac{c_{1,Ol-(1),N}}{c_{1,Ol-(1),N} + c_{1,EP,N}} = \cfrac{1}{1 + \cfrac{c_{1,EP,N}}{c_{1,Ol-(1),N}}} \tag{3}$$

with: $\dfrac{c_{1,EP,N}}{c_{1,Ol-(1),N}} =$

$$\cfrac{\left(\dfrac{k_{d,Ol-(1)}}{k_{d,P}}\right)^{-1}\left(k_a \cdot \dfrac{m_{Me}}{\dot{V}_g} + k'_a \cdot \dfrac{m_{Me}}{\dot{V}_g}\left(1 - \cfrac{1 - \left(\dfrac{k'_{d,Ol-(1)}}{k_{d,Ol-(1)}}\right)\left(\dfrac{k_{d,Ol-(1)}}{k_{d,P}}\right)\left(\dfrac{k'_{d,EP}}{k_{d,P}}\right)^{-1}}{1 + \left(\dfrac{k'_{d,Ol-(1)}}{k_{d,Ol-(1)}}\right)\left(\dfrac{k_{d,Ol-(1)}}{k_{d,P}}\right)\left(\dfrac{k'_{d,EP}}{k_{d,P}}\right)^{-1}}\right) + \dfrac{\dot{V}_1}{\dot{V}_g} + \dfrac{1}{K_N}\right)}{\dfrac{\dot{V}_1}{\dot{V}_g} + \dfrac{1}{K_N}}$$

Figure 30 Equations to calculate effective chain growth probabilities and α-olefin contents in linear hydrocarbon fractions. Olefin readsorption model accounting for chain length dependent solubility effects [121]

A combination of steady state mass conservation equations around the surface species Sp and Sp', as well as alpha olefins Ol-(1) and end products EP (see Fig. 29) was used to define a set of intensive dimensionless parameters which allow calculation of effective (i.e. accounting for secondary olefin reactions) chain growth and desorption probabilities. These effective probabilities – in analogy to the description for carbon number dependent chain growth in Fig. 10 (Eq. 10.9) – can be used to calculate product distributions by multiplication of probabilities of individual reaction steps, as well as carbon number dependent olefin contents accounting for secondary olefin reactions (see Fig. 30). As in the basic model with no olefin readsorption, the formation rate of all products can be normalised and linked to any kinetic equation describing the rate of synthesis gas consumption (see Section 2). The intensive dimensionless parameters of the model and their meanings are:

$\dfrac{k_d}{k_g}$ fixes the 'primary chain growth probability'

$$(p_g = \frac{k_g}{k_g + k_d} = \frac{1}{1 + k_d/k_g})\ \text{without olefin readsorption}$$

$\dfrac{k_{d,Ol-(1)}}{k_{d,EP}}$ expresses the 'primary selectivity of α-olefins in hydrocarbon

fractions'$= \dfrac{k_{d,Ol-(1)}}{k_{d,Ol-(1)} + k_{d,EP}} = \dfrac{1}{1 + (k_{d,Ol-(1)}/k_{d,EP})^{-1}}$, which is believed to be carbon number independent and typically lies in the range 70-90 mol-% [59].

$k_a \cdot \dfrac{m_{Me}}{\dot{V}_g}$ determines the rate of olefin-(1) readsorption to form surface species Sp_N

$k'_a \cdot \dfrac{m_{Me}}{\dot{V}_g}$ determines the rate of olefin-(1) readsorption to form surface species Sp'_N, that do not participate in chain growth

$\dfrac{\dot{V}_l}{\dot{V}_g}$ ratio of volumetric flow rates (at reaction conditions) of liquid and gas phase exiting the reactor

All these parameters are carbon number independent, the carbon number dependency in the model is introduced by the carbon number dependent partition ratio K_N. To account for the higher reactivity of ethene as observed in olefin co-feeding experiments [101] the ratio $k_{a,2}/k_a$ lies between 10 and 25.

The effect of olefin readsorption on the product distribution, effective chain growth probability and α-olefin content as predicted by the model is shown in Fig. 31. The parameter $k_a \cdot m_{Me}/\dot{V}_g$ has been varied over a wide range; a value of zero is equivalent to no secondary olefin reactions occuring, i.e. primary selectivity (dashed lines). The 'primary chain growth probability' was chosen to be 0.7 (i.e. $k_d/k_g = 0.43$) and a 'primary alpha-olefin content' of 80 mol-% was assumed (i.e. $k_{d,Ol-(1)}/k_{d,EP} = 4$). In accordance with typical experimental values a ratio of $\dot{V}_l/\dot{V}_g = 0.0001$ was assumed; the values for the solubility (K_N) refers to a reaction temperature of 250°C (see above). The formation of species Sp', which do not participate in chain growth is not taken into account here ($k'_a \cdot m_{Me}/\dot{V}_g = 0$).

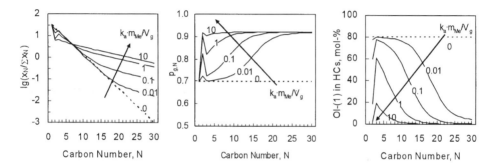

Figure 31 Calculations with the extended kinetic model taking into account carbon number dependent product solubilities; effect of olefin-(1) readsorption as surface species Sp_N. (Dashed lines: without readsorption, solid lines: with readsorption). (Model parameters: $k_d/k_g=0.43$, $k_{d,Ol-(1)}/k_{d,P}=4$, $k'_{d,Ol-(1)}/k_{d,Ol-(1)}=1$, $k'_{d,EP}/k_{d,P}=1$, $k_a \cdot \mathbf{m}_{Me}/\dot{V}_g$ =varied, $k'_a \cdot m_{Me}/\dot{V}_g$ =0, $k_{a,2}/k_a$=10, $\dot{V}_l/\dot{V}_g$ =0.0001). Left: ASF-plot (normalised for $\sum_{N=1}^{30} x_N$); middle: chain growth probability $p_{g,N}$; right: molar olefin-(1) content in hydrocarbon fractions

Readsorption of olefins as described by the model result in non-ideal distributions with typical carbon number dependent deviations with high methane contents, low values of molar C_2 concentrations (due to preferred incorporation of the highly reactive ethene; $k_{a,2}/k_a = 10$), a curvature in the ASF plots approaching constant values and the carbon number dependent deviations in the molar olefin contents. These deviations are more pronounced with increasing values of $k_a \dot{} m_{Me} / \dot{V}_g$, which for example can be interpreted as faster olefin readsorption rates but also as slower convective removal of the olefins via the reactor gas phase. The ratio $\dot{V}_l / \dot{V}_g$ is normally very low and can often be neglected. An increase of this value, which could for example be achieved by recirculation of a liquid, would have the same effect as lowering the parameter $k_a \dot{} m_{Me} / \dot{V}_g$. Lowering product solubility would also lead to a lower extent of secondary olefin reactions.

Olefin co-feeding experiments conducted at different reaction conditions [39, 101] showed that generally large amounts of olefins are directly hydrogenated to the paraffins of the corresponding carbon number. Although readsorption via a surface species Sp does account for some direct hydrogenation, the large yields of direct hydrogenation of α-olefins as found in most olefin co-feeding experiments in literature (see also Fig. 26 right) can only be explained by additional *direct* hydrogenation of these olefins. Furthermore isomerisation of primarily formed α-olefins via double bond shift can occur, which like the hydrogenation, leads to a product which has no further impact on the product distribution (as proven via co-feeding of the respective components, e.g. [101]). The readsorption of these end-products EP is therefore neglected in the model. The steps of direct hydrogenation and double bond shift isomerisation are included in the extended model by introducing the possibility of readsorption as species Sp'. This route, which is taken into account with the parameter $k'_a m_{Me} / \dot{V}_g$, does not imply or exclude whether these reactions take place at different kinds of catalytic sites. Fig. 32 demonstrates the effect of this route, showing an even further decrease of α-olefin contents in respective fractions of end products (including olefins with internal double bonds), while deviations of the product distributions in the ASF plots are much less pronounced.

Examples of experimental data obtained with a cobalt-based and a potassium promoted iron-based catalyst fitted with the model are given in Fig. 33. The experiments were conducted in a laboratory scale stirred slurry reactor, in which the reaction conditions were kept constant for over 60 days, so that product distributions are reliable even at high carbon numbers. The fitting with the model is done by fixing the initial slope of the ASF curve with the parameter k_g / k_d ('primary chain growth') and varying the parameters $k_a \dot{} m_{Me} / \dot{V}_g$ and $k'_a m_{Me} / \dot{V}_g$ until the best possible fit is obtained. In all cases a primary olefin content of 80-mol% ($k_{d,Ol-(1)} / k_{d,EP} = 4$) and a 10 fold higher reactivity of ethene compared to other α-olefins was assumed ($k_{a,2} / k_a = 10$).

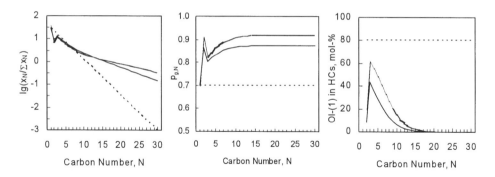

Figure 32 Calculations with the extended kinetic model taking into account carbon number dependent product solubilities; additional effect of olefin-(1) readsorption as surface species Sp'$_N$. (Dashed lines: without readsorption, solid black lines: with readsorption as species Sp$_N$, grey solid lines: with readsorption as species Sp$_N$ and Sp'$_N$).
(Model parameters: k_d/k_g=0.43, $k_{d,Ol-(1)}/k_{d,P}$=4, $k'_{d,Ol-(1)}/k_{d,Ol-(1)}$=1, $k'_{d,EP}/k_{d,P}$=1, $k_a \cdot m_{Me}/\dot{V}_g$=1, $k'_a \cdot m_{Me}/\dot{V}_g$=1, $k_{a,2}/k_a$=10, $k'_{a,2}/k'_a$=10, $\dot{V}_l/\dot{V}_g$=0.0001); Left: ASF-plot (normalised for $\sum_{N=1}^{30} x_N$); middle: chain growth probability $p_{g,N}$; right: molar olefin-(1) content in hydrocarbon fractions

Generally the model describes the experimental data and the carbon number dependencies very well. A second example of applying the model to fit experimental data, which deal with the effect of water partial pressure on FT product distributions with a cobalt catalyst is shown in Fig. 34. The model parameters give an indication of the extent and selectivity of secondary olefin reactions occurring on different catalysts or at different reaction conditions and allow for a direct comparison. Intensive model parameters used to fit the data are listed in Table 1.

A fairly high extent of secondary olefin reactions (larger values of readsorption parameters) is for example obtained at low water partial pressures with the cobalt catalyst, indicating water to strongly inhibit secondary olefin reactions. A similar effect is observed when comparing a potassium promoted catalyst with a non promoted precipitated iron catalyst.

650

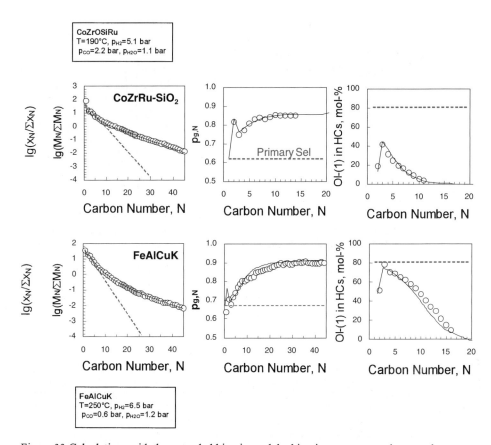

Figure 33 Calculations with the extended kinetic model taking into a⁀ ϽϽunt carbon number dependent product solubilities; fitting of experimental data with the catalysts: 100Co/15ZrO₂/100SiO₂/0.66Ru (top), 100Fe/13Al₂O₃/10Cu/5K (bottom). (Dashed lines: without readsorption, solid lines: with readsorption as species Sp_N and Sp'_N).

(Model parameters: see table 1). Left: ASF-plot (normalised for $\sum_{N=2}^{45} x_N$); middle: chain growth probability $p_{g,N}$; right: molar olefin-(1) content in hydrocarbon fractions

Besides suppressing secondary olefin hydrogenation, an important role of potassium promotion in iron-based FT synthesis is to compensate the acidity of the catalyst as for example introduced by structural promoters / supports such as Al_2O_3 and therefore forcing back double bond shift isomerisation which readily occurs on acidic sites. As a result fairly high contents of α-olefins can be obtained with potassium or generally alkali promoted iron catalysts.

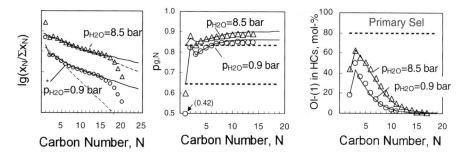

Figure 34 Calculations with the extended kinetic model taking into account carbon number dependent product solubilities; fitting of experimental data with the catalyst 100Co/15ZrO$_2$/100SiO$_2$/0.66Ru at two different water partial pressures (reaction conditions: T=190°C, p$_{H2}$≈5.8 bar, p$_{CO}$≈2.9 bar, τ≈27 min [121,137]). (Dashed lines: without readsorption, solid lines: with readsorption as species Sp$_N$ and Sp'$_N$).

(Model parameters: see table 1); Left: ASF-plot (normalised for $\sum_{N=2}^{15} M_N$, arbitrary scale); middle: chain growth probability p$_{g,N}$; right: molar olefin-(1) content in hydrocarbon fractions

Table 1

Model parameters used to fit experimental FT product distributions shown in figures 33 and 34 and corresponding selectivity of olefin incorporation. (k$_{a,2}$/k$_a$=10, k'$_{a,2}$/k'$_a$= k$_{a,2}$/k$_a$, k'$_{d,Ol-(1)}$/k$_{d,Ol-(1)}$=1 and k'$_{d,EP}$/k$_{d,P}$= k'$_{d,Ol-(1)}$/k$_{d,Ol-(1)}$; for reaction conditions with catalysts CoZrSiRu and FeAlCuK see figures 33 and 34, catalyst 100Fe/13Al2O3/10Cu: T=250°C, p$_{H2}$=5.7 bar, p$_{CO}$=2.8 bar, p$_{H2O}$=0.5 bar)

	Catalyst			
	FeAlCuK	FeAlCu	CoZrSiRu	
			p$_{H2O}$=0.9 bar	8.5 bar
k$_d$/k$_g$	0.49 (67%)[a]	0.90 (53%)	0.55 (65%)	0.20 (83%)
k$_{d,Ol}$/k$_{d,P}$	4 (80%)[b]	4 (80%)	4 (80%)	4 (80%)
k$_a$·m$_{Me}$/V$_g$	0.12	1.0	0.75	0.07
k$_a$'·m$_{Me}$/V$_g$	0.07	1.9	0.50	0.25
S$_{Olef. Incorp.Model}$	38%	29%	54%	21%
S$_{Olef. Incorp.Exp.}$	≅20%	≅15%	≅20%	n.a.

a. 'primary chain growth probability', b. 'primary olefin selectivity', c. from olefin co-feeding

As an overall effect of inhibitions of secondary reactions with the potassium promoted iron catalyst, the impact of reinsertion as implied in the model only comes into effect at higher carbon numbers compared to product distributions obtained on non-promoted iron catalysts (not shown) or typical cobalt catalysts (see Fig. 33). Furthermore the potassium promotion leads to increased primary chain growth probability (i.e. 67% as opposed to 53% with the non potassium promoted iron catalyst); increased water partial pressures in cobalt-based FT synthesis seems to have a similar effect.

When comparing the selectivity of olefin re-incorporation as predicted by the model with that obtained from experimental data from olefin co-feeding experiments it can be noted that the model seems to overpredict the effect of olefin incorporation. This may be due to the choice of a fixed value of the primary olefin selectivity, a parameter, which has been shown to have a pronounced effect on the fitting parameters. The true value of the primary olefin selectivity, which itself should be dependent on the catalysts used and the reaction conditions, is however not known. It was further noted by Kuipers et al. [116] that different from cofed olefins, an olefin produced on a growth site might retain physisorbed interactions with the catalyst surface, at least for some time, so that the reactivity of co-fed olefins may differ from the adsorbed olefin intermediate products. This scenario would – similar to diffusional effects – suggest a transport limitation, which are not in agreement with the co-feeding studies in which longer chain α-olefins were preferentially converted via hydrogenation, isomerisation and incorporation (see Fig. 26).

Deviations from ideal ASF distributions have also been ascribed to the presence of different kinds of chain growth sites or the co-occurrence of different chain growth mechanisms (see Section 3.6). Models accounting for this do not address chain length dependencies of olefin contents; it should be noted, that the carbon number trends of olefin contents could also be explained with the above-described model namely by assuming exclusive direct secondary hydrogenation and isomerisation, i.e. readsorption via Sp' only and therefore not affecting chain growth.

Another discrepancy of model predictions with real distributions is the methane content. Although, by allowing for secondary olefin incorporation, the model predicts methane contents higher than to be expected based on ideal ASF kinetics, the methane contents in the real products are often even larger. This is particularly the case with cobalt-based catalysts (see Fig. 34). This high methane selectivity has been suspected to be due to formation of additional methane on different sites, which do not promote chain growth [79, 94] and could be taken into account in a model by assuming higher desorption probabilities of C_1 species compared to other surface species (as assumed in the 'diffusion enhanced readsorption model'). Higher methane selectivities may also result from hydrogenolysis of longer chain products, particularly olefins. This reaction is however strongly inhibited by carbon monoxide and is therefore considered to be negligible at normal FT reaction conditions [115]. In fact co-feeding of olefins at reaction conditions have often led to decreased formation rates of methane [104, 105, 107-112].

Results from fitting experimental data and olefin co-feeding studies show that secondary reactions of α-olefins, as well as primary probability of chain growth and olefin contents are strongly dependent on the catalysts used and the reaction conditions. For a predictive model all intensive parameters of the model (as listed in Table 1) as function of reaction conditions (p_{H2}, p_{CO}, p_{H2O} and temperature) would have to be determined for a given catalyst. The qualitative effects of process parameters on secondary olefin reactions are briefly discussed in Section 3.3.1.3 and Chapter 3.

Analogue Models

Similar models taking carbon number dependent effects of secondary olefin reactions on product distributions and olefin contents have been developed by Shell researchers [116, 119, 120], and van der Laan and Beenackers [123] and Zimmerman et al. [122].

Geerlings et al. [120] assumed desorption of linear α-olefins and paraffins only and included a direct hydrogenation route for secondary olefin hydrogenation. Solubility of olefins in liquid FT-product was considered the only chain length dependent parameter, and effects of physisorption were shown to possibly account for chain length dependencies at dry operation. Although good agreement with experimental data on cobalt catalysts was obtained, the authors mention that deviations in ASF distributions might not exclusively be due to olefin readsorption.

Van der Laan and Beenackers [123] used the same kinetic scheme with desorption of α-olefins and paraffins, with no direct hydrogenation route. Solubility and physisorption were lumped together as the only chain length dependent parameter of their model (proportional to $e^{0.29 \pm 0.07N}$). Comparison of the model with experimental data obtained with a Ruhrchemie-type iron-based catalyst (Fe-Cu-K-SiO$_2$) were satisfactory and effects of process variables on model parameters are included in the model (readsorption rate constant $\propto$ $p_{H2}^{1.4} p_{CO}^{-0.49}$; olefin termination probability $\propto p_{H2}^{-0.5}$; chain growth probability $\propto$ $p_{H2}^{-0.47} p_{CO}^{0.43}$).

The model by Zimmerman et al. [122] also does not take direct hydrogenation into account and only solubility was assumed to affect the product distribution and olefin contents. Only a fairly poor fit with experimental data obtained with an iron-based catalyst was achieved. This may be due to the fact that the additional route of direct hydrogenation and also the possibility of double bond isomerisation of primary α-olefins were neglected.

Puskas et al. [134, 135] only considered incorporation of light olefins (C$_2$ to C$_4$) to account for relatively low molar contents of these products in this carbon number range of ASF plots as experimentally obtained by them. No physical carbon number dependent effects such as solubility or diffusivity were considered by the authors and olefin contents were not described by their model. It was argued that – in contrast to experimental findings [101] - incorporation of long chain α-olefins would not play a role. Deviations from ideal ASF kinetics were explained by multiplicity of chain growth probabilities as a consequence of "different microscopic environments" on catalyst sites (see also Section 3.6). In other words the changing temperatures and concentrations in a reactor and/or a catalyst pore result in a range of growth probabilities. Although reaction local reaction conditions undoubtedly can play a role, it should be noted that non-ASF product distributions have also been obtained in FT synthesis with small particles (no pronounced transport effects) and gradientless reactors e.g. Refs. [121, 136], so the deviations can therefore not exclusively be attributed to 'different microscopic environments'.

In conclusion: It is evident that primarily formed α-olefins undergo secondary reactions during Fischer-Tropsch synthesis. Chain length dependent physical effects such as diffusivity and solubility cause an increase of residence time of long chain products in the liquid filled pores of a catalyst and the reactor, therefore resulting in preferred secondary conversion of long chain olefins and decreasing olefin contents in higher carbon number fractions. Olefin re-incorporation can – at least partially - account for 'positive deviations' from ideal ASF-kinetics. Since α-olefins are the main primary organic products of Fischer-Tropsch synthesis, the extent and the selectivity of secondary olefin reactions have a pronounced effect on the final overall product composition of a FT-process. Reaction and process conditions as well as catalyst properties can affect primary formation of olefins as well as the extent of the secondary reactions (re-incorporation, hydrogenation, isomerisation) and hence the selectivity of the olefins (see Section 3.3.1.3).

3.3.1.3   Effects of process parameters on olefin reactions

The above model descriptions including chain length dependent product diffusivity and solubility have shown that the incorporation of α-olefins, which are the main primary products of FTS can – at least partially – account for deviations leading to increased formation of long chain products. The knowledge of process parameters which enhance this reaction is therefore of high importance, when aiming at high selectivity of diesel and wax as in a GTL process. α-olefins are however also valuable products. If chemicals rather than long chain paraffins are the desired products of a FT process an effective suppression of secondary olefin reactions is required.

Little is known about the effect of reaction conditions on the **primary olefin content** in FT synthesis. Higher temperatures seem to favour the formation of olefins compared to paraffins [59], and according to the kinetic scheme of the desorption of an alkyl surface species (Fig. 13 left) an increase of the H_2/CO ratio can be expected to result in a decrease of primary olefin selectivity. Effects of catalyst composition and product water on the primary olefin to paraffin ratio have not been recorded in the open literature.

Secondary olefin reactions are generally inhibited by carbon monoxide and product water [100, 101, 121], whereas increased hydrogen concentrations or H_2/CO ratios respectively and higher reaction temperature lead to kinetically enhanced secondary olefin consumption [101]. Iglesia et al. [100,102] assume an almost complete selective inhibition of olefin hydrogenation to the corresponding paraffin at FT-typical reaction conditions due to inhibition by CO and product water, while olefin incorporation can still occur. Olefin co-feeding during FTS with a cobalt catalysts at varied CO partial pressure showed that hydrogenation can however only be completely suppressed at very high CO partial pressures (e.g. $p_{CO} > 15$ bar at $p_{H2}=5$ bar, $p_{H2O}=1$ bar, T=190°C, FT-synthesis in slurry reactor, catalyst $CoZrRu-SiO_2$ [94, 101]), isomerisation via double bond shift seems to be enhanced, while incorporation of olefins was hardly affected with increasing CO partial pressure (see also Section 4). Water partial pressure has been reported to generally suppress secondary olefin reactions [121, 137-139], according to Iglesia et al. [100, 102] the selectivity of olefin re-incorporation is however enhanced by water; Potassium promotion in iron based FTS seems to have a similar overall effect [121]. Reaction temperature enhances secondary olefin conversion, but has no pronounced effect on the selectivity of secondary olefin reactions [101]. All these effects are of particular importance in fixed bed FT processes where temperature and concentration gradients along a catalyst bed and within a catalyst pellet can be present.

Optimising olefin re-incorporation in order to achieve maximum C_{5+} selectivity can therefore be done by taking the above effects into account (e.g. high CO and water partial pressure throughout the reactor), furthermore intermediate catalyst pellet sizes or eggshell-type catalysts with the FT active metal (e.g. Cobalt) only present in the outer layer may have a positive effect on C_{5+} selectivity as a result of diffusion enhanced readsorption of olefins (parameter χ see above). The latter type of catalysts (eggshell) would allow the utilisation of large catalyst pellets for fixed bed operation in order to minimise pressure resistance across the catalysts bed [100, 140, 141].

An increase of the selectivity of long chain products might also be obtainable via recycling olefin rich product cuts [142, 143].

A **suppression of secondary olefin reactions** can generally be achieved by shortening the residence time of the reactive olefins in the pore system of the catalysts and or the reactor. This can be either obtained by increasing space velocity, therefore shortening reactor residence time, which is often impractical if high per pass conversion is wanted, and/or using small catalyst particles in order to force back diffusion enhanced olefin readsorption effects. (For the effect of reaction conditions on suppression of secondary olefin reactions also see above.)

As carbon number dependent effects on secondary olefin reactions are due to diffusivity and/or solubility, it has been suspected that altering the physical properties of the liquid phase in the catalyst pores e.g. by conducting FT synthesis in a **supercritical** phase, which would increase diffusivity and decrease solubility of olefins in the fluid phase, might have force back secondary olefin reactions and therefore impact on the product composition [101, 126, 144]. However, a theoretical evaluation of this scenario assuming realistic Fischer-Tropsch conditions (T = 520 K, p = 60 bar including partial pressure of super-critical fluid) and n-hexane as a supercritical fluid shows that regardless of the concentration of n-hexane in the feed (H_2:CO:n-hexane = 2:1:y) two phase systems are to be expected to be present at chain growth probabilities above 0.7 (see Fig. 35).

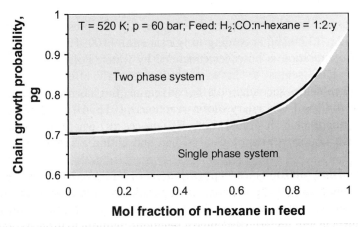

Figure 35 Phases present in FT synthesis as function of chain growth probability and molar fraction of n-hexane; calculation using Peng-Robinson equation of state [145]

In commercial FT systems chain growth probabilities are generally larger than 0.7, and a liquid phase should therefore still be present when trying to conduct super-critical FT synthesis. The reported increase of contents of long chain olefins when adding a supercritical fluid during FT synthesis [126, 144] is therefore more likely due to a more efficient extractive removal of the olefins from the liquid filled catalyst pore system rather than caused by faster diffusion rates. Overall however, adding a super-critical fluid during FT-synthesis can help to maximise the production of chemicals such as olefins and oxygenates.

3.4 Formation and readsorption of oxygenates

Oxygenates formed in Fischer-Tropsch synthesis are mainly n-aldehydes and n-alcohols-(1) and to a smaller extent methyl ketones, secondary alcohols, carboxylic acids and esters [146, 147]. In commercial iron based FT-processes ca. 6-12 wt.% of the product are oxygenates [148]. Little is known about their formation routes. According to Pichler and Schulz [46] an oxygen containing surface species can be formed via CO insertion into an alkyl-metal bond ('CO insertion mechanism', see Section 1); Johnston and Joyner [44] postulated, that the same surface species might be formed via addition of an hydroxyl group to an alkylidene species (see Fig. 36). The desorption of such a species would then result in primary formation of e.g. alcohols and aldehydes.

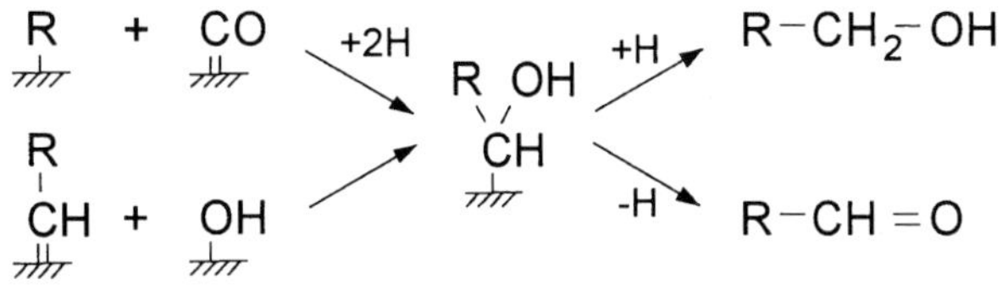

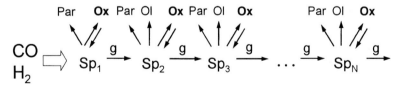

Figure 36   Formation routes of oxygenates (alcohols-(1) and aldehydes) in FT synthesis.

An additional route involving the CO insertion is that of hydroformylation of olefins, which has been proven to occur by means of olefin co-feeding experiments with cobalt [105], ruthenium [107-109] and iron catalysts [101]. However, exclusive formation of oxygenates via hydroformylation of olefins cannot explain the existence of C_1 and C_2 oxygenates in the FT product.

The primary character of linear aldehydes and alcohols was for example noted by Schulz et al. [62], who observed parallel lines of these components and hydrocarbons in ASF plots. In analogy to olefins oxygenates can readsorb on a catalyst surface and undergo secondary reactions, namely hydrogenation to form a paraffin and incorporation into growing chains as shown in co-feeding experiments (e.g. [58, 149-153]).

The kinetic scheme accounting for the formation and readsorption of oxygenates can therefore be extended as follows:

Figure 37   Kinetic scheme accounting for formation of paraffins, olefins and oxygenates and their readsorption

Davis et al. [58, 153] reported the reactivity for incorporation of ethanol added during FT-synthesis with an iron based catalyst as being 50-100 times larger than that of the corresponding olefin, ethene. Secondary reactions of oxygenates (alcohols and aldehydes) might therefore – in addition to olefins – also impact on deviation from ideal ASF distributions. The formation and readsorption of oxygenates can be mathematically introduced into a product distribution model in the same manner as olefin reactions (see Section 3). The effect of process variables on oxygenates formation and secondary consumption and strategies for e.g. maximising yields of these valuable products should be qualitatively similar to those regarding olefins (see Section 3.3.1.3).

Fig. 38 shows molar aldehyde plus alcohol-(1) contents in the fraction of the corresponding carbon number fractions of linear organic products as obtained in fixed bed FT experiments with a non-alkalised and a potassium promoted modified iron catalyst (100g Fe : 13 Al_2O_3 : 10 Cu : (5 K)), tested at the same reaction conditions [84, 129].

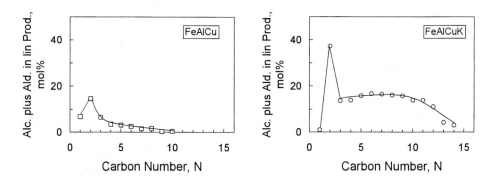

Figure 38 Molar contents of alcohols and aldehydes in respective product fractions as function of carbon number. FT-experiments in fixed bed reactor using catalysts 100Fe/13Al_2O_3/10Cu (left) and 100Fe/13Al_2O_3/10Cu/5K (right) (T=250°C, p_{total}=10 bar, $(H_2/CO)_{in}$=2.0, SV=30 ml(NTP)/min/g_{Fe})

Whereas only very small amount of oxygenates were found with the non-alkalised iron catalyst, most likely due to extensive secondary reactions, fairly large amounts of oxygenates were obtained with the potassium promoted catalyst. With the exceptions of the C_1 and the C_2 fractions (for possible reasons see Refs. [79, 154]), constant molar oxygenate contents were obtained with potassium promoted catalyst. This result - in analogy to constant molar olefin contents – might be interpreted as primary selectivity of these compounds, which is not affected by secondary reactions. The rapid decline of oxygenate contents at higher carbon number might then again be attributed to carbon number dependent effects such as solubility. As with secondary olefin reactions an important function of potassium promotion in iron based FT catalysts is forcing back secondary reactions.

3.5   Formation of branched products

In addition to linear compounds, branched compounds are also found in the product of FT synthesis. Whereas in low temperature processes the amount of branched compounds is typically very low (i.e. < 5%), in the commercial high temperature FT process at Sasol more than 40% of all paraffins are branched (see Chapter 3 and [148]). Mainly mono-methyl branched compounds are formed, only small amounts of di- and tri-methyl and ethyl-branched compounds are found, mainly in high temperature FT processes [147, 155]). Remarkably, no components with quaternary carbon atoms have been found in FT products.

The formation of branched products is believed to occur involving a surface alkyl species which bonds to the catalyst surface with the penultimate carbon atom (see Fig. 3, [61]). This may proceed via both, primary and secondary routes; the secondary route, which includes readsorption of olefins, has for example been experimentally been proven by Schulz et al. [39] through co-feeding of ^{14}C-labelled propene during FT synthesis.

First attempts to account for the formation of branched compounds in FT synthesis were undertaken by Anderson et al. [45, 85, 156], who assumed carbon number independent chain branching probabilities. Based on detailed product analyses [78], which typically shows a decline of homologous series of the different mono-methyl branched isomers approaching constant values at high carbon numbers (see Fig. 40, right for a real distribution of these isomers), Schulz et al. [61] questioned Anderson's postulate and included an exponentially decreasing chain branching rate constant in a extended model which accounts for branching steps from an Sp$_3$ surface species on. This model is shown in Fig. 39, the formation of individual product classes (paraffins, olefins or oxygenates) are not explicitly included here.

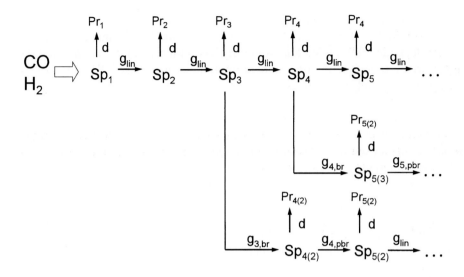

Figure 39   Kinetic scheme accounting for chain branching [61]

The kinetic scheme implies the assumption that rate constants of linear chain growth and desorption are carbon number independent, and that the chemisorbed species $Sp_{N(j)}$ (with j indicating the position of the methyl group) have the same growth and desorption probabilities as Sp_N. Only the branching probability was assumed to be carbon number dependent, namely exponentially deceasing with increasing carbon number:

$$k_{g,br,N} = k_{g,br,4} \cdot f^{(N-4)}, \tag{30}$$

($k_{g,br,3}$ can be chosen individually)

In analogy to the model with formation of on sorts of products only (see Section 3.2) probabilities of linear growth ($p_{g,lin,N}$), growth with chain branching ($p_{g,br,N}$) and product desorption ($p_{d,N}$) can be defined as follows:

$$p_{g,lin,N} = \frac{k_{g,lin,N}}{k_{g,lin,N} + k_{g,br,N} + k_{d,N}} \tag{31}$$

$$p_{d,N} = \frac{k_{d,N}}{k_{g,lin,N} + k_{g,br,N} + k_{d,N}} \tag{32}$$

$$p_{g,br,N} = \frac{k_{g,br,N}}{k_{g,lin,N} + k_{g,br,N} + k_{d,N}} \qquad (33)$$

with

$$p_{g,lin,N} + p_{d,N} + p_{g,br,N} = 1 \qquad (34)$$

Furthermore, after a branching step ('post-branching', pbr) the probability of forming a further branch is zero (no quaternary C-atoms in product), so that now only linear chain growth or desorption are possible:

$$p_{g,pbr,N} = \frac{k_{g,lin,N}}{k_{g,lin,N} + k_{d,N}} \qquad (35)$$

$$p_{d,pbr,N} = \frac{k_{d,N}}{k_{g,lin,N} + k_{d,N}} \qquad (36)$$

Analogue to the simple model with 'formation of one sort of products' only (see Section 3.2), the molar content of an individual product can then be calculated by multiplying the individual reaction probabilities of all steps involved in the formation of the component[2], for example:

Linear compounds:

$$x_{lin,N} = p_{g,lin,1} \cdot p_{g,lin,2} \cdot \ldots \cdot p_{g,lin,N-1} \cdot p_{d,N} \qquad (37)$$

2-methyl-branched compounds:

N=4 $\quad x_{4(2)} = p_{g,lin,1} \cdot p_{g,lin,2} \cdot p_{g,br,3} \cdot p_{d,4} \qquad (38)$

N=5 $\quad x_{5(2)} = p_{g,lin,1} \cdot p_{g,lin,2} \cdot p_{g,br,3} \cdot p_{g,pbr,4} \cdot p_{d,5}$

$\qquad\qquad + p_{g,lin,1} \cdot p_{g,lin,2} \cdot p_{g,lin,3} \cdot p_{g,br,4} \cdot p_{d,pbr,5} \qquad (39)$

Note that mono-methyl branched compounds can be formed via routes, this is however not the case for the first monomer of a homologous series.

---

[2] Note: for this the sum of all molar product formation rates have to be normalised and steady state conditions are assumed

This model was successfully used to fit experimentally observed distributions of the methyl branched compounds; an example for a model fit is given in Fig. 40 right (FT synthesis with Mn modified iron catalyst from [61]). The corresponding chain branching probability as function of carbon number is shown in Fig. 40 middle. It shows a decline from C_4 onwards; correspondingly the probability of chain growth goes through a local minimum at carbon numbers C_3 to C_4 and then approaches constant values at high carbon numbers. It is interesting to note, that the resulting calculated ASF distribution of the linear compounds shows a slight 'positive deviation'.

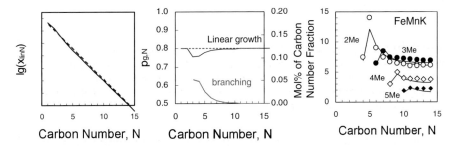

Figure 40 Model predictions accounting for chain branching in FT synthesis. Left: ASF distributions, middle: probabilities of linear chain growth and branching probability, right: Distribution of mono-methyl compounds. Dashed lines: no chain branching. Experimental data points from [61]

In return the model has also been used to calculate probabilities of chain branching, linear chain growth and desorption from experimental product distributions [79, 156, 157-160]. For this it is assumed that the chain branching probability at high carbon numbers approaches zero or constant values and that the desorption probability at high carbon numbers is also constant. The probabilities can then be calculated starting at high carbon numbers using the equations given in Fig. 41.

$$p_{g,lin,N} = \cfrac{1}{1 + \cfrac{\cfrac{p_{d,N+1}}{\dot{n}_{lin,N+1}}}{\dot{n}_{lin,N}} + (1 - p_{g,br,N+1})\left(\cfrac{\dot{n}_{N+1(2)}}{\dot{n}_{lin,N+1}} - \cfrac{\dot{n}_{N+2(2)}}{\dot{n}_{lin,N+2}} + \cfrac{p_{g,br,N+1}}{p_{g,lin,N+1}(1 - p_{g,br,N+2})} \right)} \qquad (1)$$

$$p_{d,N} = p_{g,lin,N} \cdot p_{d,N+1} \cdot \frac{\dot{n}_{lin,N}}{\dot{n}_{lin,N+1}} \qquad (2)$$

$$p_{g,br,N} = 1 - p_{g,lin,N} - \cdot p_{d,N} \qquad (3)$$

The algorithm is to be started at high carbon numbers where chain linear chain growth and chain branching are constant (starting value for $p_{g,br}$ via iteration). The total carbon number distribution can again be derived by multiplication of the single (effective) reaction probabilities, that lead to the formation of a regarded product (see figure 10 eq.(10.9)).

Figure 41 Algorithm to calculate probabilties of linear chain growth, desorption and chain branching from experimental data ($\dot{n}_N$, molar formation rate)

Schulz et al. [161] have for example used the model to investigate the effect of process parameters such as hydrogen and CO partial pressure and temperature on chain branching on cobalt and iron based catalysts (see Section 4). Furthermore chain branching reactions have been identified to be an ideal 'tool' to probe spatial constraints that prevail on the surface of working FT catalysts. For example, Fig. 42 shows the change of chain branching probabilities obtained in early stages of a FT-experiment with a modified cobalt catalyst and probabilities obtained at steady state.

Whereas at steady state a decline of chain branching probability towards almost zero has been obtained, which would indicate a sterical hindrance of formation of long chain branched compounds, at initial stages of an experiment much higher branching probabilities, which even increased with increasing carbon number towards constant values, were found. The authors suggested that this is most likely due to the fact that at the beginning spatial constraints are less pronounced and furthermore chain length dependent readsorption of olefins in position 2 (see Fig. 3) might account for the increase of chain branching probability with increasing carbon number.

664

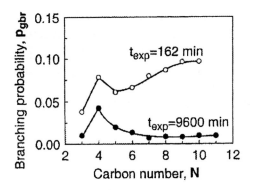

Figure 42 Branching probability as function of carbon number at the beginning of an FT
experiment and at steady state. (Reaction conditions: $100Co/15ZrO_2/100SiO_2$,
$T=190°C$, $p_{total}=5$ bar, $(H_2/CO)_{in}=1.9$, SV=30 ml(NTP)/min/g_{Co}, fixed bed reactor)

A **complete description of a FT product might** be obtained by
combining the above models taking into account:
- Desorption of paraffins
- Reversible desorption of olefins and oxygenates (chain length dependent
  solubility and diffusivity)
- Formation of branched compounds

This would still not include the formation of cyclic compounds, naphthenic and
aromatic, which are formed in high temperature FT processes (up to 15 wt.% in
carbon number range C_{11}-C_{14} [147, 148]). It is likely that these compounds are
formed in secondary reactions from olefins at the severe conditions of a high
temperature process [157].

3.6 Non ideal distributions due to several growth sites or mechanisms

Noting the deviations from ideal ASF distributions, which often show a
sudden change in slope at carbon numbers C_8 to C_{12}, prompted a number of
researchers [136, 162-166] to fit the distribution with two straight lines
representing two different chain growth probabilities, that might originate from:
- two different growth mechanisms occurring in parallel
- two different catalytic sites[3]

---

[3] As an alternative to the two site model, Stenger et al. [167] suggested a 'distributed site'
model, which was shown to be equivalent.

Such a bimodal distribution can mathematically then be simply described by superimposition of two ideal ASF distributions with different chain growth probabilities (p_{g1} and p_{g2}):

$$x_N = A \cdot p_{g1}^{N-1} + B \cdot p_{g2}^{N-1}$$ (40)

The result of a calculated bimodal distribution is given in Fig. 43.

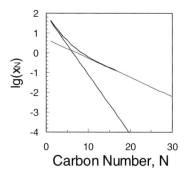

Figure 43:   Bimodal distribution using Eq. (43)

It should be emphasised that there is no direct proof for the co-occurrence of two mechanisms or the existence of two different catalytic sites. As 'bimodal distributions' have been found by authors with many different catalysts, Dictor and Bell [132] suggested that the deviations in ASF distributions are rather due to physico-chemical effects (see models dealing with olefin re-insertion above). In order to support the concept of two different growth mechanisms, Patzlaff et al. [166] argued that effects of olefin re-incorporation would only account for a slight modification of the ASF distribution in FT synthesis with cobalt catalysts and olefin re-incorporation on iron based catalysts would generally only play a negligible role. It should be noted that the above model does only address the total hydrocarbon product distributions; it does for example not explain decreasing olefin contents in higher carbon number fractions, which can accurately be described with olefin re-adsorption followed by hydrogenation or chain start, and which always coincides with a change in the slope of the corresponding ASF distribution, strongly suggesting a crucial role of olefin reactions on the distribution. Furthermore, at high carbon numbers, where no olefins are present and the product is almost exclusively paraffinic generally no further changes of the slope in the ASF diagrams are found. A straight line in the ASF diagram has also been reported for the product of the commercial high temperature process [130]. Correspondingly the olefin content in the reported carbon number range (up to C_{12}) is almost carbon number independent; it may be suspected that secondary olefin reactions and possible resulting changes in the ASF distribution only take place at very high carbon numbers. The exact origin of the curvature of logarithmic molar FT product distributions is however still being debated.

4. CONTROLLING SELECTIVITY IN THE FT SYNTHESIS

The Fischer-Tropsch synthesis is a polymerisation reaction with some marked deviations, viz. higher than expected methane content in the fraction of organic product compounds, a lower than expected C_2 content, and an increase in the chain growth probability for high carbon numbers (see Section 3). In polymerisation reactions selectivity can only be controlled in a limited manner, i.e. over the chain growth probability. Chain growth in the FT synthesis is terminated in various ways yielding in a certain carbon number a variety of products, viz. olefins, paraffins and oxygenates. These products are of high economic value and the Fischer-Tropsch process might be targeting maximisation of a certain product class, e.g. α-olefins.

In production the yield of a certain product group is of interest, e.g. in the GTL process the maximisation of production of liquid fuels is of interest. Thus, it is of economic interest to maximise the weight of liquid products (usually characterised by C_{5+} yield). The kinetic expressions are typically given in terms of a rate with which carbon is incorporated in organic products. The selectivity of the FT synthesis is governed by the Anderson-Schulz-Flory distribution, which gives the molar content of certain carbon numbers in the organic product. The rate of formation of C_{5+} on carbon or weight basis is then given by:

$$r_{C,C_{5+}} = r_{C,org} \cdot \frac{\sum\limits_{n=5}^{n=\infty} n \cdot (1 - p_{g,n}) \cdot \prod\limits_{i=1}^{i=n-1} p_{g,i}}{\sum\limits_{n=1}^{n=\infty} n \cdot (1 - p_{g,n}) \cdot \prod\limits_{i=1}^{i=n-1} p_{g,i}} \tag{41}$$

which simplifies for constant chain growth probability

$$r_{C,C_{5+}} = r_{C,org} \cdot \frac{\sum\limits_{n=5}^{n=\infty} n \cdot p_g^{n-1} \cdot (1 - p_g)}{\sum\limits_{n=1}^{n=\infty} n \cdot p_g^{n-1} \cdot (1 - p_g)} \tag{42}$$

$$r_{C,C_{5+}} = r_{C,org} \cdot \left(1 - (1 - p_g)^2 - p_g \cdot (1 - p_g)^2 - p_g^2 \cdot (1 - p_g)^2 - p_g^3 \cdot (1 - p_g)^2 \right)$$

$$\tag{43}$$

Knowledge of the dependency of the rate of formation of organic product compounds on carbon basis **and** the chain growth probability on the kinetically determining parameters (such as partial pressures and temperature) allows optimisation of the desired and undesired products formed in the FT synthesis. Typically, methane formation is to be minimised. The rate of methane formation is given by:

$$r_{CH_4} = r_{C,org} \cdot \frac{\left(1 - p_{g,1}\right)}{\sum\limits_{n=1}^{n=\infty} n \cdot \left(1 - p_{g,n}\right) \cdot \prod\limits_{i=1}^{i=n-1} p_{g,i}} \tag{44}$$

which simplifies for constant chain growth probability

$$r_{CH_4} = r_{C,org} \cdot \left(1 - p_g\right)^2 \tag{45}$$

The simplistic conclusion is that the minimisation of the rate of methane formation is not possible. The constraint on the FT synthesis is, however, different. In the FT synthesis the rate of formation of C_{5+} product formation needs to be maximised relative to the methane formation. This can be achieved by maximising the chain growth probability.

Typically, more methane is formed than expected based on the polymerisation characteristic of the reaction (see e.g. Fig. 44). The assumption of constant carbon number independent chain growth probability is therefore not valid. Some explanations have been offered in literature to explain the 'excess' methane formation. However, it must be realised that the 'excess' methane formation is much larger than typically thought by researchers in the field of the FT synthesis. Surface species with more than one carbon number have additional desorption pathways, e.g. β-hydrogen elimination yielding olefins. This desorption pathway is not open for C_1 surface species. Hence, chain growth for the C_1 surface species should be larger than that for the higher carbon numbers, which have more pathways for chain growth termination. The excess methane formation is better understood by plotting the n-paraffins in the Anderson Schulz-Flory plot (see Fig. 44). The large excess of methane is easily recognised for cobalt-catalysed Fischer-Tropsch synthesis. In iron-catalysed Fischer-Tropsch synthesis, methane content seems to fall almost at the expected value if the total organic product is considered. However, it can be clearly seen that methane is also in large excess in iron-catalysed Fischer-Tropsch synthesis if only paraffins are considered.

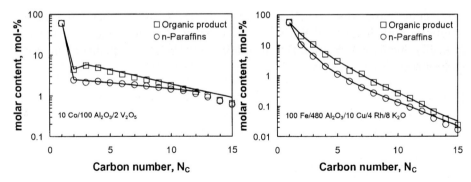

Figure 44 Anderson-Schulz-Flory plot for the formation of organic product compounds and
n-paraffins in Fischer-Tropsch synthesis in a fixed bed reactor
Left: catalyst 10 Co/100 Al$_2$O$_3$/2 V$_2$O5; T=220°C; p=20 bar, H$_2$/CO=2; SV=1.5
          l/g/hr [168]
Right: catalyst 100 Fe/480 Al$_2$O$_3$/10 Cu/4 Rh/8 K$_2$O; T=260°C; p=20 bar;
          H$_2$/CO=2; SV=5 l/g/hr [169]

Up to now the excess formation of methane has not been explained
satisfactorily. Some researchers view methane formation as being different from
the formation of the long chain hydrocarbons in the FT synthesis [52, 63] with
methane formation through hydrogenation of a CH$_3$ surface species, which does
not take part in the chain growth scheme. Alternatively, additional routes may
exist for the formation of methane (e.g. direct hydrogenation of CH$_2$ surface
species in addition to hydrogenation of a CH$_3$ surface species). Furthermore, a
contribution of hydrogenolysis to methane formation cannot be completely
excluded under all Fischer-Tropsch conditions, although this reaction seems to
be strongly inhibited by carbon monoxide [46, 54].

Schulz et al. [94, 158] proposed the formation of methane on non-specific
sites on the Fischer-Tropsch catalyst. These sites have as yet not been identified.

Zheng at al. [30] showed using extended Hückel-calculations, that C$_1$
surface species on Co(0001) carry a negative charge. Hence, the approach of a
CH$_2$ surface species to a CH$_3$-surface species is hindered through electrostatic
repulsion. Longer chain surface alkyl species may have a reduced charge on the
carbon atom bonded to the metal surface, which may result in an enhanced rate
of growth for higher carbon number surface alkyl species. The charge on the C$_1$
surface species is reduced going from right to left in the periodic table (e.g. Ni
→Fe), which coincides with the observed lowering of the molar content of
methane in the organic products going from nickel to iron.

The formation of methane can be controlled kinetically. Schulz et al. [79] reported the chain growth probability for C_1 surface species calculated from the product spectrum as a function of the partial pressures of hydrogen, carbon monoxide and water. The rate of desorption of the C_1-surface species relative to the rate of chain growth of the C_1 surface species is shown in Fig. 45. The desorption of a C_1 surface species is favoured by hydrogen and inhibited by carbon monoxide. Water partial pressure does not affect the rate of desorption relative to the rate of chain growth.

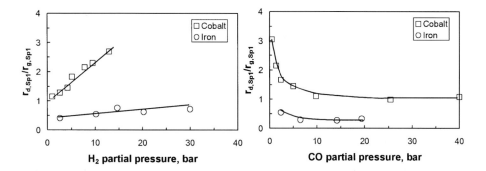

Figure 45 Rate of desorption relative to the rate of chain growth of a C_1-surface species as function of hydrogen partial pressure (left) and carbon monoxide partial pressure (right) [79]

Similar relationships have been reported for the chain growth probability at higher carbon numbers as a function of partial pressures. In general it has been found that the rate of desorption relative to the rate of chain growth is inhibited by carbon monoxide and favoured by hydrogen (see Table 2). This type of kinetic dependency might have been expected a priori, since desorption yielding paraffins involves hydrogen and chain growth involves the addition of a carbon unit. Furthermore, it is well known that the rate of desorption is strongly favoured by an increase in the reaction temperature.

Table 2

Kinetic dependency of chain growth probability (p_g or α) or the rate of chain desorption relative to the rate of chain growth ($\frac{(1-\alpha)}{\alpha}$)

Catalyst	T, K		Source
Fe-Mn	322-562	$\alpha \propto p_{CO}^{0.11}$	[170]
Fe$_2$O$_3$	480	$\frac{(1-\alpha)}{\alpha} = a + b \cdot \frac{p_{H_2}}{p_{CO}}$	[171]
Fe$_3$C, Fe	582	$\alpha = \dfrac{1}{a + b \cdot \dfrac{p_{H_2}^{0.11}}{p_{CO}^{0.25}}}$	[171]
Fe	523-573	$\frac{(1-\alpha)}{\alpha} = b \cdot \frac{p_{H_2}^{0.4}}{p_{CO}^{0.16}}$	[164]
Ru	500-573	$\frac{(1-\alpha)}{\alpha} = \frac{b}{p_{CO}^{0.33}}$	[164]

Knowing the dependency of the chain growth probability and the rate of formation of all organic product compounds on carbon basis, the productivity of a catalyst for the production of e.g. liquid products (C_{5+}) can be optimised. Fig. 46 shows the principal relationship between the productivity for the formation of liquid products or C_{5+}-selectivity and the total pressure using a feed of $H_2/CO = 2$ and $X_{CO}=50\%$. It is assumed that the rate expression developed by van Steen and Schulz [77] is valid, chain growth probability is carbon number independent, and the rate of desorption relative to the rate of chain growth ($\frac{(1-\alpha)}{\alpha} = \frac{(1-p_g)}{p_g}$) is proportional to $\frac{p_{H_2}^{0.4}}{p_{CO}^{0.16}}$. The productivity passes a maximum, since the rate of formation of organic product compounds passes a maximum. The selectivity for the formation of liquid products decreases steadily with increasing pressure, if the assumed correlation for the rate of desorption relative to the rate of chain growth is correct.

It is however well established that the chain growth probability is not constant (see e.g. Refs. [62, 79] and Section 3). The above outlined expression merely describes an ideal situation, to which reincorporation of reactive product compounds needs to be taken into account.

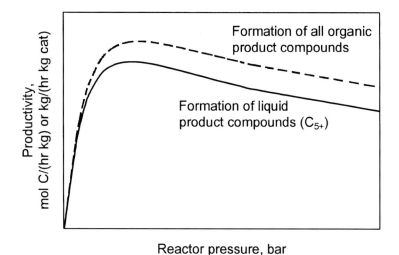

Figure 46 Predicted productivity of liquid hydrocarbons (C_{5+}) in the Fischer-Tropsch synthesis at $X_{CO}=50\%$ assuming that that the rate expression developed by van Steen and Schulz [77] is valid, chain growth probability is carbon number independent, and the rate of desorption relative to the rate of chain growth $\left(\frac{(1-\alpha)}{\alpha} = \frac{(1-p_g)}{p_g} \right)$ is proportional to $\frac{p_{H_2}^{0.4}}{p_{CO}^{0.16}}$

4.1 Branched compounds

The FT synthesis produces linear and branched hydrocarbons. The distribution of branched product compounds has been mathematically modelled by Friedel and Anderson [85], Taylor and Wojciechowski [172] and Schulz et al. [61]. These models describe the principal characteristic of the distribution of branched product compounds well. The detailed description of the distribution of branched product compounds is, however, not entirely accurate [173].

Branched hydrocarbons can be formed primarily [60-62] or by readsorption of olefins and successive chain growth [39]. The distribution of branched product compounds has been described using a chain branching probability, which declines exponentially with carbon number [62]. The dependency of the chain branching probability on the reaction parameters has been given by Schulz et al. [79]. The chain branching probability was related to the hydrogen availability on the surface and the likelihood for secondary reactions. Under hydrogen-poor conditions the primary formation of branched compounds becomes favourable, whereas under hydrogen-rich conditions the secondary formation of branched compounds becomes feasible.

Fig. 47 shows as an illustration the rate of formation of iso-C_5 relative to the rate of formation of n-C_5 as a function of the reaction parameters on cobalt- and iron-based catalysts (data from [173]). With cobalt-based catalysts the secondary formation of branched compounds is expected to dominate [39], whereas with iron-based catalysts the primary formation of branched compounds is expected [60]. The amount of branched compounds formed relative to the amount of linear compounds in with cobalt-based catalysts generally much less than with iron-based catalysts. Hydrogen does not seem to affect the formation of branched product compounds that much, although with the cobalt-based catalyst a slight decline in the formation of branched compounds was observed. This might be attributed to the decline of olefinic precursors in the reactor, which are necessary for the secondary formation of branched compounds.

The formation of branched compounds over iron-based catalysts is favoured with increasing partial pressure of carbon monoxide. An increase in the formation of branched product compounds is also observed with a cobalt-based catalyst at low reaction temperatures. The enhanced rate of formation of branched product compounds with increasing partial pressure of carbon monoxide might be related to the primary formation of branched compounds. Increasing the partial pressure of carbon monoxide will yield hydrogen-poor conditions at the catalyst surface, which may favour the formation of branched compounds. At higher reaction temperatures a decrease in the formation of branched compounds relative to the formation of linear compounds with increasing partial pressure of carbon monoxide is observed. This is likely to be related to the inhibition of the re-incorporation of reactive product compounds in the chain growth pathway leading to the formation of branched compounds.

With increasing temperature the rate of formation of branched compounds relative to the rate of formation of linear compounds rises rapidly over an iron-based catalyst (it should however be noted that this catalyst was tested under atypical Fischer-Tropsch conditions). With a cobalt-based catalyst the relative rate of formation of iso-C_5 seems to pass a minimum. Both these effects can be explained in terms of enhanced secondary formation of branched compounds (the initial decrease must be attributed to a decrease in the primary formation of branched compounds).

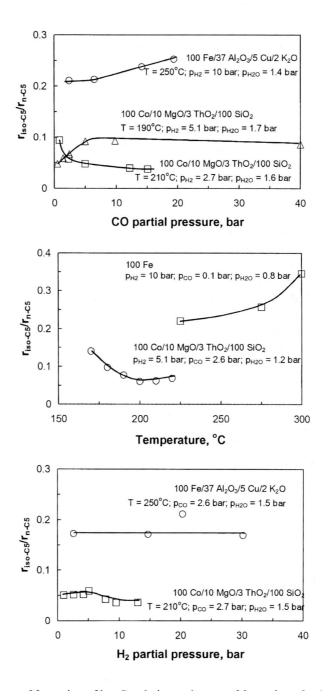

Figure 47 The rate of formation of iso-C$_5$ relative to the rate of formation of n-C$_5$ as a function of the reaction parameters (data from [173]).

4.2 Formation of specific product classes

Additional value can be obtained from the Fischer-Tropsch process by the separation of valuable chemical compounds from the complex product stream. Hence, it is of interest to establish definite relationships between the synthesis conditions and the rate at which these valuable chemicals are formed. Valuable chemicals are for instance olefins and oxygenates. These product compounds can undergo secondary reactions under Fischer-Tropsch conditions (see Section 3). The interpretation of kinetic data for the formation of these compounds must take the possible secondary reaction of these compounds into account.

Propene is the least reactive olefin. Fig. 48 shows the paraffin to olefin ratio in C_3 as a function of the reaction conditions. The rate of formation of propane relative to the rate of formation of propene strongly increases with increasing partial pressure of hydrogen. The increase in paraffin formation is expected for both the primary reaction (based on the alkyl-mechanism – see Section 1.2.1 – and on the CO-insertion mechanism – see Section 1.2.4) and for the secondary conversion of propene.

Olefin formation is favoured at high partial pressures of carbon monoxide. It must be stressed that olefins and paraffins are always formed (olefin content in the fraction of linear hydrocarbons seems to be limited to 70 – 90% depending on the reaction conditions).

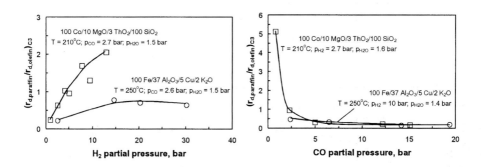

Figure 48   The rate of formation of propane relative to the rate of formation of propene as a function of the reaction parameters (data from [173])

The formation of oxygenates is poorly described in the literature, despite the high value these compounds may have as chemicals. This might be attributed to the difficulty of separation and analysis for these compounds. The distribution of oxygenates in the Fischer-Tropsch synthesis follows the normal Anderson-Schulz-Flory distribution [7, 173-176]. Oxygenates are rather reactive and the observed oxygenate content is influenced by the primary formation of these compounds and the secondary consumption in both the oxygenate fraction and the hydrocarbon fraction.

Fig. 49 shows the oxygenate content in C_2 (defined as the number of moles of ethanol plus acetaldehyde relative to all C_2 product compounds). With increasing partial pressure of hydrogen the oxygenate content passes a maximum with the iron-based catalyst, whereas the oxygenate content seems to increase with the cobalt based catalyst (albeit with a lot of scatter in the data). The observed dependency must be ascribed to the interplay between primary formation, secondary consumption of oxygenates and secondary consumption of the olefin.

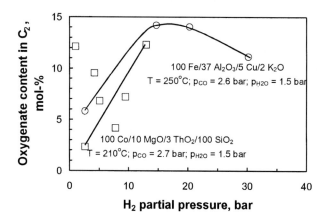

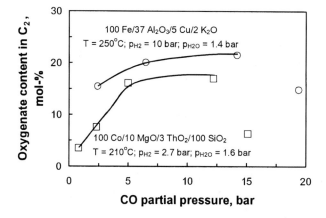

Figure 49 Oxygenate content in the fraction of C_2-compounds as a function of the reaction parameters (data from [173])

The oxygenate content in C_2 increases initially with increasing partial pressure of carbon monoxide. This behaviour would have been expected if oxygenates are formed according to the CO-insertion mechanism (see Section 1.2.4) and hydrocarbons are mainly formed through another reaction pathway. The decrease in the oxygenate content at high partial pressure of carbon monoxide might be attributed to the way these experiments were performed.

The detailed reaction pathway of the reactive product compounds must be known to describe the formation of the desired products accurately. This often leads to a complex interplay of factors, such as the primary reaction pathway and the relative reactivity of the compounds under consideration.

REFERENCES

[1]   R. Reuel and C. Bartholomew, J. Catal., 85 (1984) 63.
[2]   H. Shi and K. Jacobi, Surface Science, 331 (1994) 289-294.
[3]   D.E. Jiang and E.A. Carter, Surface Science, 547 (2003) 85-90.
[4]   D. Klinke and L.J. Broadbelt, Surface Science, 429 (1999) 160-177.
[5]   M. Watanabe and P. Wissmann, Catal. Lett., 7 (1990) 15.
[6]   G. Wedler, Chem.-Ing.-Techn., 47 (1975) 1005.
[7]   H. Schulz, Erdöl und Kohle, 30 (1977) 123.
[8]   U. Roland, U., T. Braunschweig, and F. Roessner, J. Mol. Catal. A, 127 (1997) 61.
[9]   G. Blyholder, J. Phys. Chem., 68 (1964) 2772.
[10] R. Hoffmann, Rev. Mod. Phys., 60 (1988) 601.
[11] M.A. van Daelen, Y.S. Li, J.M. Newsam, and R.A. van Santen, Chem. Phys. Lett., 226 (1994) 100.
[12] B. Hammer, Y. Morikawa, and J.K. Nørskov, Phys. Rev. Lett., 76 (1996) 2141.
[13] R. Eischens and A. Pliskin, Adv. Catal., 10 (1958) 1.
[14] J. Lahtinen, J. Vaari, K. Kauraala, E.A. Soares, and M.A., Van Hove, Surface Science, 448 (2000) 269-278.
[15] G. Michalik, W. Moritz, H. Pfnür, and D. Menzel, Surface Science, 129 (1983) 92.
[16] S.D. Kevan, R.F. Davis, D.H. Rosenblatt, J.G. Tobin, M.G. Mason, D.A. Shirley, C.H. Li, and S.Y. Tong, Phys. Rev. Lett., 46 (1981) 1629.
[17] S. Sung and R. Hoffmann, J. Am. Chem. Soc., 107 (1985) 578.
[18] Y. Morikawa, J.J. Mortensen, B. Hammer, and J.K. Nørskov, Surface Science, 386 (1987) 67.
[19] I.M. Ciobîcă and R.A. van Santen, J. Phys. Chem. B, 107 (2003) 3808.
[20] G. Blyholder and M. Lawless, Langmuir 7 (1991), 140.
[21] R.B. Anderson, in Catalysis, Vol. IV, P. Emmett (ed.), Reinhold Publ., New York, 1956, p.1
[22] O. Malan, J. Louw, and L. Ferreira, Brennstoff-Chem., 42 (1961) 209.
[23] J. Reymond, P. Meriandeau, and S. Teichner, J. Catal., 75 (1982) 39.
[24] H.P. Bonzel and H.J. Krebs, Surface Science, 91 (1980) 499.
[25] D.J. Dwyer and J.H. Hardenbergh, J. Catal., 87 (1984) 66.
[26] M. Kaminsky, N. Winograd, G., Geoffroy, and M.A. Vannice, J. Am. Chem. Soc., 108 (1986) 1315.

[27] H. Yamasaki, Y. Kobori, S. Naito, T. Onishi, and K. Tamaru, J. Chem Soc. Faraday Trans 1, 77 (1981) 2913.

[28] W. Erley, P. McBreen, and H. Ibach, J. Catal., 84 (1983) 229.

[29] C.J. Wang and J.G. Ekerdt, J. Catal., 86 (1984) 239.

[30] C. Zheng, Y. Apeloig, and R. Hoffmann, J. Am. Chem. Soc., 110 (1988) 749

[31] Q. Ge, M. Neurock, H.A. Wright, and N. Srivinas, N., J. Phys Chem. B, 106 (2002) 2826.

[32] W. van Barneveld and V. Ponec, Ind. Eng. Chem. Prod. Res. Dev., 18 (1979) 268.

[33] W. van Barneveld and V. Ponec, J. Catal., 88 (1984) 382.

[34] R. Brady and R. Pettit, J. Am. Chem. Soc., 102 (1980) 6181.

[35] R. Brady and R. Pettit, J. Am. Chem. Soc., 103 (1981) 1287.

[36] G.A. Somorjai, M.A. van Hove, and B.E. Bent, J. Phys. Chem., 92 (1988) 973.

[37] A.B. Anderson and S.J. Choe, J. Phys. Chem. B, 93 (1989) 6145.

[38] F. Calvalcanti, D. Blackmond, R. Oukaci, A. Sayari, A., Erdem-Senatalar, and I. Wender, J. Catal., 113 (1988) 1.

[39] H. Schulz, B.R. Rao, and M. Elstner, Erdöl und Kohle, 22 (1970) 651.

[40] R. Ugo, Catal. Rev.-Sci. Eng., 11 (1975) 225.

[41] P.M. Maitlis, H.C. Long, R. Quyoum, M.L. Turner, and Z.-Q. Whang, J. Chem. Soc. Chem. Commun. (1996) 1.

[42] e.g. D.A. Wessner, F.P. Coenen, and H.P. Bonzel, Langmuir, 1 (1985) 478.

[43] M.T.M. Koper and R.A. van Santen, J. Electroanal. Chem. 472 (1999) 126.

[44] O. Johnston and R. Joyner, Stud. Surf. Sci. Catal., 75 (1993) 165.

[45] H.H. Storch, N. Golumbic, and R.B. Anderson, The Fischer-Tropsch and Related Synthesis, John Wiley & Sons, New York, 1951.

[46] H. Pichler and H. Schulz, Chem.-Ing. Techn., 42 (1970) 1162.

[47] R. George, J.-A.M. Andersen, and J.R. Moss, J. Organomet. Chem., 505 (1995) 131.

[48] F. Fischer and H. Tropsch, Brennstoff-Chem., 7 (1926) 97.

[49] S.R. Craxford and E. Rideal, Brennstoff-Chem., 20 (1939) 263.

[50] J.T. Kummer, T.W. de Witt, and P.H. Emmett, J. Am. Chem. Soc., 70 (1948) 3632

[51] A. de Koster and R.A. van Santen, J. Catal. 127 (1991) 141.

[52] I.M. Ciobîcă, G.J. Kramer, Q. Ge, M. Neurock, and R.A. van Santen, J. Catal., 212 (2002) 136.

[53] M.E. Dry, Appl. Catal. A, 138 (1996) 319.

[54] H. Kölbel, H.B. Ludwig, and H. Hammer, H., J. Catal., 1 (1962) 156.

[55] H. Pichler, H. Schulz, and F. Hojabri, Brennstoff-Chem., 45 (1964) 215.

[56] D. Bukur, D. Mukesh, and S. Patel, Ind. Eng. Chem. Res., 29 (1990) 194.

[57] R. Madon, S. Reyes, and E. Iglesia, E., J. Phys. Chem., 95 (1991) 7795.

[58] L. Tau, H. Dabbagh, and B.H. Davis, Energy & Fuels, 4 (1990) 94.

[59] H. Schulz and H. Gökcebay, in Catalysis of Organic Reactions, J. Kosak (ed.), Marcel Dekker, New York, 1984, p.153.

[60] C. Lee and R. Anderson, Proc. 8th Int. Congr. on Catalysis, Verlag Chemie, Weinheim, 1984, Vol. 2, p.15.

[61] H. Schulz, K. Beck, and E. Erich, Proc. 9th Int. Congr. on Catalysis, M. Philips and M. Teman (eds.), The Chemical Institute of Cananda, Ottawa, 1988, Vol. 2, p.829

[62] H. Schulz, E. Erich, H. Gorre, and E. van Steen, Catal. Lett., 7 (1990), 157

[63] P.M. Maitlis, R. Quyoum, H.C. Long, and M.L. Turner, Appl. Catal. A, 186 (1999) 363.

[64] M.L. Turner, N. Marsih, B.E. Mann, R. Quyoum, H.C. Long, and P.M. Maitlis, J. Am. Chem. Soc., 124 (2002) 10456.

[65] R. Balaji Gupta, B. Viswanathan, and M.V.C. Sastri, J. Catal., 26 (1972) 212.

678

[66] H. Kölbel, G. Patzschke, and H. Hammer, Brennstoff-Chem., 47 (1966) 4.
[67] A. Sternberg and J. Wender, Proc. Intern. Conf. Coordination Chem., The Chemical Society, London, (1959) 53.
[68] S. Roginski, Proc. 3rd Congr. on Catalysis, Amsterdam, 1965, p.939.
[69] K.G. Anderson and J.G., Ekerdt, J. Catal., 95 (1985) 602.
[70] M.E. Dry, Catal. Today, 6 (1990), 183.
[71] M.A. Vannice, Catal. Rev.-Sci. Eng., 14 (1976) 153.
[72] G. Huff jr. and C.N. Satterfield, Ind. Eng. Chem. Proc. Des. Dev., 23 (1984) 696.
[73] G.P. van der Laan and A.A.C.M. Beenackers, A.A.C.M., Catal. Rev.-Sci. Eng., 41 (1999) 255.
[74] I. Yates and C.N. Satterfield, Energy & Fuels, 5 (1991) 168.
[75] A. Rautavuoma and H. van der Baan, Appl. Catal., 1 (1981) 247.
[76] E.F.G. Herrington, Chem. Ind., 65 (1946) 346.
[77] E. van Steen and H. Schulz, Appl. Catal. A: Gen., 186 (1999) 309.
[78] H. Pichler, H. Schulz, and M. Elstner, Brennstoff-Chem., 48 (1967) 78.
[79] H. Schulz, E. van Steen, and M. Claeys, Stud. Surf. Sci. Catal., 81 (1994) 455.
[80] M. Claeys and E. van Steen, Catal. Today, 71 (2002) 419.
[81] M.E. Dry, Ind. Eng. Chem. Res., 15 (1976) 282.
[82] E.S. Lox and G.F. Froment, Ind. Eng. Chem. Res., 23 (1993) 71.
[83] G.P. van der Laan and A.A.C.M. Beenackers, Appl. Catal. A, 193 (2000) 39.
[84] M. Claeys, PhD thesis, University of Karlsruhe (1997)
[85] R.A. Friedel and R.B. Anderson, J. Amer. Chem. Soc., 72 (1950) 121, 2307.
[86] H. Schulz, K. Beck, and E. Erich, Stud. Surf. Sci. Catal., 36 (1988) 457.
[87] G.V. Schulz, Z. Physik. Chem., B30 (1935) 379.
[88] P. Flory, J. Amer. Chem. Soc., 58 (1936) 1877.
[89] J. Eilers, S.A. Posthuma, and S.T. Sie, Catal. Letters 7 (1990) 253.
[90] S.T. Sie, M.M.G. Senden, and H.M.H. van Wechem, Catal. Today 8 (1991) 371.
[91] M.E. Dry, Encyclopaedia of Catalysis, I.T. Horvath (ed.), John Wiley & Sons, N.Y. USA, 2003, Vol. 3, p.347.
[92] S.T. Sie and R. Krishna, Appl. Catal. A: Gen., 186 (1999) 55.
[93] M.E. Dry, T. Shingles, L.J. Boshoff, and G.J. Oosthuizen, J. Catal., 15 (1969) 190.
[94] H. Schulz, Zh. Nie, and F. Ousmanov, Catal. Today, 71 (2002) 351.
[95] H. Schulz, K. Beck, and E. Erich, Fuel Proc. Techn. 18 (1988) 293.
[96] D.R. Smith, C.O. Hawk, and P.L. Golden, J. Amer. Chem. Soc., 52 (1930) 3221.
[97] S.R. Craxford, Trans. Faraday Soc., 35 (1939) 946.
[98] Y.T. Eidus, N.D. Zelinskii, and N.I. Ershov, Dok. Akad. Nauk. SSSR, 60 (1948) 599.
[99] J.T. Kummer and P.H. Emmett, J. Amer. Chem. Soc. 73 (1951) 564.
[100] E. Iglesia, S.C. Reyes, and R.J. Madon, and S.L. Soled, Adv. Catal., 39, 2 (1993) 221.
[101] H. Schulz and M. Claeys, Appl. Catal. A: Gen., 186 (1999) 71.
[102] R.J. Madon, S.C. Reyes, and E. Iglesia, J. Phys. Chem., 95 (1991) 7795.
[103] W.K. Hall, R.J. Kokes, and P.H. Emmett, J. Amer. Chem. Soc., 82 (1960) 1027.
[104] D.H. Dwyer and G.A. Somorjai, J. Catal., 56 (1979) 249.
[105] C. Kibby, R. Pannell, and T. Kobylinski, Prepr. ACS Div. Petr. Chem., 29 (1984) 1113.
[106] A.A. Adesina, R.R. Hudgings, and P.L. Silveston, Appl. Catal. 62 (1990) 295.
[107] D.S. Jordan and A.T. Bell, J. Phys. Chem., 90 (1986) 4797.
[108] D.S. Jordan and A.T. Bell, J. Catal., 107 (1987) 338.
[109] D.S. Jordan and A.T. Bell, J. Catal., 108 (1987) 63.
[110] R. Snel and R.L. Espinoza, C$_1$ Mol. Chem., 1 (1986) 349.
[111] J.H. Boelee, J.M.G. Cüsters, and K. van der Wiele, Appl. Catal., 53 (1999) 1.

[112] R.T. Hanlon and C.N. Satterfield, Ind. Eng. Chem. Res., 27 (1988) 162.
[113] K. Fujimoto, Topics in Catalysis 2 (1995) 259.
[114] O Roelen, DRP 849,548 (1938)
[115] R.J. Madon and E. Iglesia, J. Catal. 149 (1994) 428.
[116] E.W. Kuipers, I.H. Vinkenberg, and H Oosterbeek, J. Catal., 152 (1995) 137.
[117] E. Iglesia, S.C. Reyes, and R.J. Madon, J. Catal., 129 (1991) 238.
[118] R.J. Madon and E. Iglesia, J. Catal., 139 (1993) 576.
[119] E.W. Kuipers, C. Scheper, J.H. Wilson, I.H. Vinkenburg, and H. Oosterbeek, J. Catal., 158 (1996) 288.
[120] J.J.C. Geerlings, J.H. Wilson, G.J. Kramer, H.P.C.E. Kuipers, A. Hoek, and H.M. Huisman, Appl. Catal. A: Gen., 186 (1999) 27.
[121] H. Schulz and M. Claeys, Appl. Catal. A: Gen., 186 (1999) 91.
[122] W.H. Zimmerman, D.B. Bukur, and S. Ledakowicz, Chem. Eng. Sci., 47 (1992) 2707.
[123] G.P van der Laan and A.A.C.M. Beenackers, Stud. Surf. Sci. Catal., 119 (1998) 179.
[124] T. Komaya and A.T. Bell, J. Catal., 146 (1994) 237.
[125] H. Raak, PhD thesis, University of Karlsruhe (1995)
[126] K. Fujimoto, M. Shimose, and Y.Z. Han, Stud. Surf. Sci. Catal., 107 (1997) 171.
[127] R.B. Anderson, B. Seligman, J.F. Shultz, R. Kelly, and M.A. Elliott, Ind. Eng. Chem., 44, 2 (1952) 391.
[128] M.F.M. Post, A.C. van't Hoog, J.K. Minderhoud, and S.T. Sie, AIChE, 35, 7 (1989) 1107.
[129] M. Claeys and H. Schulz, to be published in Prepr. ACS Petr. Div., ACS Meeting Anaheim (2004).
[130] R.J. Madon, E. Iglesia, and C.R. Reyes, in Selectivity in Catalysis (eds. S.C. Suib, N.E. Davis), ACS Symposium Series 517 (1993) 385.
[131] B. Jager and R. Espinoza, Cat. Today, 23 (1995) 17.
[132] R. Dictor and A.T. Bell, J. Catal., 97 (1986) 121.
[133] R. Dictor and A.T. Bell, Appl. Catal., 20 (1986) 165.
[134] I. Puskas, R.S. Hurlbut, and R.E. Pauls, J. Catal., 139 (1993) 591.
[135] I. Puskas and R.S. Hurlbut, Catal. Today, 84 (2003) 99.
[136] G.A. Huff and C.N. Satterfield. J. Catal., 85 (1986) 370.
[137] H. Schulz, M. Claeys, and S. Harms, Stud. Surf. Sci. Catal. 107 (1997) 193.
[138] C.N. Satterfield, R.T. Hanlon, S.E. Tung, Z. Zou, and G.C. Papaefthymion, Ind. Eng. Chem. Prod. Res. Dev., 25 (1986) 407.
[139] M. Claeys and E. van Steen, Catal. Today, 71 (2002) 395.
[140] E. Iglesia, S.L. Soled, J.E. Baumgartner, and S.C. Reyes, J. Catal., 153 (1995) 108.
[141] E. Iglesia, S.L. Soled, J.E. Baumgartner, and S.C. Reyes, Topics in Catalysis, 2 (1995) 17.
[142] H. Kölbel and E. Ruschenberg, Brennstoff-Chemie, 55 (1954) 161.
[143] E. Iglesia and R.J. Madon, (Exxon Res. Eng. Co.) U.S. Pat. 4,754,092
[144] G. Jakobs, K. Chaudhari, D. Sparks, Y. Zhang, B. Shi, R. Spicer, T. Das, J. Li, and B.H. Davis, Fuel, 82 (2003) 1251.
[145] L. Biquizo, M. Claeys, and E. van Steen, unpublished
[146] H. Schulz and A. Zein el Deen, Fuel Proc. Techn., 1 (1977) 247.
[147] C.D. Frohning, H. Kölbel, M. Ralek, W. Rottig, F. Schnur, and H, Schulz, in Chemierohstoffe aus Kohle, J. Falbe (ed.), Georg Thieme Verlag, Stuttgart, 1977, p.219.
[148] B. Jager, Stud. Surf. Sci. Catal., 107 (1997) 219.
[149] Ya. T. Eidus, Russ. Chem. Rev., 36, 5 (1967) 338.

[150] J.T. Kummer, P.H. Emmett, H. Podgurski, W. Spencer, and P.H. Emmett, J. Amer. Chem. Soc., 73 (1951) 564.

[151] J.T. Kummer, J. Amer. Chem. Soc., 75 (1953) 5177.

[152] R.J. Kokes, W.K. Hall, P.H. Emmmett, J. Amer. Chem. Soc., 79 (1957) 2989.

[153] B.H. Davis, Proc. Int. Conf. on Catalysis and Catalytic Processing, Cape Town, (1993) 305.

[154] H. Schulz, G. Schaub, M. Claeys, and T. Riedel, Appl. Catal A: Gen. 186 (1999) 2115.

[155] H. Pichler, H. Schulz, and P. Kühne, Brennstoff-Chemie, 49 (1968) 36.

[156] R. B. Anderson, L. Hofer, and H. Storch, Chem.-Ing. Tech., 30 (1958) 560.

[157] H. Schulz and Zh. Nie, Stud. Surf. Sci. Catal., 136 (2001) 159.

[158] H. Schulz, Zh. Nie, and F. Ousmanov, Catal. Today, 71 (2002) 351.

[159] Zh. Nie, PhD thesis, University of Karlsruhe (1997).

[160] M. Claeys, R. Cowan, H. Schulz, Topics in Catalysis 26 (2003) 139.

[161] H. Schulz, E. van Steen, and M. Claeys, Stud. Surf. Sci. Catal. 81 (1994) 455.

[162] B. Sarup and B.W. Wojciechowski, Can. J. Chem. Eng., 66 (1988) 831.

[163] N.O. Egibor, W.C. Cooper, and B.W. Wojciechowski, Can. J. Chem. Eng., 63 (1995) 826.

[164] A.C. Vosloo, PhD thesis, University of Stellenbosch (1989).

[165] L. König and J. Gaube, Chem.-Ing. Tech., 55 (1983) 14.

[166] J. Patzlaff, Y. Liu, C. Graffmann, and J. Gaube, Appl. Cat. A: Gen. 186 (1999) 109.

[167] H. Stenger, J. Catal., 92 (1985) 426.

[168] S. Zwane, MSc-thesis, University of Cape Town (2003).

[169] G. Joorst, MSc-thesis, University of Cape Town (2000).

[170] G. Bub and M. Baerns, Chem. Eng. Sci., 35 (1980) 348.

[171] R.L. Dictor, PhD thesis, University of California (Berkeley), (1984).

[172] P. Taylor, and B. Wojciechowski, Can. J. Chem. Eng. 61 (1983) 98.

[173] E. van Steen, PhD thesis, University of Karlsruhe (1993).

[174] G.A. Huff, jr. and C.N. Satterfield, J. Catal. 85 (1984) 370.

[175] R.A. Dictor and A.T. Bell, Appl. Catal. 20 (1986) 145.

[176] E. Erich, PhD thesis, University of Karlsruhe (1990).

Index

STUDIES IN SURFACE SCIENCE AND CATALYSIS

Advisory Editors:
B. Delmon, Université Catholique de Louvain, Louvain-la-Neuve, Belgium
J.T. Yates, University of Pittsburgh, Pittsburgh, PA, U.S.A.

694

696

700